B. Dietrich, P. Viout, J.-M. Lehn

Macrocyclic Chemistry

Distribution:
VCH, P.O. Box 101161, D-6940 Weinheim (Federal Republic of Germany)
Switzerland: VCH, P.O. Box, CH-4020 Basel (Switzerland)
United Kingdom and Ireland: VCH (UK) Ltd., 8 Wellington Court, Cambridge CB1 1HZ (England)
USA and Canada: VCH, 220 East 23rd Street, New York, NY 10010-4606 (USA)

ISBN 3-527-28330-7 (VCH, Weinheim) ISBN 1-56081-164-1 (VCH, New York)

B. Dietrich, P. Viout, J.-M. Lehn

Macrocyclic Chemistry

Aspects of Organic and Inorganic Supramolecular Chemistry

VCH

Weinheim · New York
Basel · Cambridge

First published under Aspects de la chimie des composés macrocycliques

Author: Dr. Bernard Dietrich
Prof. Dr. Jean-Marie Lehn
Université Louis Pasteur, Laboratoire de Chimie
Supramoléculaire
4. Rue Blaise Pascal
F-67000 Strasbourg
Dr. Paulette Viout
CNRS
2rue Henri-Dunant
F-94320 Thiais

Published jointly by VCH Verlagsgesellschaft mbH, Weinheim (Federal Republic of Germany) VCH Publishers, Inc., New York, NY (USA)

Translator: Dr. William E. Douglas
Copy Editor: Karin von der Saal
Production Manager: Dipl.-Ing. (FH) Hans Jörg Maier

Library of Congress Card No. applied for

British Library Cataloguing-in-Publication Data: A catalogue record for this book is available from the British Library

Deutsche Bibliothek Cataloguing-in-Publication Data

Dietrich, Bernard:
Macrocyclic chemistry: aspects of organic and inorganic supramolecular chemistry / B. Dietrich; P. Viout; J.-M. Lehn. [Transl.: William E. Douglas]. – Weinheim; New York; Basel; Cambridge: VCH, 1992
Einheitssacht.: Aspects de la chimie des composés macrocycliques ⟨dt.⟩
ISBN 3-527-28330-7 (Weinheim ...)
ISBN 1-56081-164-1 (New York)
NE: Viout, Paulette:; Lehn, Jean-Marie:

Printed on acid-free and chlorine-free paper.

Composition: Filmsatz Unger & Sommer GmbH, D-6940 Weinheim. Printing: betz-druck gmbh, D-6100 Darmstadt 12. Bookbinding: Verlagsbuchbinderei Georg Kränkl, D-6148 Heppenheim.
Printed in the Federal Republic of Germany

Preface

The present work is based on the series of lectures in macrocyclic chemistry which I delivered at the Collège de France some years ago as holder of the Chair of Chemistry of Molecular Interactions.

The aim was to cover extensively, but not however exhaustively, the vast panorama of macrocyclic and macropolycyclic chemical compounds. This field has very greatly advanced over the last twenty years mainly with the goal of developing complexing-agents, or molecular receptors, capable of binding very efficiently and very selectively a given substrate. Thus, by taking into account specific molecular interactions and the relationship between geometrical structure and binding sites, the chemistry of molecular recognition has been developed, this being a major part of **supramolecular chemistry.** A particular attraction is its strongly interdisciplinary nature involving as it does organic, inorganic, physical and analytical chemistry as well as the numerous analogies with biological phenomena. Many laboratories have become increasingly interested in this new field and are intensely active in it.

The intention in these lectures was to present the very great variety of structures and properties of macrocyclic and macropolycyclic compounds ranging from the ligand to the complex and from the receptor to the supermolecule, involving increasing complexity of structural type, and starting with synthetic routes and ending with the ability to bind different substrates. It was more a matter of showing the immense diversity of the field covered and of illustrating major aspects and indicating directions of developments, rather than giving an exhaustive description.

Throughout the work, various digressions are made from the principal theme. The purpose is to generalize certain phenomena which are manifested in a specific way in the case of macrocyclic compounds.

Starting with this framework, Bernard Dietrich and Paulette Viout have produced a much richer and more complete work than the lecture course on which it is based could ever be. They have developed, completed and brought up to date the lecture themes to such an extent that the final result has become their very own creation. I wish to thank them very sincerely since without their impressive efforts, the initital work would have remained what it was, that is to say mere lecture notes.

Jean-Marie Lehn

The authors thank warmly Mrs. Cécile Bertrand for typing the manuscript so patiently and competently. They also wish to thank Mrs. Françoise Le Goff for her numerous excellent drawings which illustrate the text.

Contents

Introduction

Ring systems, cyclo $(CH_2)_n$, can be classified into four groups: *small rings* (n = 3, 4), *normal rings* (n = 5, 6, 7), *medium rings* (n = 8–11), and *large rings* (n ⩾ 12). This classification arises in part from the historical observation that synthetic access to each group requires either modification of a method well-adapted to one of the other categories or the development of a specific procedure.

Normal rings date back to the beginning of organic chemistry and have been very widely studied.

The persistent fascination in small rings should be seen in relation to the development of bonding theories. The synthesis of "strained" rings has been a constant challenge for existing theories.

The synthesis of medium and large rings was quite well studied about 50 years ago and certain basic rules in this area were established at that time. The names of Ruggli, Ruzicka, Stoll, Lüttringhaus, Prelog and Ziegler should be cited. In the following decades interest in macrocycles was more sporadic.

The second half of the 1960s marked the beginning of current macrocyclic chemistry with the realization that macrocyclic structures can give rise to particularly effective and selective complexing agents. This is the case for various natural macrocycles, e. g., those of a peptide or depsipeptide type such as valinomycin which will be extensively discussed later in the work.

Many natural substances have a macrocyclic structure whether or not they exhibit complexing properties. A few examples will illustrate the variety.

Cyclic antibiotics constitute an important class containing an impressive number of members (Fig. 0-1). Those exhibiting complexing properties towards cations will be mentioned in Part II.

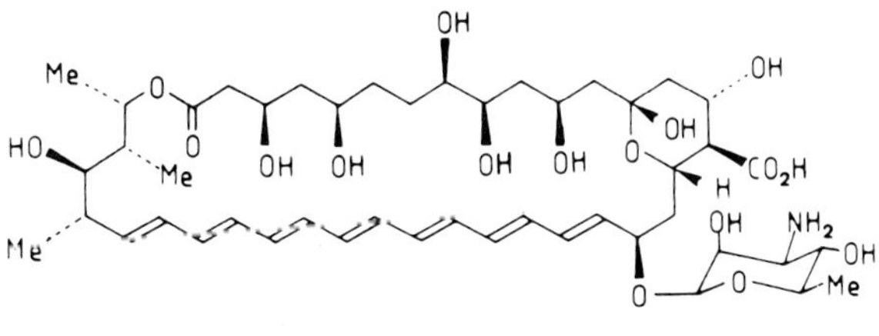

Fig. 0-1. Amphotericin B.

Cyclodextrins and certain cyclodextrin derivatives can bind various molecular substrates thus giving rise to reactions which can serve as models for certain enzymatic processes (Fig. 0-2).

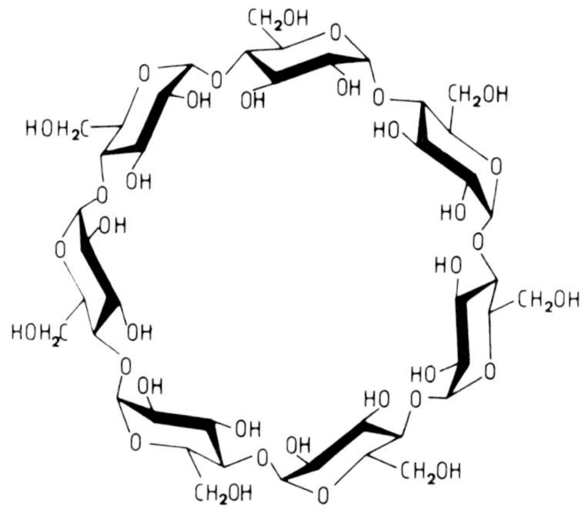

Fig. 0-2. *β*-Cyclodextrin.

The above are the two classes which have been most extensively studied but many other macrocycles are present in living systems. Certain alkaloids are macrocyclic (Fig. 0-3).

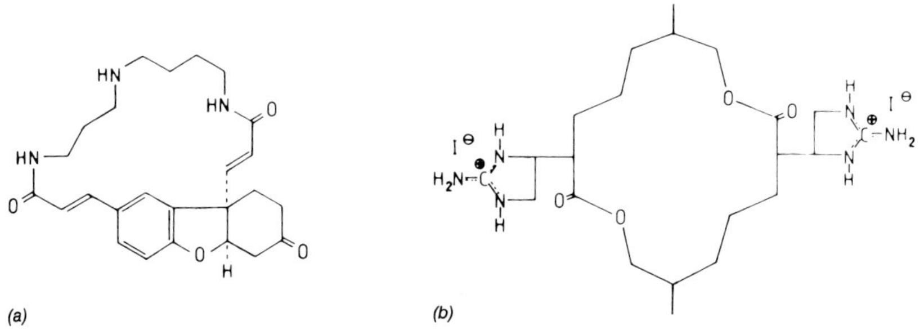

Fig. 0-3. (a) Lunarine, (b) Chaksine.

Very large macrocycles with bolaamphiphile properties are found in lipidic membranes of certain bacteria existing under somewhat severe conditions (pH ~ 2, $T = 85\,°C$; Fig. 0-4).

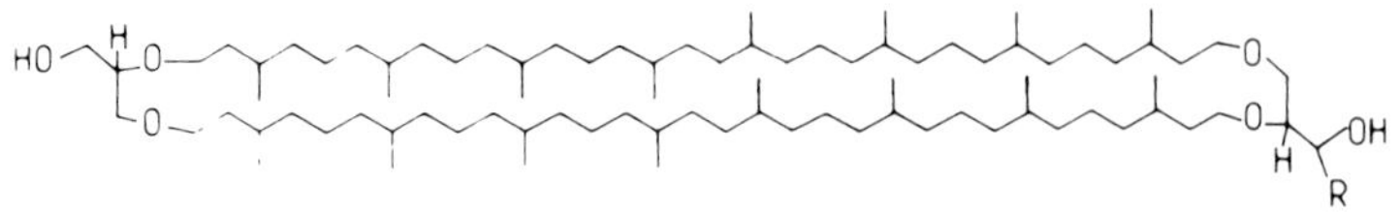

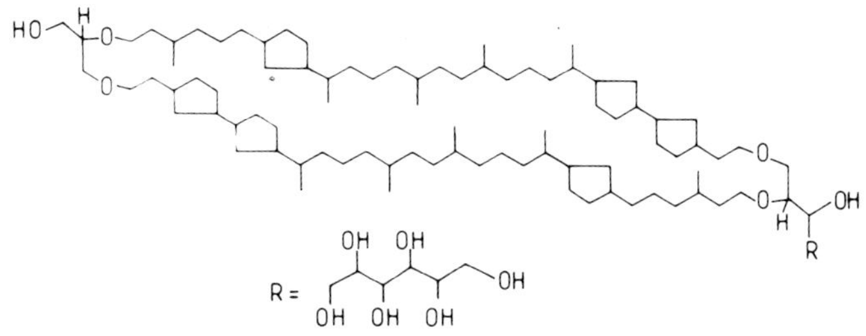

Fig. 0-4. Bolaamphiphiles.

Marine chemistry has revealed a large number of macrocycles some of which show interesting antineoplastic properties. Their structures are sometimes surprising (Fig. 0-5).

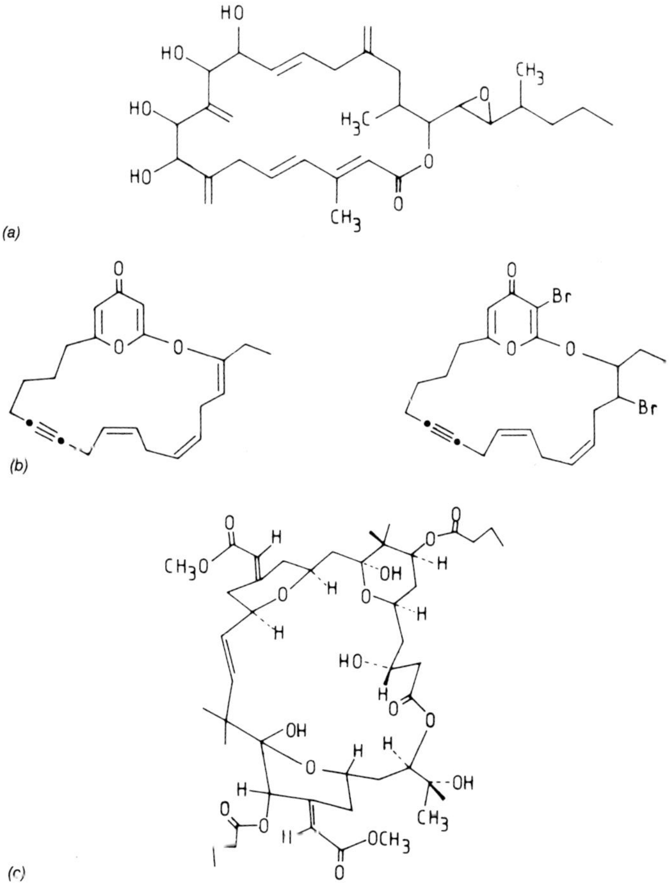

Fig. 0-5. (a) Amphidinolide A, (b) Macrocycles from red algae *Phacelocaspus labillardieri,* *(c) Bryostatin 8.*

Some natural *macrobicycles* also exist such as phalloidin obtained from the poisonous toadstool, *Amanita phalloides.* Other examples are also known (Fig. 0-6).

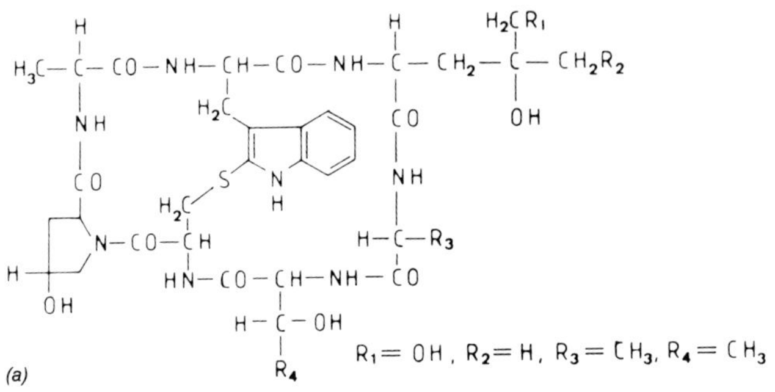

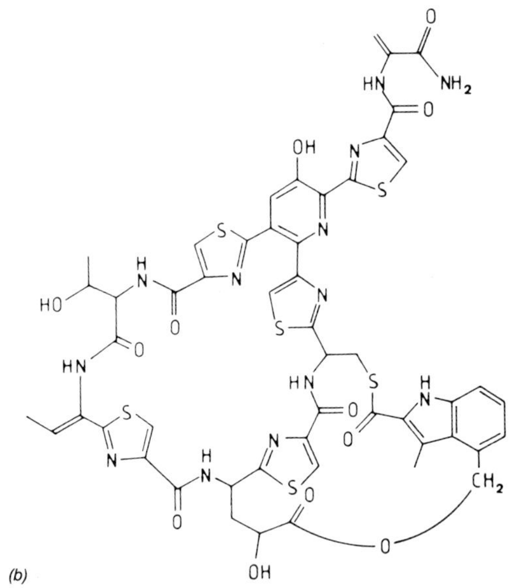

Fig. 0-6. (a) Phalloidin, (b) Nosiheptide.

Finally, we should note the immense circular DNA macrocycles (Fig. 0-7).

In parallel with the discovery of numerous natural macrocycles, the chemical synthesis of macrocycles has been developed. A large number of cyclopeptides based on natural amino acid constituents have been synthesized and studied. Their applications, both fundamental (conformation, ion transport, etc.) and practical (biological activity) are very important. Amongst the impressive amount of work devoted to natural macrocyclic systems we shall mention only a few very specific points.

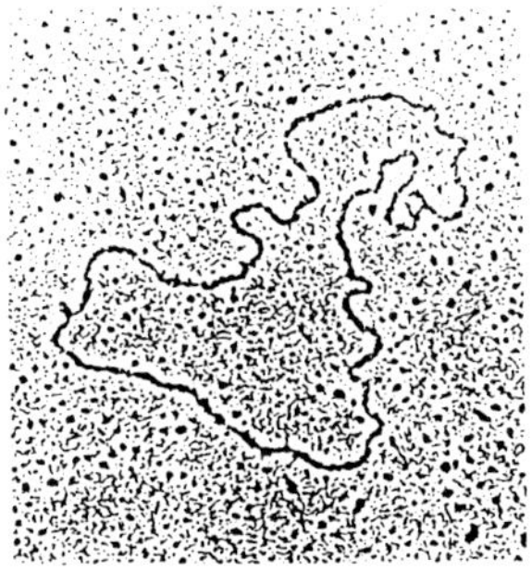

Fig. 0-7. Circular DNA.

Since the macrocyclic structure is so widespread in nature, one can conclude that it offers over the linear arrangement certain "useful" properties with respect to the functions performed. This fundamental observation has been incorporated in the chemistry of molecular recognition ever since the latter was first developed. The *organization* of the ligand is a basic concept in supramolecular chemistry.

Thus, from linear to macrocyclic to macropolycyclic structures, command of molecular shape is refined so allowing an increasingly perfect correspondence between receptor and substrate, or between lock and key in the image of Emil Fischer. It is this path leading in the direction of increasing complexity which will be followed here.

Note Concerning Nomenclature

Because of the rapid increase in the number of known macrocyclic compounds, both natural and synthetic, the difficult problem of *nomenclature* of these substances has arisen.

In the case of natural macrocycles, tradition has it that the group having isolated the substance and/or elucidated its structure should propose a name which often reflects either the natural origin of the product (e.g., phalloidin comes from *Amanita phalloides*) or else a structural characteristic as well (e.g., valinomycin contains only one type of amino acid, valine, and is extracted from *Streptomyces fulvissimus*).

The situation is more delicate for synthetic macrocyclic and macropolycyclic systems and is, it should be admitted, still extremely confused. It is clear that the complexity of the compounds makes official nomenclature (IUPAC) unusable in practice. Research workers use the expedient of coining a name for each new class of macrocycles. Thus, the following were conceived: crown ether, cryptand, spherand, cavitand, carcerand, sepulchrate, coronand, podand, calixarene, etc. This is a long list which continues to lengthen rapidly.

Simplified nomenclature, used in publications, varies according to authors (e.g., [18]crown-6, [18]C-6, 18$\langle O_6$coronand-6$\rangle$, 18-0-6, etc.). Because of this diversity we have not attempted to impose a strict system. The nomenclature, or rather nomenclatures, used in this work reflect the variety found in the literature. On the other hand, the symbolism used is, of course, carefully specified in each case.

Part I

Macrocycle Synthesis

Chapter 1

Synthetic Principles

Formation of macrocycles can be carried out starting from a chain or a group of linear fragments. Whatever the nature of the starting materials or the method used, the last step is almost invariably connexion of the two ends of a chain. However, this chain can either be the starting material itself or result from the *in situ* assembly of bifunctional units.

Several ways of obtaining macrocycles can be envisaged: (a) simple cyclization, (b) cyclization in conjunction with another molecule (capping), (c) condensation of two or four either identical or different units (Fig. 1-1).

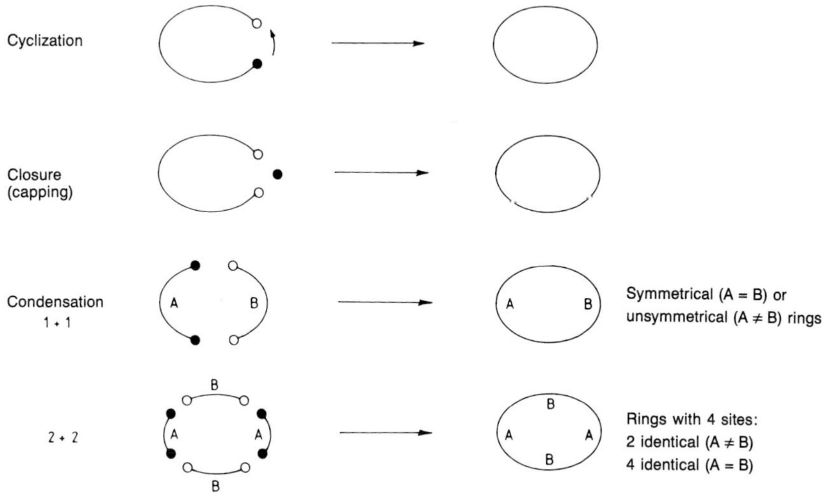

Fig. 1-1. Some methods of cyclization.

The fundamental problem in the synthesis of macrocycles from chains is to favor cyclization, that is to say the intramolecular reaction (a), over end to end linking giving polycondensation, this being the intermolecular reaction (b) (Fig. 1-2).

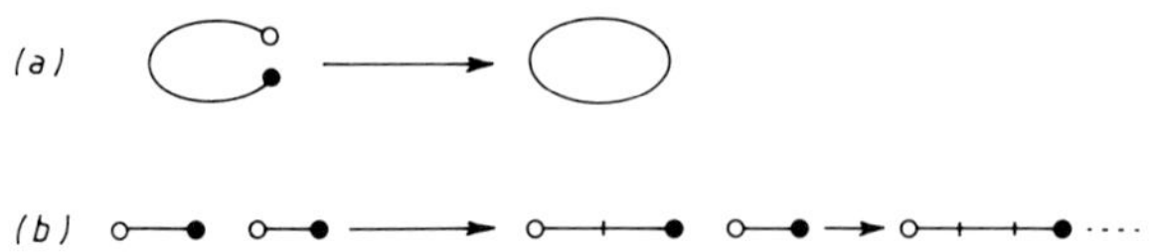

Fig. 1-2. (a) Cyclization and (b) polycondensation of a chain.

It is therefore necessary to analyse the factors favoring intramolecular reaction: 1) chain length; 2) nature of the atoms and groups from which the chain is made up; 3) type of cyclization reaction, and hence the nature of the end groups; 4) experimental technique.

The initial work of Ruggli, Ruzicka, Ziegler, Stetter, Lüttringhaus, and Prelog (1910 to 1940) [1] gave rise to a body of experimental results which can be related to the thermodynamic and structural properties of macrocyclic compounds.

Ring size is defined in relation to the number of atoms in the ring:

$n = 3,4$ *small* rings
$n = 5,6,7$ *normal* rings
$n = 8-11$ *medium* rings
$n \geqslant 12$ *large* rings

1.1 Effect of Chain Length on Cycloalkane Formation

The tendency towards cyclization depends on interactions and on the ensuing entropy change $\Delta S < 0$, $\Delta S^{\neq} < 0$.

Ring strain can be expressed in terms of the heats of combustion of cycloalkanes per CH_2 group, cyclohexane acting as reference in this case (Table 1-1).

Table 1-1. Heat of combustion (H_c/n) per CH_2 group and ring strain ((H_c/n)-157.4) with respect to cyclohexane.

n	H_c/n	(H_c/n)-157.4	n	H_c/n	(H_c/n)-157.4
3	166.6	9.2	11	158.4	1.0
4	163.95	6.55	12	157.7	0.3
5	158.7	1.3	13	157.8	0.4
6	157.4	0.0	14	157.4	0.0
7	158.3	0.9	15	157.5	0.1
8	158.6	1.2	16	157.5	0.1
9	158.8	1.4	17	157.2	−0.2
10	158.6	1.2	∞	157.4	0.0

(Taken from E. L. Eliel, *Stereochemistry of Carbon Compounds,* McGraw-Hill Book Company, Inc., 1962, p. 189.)

It is at a maximum for small rings as a result of angular strain, and at a minimum for n = 6. It increases strongly from n = 8 to 11 and then decreases for n = 12 and 13 becoming approximately zero for the largest rings (n > 14). A wide-ranging discussion of ways of expressing ring strain is to be found in reference [2b].

The degree of strain in the ring to be formed has a direct effect on the ease of preparing it.

In particular, *cyclization yields* have been determined for two cyclization methods, one based on distillation of thorium salts of α, ω-diacids (Ruzicka) [3] (Fig. 1-3), and the other being the cyclization of α, ω-dinitriles (Ziegler) [4] (Fig. 1-4).

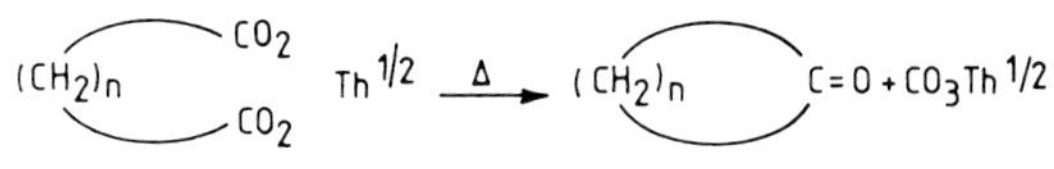

Fig. 1-3. Cyclization of thorium salts of α, ω-diacids.

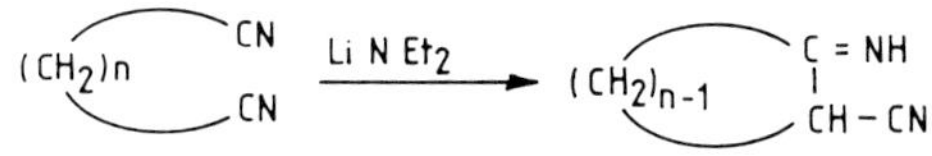

Fig. 1-4. Cyclization of α, ω-dinitriles.

The plot of yield as a function of n (Fig. 1-5) exhibits two different alternations, one being an initial maximum for n = 5 to 8, a minimum for n = 9 to 12 and a second maximum for n = 14 followed by more or less constant yields. This alternation is especially marked in the case of curve I.

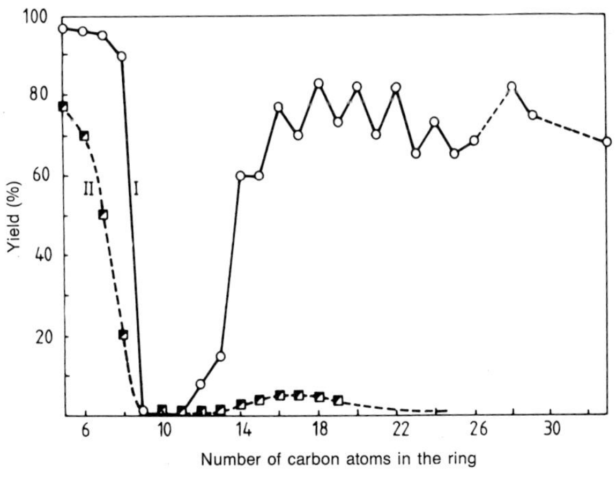

Fig. 1-5. Profile for cyclization yield; I: α, ω-dinitriles; II: α, ω-diacids (taken from K. Ziegler, *Methoden der Organischen Chemie,* Georg Thieme Verlag, Stuttgart, 4/2, 1955, p. 729).

This variation parallels that of the heats of combustion and ring strains, and the minimum observed for medium rings originates at least in part from transannular repulsive interactions (Fig. 1-6).

It can also be noted (curve I) that formation of rings with an even number of members is more facile than that of rings with an uneven number. This odd-and-even alternation is similar to that for melting points and is related to chain entropy.

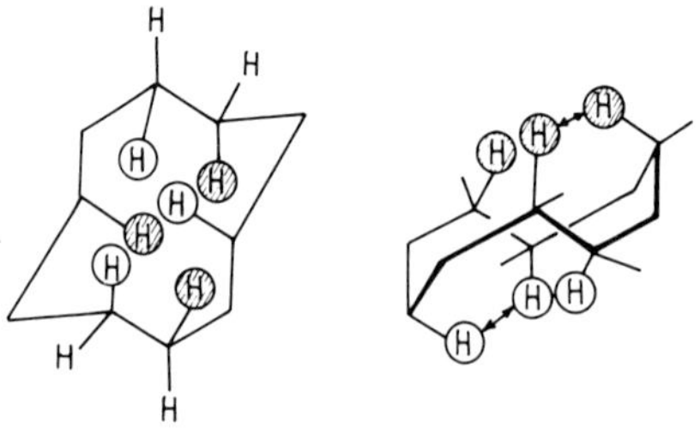

Fig. 1-6. Repulsive interaction in medium rings: e.g., cyclodecane.

The above results have been recently confirmed by Illuminati, Mandolini et al. [5]. But first an important concept should be introduced.

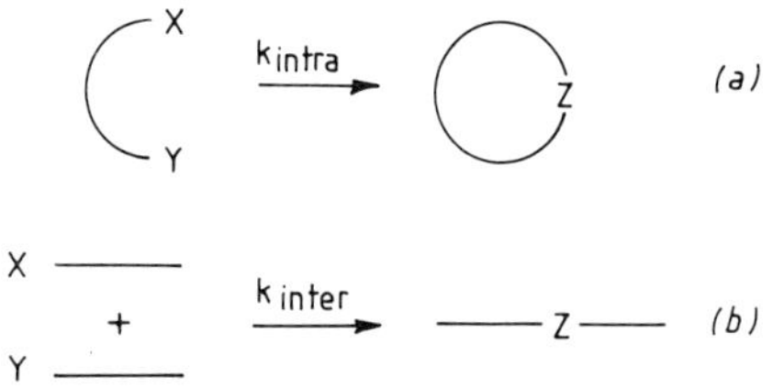

Fig. 1-7. (a) Intramolecular cyclization and (b) intermolecular reaction between two model reactants.

In order to assess the tendency for a system to cyclize and to obtain an order of reactivity, these authors used extensively the concept of *effective molarity,* EM. The rate of the 1st order cyclization reaction in Fig. 1-7a is $v = k_{intra} \cdot c\,(\text{X-Y})$. Furthermore, the rate of the intermolecular reaction (2nd order) in Fig. 1-7b taking place between two reactants with corresponding structural characteristics to those of X-Y is $v' = k_{inter} \cdot c\,(-\text{X}) \cdot c\,(\text{Y}-)$. The ratio k_{intra}/k_{inter} defines the effective molarity (units: mol 1^{-1}). It should be noted that the effective molarity is directly analogous to the cyclization constant introduced by Stoll et al. in 1934 [6]. A review concerning the effective molarity of intramolecular reactions has been published [7]. The number EM is of great practical use. For any system X-Y there is a concentration $c\,(M)$ for which

$$v = v' = k_{intra} \cdot c\,(M) = k_{inter} \cdot c^2\,(M)$$

hence

$$k_{intra}/k_{inter} = c\,(M) = \text{EM}$$

EM ($c\,(M)$) is the concentration necessary in order for the intermolecular reaction to occur at the same rate as the intramolecular reaction. Thus, at this concentration, $c\,(M)$, equal amounts of cyclic and open product are formed. A significantly more dilute solution will therefore give predominantly the desired macrocycle. Conversely, a solution which is too concentrated will afford almost exclusively polymers. This practical aspect has been emphasized by Mandolini in a critical article surveying the literature results [8].

The kinetic study of (a) lactonization of ω-bromoalkanoate ions compared to (b) the corresponding intermolecular reaction (Fig. 1-8) illustrates the concept of effective molarity [9].

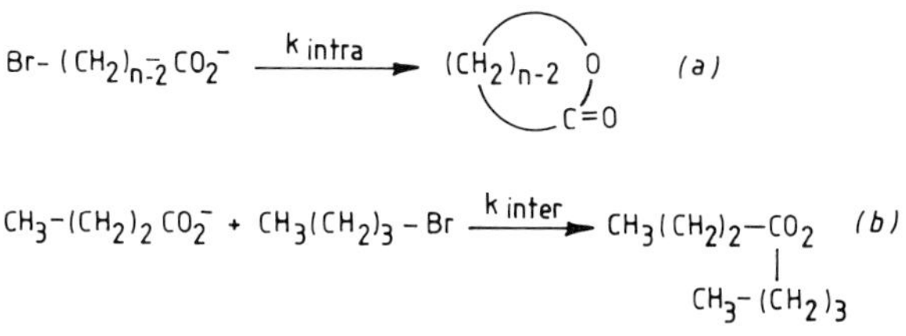

Fig. 1-8. (a) Intramolecular cyclization of ω-bromoalkanoate ions to give lactones, (b) intermolecular condensation of an alkanoate ion with a bromo derivative.

In Table 1-2, it can be seen that the value of EM is high for n = 4,5,6. It should be stressed here that this effective molarity should not be considered as a concentration in the usual sense of the term (1.6×10^3 mol l^{-1} is a concentration impossible to achieve), but rather it simply denotes exclusive formation of the cyclic product since it represents the ratio k_{intra}/k_{inter} (see [7]). For n = 7, the value of EM is low, and this ring could only be obtained at low concentrations of the bifunctional reactant. For n = 8 to 12, the values of EM observed are the lowest (great difficulty in obtaining these rings). From n = 13 upwards, EM increases again to reach a more or less constant value with increasing n.

Table 1-2. Lactonization. Effective molarity, EM (t = 50 °C).

n	EM (M)	n	EM (M)
3	1.23×10^{-2}	11	4.82×10^{-3}
4	13.5	12	6.02×10^{-3}
5	1.6×10^{3}	13	1.82×10^{-2}
6	14.5	14	2.37×10^{-2}
7	5.51×10^{-2}	15	2.56×10^{-2}
8	5.66×10^{-4}	16	2.94×10^{-2}
9	6.33×10^{-4}	18	2.90×10^{-2}
10	1.90×10^{-3}	23	3.42×10^{-2}

(Taken from G. Illuminati et al. (9)).

In Fig. 1-9a, it can be seen that the rate constant profile for lactone formation from the parent ω-bromoalkanoate ions shows a very marked maximum for n = 5. Thereafter, k_{intra} decreases until n = 8,9 followed by an increase to reach a constant value for larger rings.

The values of the enthalpy of activation (Fig. 1-9b) indicate maximum ring strain for 3- and 8-membered rings. On the other hand, for 18- to 23-membered rings, the enthalpy of activation approaches that for the corresponding intermolecular reaction (low ring strain).

The entropy of activation (Fig. 1-9c) decreases with increasing chain length (decrease in probability of the two chain ends meeting) and tends towards a minimum for long chains. The profile for the entropy of activation shows maxima at n = 8, 12 and 16, just as does that for the enthalpy of activation.

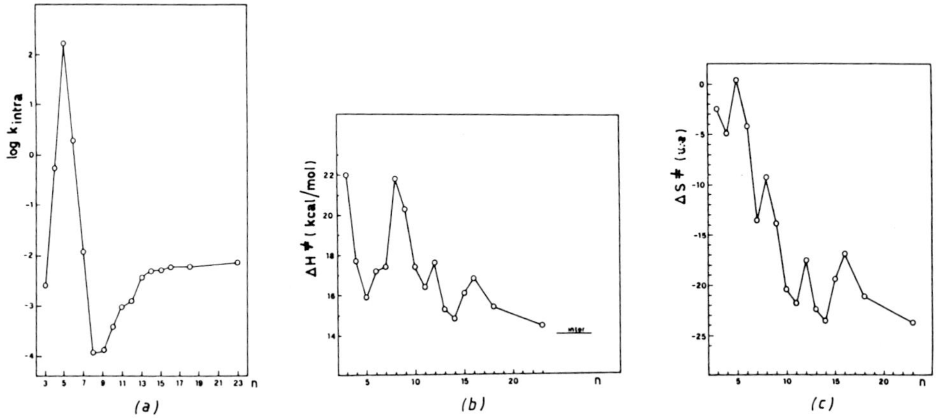

Fig. 1-9. Intramolecular lactone formation profiles for (a) the rate constant, (b) the enthalpy of activation, (c) the entropy of activation; N.B.: n is the size of the ring obtained from Br $(CH_2)_{n-2}$ $CO_2^{\ominus}$ (taken from L. Mandolini et al. *Acc. chem. Res.* **14**, 95, 1981; L. Mandolini, *Adv. Phys. Org. Chem.,* **22**, 1, 1986).

1.2 Effect of the Nature of the Chain Atoms and Groups on Ring Formation

1.2.1 Heteroatoms

The presence of heteroatoms decreases transannular interactions: –O–, –S–, and –N–H are groups which are less bulky than is CH_2 (Fig. 1-10).

Heteroatoms favor formation of rings in which transannular interactions are high [10]. Thus, the minimum in the yield profile vs. ring size for cycloalkane formation almost disappears for oxygen-containing medium rings.

The examples in Fig. 1-11 illustrate the effect of an oxygen atom in the chain [11]. Methylene group replacement by an oxygen atom greatly favors cyclization. Another example illustrating the role of the oxygen atom will be given later.

However, in large rings the effect of heteroatoms is much less clear-cut (Fig. 1-12) since even in their absence transannular interactions are weak or non-existent [12-15]. Both with and without oxygen atoms, the yields for these two reactions were comparable, being of the order of 75%.

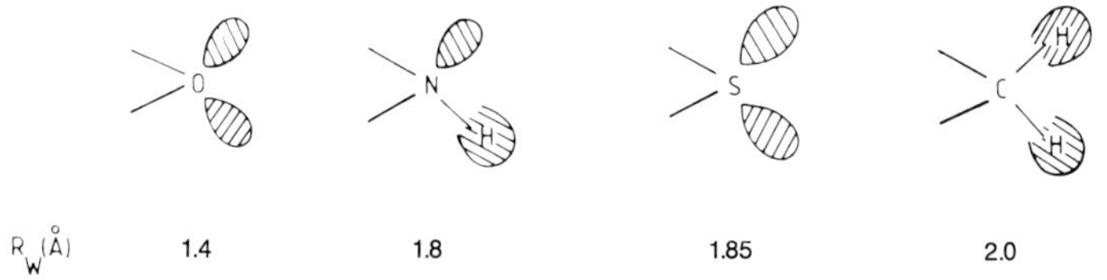

Fig. 1-10. Schematic representation of the bulkiness of some heteroatoms and of the methylene group.

Fig. 1-11. Effect of CH_2 replacement by an oxygen atom on dinitrile cyclization yield (medium rings).

Dinitrile	Ring yield
$N \equiv C-(CH_2)_9-C \equiv N$	0,5%
$N \equiv C-(CH_2)_4-O-(CH_2)_4-C \equiv N$	5%
$N \equiv C-(CH_2)_{12}-C \equiv N$	12%
$N \equiv C-(CH_2)_4-O-CH_2CH_2-O-(CH_2)_4-C \equiv N$	70%

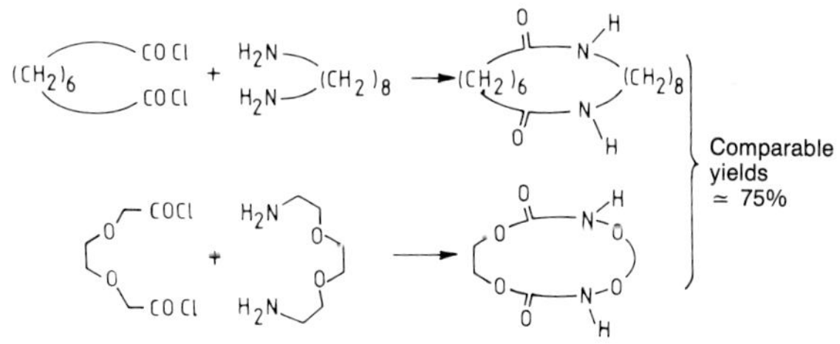

Fig. 1-12. Absence of heteroatom effects on yield in large ring formation.

1.2.2 Rigid Groups

Because of the flexibility of the chains and the greater number of possible conformations with increasing chain length, the probability of intramolecular encounter of the two chain ends decreases as chain length and internal entropy increase. Decrease in internal entropy should therefore favor ring closure, hence the idea of incorporating rigid groups in order to increase the probability of intramolecular reactions.

This principal was initially proposed by Baker et al. [16], and then expanded and established as a rule by Ziegler [1]. A more quantitative study has been recently carried out by Illuminati et al. [17-19a].

The effect of rigid (aromatic) groups on cyclization yield has been demonstrated in the case of medium rings for cyclic mono-and di-ether formation from phenol halogeno derivatives (Fig. 1-13). The results of this study are given in Table 1-3.

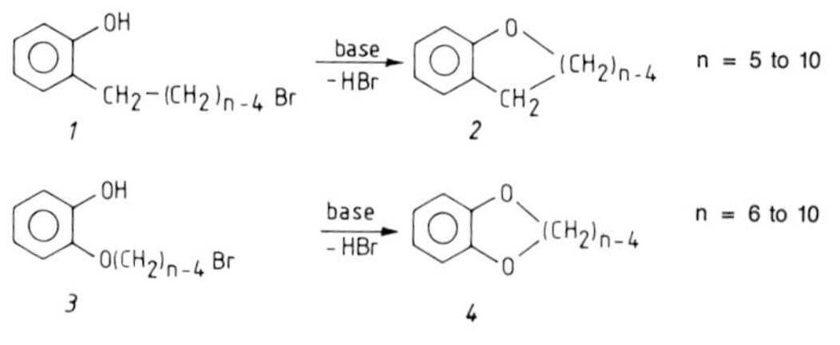

Fig. 1-13. Rigid group effect in medium ring formation.

Table 1-3. Rigid group effect on cyclization yield of phenol compounds.

Cyclization Product	n (Ring size)					
	5	6	7	8	9	10
2	>96	>96	>96	64	78	85
4		>96	>96	85	95	95

(Taken from G. Illuminati et al. [17, 18]).

The rigid group effect is quite clear-cut; the yield showed no pronounced minimum. Replacement of CH_2 by O resulted in an increased yield of cyclic product especially when the ring was more strained (n = 8 and 9) in agreement with the results described above.

Double and triple bonds also have the effect of making the chain more rigid (decrease in the degrees of freedom). Thus ring closure for a C_{10} chain with double and triple bonds in the 3 and 8 positions gave acceptable yields (40–70%) of the [10]-monocycle [19b].

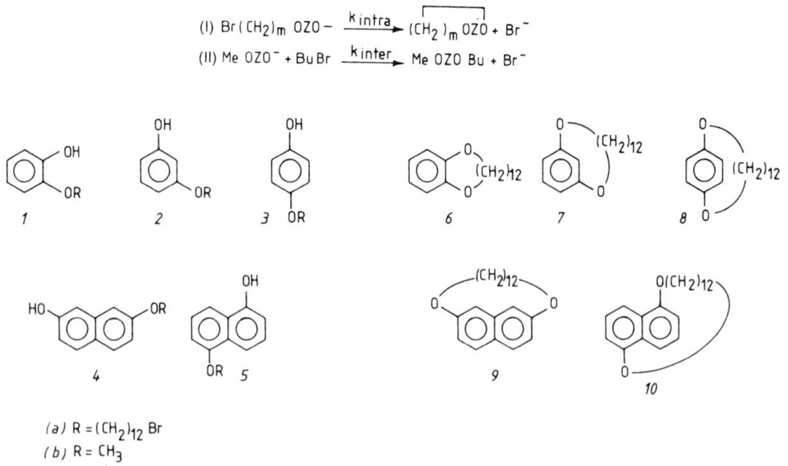

Fig. 1-14. Rigid group effect in large ring formation. Compounds *6-10* are macrocycles formed by intramolecular reaction of *1a-5a*. In reactions (I) and (II), Z represents the aromatic part of compounds *1-10* (taken from L. Mandolini, *Adv. Phys. Org. Chem.,* **22**, 1, 1986).

In order to determine a possible rigid group effect in the formation of large rings, a kinetic study of ring closure reactions of compounds 1a–5a (Fig. 1-14) and intermolecular models 1b-5b has been performed [19a]. It can be seen (Table 1-4) that in the case of large rings, the rigid group effect was slight or non-existent. The ratio k_{intra}/k_{inter} = EM (effective molarity) varied very little despite the differences in size and geometry of the groups introduced.

Table 1-4. Kinetic results for (a) ring closure of phenol compounds, (b) the corresponding intermolecular model in aqueous DMSO at 25 ± 0.2 °C.

Compound	$10^3\ k_{intra}$ sec^{-1}	10 k_{inter} M^{-1} sec^{-1}	10^2 EM M	Log EM
1	9.04 ± 0.11	2.95 ± 0.03	3.06	−1.51
2	3.56 ± 0.04	1.53 ± 0.04	2.33	−1.63
3	7.22 ± 0.13	4.68 ± 0.20	1.54	−1.81
4	1.59 ± 0.04	1.07 ± 0.06	1.49	−1.83
5	1.59 ± 0.01	0.84 ± 0.01	1.90	−1.72

(Taken from L. Mandolini et al., *J. Org. Chem.* **42**, 3733, 1977).

1.2.3 *Gem*-dimethyl Substituents

The effect of *gem*-dimethyl substituents has been observed for both dinitrile and diester ring closure [20a].

It should be noted that a) both methods (X = CN or COOR) gave the same products (Fig. 1-15), and b) in certain cases an appreciable quantity of the diketone twice the size of the monoketone was formed.

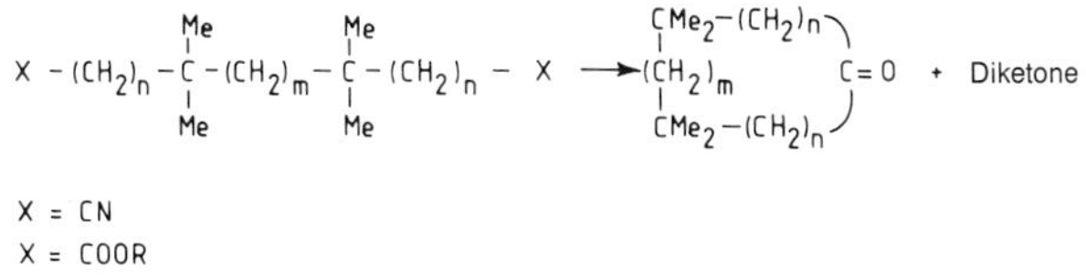

Fig. 1-15. *Gem*-dimethyl substituent effects on dinitrile and diester ring closure.

These reactions have been compared to the ring closure reaction carried out in the case of the chains without *gem*-dimethyl substituents.

For Ziegler ring closure, a negative effect was observed in all cases (Table 1-5). Both with and without *gem*-dimethyl substituents this type of ring closure gave (with rare exceptions) mainly the 18- to 20-membered cyclic diketone, but here too dimethyl groups had a negative effect.

In the case of the Dieckmann cyclization, *gem*-dimethyl groups favored formation of 9-, 11- and 12-membered rings, and disfavored or only slightly favored 8- and 10-membered ring formation. Diketone yields were generally relatively little affected.

Table 1-5. (a) Ziegler cyclization of dinitriles, (b) Dieckmann cyclization of diesters. Cyclization yield as a function of ring size.

m	n	Ring Size (mono-ketone)	Monoketone yield (%)	Yield (%) from unsub-stituted compound	Ring size (diketone)	Diketone yield (%)	Yield (%) from un-substituted compound	
2	1	7	traces	96	14	0	0	
2	2	9	2,4	2.8	18	18	62	(a)
3	2	10	0	0.4	20	24	70	
2	1	7	30	47	14	0	0	
3	1	8	0	15	16	0	11	
2	2	9	30	0	18	7	25	(b)
3	2	10	0.4	0	20	27	12	
2	3	11	13	0.5	22	15	23	
3	3	12	8	0.5	24	10	16	

(Taken from J. Dale et al. [20a]).

A kinetic study of cyclization of ω-bromoalkanoates with a *gem*-dimethyl group in the chain showed a positive effect for 9-membered ring formation. In the case of the larger rings studied (10-, 11- and 16-membered) the effects were slight [20b].

The *gem*-dimethyl effect which is well established for small and normal rings is less clear-cut in the case of macrocycle formation.

1.3 Influence of Cyclization Type on Macrocycle Formation

The type of reaction also affects ring closure yield since the stereochemistry as well as particular reaction features (reactant nature, temperature) are involved. The combination of several effects is possible.

The number of units is important since it determines the number of condensation steps. As far as possible, macrocycle formation should be restricted to cyclization involving a single bifunctional unit rather than the combination of several units.

One component cyclization is intramolecular (Fig. 1-16); (a)). An intermolecular reaction may occur simultaneously giving larger cyclization products (Fig. 1-16; (b)).

Fig. 1-16. One component cyclization.

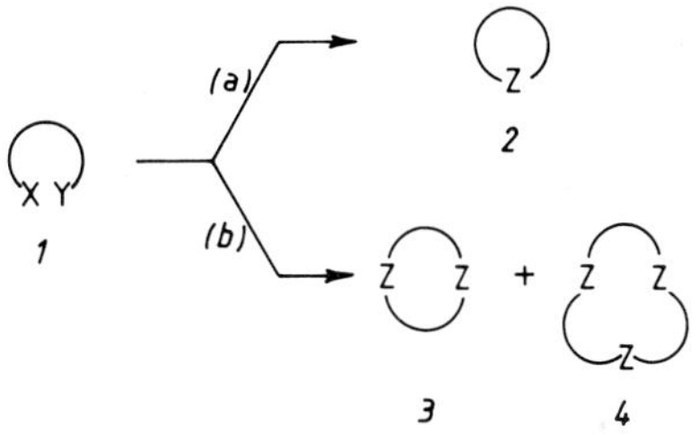

The following are examples of intramolecular one component cyclizations; the Wurtz reaction using a dibromo derivative, the acyloin condensation of a diester, dinitrile condensation, and amide formation from a carboxylic acid and a terminal amine.

Two component cyclization takes place in two steps: (a) intermolecular condensation, followed by (b) intramolecular cyclization (Fig. 1-17). It can also be accompanied by larger ring formation (c).

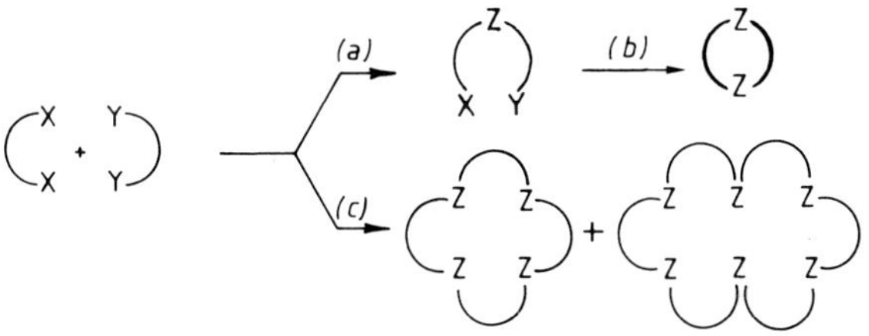

Fig. 1-17. Two components cyclization.

This type of reaction has been much used. Examples are the Wurtz reaction (X = Na, Y = Br) and especially the formation of dilactams (X = COC1, Y = NH_2) or cyclic amines (e.g. X = NHTs, Y = Br).

N.B., the cyclization reactions in Figures 1-16 and 1-17 all take place, of course, in competition with polymerization reactions.

Various Effects

Various factors influence the course of ring closure reactions. The solvent affects the distribution of conformations and possibly also the reaction rate and interactions between the chain ends. The viscosity can also have an effect. Ziegler gave an example where the yield increased from 8.3% in benzene to 52.9% in ethanol for 10-membered ring formation [1].

The effect of temperature on the course of macrocyclization reactions has until now been little studied. However, it has been shown in one case that there was an optimum temperature (60 °C in this case) for which the yield almost doubled with respect to that obtained at 20 °C [21a]. In another case where both a cyclic monomer and a cyclic dimer could be formed, the ratio of the two was found to be very sensitive to variations in temperature [21b].

Chapter 2

Cyclization

2.1 Principles and Concepts

The term cyclization taken in its broadest sense defines any interaction or reaction taking place between two ends of a chain [5, 22, 23] (Fig. 2-1). It depends on chain end proximity.

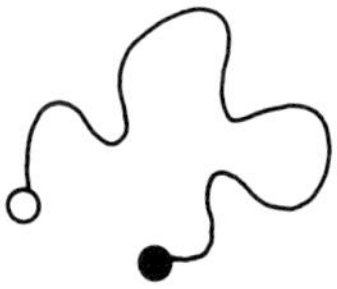

Fig. 2-1. Interaction between chain ends by proximity effect.

However, the definition of "proximity" depends on the ability of the two ends to undergo the process envisaged and thus varies with the type of reaction. Cyclization may occur over a short distance (bond formation and therefore ring closure), a medium distance up to 10 Å (electron transfer), or long distance up to 100 Å (energy transfer).

There are several modes of cyclization according to the position of the reacting groups, that is end-to-end, end-to-backbone, backbone-to-backbone (Fig. 2-2).

Fig. 2-2. Types of cyclization according to the position of the reacting groups: (a) end-to-end, (b) end-to-backbone, (c) backbone-to-backbone.

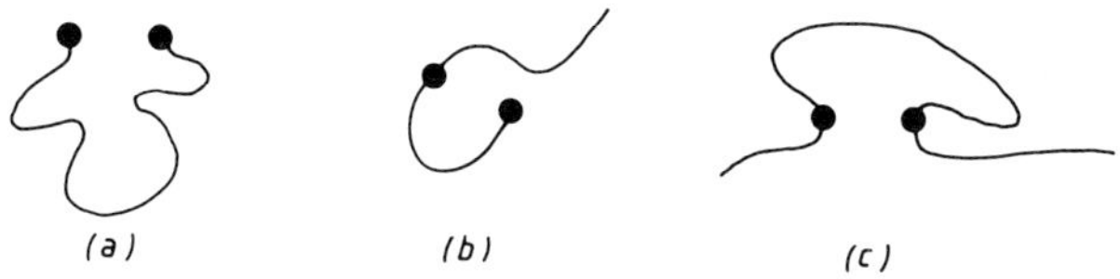

The properties of chains (conformation, mobility) influence the probability of cyclization. The kinetics can be for either *diffusion controlled* or *conformationally controlled* reactions.

In the first case the reaction depends on the probability of the two chain ends meeting, energy transfer reactions between the two ends being an example.

In the second case, the reaction is much slower than diffusion, this being chiefly so for covalent bond formation.

2.2 Diffusion Controlled Cyclization

For the reaction to be under diffusion control, it should be fast with respect to chain movements. This is so mainly for physical interactions such as those giving rise to energy transfer, donor-acceptor interactions or luminescence quenching.

Since these processes are very fast with respect to conformational movements, they allow study of (a) conformational populations, (b) the average distance between chain ends, and (c) chain dynamics. This has been achieved for polymeric chains thanks to the introduction of photosensitive probes at the chain ends. In our case we are principally interested in the intramolecular process in systems of type (a) (Fig. 2-2) for which the chain ends are chromophores. Several types of study have been carried out, including especially (a) the observation of charge transfer bands in systems incorporating in the same molecule both the donor and the acceptor, (b) excimer fluorescence detection, and (c) fluorescence quenching *via* energy or electron transfer.

2.2.1 Intramolecular Donor-Acceptor Interactions

Compounds containing in the same molecule both an acceptor and a donor group (Fig. 2-3), show D-A polarization interactions in the ground state, and charge transfer bands when photoexcited.

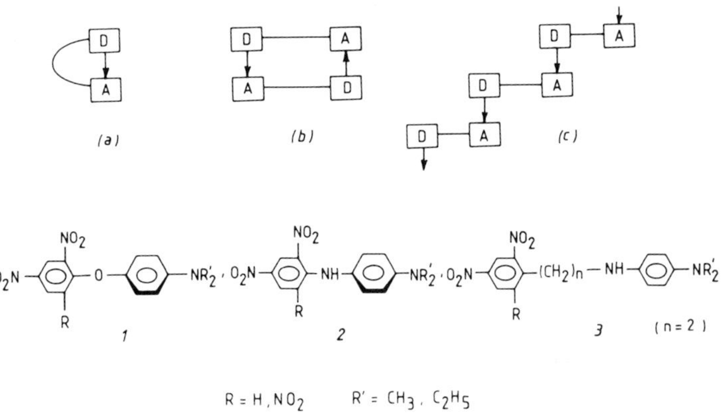

Fig. 2-3. Donor-acceptor complex molecules: (a) intramolecular, (b) bimolecular, (c) intermolecular.

Short chain compounds such as *1* and *2* give only (b) bimolecular or (c) intermolecular charge transfer complexes (CTC). Compound *1* is colored in the solid state but in solution the solvent acts as a screen and separates the donor and acceptor groups on neighboring molecules.

In the case of the longer chain compounds *3*, charge transfer complexes may be formed both in the solid state and in solution, the band intensity depending on the probability of the two chain ends meeting (Fig. 2-3 a). Compound *3* thus gives an intense charge transfer band both in the solid state and in solution. Such phenomena have been studied by many workers (e.g., see [24]). These CTC, or self-complexes, have organic semi-conductor properties and are photosensitive in the visible and near-UV regions.

In another example [25], the charge transfer bands obtained when the donor and acceptor were either two independent molecules or both contained in the same molecule, were studied.

When the donor and acceptor were different molecules, intermolecular charge transfer bands appeared in *concentrated solution* or in the *solid*. Thus, intermolecular complex formation was observed between methyl trinitrobenzoate (TNMB) and pyrene or anthracene (Fig. 2-4).

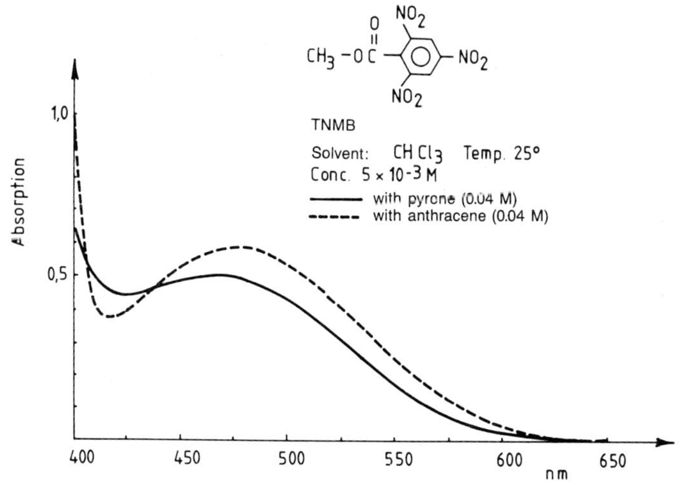

Fig. 2-4. Charge transfer bands for the methyl trinitrobenzoate-pyrene (or anthracene) complex (taken from H.A.H. Craenen et al., *Tetrahedron,* **27**, 2561, 1971).

When the D and A groups were contained in the same molecule (Fig. 2-5), a charge transfer band appeared at concentrations for which it was not detected in the case of intermolecular interactions. There were also chain length and conformation effects, the charge transfer band being most intense for 3-membered chains (n = 1) (Fig. 2-5).

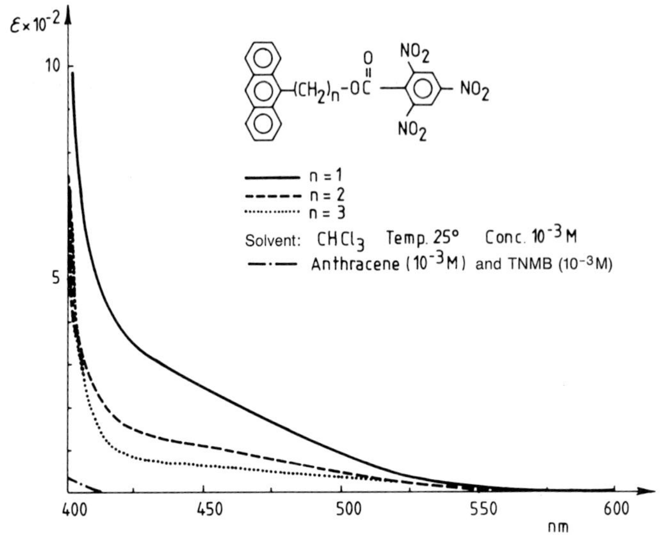

Fig. 2-5. Intramolecular charge transfer bands for three anthrylalkyl trinitrobenzoates. It should be noted that at the same concentration, the intermolecular complex exhibits a very weak band at ca. 400 nm.

When the donor was pyrene and n = 3 (Fig. 2-6), the general appearance of the curve was very similar to that for the intermolecular complex (cf. Fig. 2-4). This indicated that in this case the molecule adopted a conformation favorable to strong interactions between D and A.

It should be noted that the various studies which we have just described mainly concerned relatively short chains.

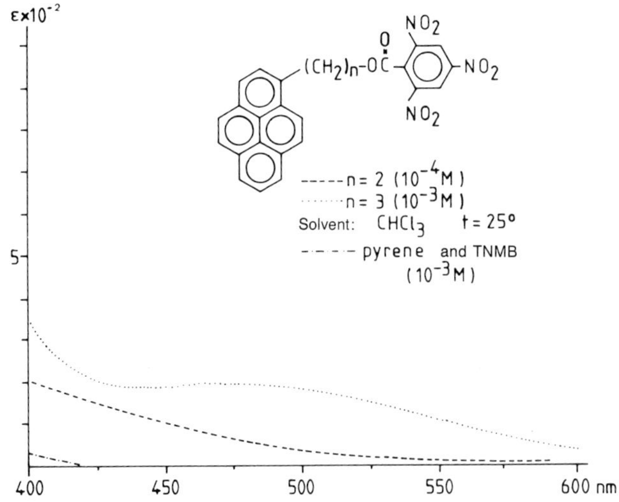

Fig. 2-6. Intramolecular charge transfer bands for two pyrenealkyl trinitrobenzoates. In the case of the intermolecular complex, the remark made in Fig. 2-5 applies here as well.

2.2.2 Excimer-Exciplex

The term *excimer* denotes a molecular complex formed between an excited molecule and another molecule of the same type in its ground state. The term *exciplex* (also sometimes called *heteroexcimer*) denotes the same type of association but between two molecules of different types one being in the excited state (Fig. 2-7). It is important to note that these complexes are formed between species which are not associated in the ground state or only very weakly so (such as AA or AB).

The fluorescence spectra of these complexes are characterized by the appearance of an emission band situated at longer wavelengths than those for emission from the monomer. A schematic representation of this modification of the fluorescence spectra is given in Fig. 2-7. This excimer (or exciplex) emission is accompanied by progressive disappearance of monomer emission when in the case of an excimer the concentration of A increases and in the case of an exciplex the concentration of the non-excited species increases.

$$A + h\nu \longrightarrow A^*$$

$$A^* + A \longrightarrow \underset{\text{excimer}}{(AA)^*} \longrightarrow A + A + h\nu'$$

$$A^* + B \longrightarrow \underset{\text{exciplex}}{(AB)^*} \longrightarrow A + B + h\nu''$$

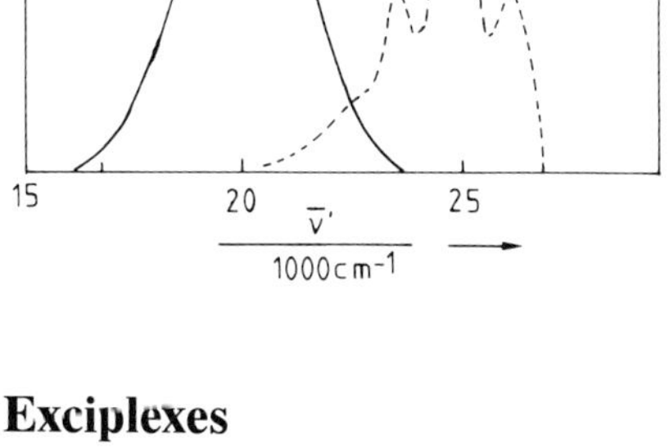

Fig. 2-7. Excimer and exciplex formation on photoirradiation: fluorescence spectra of a monomer, A*, and an exciplex, (AB)* (or an excimer (AA)*).

2.2.3 Intramolecular Excimers and Exciplexes

Intramolecular excimer formation was studied initially by Hirayama for $Ph-(CH_2)_n-Ph$ [26]. Since then, numerous symmetrical excimers, $Ar-(CH_2)_n-Ar$, and heteroexcimers, $Ar-(CH_2)_n-Ar'$, have been studied [23].

According to the *Hirayama rule,* excimer fluorescence is only intense for a trimethylene chain (n = 3). However, other studies have shown that although the optimum chain length is when n = 3, intramolecular complexation of excited states also occurs for other chain lengths.

1-Methylpyrene was found to exhibit blue fluorescence, I_{Mono}, at 400 nm in dilute solution (10^{-5} mol/l) (Fig. 2-8a). In more concentrated solution ($\sim 10^{-3}$ mol/l) an excimer was formed with green fluorescence, I_{Ex}, at 475 nm as a result of the pyrene molecules being face-to-face (Fig. 2-8b).

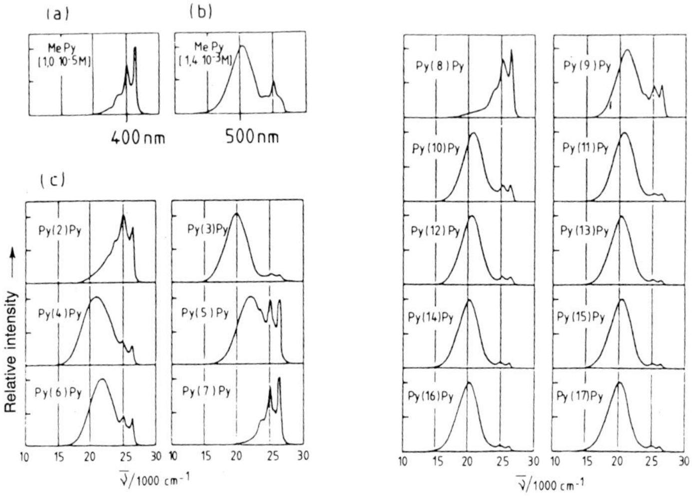

Fig. 2-8. Fluorescence spectra of (a) 1-methylpyrene monomer, (b) 1-methylpyrene excimer, (c) dipyrenylalkanes (taken from K. Zachariasse et al., *Z. Phys. Chem.,* **101,** 267, 1976).

For di(1-pyrenyl)alkanes, $Py(CH_2)_nPy$, at a concentration of 10^{-5} mol/l, both I_{Mono} and I_{Ex} were observed (Fig. 2-9) with very different relative intensities depending on m (Fig. 2-8c).

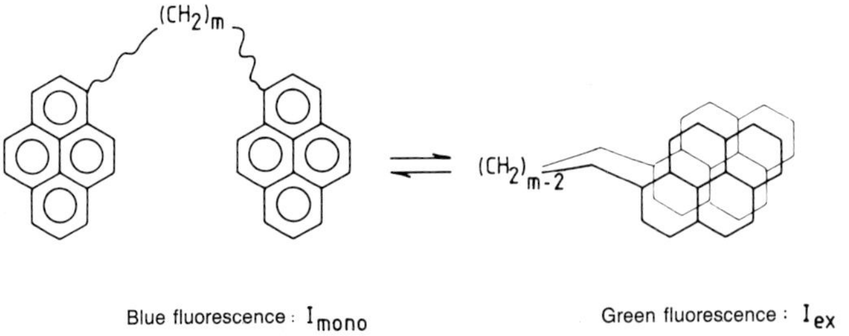

Fig. 2-9. Fluorescence of the separated (monomer) and associated (excimer) forms of dipyrenylalkanes.

This superposing of the monomer and excimer spectra allowed direct observation of the proportion of "ring"-type conformations as a function of chain length [27a]. Depending on the value of m, excimer formation was facilitated or inhibited. It can be seen in Fig. 2-10 that I_{Ex}/I_{Mono} was at a maximum when m = 3 (the Hirayama rule), went through a minimum for m = 7, 8 and increased again to reach a second maximum for m = 13.

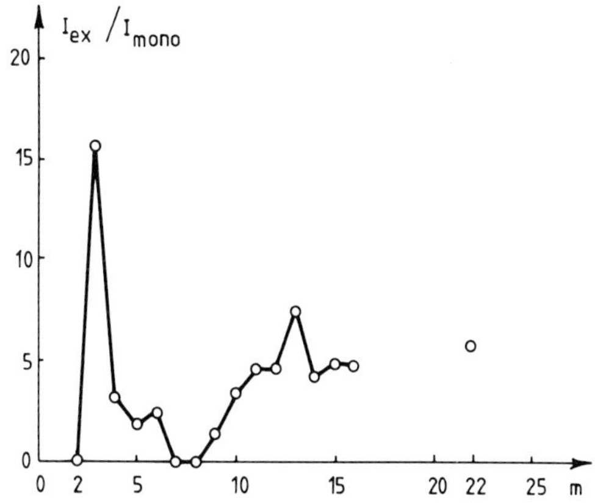

Fig. 2-10. Dependence of I_{ex}/I_{mono} on m for dipyrenylalkanes Py–$(CH_2)_m$–Py.

The strongest fluorescence was obtained for m = 3, this being equivalent to formation of a virtual 6-membered ring (Fig. 2-11).

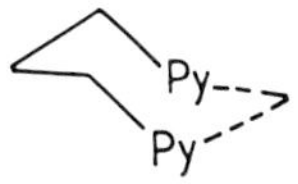

Fig. 2-11. "Virtual" 6-membered ring for dipyrenylpropane.

It was therefore necessary to consider ring size as being in fact n = m + 3. The minimum then occurred around (7, 8) + 3 = (10, 11) and the second maximum at about 16.

The plot of variation of I_{Ex}/I_{Mono} as a function of m recalls that of Ziegler and gives directly the conformation distribution as a function of m.

In a recent review, De Schryver et al. have analyzed the factors influencing intramolecular excimer formation, and in particular the conformations and configurations of bichromophoric systems [27b].

2.2.4 Fluorescence Quenching by Intramolecular Energy Transfer (Internal Conversion)

NR_3 tertiary amines are fluorescent when UV photoexcited (λ_{ex} = 240–250 nm) in alkane solution. The excited species NR_3^* can undergo fluorescence quenching by ground state NR_3 (self-quenching: s. q.):

$$R_3N + R_3N^* \xrightarrow{\text{s.q.}} 2\,R_3N$$

The lifetime of NR_3^*, when isolated, is ca. 25 ns.

Halpern has applied this effect to the study of fluorescence properties of $(CH_3)_2N-(CH_2)_m-N(CH_3)_2$ molecules (denoted TM_m) in which one of the amine functions was photoexcited [28]. The fluorescence spectra (Fig. 2-12) of the monoamine compound DM_{12} and the products TM_2 to TM_5 showed:

- for DM_{12}, a single band at 292 nm and no intermolecular excimer formation even at high concentration,
- for TM_3, a strong emission at 365 nm corresponding to excimer formation (the Hirayama rule),
- for TM_4, emission from both the monomer (292 nm) and the excimer (365 nm),
- for TM_2 and TM_5, the spectra corresponded to emission from an equilibrium mixture of monomer and excimer (with, however, very little excimer in the case of TM_5).

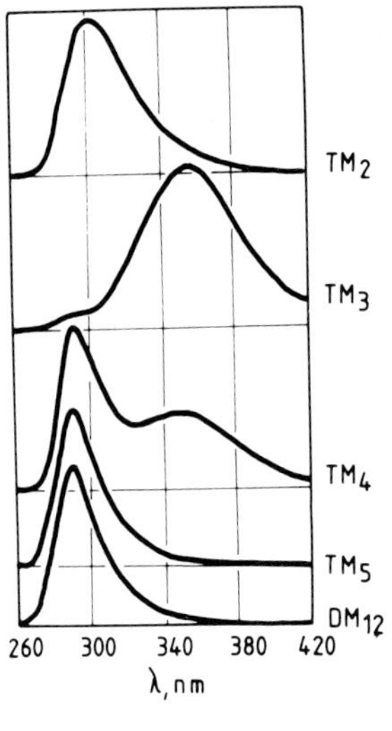

Fig. 2-12. Fluorescence spectra of $(CH_3)_2N(CH_2)_mN(CH_3)_2$ (TM_m) and a monoamine $CH_3-(CH_2)_{11}-N(CH_3)_2$ (DM_{12}) (taken from A.M. Halpern et al., *J. Am. Chem. Soc.,* **101,** 5736, 1979).

For diamines with m greater than 5, the fluorescence spectra were identical to that of the monoamine DM_{12} but the quantum yields were lower and the lifetimes shorter. This was explained by intramolecular quenching. The intramolecular quenching rate constants k_{iq} were determined. Their values are gathered together in Fig. 2-13. It can be seen that for TM_7 to TM_9 (9, 10, 11-membered rings) the values of k_{iq} were the lowest. It can also be noted that the general shape of the curve is similar to that of the other cyclization curves shown above.

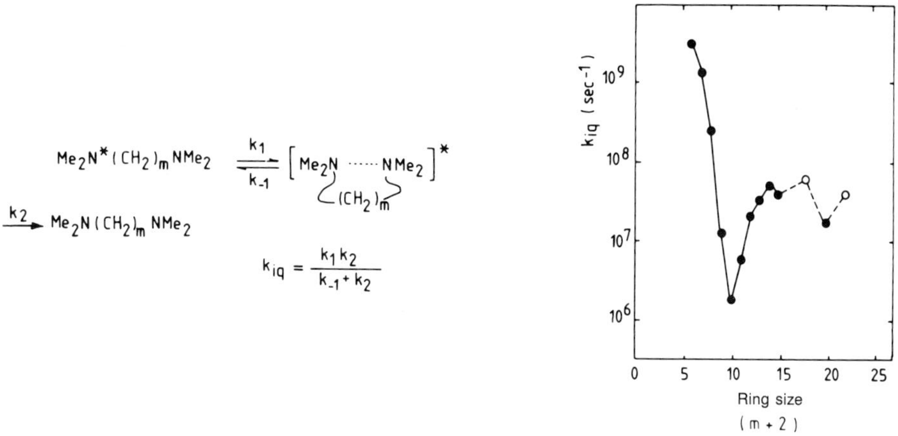

Fig. 2-13. Scheme showing intramolecular self-quenching mechanism for T_m compounds, and plot of self-quenching rate constant vs. ring size. The rate constant TM_3 is too high and was not measured.

2.2.5 Fluorescence Quenching by Intramolecular Electron Transfer

This effect is brought about by charge transfer leading to fluorescence quenching [29]. The various possible processes are portrayed in Fig. 2-14.

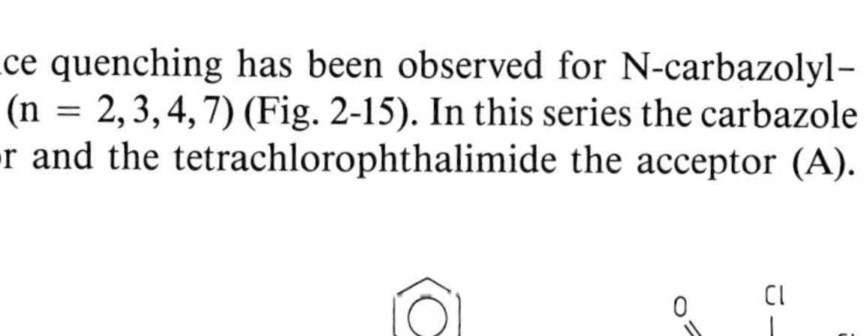

Fig. 2-14. Mechanisms of formation and deactivation of excited states (D*, A) and $(D^{+}A^{-})^{*}$.

This type of fluorescence quenching has been observed for N-carbazolyl-$(CH_2)_n$-tetrachlorophthalimides (n = 2, 3, 4, 7) (Fig. 2-15). In this series the carbazole unit (D*) was the electron-donor and the tetrachlorophthalimide the acceptor (A).

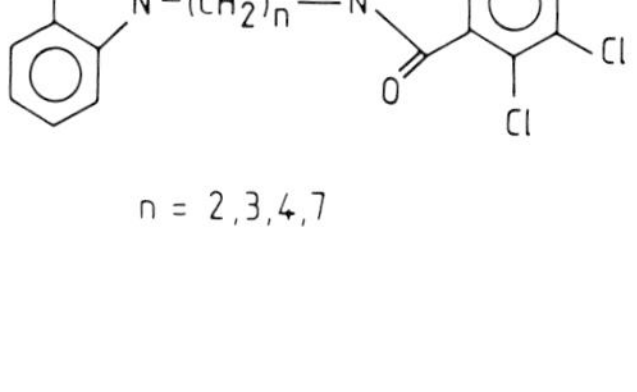

Fig. 2-15. N-carbazolyl-$(CH_2)_n$-tetrachlorophthalimides.

The carbazole emission fluorescence was completely quenched by intramolecular one electron transfer with formation of a dipolar excited state which itself emitted (Fig. 2-16).

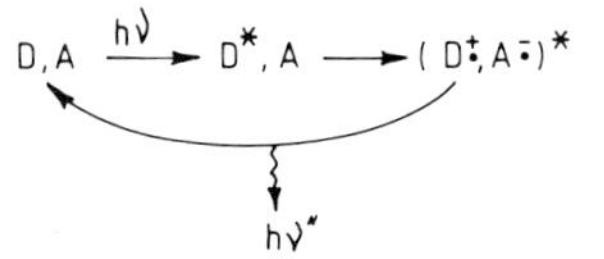

Fig. 2-16. Deactivation by emission from the excited dipolar species.

For n = 7, a weak fluorescence remained, however, due to emission from the carbazole group. In this case too, the shortest chains were those best adapted to the observed reaction. Numerous examples of intramolecular quenching by electron transfer are given in reference [35].

2.3 Cyclizations under Conformational Control

Cyclization reactions, when slower than chain movements, are controlled by conformational effects. It should be noted that distinction between the two types of cyclization, that is diffusion and conformationally controlled, is not always obvious. In certain cases, very detailed studies are necessary to distinguish between the two possibilities (a wide-ranging discussion concerning this point is to be found in reference [23]).

2.3.1 Electron Transfer

Szwarc et al. have studied the intramolecular electron exchange kinetics for the radical anion of the symmetrical species $X-(CH_2)_n-X$ (X = naphthalene or phthalimide). The process and the compounds studied are given in Fig. 2-17.

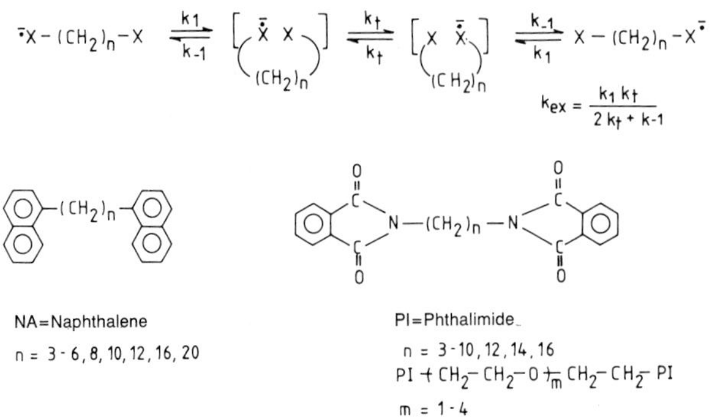

Fig. 2-17. Intramolecular electron exchange in the radical anions $NA(CH_2)_nNA^{\bullet-}$, $PI(CH_2)_nPI^{\bullet-}$ and $I(CH_2CH_2O)_mCH_2CH_2PI^{\bullet-}$.

The EPR spectrum of the radical anion of a symmetrical species $X-(CH_2)_n-X^{\overline{\cdot}}$ depends on the rate of exchange of the electron between the two chain ends. When exchange is slow, the radical is localized, and when it is fast, the radical is delocalized. In intermediate cases, spectral analysis allows one to deduce the rate of transfer. This has been done for various chain lengths at different temperatures [30] (EPR measurements were used to determine the values of $k_{exchange}$).

Thus, for example, in the system $NA-(CH_2)_n-NA^{\overline{\cdot}}$ if the electron is localized on only one of the naphthalene groups, its spin is coupled by hyperfine interactions with those of the 7 protons, and exchange is slow. If, on the other hand, it is coupled with those of the 14 protons of both naphthalene groups, exchange is fast. Spectral analysis thus allows the rate of transfer to be determined.

In the case of the dinaphthyl system, a regular decrease in the intramolecular electron exchange frequency $P\,(s^{-1})$ with increasing chain length was observed, there being no minimum (Fig. 2-18).

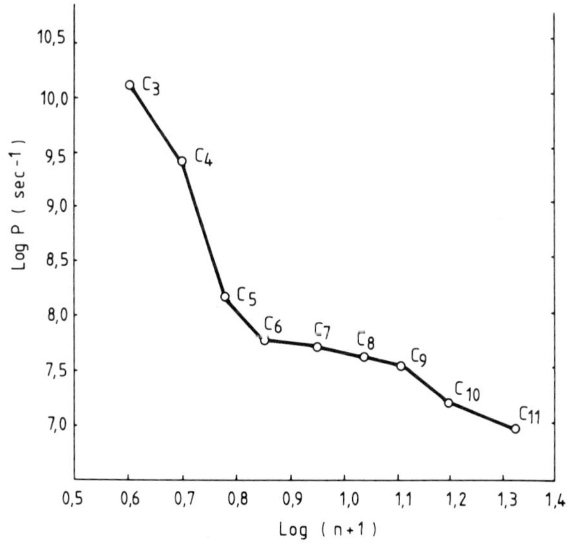

Fig. 2-18. Dependence of intramolecular electron exchange frequency on n for $NA(CH_2)_nNA^{\overline{\cdot}}$ (taken from M. Szwarc et al., *J. Am. Chem. Soc.*, **97**, 3313, 1975).

This absence of a minimum has been attributed to the fact that electron transfer could occur over a great distance of approach between the two naphthalene groups (8–10 Å) and as a result no actual cyclization was required. The strain for medium ring formation was therefore not observed.

In compounds with phthalimide-type end-groups, electron exchange occurred over a shorter distance and the minimum which is characteristic of medium rings was once again apparent (Fig. 2-19).

In the series of compounds containing oxygen atoms, the rate of electron transfer for a given compound is generally greater than that observed for the analogous compound with a hydrocarbon chain of the same length (heteroatom effect).

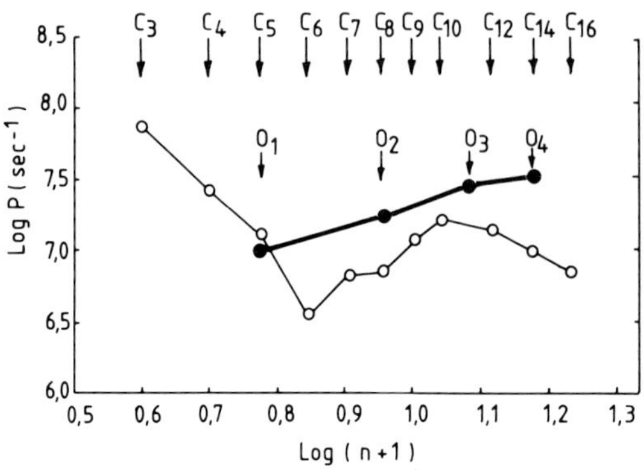

Fig. 2-19. Dependence of the frequency of intramolecular electron transfer on n, the number of atoms separating the PI moieties.
(o) $PI(CH_2)_nPI^{\bar{\cdot}}$;
(●) $PI(CH_2CH_2O)_mCH_2CH_2PI^{\bar{\cdot}}$
(taken from M. Szwarc et al., *J. Am. Chem. Soc.*, **97**, 5834, 1975).

2.3.2 Phosphorescence Quenching

Phosphorescence quenching for benzophenone in the triplet state by a monosubstituted alkene, $RCH = CH_2$, is 100- to 1000-fold slower than is diffusion. Consequently, the intramolecular reaction is under conformational control for the molecules with an alkene-terminated chain attached to a benzophenone group (Fig. 2-20) [31–33].

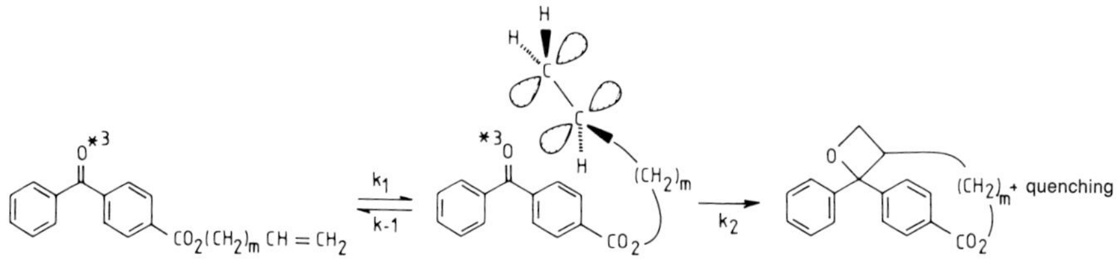

Fig. 2-20. Intramolecular phosphorescence quenching of triplet state benzophenone by an alkene group.

The rate constant for total intramolecular quenching, k_{iq}, can be calculated from the values for the triplet state lifetimes extrapolated to infinite dilution, of on the one hand the long chain ester derivative of benzophenone, τ^o_m, and on the other the corresponding methyl ester, τ^o_{Me} (τ^o_{Me} allows estimation of the intermolecular contribution to quenching).

$$k_{iq} = 1/\tau^o_m - 1/\tau^o_{Me} = (k_1/k_{-1})\, k_2$$

When m is greater than 10, this constant k_{iq} has to be corrected because of quenching due to hydrogen abstraction from the chain by the ketone:

$$k_o\,(m) = k_{iq}\,(m) - k_H\,(m)$$

One thus obtains the rate constant for intramolecular phopshorescence quenching of the ketone excited state by the terminal alkene group of the long chain ester. This constant can be expressed in terms of the effective concentration (or effective molarity):

$$C_{\text{eff}}\,(\text{m}) = K_{\text{o}\,(\text{m})}/k_{\text{q}}^{(2)}$$

where $k_{\text{q}}^{(2)}$ is the rate constant for the bimolecular reaction between the methyl ester and 1-pentene. Fig. 2-21 shows the dependence of C_{eff} on m.

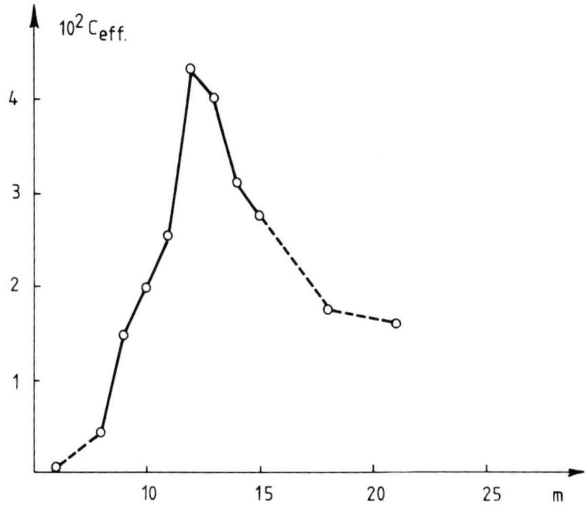

Fig. 2-21. Plot of effective concentration for intramolecular quenching of the benzophenone triplet vs. chain length (taken from M. A. Winnik et al., *Chem. Phys. Letters,* 77, 73, 1981).

For short chains, m < 8, there is no intramolecular phosphorescence quenching because the double bond cannot get sufficiently near to the carbonyl group.

The effective concentration is at a maximum for m = 12. When m > 12, unfavorable entropy effects (increase in the number of conformations) leads to a decrease in C_{eff}.

The few examples given above show the importance of photochemical methods applied to the study of chain flexibility. These methods are developing fast [34, 35], and have been now extended to cyclization of polymeric chains [36].

2.4 Internal Ring Closure. Remote Functionalization

Formation of a ring or pseudo-ring from linear chains can be used to direct chemical reactions towards specific sites of a molecule remote from the functional groups present. The basic principal, which is schematically portrayed in Fig. 2-22, includes the following steps: fixing a carrier to the molecule, reaction leading to functionalization, release of the product with possible recycling of the carrier.

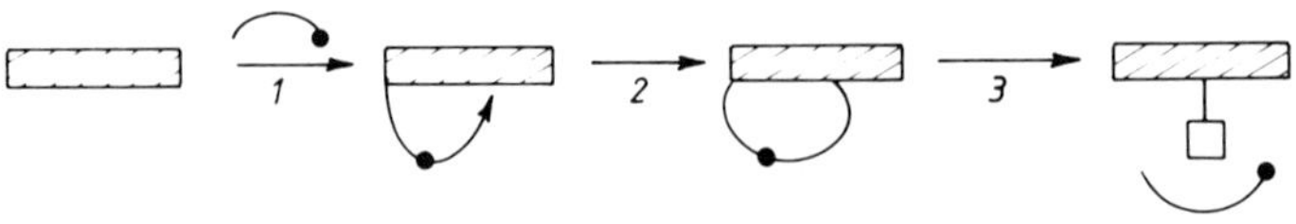

Fig. 2-22. Schematic representation of remote functionalization.

2.4.1 Chromic Anhydride Oxidations

Oxidation by CrO_3 of fatty acid esters and of esters of long chain alcohols led to monoketo-esters in reasonable yields [37a] (Fig. 2-23).

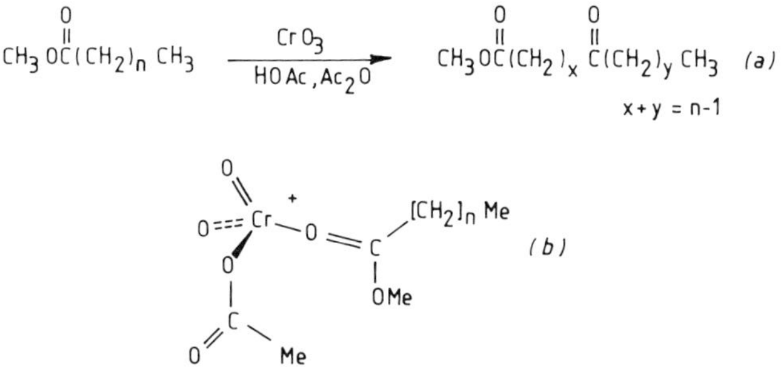

Fig. 2-23. (a) Monoketo-ester formation by oxidation of fatty acid esters, (b) presumed structure of the reactive intermediate.

The distribution of carbonyl groups along the chain (Fig. 2-24) was incompatible with random bimolecular attack by CrO_3. Thus, it should be noted that the CH_2 groups 6 to 9 atoms distant from the ester function were those most affected.

A possible explanation was $CrO_2(OAc)_2$ formation in the medium followed by substitution of one of the two acetate groups by the ester function of the fatty acid to give the reactive intermediate portrayed in Fig. 2-23b. Intramolecular oxida-

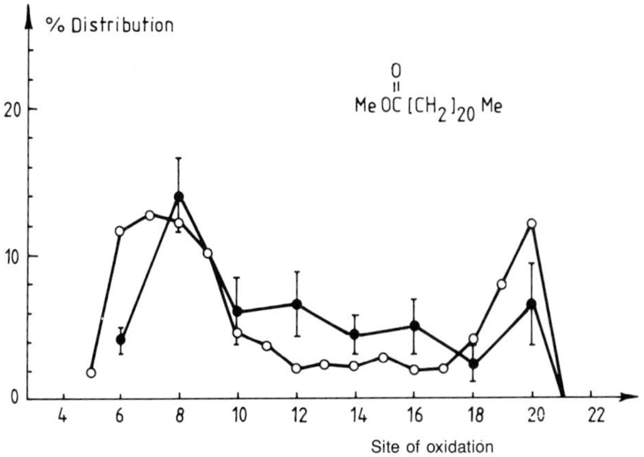

Fig. 2-24. CrO_3 oxidation of methyl docosanoate: oxidation site distribution; (o) experimental data, (●) calculated results (taken from M. A. Winnik et al., *J. Chem. Soc., Chem. Comm.,* 156, 1976).

tion of the methylene groups would then occur within this intermediate [37b]. On the basis of this reactive intermediate, the probabilities of oxidation of the various CH_2 groups in the molecule have been calculated (see Fig. 2-24).

In the case of macrolides, CrO_3 oxidation gave a narrower distribution of ketone products than for the reaction on an open chain [38] (Fig. 2-25).

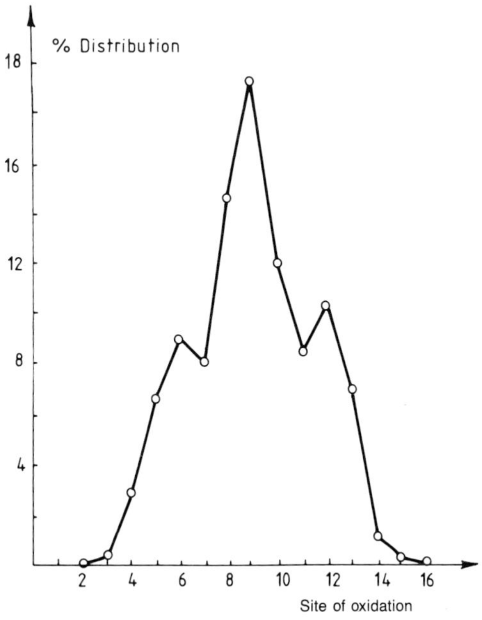

Fig. 2-25. CrO_3 oxidation of hexadecanolide (n = 14, see Fig. 2-26): oxidation site distribution (taken from T. Money et al., *J. Chem. Soc., Chem. Comm.*, 561, 1976 and *J. Chem. Soc. Perkin I.*, 896, 1979).

The illustrated distribution curve concerns hexadecanolide oxidation, an example for which the preferred distance between the ester group and the oxidation site is between 8 and 10 carbon atoms. Depending on macrolide size, there are slight variations in this distance.

The postulated reactive intermediate (Fig. 2-26) provided an explanation for the narrower distribution of the products formed.

The ester functional group was utilized in the intermediate as an internal template (Fig. 2-27).

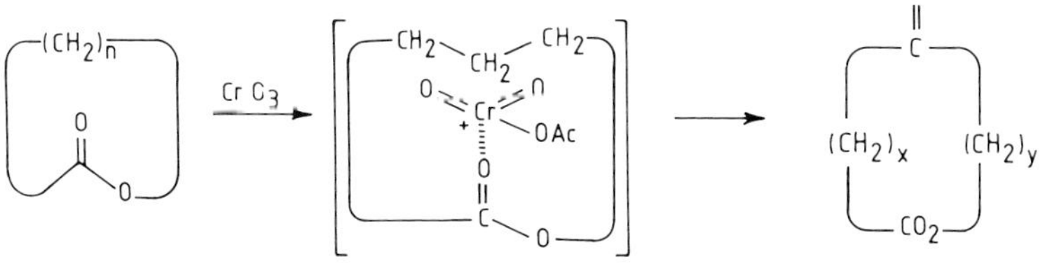

Fig. 2-26. Proposed reaction mechanism for CrO_3 oxidation of macrolides.

Fig. 2-27. Schematic representation of the role of the ester group as an internal template for the oxidizing species $[O_2Cr(OAc)]^+$.

2.4.2 Photochemical Functionalization by Benzophenone

The photophysical properties of benzophenone have been used by Breslow [39–43] to bring about remote functionalization. Very rapid singlet-triplet intersystem crossing is characteristic of this system. The lifetime τ of a benzophenone triplet depends on several factors:

$$^3Ph_2CO \xrightarrow{k_d} Ph_2CO \qquad \text{radiationless decay}$$

$$^3Ph_2CO \xrightarrow{k_P} Ph_2CO + h\nu \qquad \text{phosphorescence}$$

$$^3Ph_2CO + RH \xrightarrow{k_r} Ph_2\dot{C}OH + R^{\cdot} \qquad \text{hydrogen abstraction}$$

$$\tau\ ^3Ph_2CO = 1/(k_d + k_P + k_q \cdot c\,(Q))$$

where $k_q \cdot c\,(Q)$ represents the rate of quenching reactions, hydrogen abstraction being included in this terms. The last process becomes important under certain conditions. Thus, the triplet state lifetime is shorter in the case of intramolecular hydrogen transfer (benzophenone derivative *1*) than for intermolecular transfer (mixture of *2* and *3*). These differences are quite apparent from the two plots in Fig. 2-28.

Benzophenone, when photoexcited to the triplet state (Fig. 2-29), can therefore attack carbon-hydrogen bonds so forming a diradical which in turn gives a new carbon-carbon bond or a double bond.

Photolysis of a series of *long chain ester* benzophenone derivatives (e.g., *3*, Fig. 2-30a) afforded cyclization products (e.g., *4*), these being the only products detected. The latter, after dehydration, were transformed into benzophenonecarboxylic acid and keto-alcohols to enable identification of the positions of attack [39, 40].

For the three esters studied (C_{12}, C_{16}, C_{20}), the distribution of oxidation sites (see plot in Fig. 2-30b) was not uniform and reflected the relative probability of cyclization onto each CH_2 group. The reaction was not very selective, except for the C_{12} chain for which 55% of the attack occurred at C-11 and about 30% at C-10. Up to chain length C_{16}, preferential attack at the penultimate carbon atom was observed.

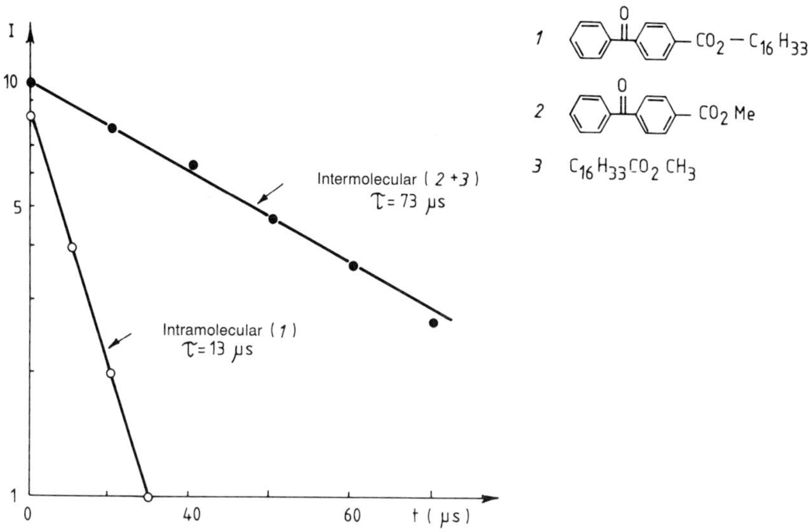

Fig. 2-28. Comparison of lifetimes of a benzophenone triplet undergoing intra- or intermolecular hydrogen transfer (taken from R. Breslow, *Chem. Soc. Rev.,* **1**, 553, 1972).

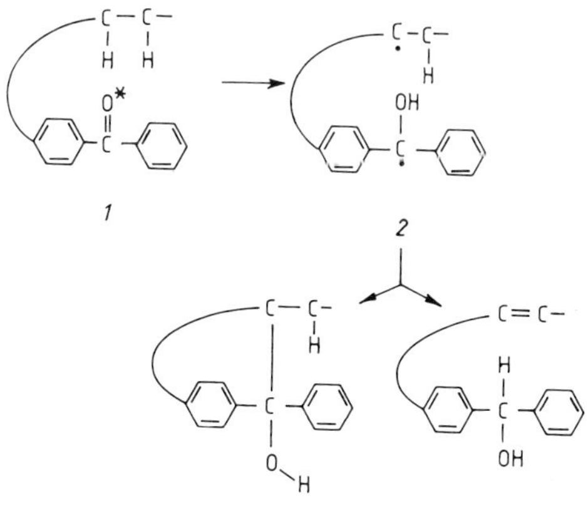

Fig. 2-29. Mechanism for remote functionalization by benzophenone in the triplet state.

This method of functionalization has been extended to *steroids.* In this case, models showed a limited number of axial hydrogen atoms as being accessible (Fig. 2-31 a), and selective attack on certain of them might be expected if the reagent were suitably chosen. This hypothesis has in fact been verified [39–41].

According to molecular models for n = 2, the carbonyl group can reach the hydrogen atoms in positions 7, 14 and 12 which could therefore be functionalized (Fig. 2-31 b). In the case of this compound, attack at C-14 was confirmed (Fig. 2-32)

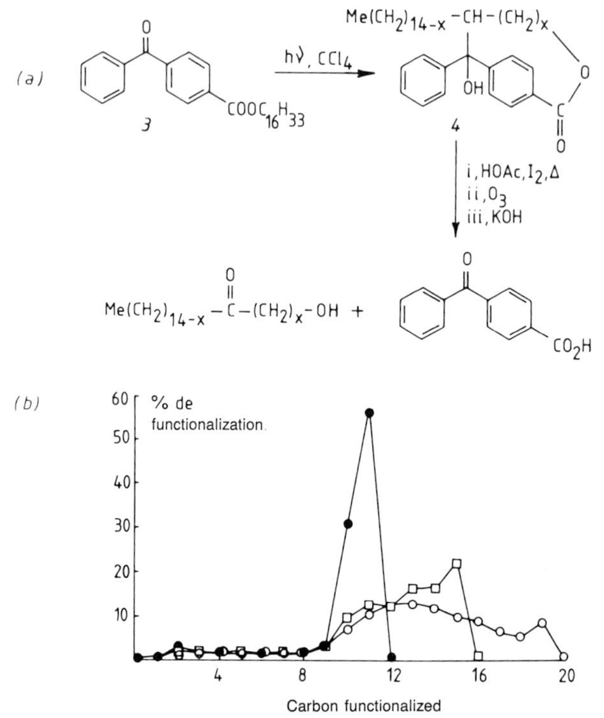

Fig. 2-30. (a) Remote functionalization of n-hexadecanol ester and method for determination of the products formed, (b) distribution curve for carbons functionalized by photolysis of: (●) n-dodecanol, (□) n-hexadecanol and (o) n-eicosanol esters (taken from R. Breslow et al., *J. Am. Chem. Soc.*, **100**, 1213, 1978).

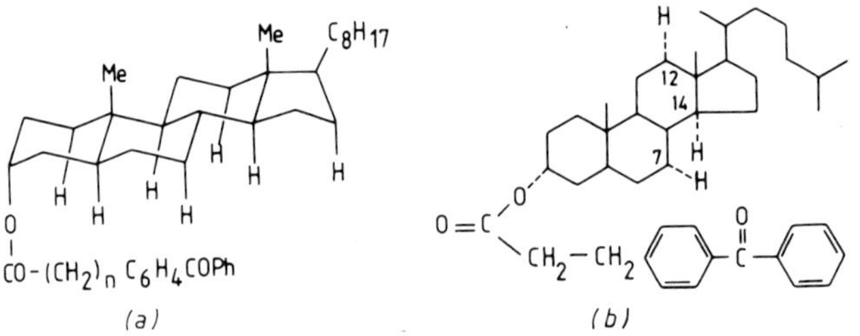

Fig. 2-31. Remote functionalization of steroids. Positions susceptible to functionalization:
(a) general case,
(b) n = 2.

by formation of a single unsaturated steroid product, the ester of Δ^{14}-cholesten-3α-ol (*1*), and formation of a new C-C bond at C-14 shown by oxidation of the alcohol *2* with lead tetraacetate to give $\Delta^{8(14)}$-cholesten-3α-ol (*3*) [44]. The scheme in Fig. 2-29 is consistent with the formation of these two products.

Attack at C-7 was also observed. A C-C bond of the type in *2* was formed at C-7. Treatment with $Pb(OAc)_4$ gave, after hydrolysis, Δ^6- and Δ^7-cholestenols [42].

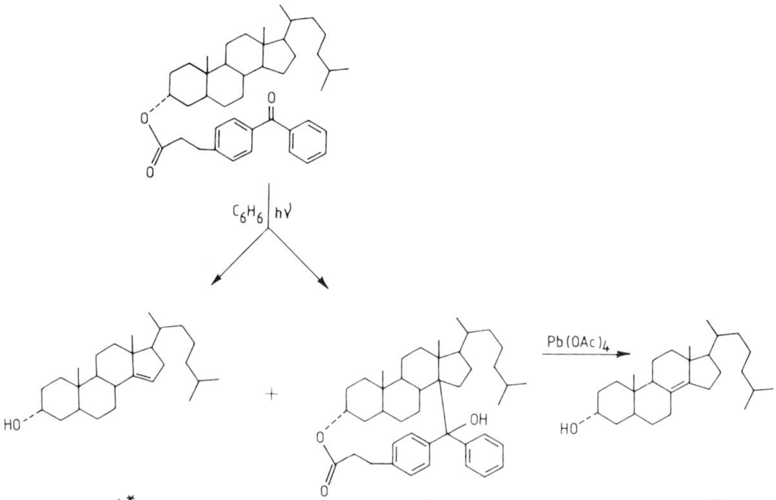

*1** *2* *3**

Fig. 2-32. Steroid functionalization at C-14 by photo-excited benzophenone. *These products were obtained after hydrolysis of the ester.

2.4.3 Remote Chlorination

Russel and Tanner have shown that dichloroiodobenzene can be used as an agent for the photochlorination of alkanes [45]. The mechanism of this photochlorination is given in Fig. 2-33a.

The PhICl˙ radicals formed selectively attacked the tertiary rather than the primary or secondary carbon-hydrogen bonds. This selectivity can be attributed to the decrease in reactivity of Cl˙ when attached to PhI, and also to the greater stability of the tertiary radical formed.

This chemical selectivity therefore reduces the number of sites of attack and can be combined with geometrical selectivity due to the relative positions of the substrate and reagent when these are fixed on the same molecule, as shown by Breslow [43]. In order to achieve these directed chlorinations, two methods have been used. One involved a radical chain reaction (b), and the other a radical-relay mechanism (c). These processes are depicted schematically in Fig. 2-33. Method (b) is equivalent to (a) applied to an intramolecular reaction. In process (c), the radical $-CH_2-C_6H_4ICl^{\cdot}$ is generated by using a radical-relay. Each of these methods will be illustrated by an example and we shall see that the combination of the two factors (chemical and geometrical selectivity) has enabled selective functionalizations to be performed.

Thus, the m-iodobenzoate ester of 3-cholestanol has been converted into the dichloro derivative *1* by Cl_2, then irradiated or treated with a radical initiator

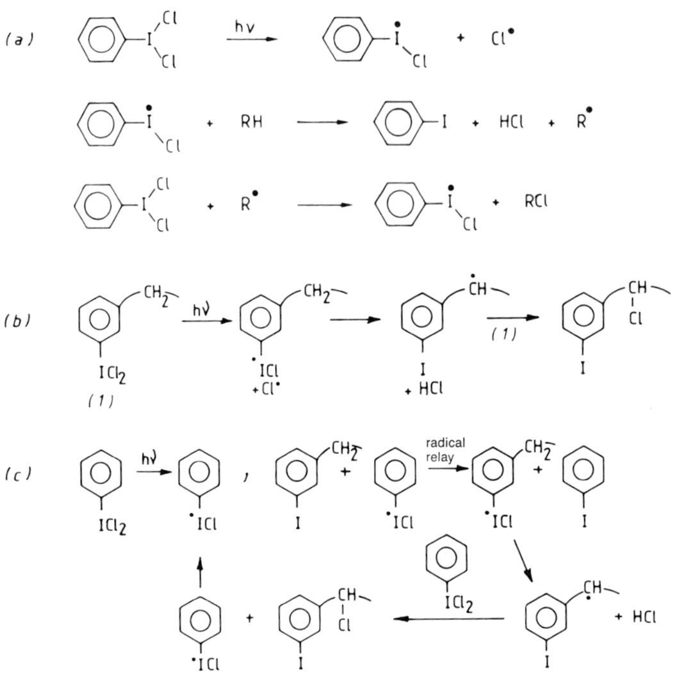

Fig. 2-33. (a) Mechanism for the chlorination of an alkane by dichloroiodobenzene, (b) remote chlorination by a radical chain reaction, or (c) by radical relay.

(Fig. 2-34). Attack occurred only at the tertiary hydrogen at C-9 as shown by exclusive formation of *5* obtained after treatment of *4* with base [46, 47].

The hydrogen abstraction was unimolecular (reaction *2* → *3*) but, in order to abstract a chlorine atom, the radical formed had to undergo a bimolecular reaction with another molecule of *1* (reaction *3* → *4*).

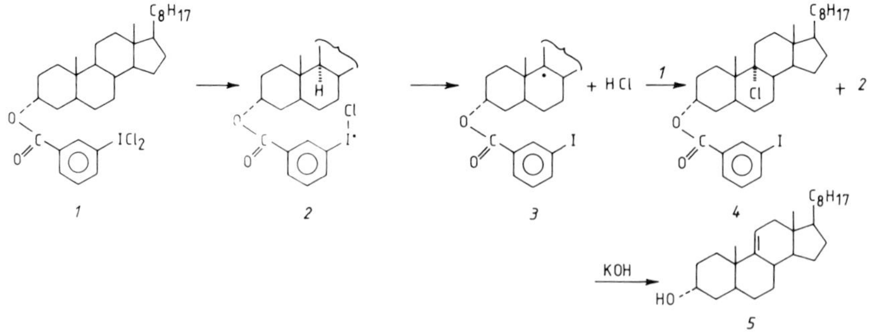

Fig. 2-34. Selective attack at C-9 of the m-iodobenzoate ester of 3-cholestanol. Compound *1* is obtained by addition of Cl_2 to the iodo precursor: $-COPhI + Cl_2 \rightarrow -COPhICl_2$.

This method was not entirely satisfactory since formation of *1* by treatment with Cl_2 might pose problems in the case of molecules containing certain fragile functional groups. *Chlorination by a radical-relay mechanism* permitted generation of the intermediate reactive radical *2* without passing via *1*, and so faster and more selective chlorinations could be achieved [47, 48]. This process has been applied to the synthesis of dihydrocortisone acetate [49] according to the scheme in Fig. 2-35.

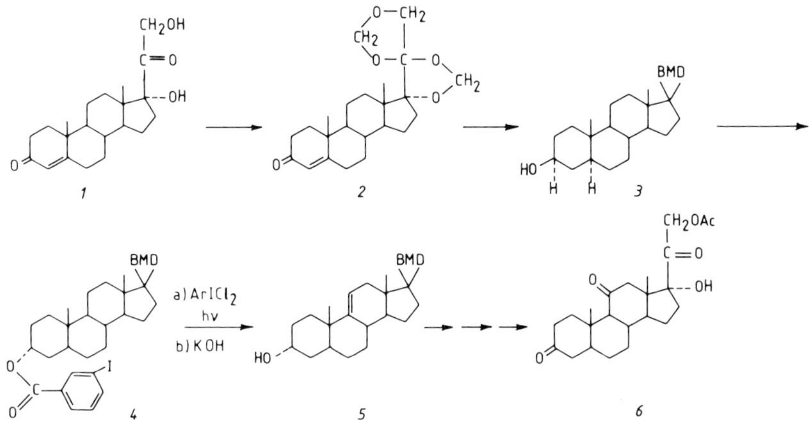

Fig. 2-35. Synthesis of dihydrocortisone acetate from cortexolone by the radical-relay method, the mechanism for which is given in Fig. 2-33.

The starting material was cortexolone (*1*) in which the dihydroxyacetone moiety was protected (*2*) before conversion of the alcohol *3* into the m-iodobenzoate ester *4*. An external ArICl˙ radical subsequently transferred a Cl atom to the iodobenzene derivative *4* and hence to the 9 position. The olefin was formed by action of KOH and then transformed into dihydrocortisone acetate, *6*, in a series of steps consisting of hydroboration, oxidation, and finally deprotection. The overall yield was 30% from cortexolone.

Diphenylsulfide could also be used for radical relay [45] in the selective chlorination of a steroid at C-14 [50].

2.4.4 Temporary Association Between the Reagent and the Substrate

In the hope of making functionalization of flexible chains in solution more selective, reactions have been carried out in organized phases such as micelles and membranes [51]. These systems were not sufficiently rigid to orient favorably the substrate and the reagent. The selectivities were therefore low.

Selective functionalization using photochemistry has been achieved by immobilizing the linear substrate through coordination of the two chain ends to a rigid reagent [52] (Fig. 2-36).

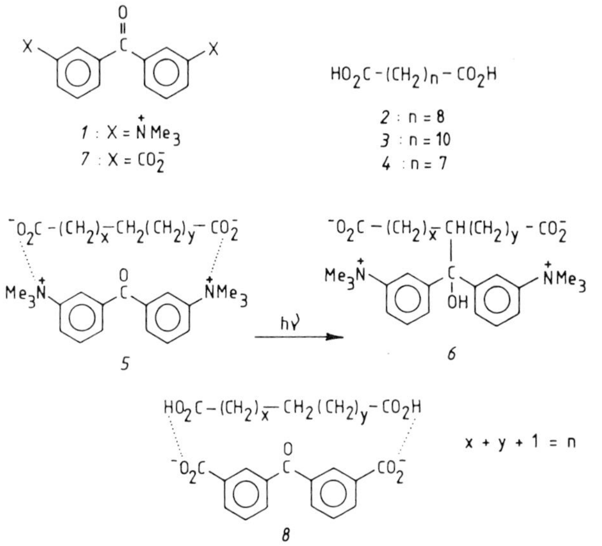

Fig. 2-36. Functionalization of flexible chains by association of the two chain ends of the substrate to a rigid reagent.

Two modes of interaction have been investigated:

- ion pairs, *5*, formed in water between a quaternary ammonium dication, *1*, and long chain dicarboxylate ions derived from diacids, *2*, *3* or *4*,
- hydrogen bonds, *8*, formed between the dicarboxylate dianion, *7*, and a dicarboxylic acid *2*, *3* or *4* in a non-polar solvent.

Table 2-1. Percent functionalization of decandioic, (*2*) dodecandioic (*3*) and nonandioic (*4*) acids at particular carbons by benzophenone derivatives *1* and *7*.

Reagent	Substrate	Carbon functionalized %				
		C-2	C-3	C-4	C-5	C-6
1	*2*	2.7	1.7	2.7	93	
1	*3*	1	0.2	3.6	62	34
1	*4*	2.6	1.4	22	74	
7	*2*	4.3	0.2	13.7	81	
7	*3*	3.8	0.5	5.6	42	49
7	*4*			49	51	

(Taken from R. Breslow et al. *J. Am. Chem. Soc.* **103**, 2905, 1981).

The results of photolysis of these different systems (Table 2-1) showed that:

- with substrate *2* (n = 8) attack was very selective at the two equivalent central CH_2 groups (93% and 81% depending on association type),
- with substrate *3* (n = 10) the product distribution was wider. The size of reagent *1* or *7* was less well matched to that of the chain. The latter was therefore not fully extended and hence it was more mobile,
- with substrate *4* (n = 7) attack by reagent *1* gave reasonable selectivity towards the (single) central CH_2 group,
- in all cases, reagent *1* (ion pair formation) led to greater selectivity than did reagent *7* (hydrogen bond formation). Control reactions with reagents having only one possible binding site turned out to be totally negative.

Chapter 3

Ring Closure Methods

In this chapter, we examine some methods for the formation of macrocycles.

3.1 High Dilution Technique

3.1.1 Principle

In order to obtain a high cyclization yield, cyclization must be favored at the expense of polycondensation.

The concept of the high dilution technique was first formulated and applied in Strasbourg in 1912 by P. Ruggli in the formation of cyclic amides [53] (Fig. 3-1).

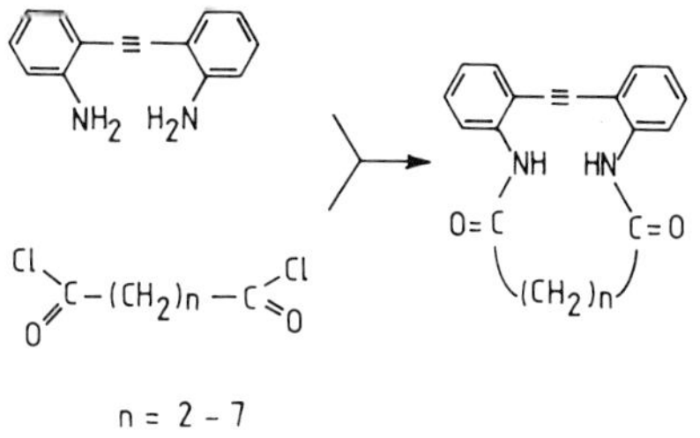

Fig. 3-1. Ruggli's method for the formation of cyclic amides (high dilution method).

It is based on the following factors. The intramolecular ring closure reaction is first order, its rate being proportional to concentration. The intermolecular condensation reaction is second order, and therefore its rate is proportional to the square of the concentration. It follows that dilution should favor the intramolecular reaction.

The two points which we have just mentioned constitute the pragmatic approach to the problem of macrocycle synthesis. The more rigorous procedure, which consists of determining the effective molarity of the reaction (k_{intra} and k_{inter}) and then using the results obtained to determine the ideal concentration, is rarely used.

The method of high dilution was developed and widely used by Ziegler and Lüttringhaus [1], and an initial theoretical analysis was established by Stoll [6] and Salomon [54] in about 1935.

This technique is surveyed in a wide-ranging review [55a]. Certain reactions are described there with complete experimental details thus giving the reader a better insight into the actual operating conditions. The review covers extensively the formation of C–S bonds without, however, neglecting other bonds (C–N, C–C, C–O). A supplement to this review has recently appeared [55b].

3.1.2 Implementation

Cyclizations using the method of high dilution can be carried out in either homogeneous or heterogeneous medium.

The method in *liquid homogeneous medium* is that most frequently used, and for a two component reaction consists of adding the two solutions dropwise into a large quantity of solvent with very vigorous stirring. The important point is not the initial concentration of the reactants but rather the stationary concentration of the intermediate which cyclizes or undergoes polycondensation. This stationary concentration should be the lowest possible. Such is the case for fast reactions, for which prolonged contact times are avoided and which are thus to be recommended.

A long-known example of the procedure in *heterogeneous medium* is the acyloin cyclization developed by Prelog [56] and Stoll [57] (Fig. 3-2).

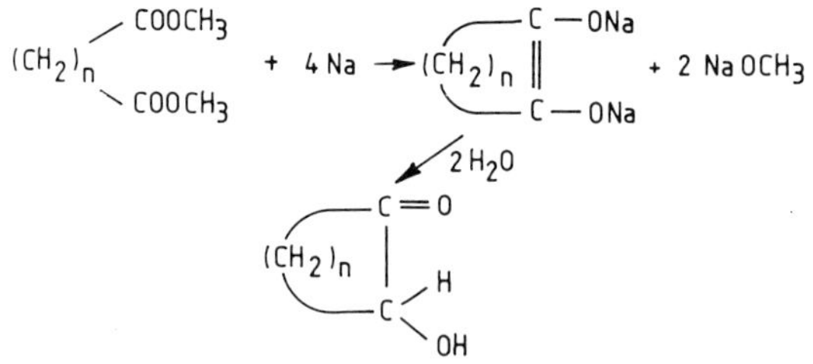

Fig. 3-2. Ring closure of diesters to give acyloins by a heterogeneous process.

The diester was added to a suspension of liquid sodium in boiling xylene with very vigorous stirring and rigorous exclusion of air and moisture. The ring closure yields were high even for medium rings, being 30% to 60% for 9- and 10-membered rings. However, 7- and 8-membered rings required efficient conditions of high dilution (very slow addition of the reactant). The high yields have been attributed in part to coordination of the diester to the metal surface, hence isolating the molecules from each other and bringing the chain ends together.

A more rational approach towards high dilution in solid-liquid heterogeneous medium is the use of a *polymer support* to bind the reactant which leads to site isolation on a solid matrix [58–60]. A "hyperentropy" effect is thus obtained through total immobilization of the reactant (Fig. 3-3).

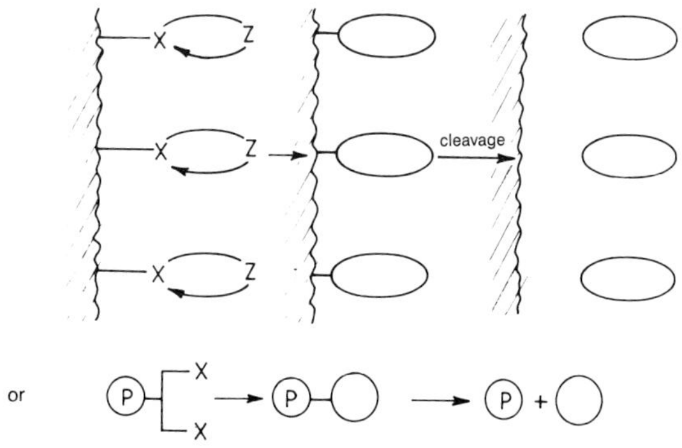

Fig. 3-3. Scheme for cyclization in solid-liquid heterogeneous medium on a polymer support.

A cyclopeptide has been synthesized on a polymer support using this method [60b]. The principle of the procedure is illustrated in Fig. 3-4. The first step consisted of binding the amino-acid *5* (which had to be trifunctional such as glutamic acid, asparagine, tyrosine, serine, cysteine etc.) onto the polymer. By using the amine and carboxylate functional groups of *5*, two linear chains were elaborated in parallel to afford as end-products polypeptides having amino-acids *1* and *9* at the two chain ends. This was followed by cyclization and then cleavage of the cyclopeptide from the polymer support.

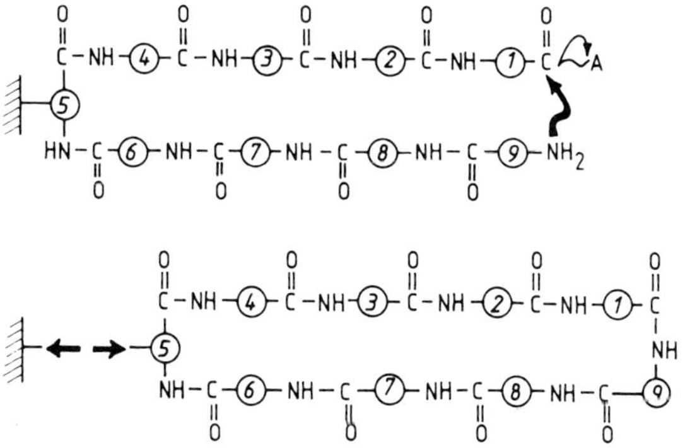

Fig. 3-4. Synthesis of a cyclopeptide on a polymer support.

However, despite the attractive nature of the idea, the results obtained are often indifferent. In fact, the sites are only imperfectly isolated and reactions between them take place.

3.2 Depolymerization

The principle of this method, which has been known since long [61], is to obtain cyclic products from linear polymers by the route shown in Fig. 3-5.

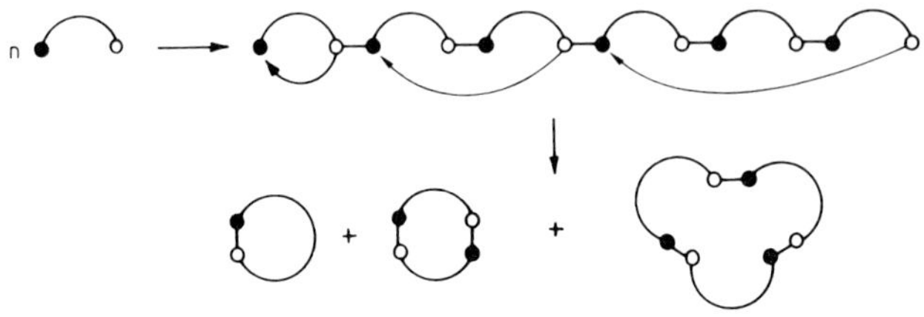

Fig. 3-5. Ring formation from linear polymers: depolymerization.

In this way, one can obtain monomers, dimers and trimers.

The method has been principally applied to depolymerization of polyesters prepared by condensation of an ω-hydroxyacid or of a diacid with a diol.

Thus, the various polyesters obtained by condensation of succinic acid with polymethylenediols ($HO-(CH_2)_m-OH$; m = 2–5 and 7–9) have been depolymerized. The reactions of intramolecular transesterification give rise to various cyclic products which are separated from the reaction mixture by distillation as they are formed.

In practice, the polymer is heated to ca. 270 °C *in vacuo* (⩽ 1 mm Hg) in the presence of a transesterification catalyst ($SnCl_2 \cdot 2H_2O$; $MnCl_2 \cdot 4H_2O$; $MgCl_2 \cdot 6H_2O$, etc.) [62]. The volatile products distil, and are recovered and analyzed. The distillate generally consists of a mixture of cyclic monomer and cyclic dimer (Fig. 3-6).

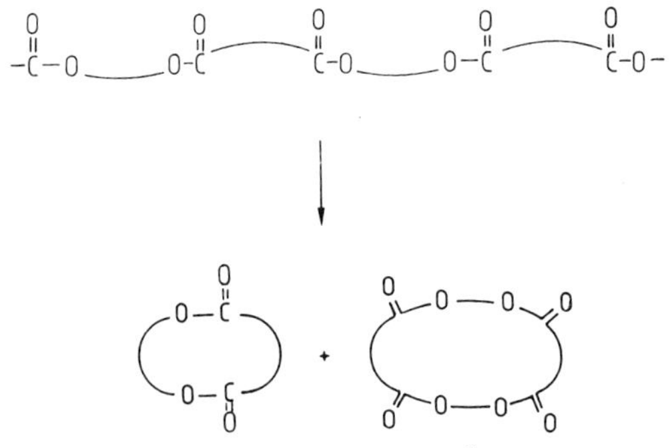

1 2

Fig. 3-6. Cyclic monomer and cyclic dimer afforded by depolymerization of polyesters.

For example, pentamethylene succinate (Fig. 3-7) afforded 46% of monomer *1* (11-membered ring; point d on the monomer plot) and 26% of dimer *2* (22-membered ring; point d on the dimer plot).

It should be noted that the 8- and 9-membered rings were formed with difficulty. The 10-membered monomer was obtained in remarkably high yield (70%). The highest yields were observed for 15- (83% monomer), 16- (95% dimer), and 18-membered (89% dimer) rings [62].

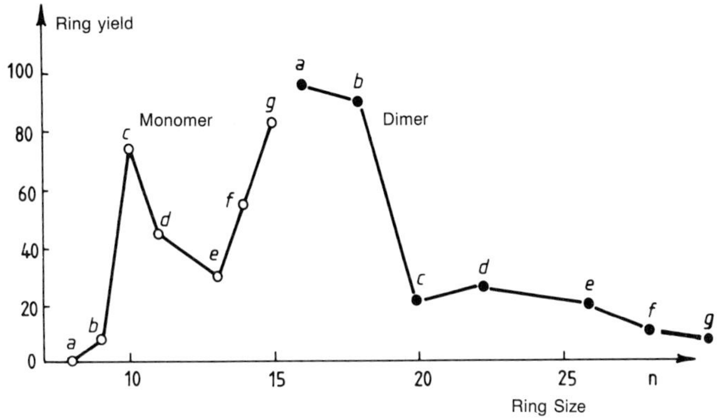

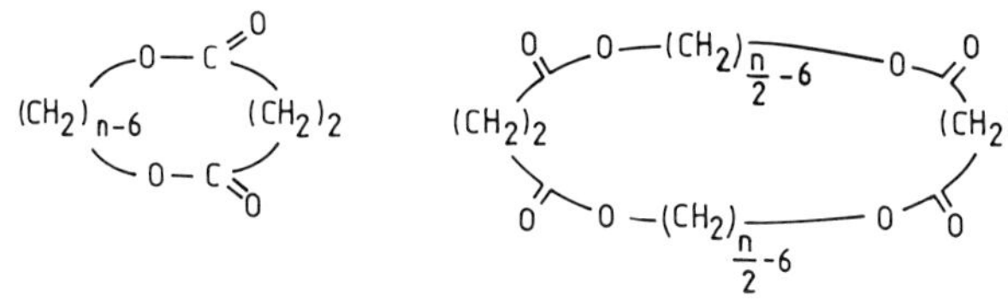

1
m - n 6

2
$m = \frac{n}{2} - 6$

m = 2-5;7-9

Fig. 3-7. Depolymerization of polyester derivatives of succinic acid (α, ω-diols + succinic acid): $-O(CH_2)_mOOC(CH_2)_2COO\,(CH_2)_mCOO-$. Each type of polyester affords a mixture of monomor and dimer. Thus, when m = 2, products (a) are formed. In the same way, m = 3 → (b), m = 4 → (c), m = 5 → (d), m = 7 → (e), m = 8 → (f), m = 9 → (g) (taken from W. H. Carothers, *J. Am. Chem. Soc.,* **57**, 929, 1935).

3.3 Templated Ring Closure

3.3.1 Principle

Many cyclization methods are based on the use of a temporary or permanent template, and on specific reactions which depend on the nature of the reacting groups between which the bond leading to ring closure is formed. Cyclization may take place using either an internal or external template (endo- or exo-template).

a) Endo-template

When the formation of a macrocycle is based on a smaller ring which is then ring-enlarged to include all the pre-existing parts of the molecule, this process can be described as occurring by the use of an endo-support. An example is the insertion of a side chain into the ring (Fig. 3-8a).

Similarly, the "zip" reaction is based on ring expansion by the process of sliding along a side chain (Fig. 3-8b).

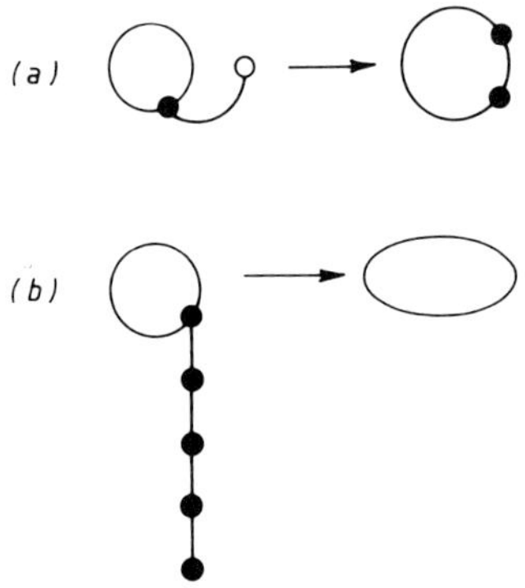

Fig. 3-8. Ring expansion using (a) an internal template or (b) successive internal templates.

Another approach is initial formation of a bicyclic system containing a cleavable bond, the subsequent breaking of which gives a macrocycle (Fig. 3-9). Bonds which have been used to this end are C–N, N–N, C–C, and C=C.

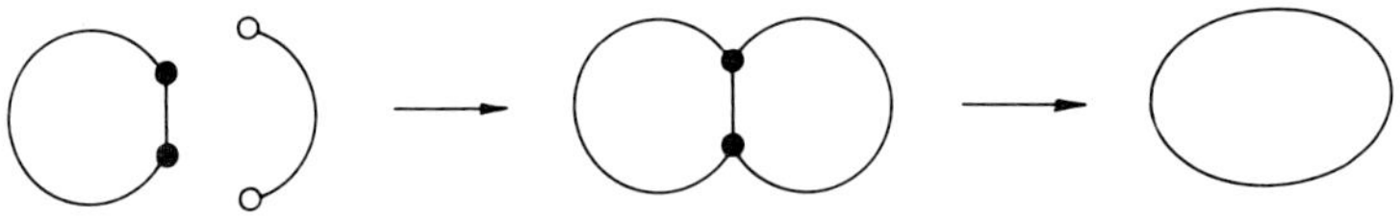

Fig. 3-9. Large ring formation by bond cleavage in a bicyclic system.

b) Exo-template

A temporary center or group, which may be either ionic or covalent, may be used in cyclization reactions. This serves as a template on which the macrocycle is assembled, and which is subsequently eliminated.

Metal cations are by far the oldest-known and most used templates. The process may occur in a single step in which assembly and connexion of the units occur simultaneously around the ion (Fig. 3-10a). Alternatively, there may be several steps in which the placing of one type of unit is followed by bridge formation by another species (Fig. 3-10b).

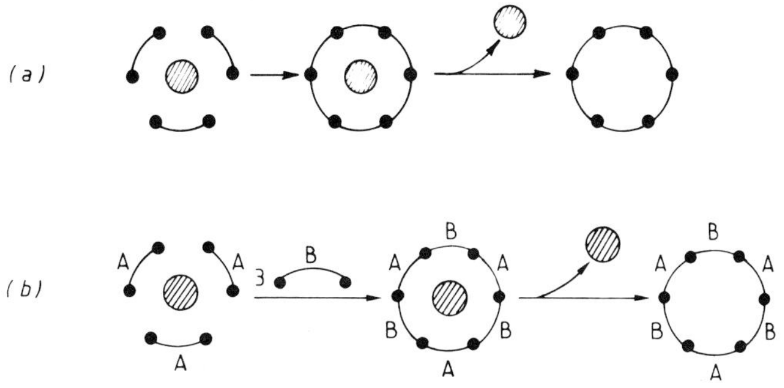

Fig. 3-10. Cyclization around a metal cation (external template).

Metalloid derivatives (Si, Sn, B) have been used as chemical templates serving both as a template for chain assembly and as a reaction site for subsequent ring closure of the chains thus locked in position (Fig. 3-11).

Fig. 3-11. Scheme for cyclization using an external template. The template ----- is called external because it is eliminated at the end of the synthesis and not incorporated into the end-product.

Certain atoms or molecular groups can also act as temporary templates. An example is sulfur which after having enabled cyclization to take place is then eliminated (Fig. 3-12).

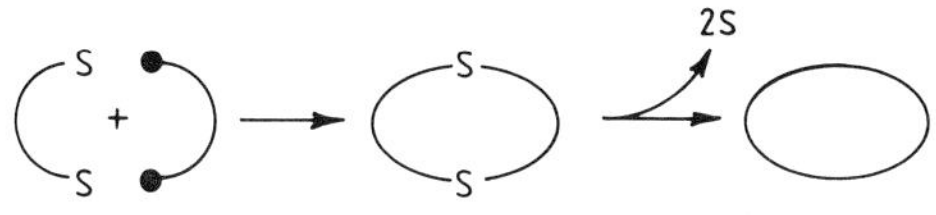

Fig. 3-12. Use of sulfur as an external template.

3.3.2 Cyclization on Endo-Templates

a) The "Zip" Reaction: Transamidation

The "zip" reaction, which results in insertion of a side chain into the ring (Fig. 3-8 b), utilizes unidirectional transamidation (Fig. 3-13) developed using linear compounds. This enables conversion of a tertiary into a secondary amide [63].

Conversion of a lactam into an amino-amide with ring enlargement has been achieved in various cases [63–65] by use of KAPA reagent (the base potassium aminopropylamide in 1,3-propanediamine) (Fig. 3-14).

An m + 2 ring gave an m + n + 3 ring. Transamidation yields were high for n = 2 or 3.

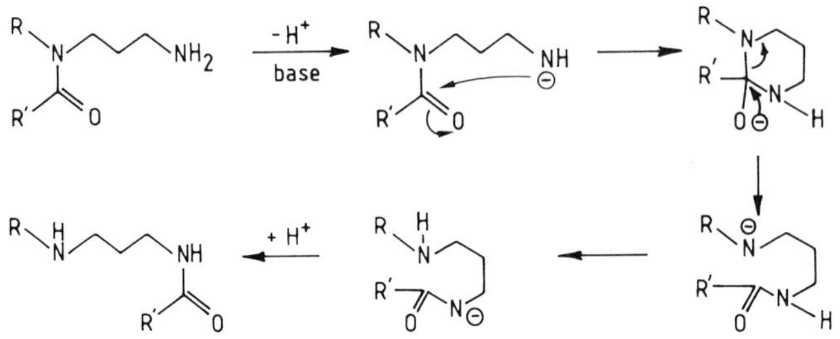

Fig. 3-13. Transamidation: conversion of a tertiary into a secondary amide.

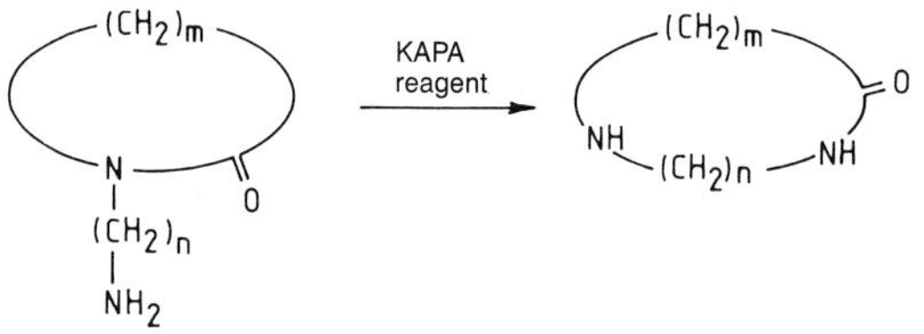

Fig. 3-14. Conversion of a lactam into an amino-amide with ring enlargement (KAPA reagent = potassium aminopropylamide in 1,3-propanediamine).

In this way, a 13-membered lactam has been transformed into a 33-membered amino-amide (Fig. 3-15a) by five successive introductions of a $CH_2CH_2CH_2N$ unit into the ring [64].

A 53-membered polyaminolactam has also been obtained from a 13-membered lactam (Fig. 3-15b) by the introduction repeated eight times of a ring-enlargement unit [65]. The yield in the last step was, however, not as good (38%) as in the previous example (85%).

The postulated mechanism for ring enlargement is given for transformation of a 13- into a 21-membered ring (Fig. 3-16).

It can be seen that initially transamidation occurred using the secondary amino group via a 6-membered ring (*1*) which was converted into *2* (17-membered ring). The latter then underwent ring enlargement by reaction of the primary amino group to give the final product *3*, a 21-membered ring [65b]. It is important to note that after each intermediate step, a new lactam group was re-formed, this being the necessary condition for insertion of the next unit. In the formation of large rings mentioned above, the course of the reactions is identical (successive insertion of all the units).

The transformation from a 13- to a 21-, 33- or 53-membered ring takes place like the opening of a zip fastener, hence the name for this type of reaction [63].

→

Fig. 3-15. Transformation of a 13-membered lactam into (a) 33-membered and (b) 53-membered polyaminolactams.

(a)

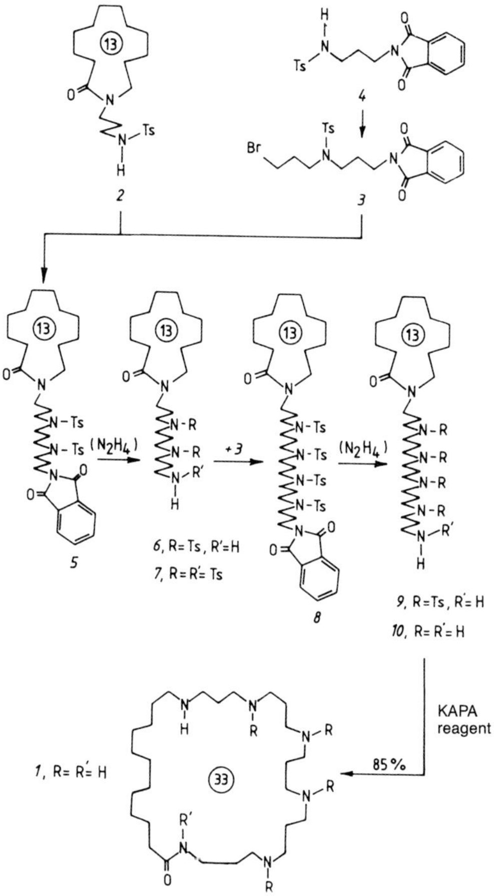
2
3
4
5
(N2H4)
+3
6, R= Ts, R'=H
7, R= R'= Ts
8
9, R=Ts, R'= H
10, R= R'= H
KAPA
reagent
85 %
1, R= R'= H

(b)

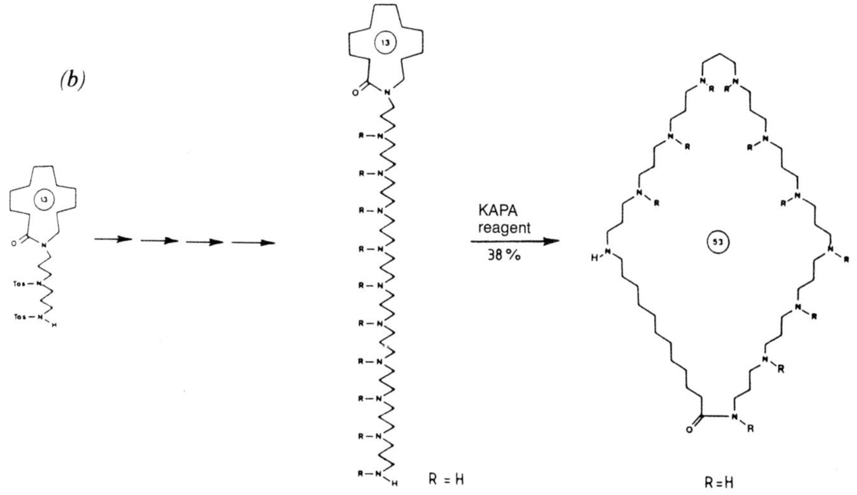
KAPA
reagent
38 %
R = H
R=H

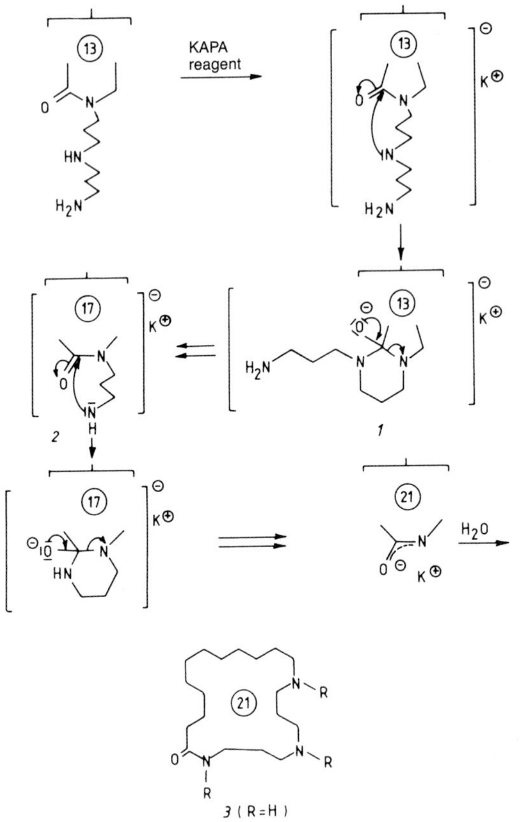

Fig. 3-16. Proposed mechanism for expansion of a 13-membered ring to give a 21-membered ring (taken from M. Hesse et al., *Helv. Chim. Acta,* **61,** 1342, 1978).

This method of ring expansion has been applied to the synthesis of a barbituric acid derivative [66]. The latter, *7*, with an azaheptylamino side chain, was obtained by means of the series of reactions shown in Fig. 3-17.

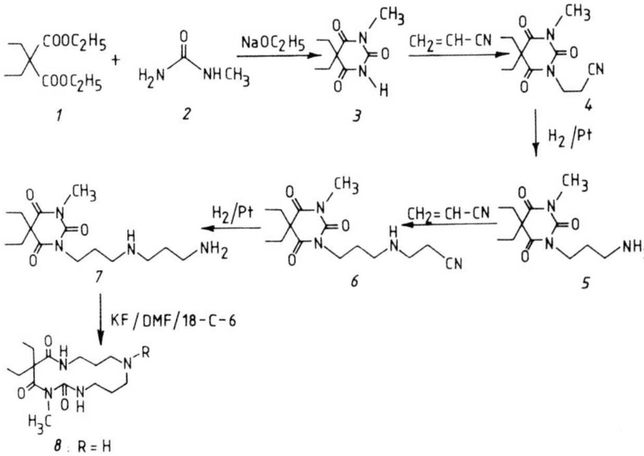

Fig. 3-17. Ring enlargement of a barbituric acid derivative using the "zip" method.

The derivative *7* when treated with KF/DMF/18-crown-6, gave the 14-membered ring *8* in good yield (83% for the last step) [66]. This synthesis demonstrated the fact that the zip reaction could be used in relatively complex systems and that its adaptation to the synthesis of a large variety of macrocyclic lactams was possible.

b) Cleavage Reactions in a Bicyclic System

These are based on the breaking of a cleavable bond serving as a temporary link: carbon-nitrogen, nitrogen-nitrogen, carbon-carbon (Fig. 3-18).

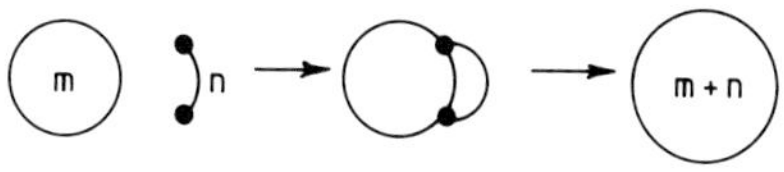

Fig. 3-18. Route to an m + n ring via formation of a bicyclic system followed by bond cleavage.

This procedure is particularly useful for forming medium rings for which any other method of synthesis is laborious.

Cleavage of carbon-nitrogen bonds. The synthesis of cyclic aminoalcohols (Fig. 3-19) has been carried out by electrolytic reduction of bicyclic α-aminoketones at a Pb electrode in 30% sulfuric acid at 60 °C [67]. Depending on the nature of the starting aminoketone, the yields for the reduction were between 40 and 75%.

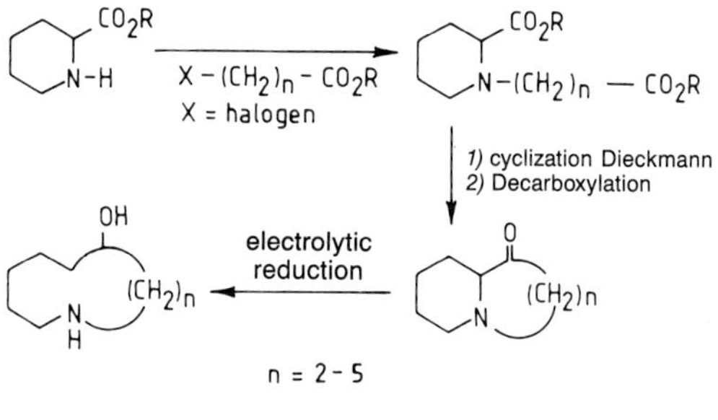

Fig. 3-19. Synthesis of cyclic aminoalcohols by electrolytic reduction of bicyclic α-aminoketones.

Cleavage of nitrogen-nitrogen bonds. A very useful synthesis of medium sized cyclic diamines has been performed [68]. The steps, which are shown in Fig. 3-20, were as follows:

- Condensation of an N,N′-diacylhydrazine *1* with a dihalogeno compound *2*.
- Treatment of the cyclic compound *4*, obtained by hydrolysis of *3*, with methyl succinate gave *5*.
- The latter was reduced to form the bicyclic dihydrazide *6*.
- Hydrogenation of *6* in the presence of Raney nickel led to N–N bond cleavage and formation of the cyclic diamine *7*.

The overall yield for the formation of *7* from the starting materials *1* and *2* was satisfactory (14% for n = 1; 20% for n = 2).

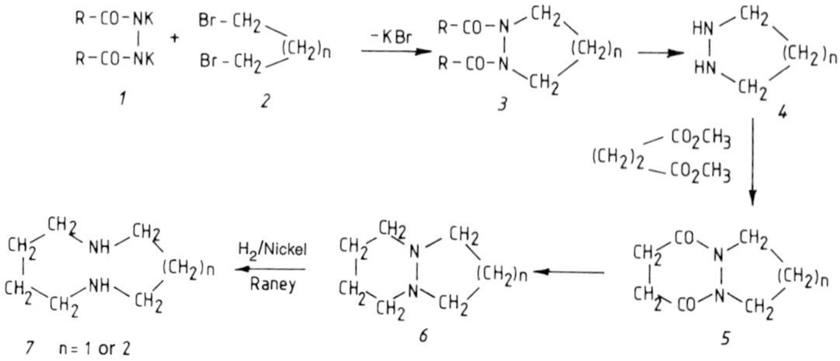

Fig. 3-20. Synthesis of cyclic diamines by N–N bond cleavage in an intermediate bicyclic compound. This method is useful for the formation of medium rings.

Cleavage of carbon-carbon bonds. This method has been used for the synthesis of a wide variety of macrocycles. Macrocyclic *lactones* were obtained as shown in the scheme in Fig. 3-21, where E is an electronegative group.

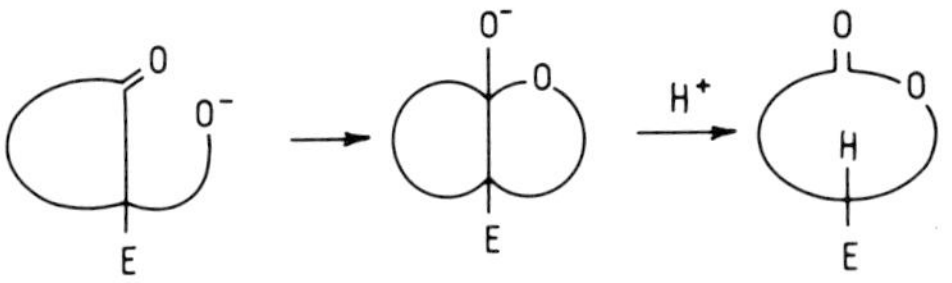

Fig. 3-21. Synthesis of macrocyclic lactones by C–C bond breaking in an intermediate bicyclic compound.

The following examples illustrate the method [69].

Sodium enolates of the 2-phenylsulfonylketones (1a-d) (Fig. 3-22) were alkylated by *cis*-4-acetoxy-1-chlorobut-2-ene in dimethoxyethane. Acid hydrolysis of the acetates *2* gave the 2-(*cis*-4-hydroxybut-2-enyl) derivatives *3* in good yield. The 3-hydroxypropyl derivative *8* was made by Michael addition of the keto-sulfone *1a* to acrolein and reduction of the aldehyde group with $NaBH_4$. The two cyclododecanone derivatives *3a* and *8* on treatment with NaH in benzene gave the lactones *4a* and *9* respectively in high yield. The sulfone group was reductively cleaved with a large excess of sodium amalgam and disodium hydrogen phosphate in MeOH-DME. The lactone *9* gave 15-pentadecanolide (exaltolide) *10,* and *4a*, after hydrogenation, gave 16-hexadecanolide (dihydroambrettolide) *6a* in overall yields of 36% and 42% respectively.

Table 3-1 gives the yields of all the intermediate products isolated during the syntheses of the macrocyclic lactones. It may be noted that the steps in which the macrocycles were formed (steps *3* → *4* and *8* → *9*) occurred in high yield.

Another more strongly electron-withdrawing group is the nitro group ($E = NO_2$, see Fig. 3-21). By use of this group, some of the above macrocycles have been obtained in even higher yields [69b].

Ketolactone macrocycles (Fig. 3-23) have been synthesized by cleavage of the C = C bond in 2,3-polymethylenebenzo- and naphthofurans, and also in 2,3-

polymethylenebenzo- and naphthopyrans, using various oxidizing agents (chromic anhydride, ozone, m-chloroperbenzoic acid) [70]. The yields of macrocyclic compounds (*2, 4, 6, 8*) were generally quite good (50–70%).

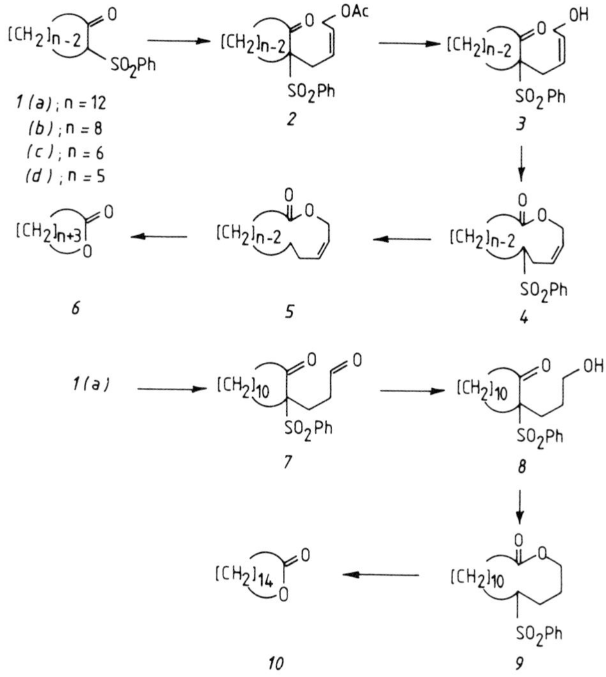

Fig. 3-22. Synthesis of dihydroambrettolide *6a* and exaltolide *10* from phenylsulfonylketones.

Table 3.1 Percentage yields of intermediate products isolated during the synthesis of lactones from the sodium enolates of 2-phenylsulfonylketones.

	(2)	*(3)*	*(4)*	*(5)*	*(6)*	*(7)*	*(8)*	*(9)*	*(10)*
a	86	97	89	82	93	61	98	90	90
b	80	98	79	–	–				
c	79	93	76	–	–				

6a) dihydroambrettolide 10) exaltolide
(Taken from V. Bhat and R. C. Cookson, *J. Chem. Soc., Chem. Comm.*, 1123, 1981.)

Macrocyclic *ketoimides* and *ketolactams* (Fig. 3-24) have been prepared by oxidation of enamine lactams, *2,* and N-acetyl-enamines, *4,* by ozone or m-chloroperbenzoic acid [71].

Cleavage of two C = C bonds has led to the synthesis of macrocyclic *tetraketones* [72]. This very elegant procedure is depicted in Fig. 3-25. Cyclopropanation of tetrakis(trimethylsilyloxy)cycloalkadienes *1* [73] gave *2* which on treatment with $FeCl_3$ in DMF afforded the tetraketones *3* in acceptable yield (33–47%). Transition metals such as Cu(II), Ni(II) and Co(II) were complexed by these tetraketones.

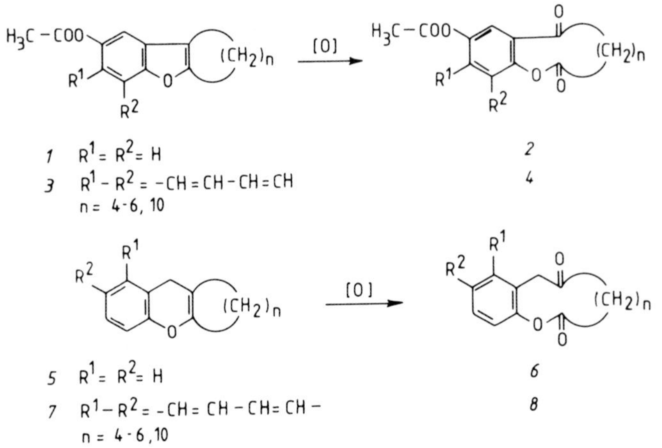

Fig. 3-23. Synthesis of macrocyclic ketolactones by cleavage of the C=C bond in 2,3-polymethylenebenzo- and naphthofurans (*1* and *3*) and in 2,3-polymethylenebenzo- and naphthopyrans (*5* and *7*).

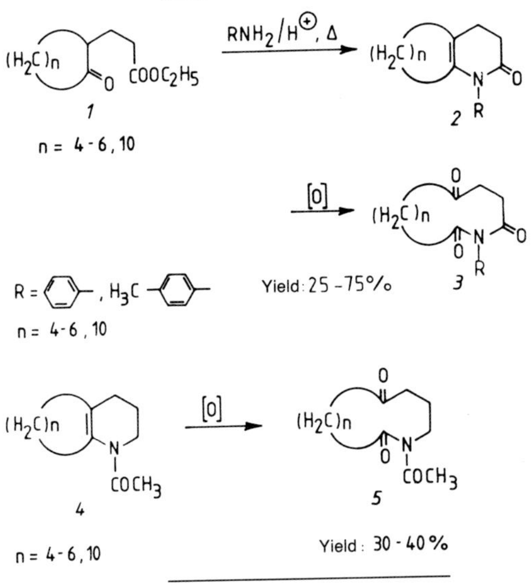

Fig. 3-24. Synthesis of macrocyclic ketoimides *3* and ketolactams *5* by oxidation of enamine lactams *2* and N-acetyl-enamines *4*.

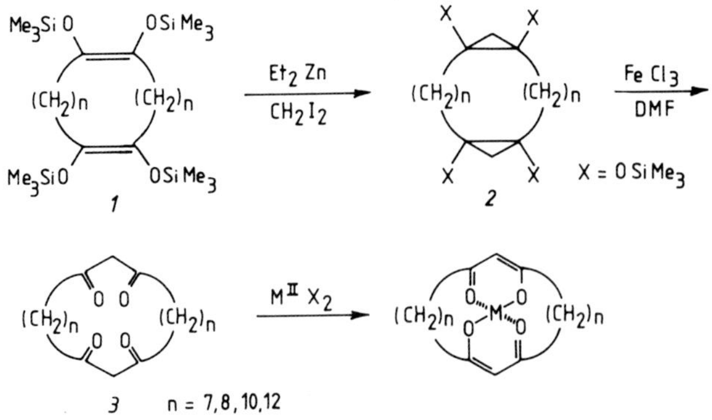

Fig. 3-25. Synthesis of macrocyclic tetraketones by cleavage of two C=C bonds.

3.3.3 Cyclization on External Templates: Metalloids

Metalloids, such as Si, Sn, and B can be used both as covalently bonded templates and also as reaction sites.

The building up of the macrocycle takes place with the aid of a supporting group, a "temporary framework", on which the desired macrocycle is assembled by means of successive condensations. It is then finally cleaved from the support. The metalloid heterocycles which are formed as intermediates serve as "chemical templates" (Fig. 3-26).

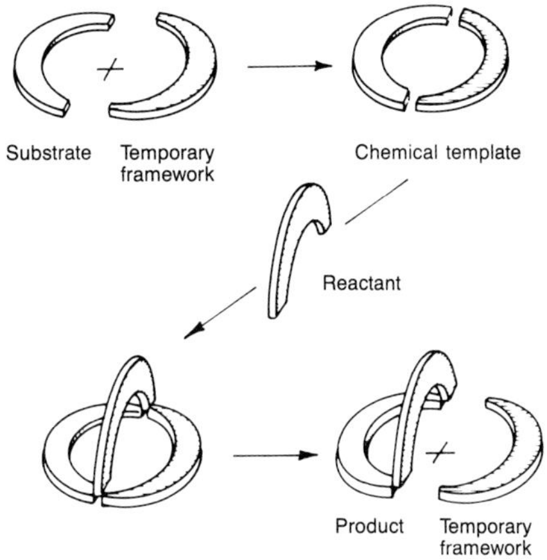

Fig. 3-26. Schematic representation of the use of "chemical templates".

The elements Si, Sn, and B have been chosen as *external templates* which are temporarily internal, because of their ability to form covalent metalloid-heteroatom bonds which can be easily cleaved by hydrolysis under mild conditions.

Thus, the metalloid rings of the type shown in Fig. 3-27 may be formed.

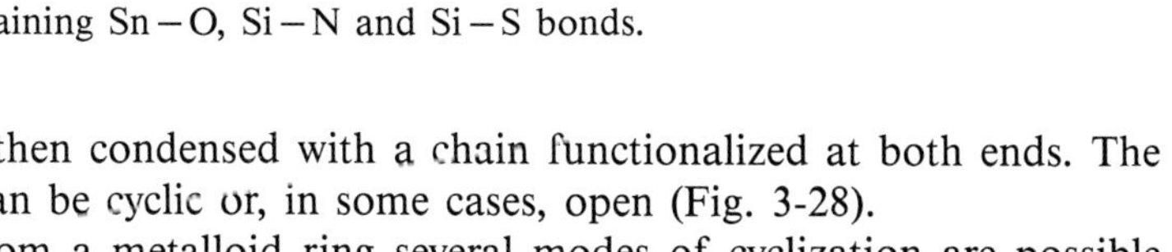

Fig. 3-27. Rings containing Sn–O, Si–N and Si–S bonds.

These are then condensed with a chain functionalized at both ends. The first intermediate can be cyclic or, in some cases, open (Fig. 3-28).

Starting from a metalloid ring several modes of cyclization are possible (Fig. 3-29).

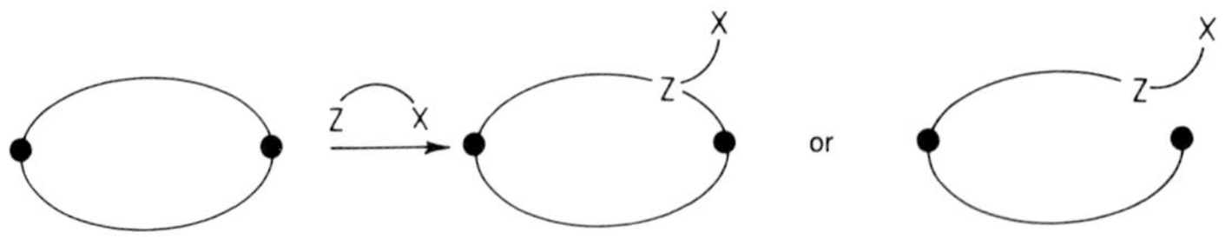

Fig. 3-28. Possible reactions of a metalloid ring.

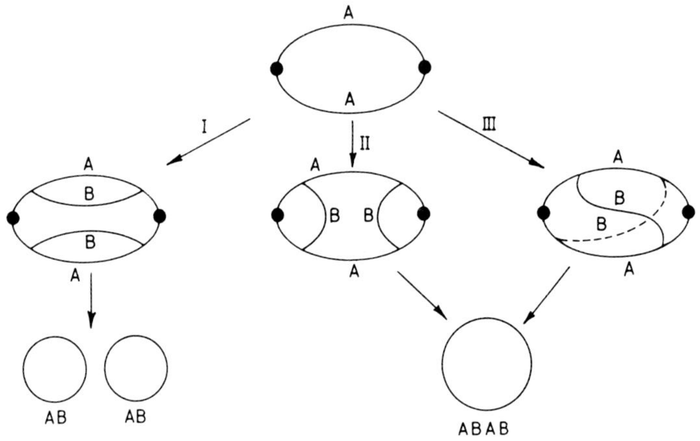

Fig. 3-29. Possible cyclization modes starting from a metalloid ring: transannular (I), lateral (II) and diagonal (III).

- Transannular (I) or diagonal (III) cyclization takes place in cases when insertion of Z deactivates the neighboring reaction site thus leading to attack by X at the other site (Fig. 3-30a).
- Lateral cyclization (II) (Fig. 3-29) occurs when X bonds to the same site, e.g., if the first step activates formation of a second bond to the same site (Fig. 3-30b).

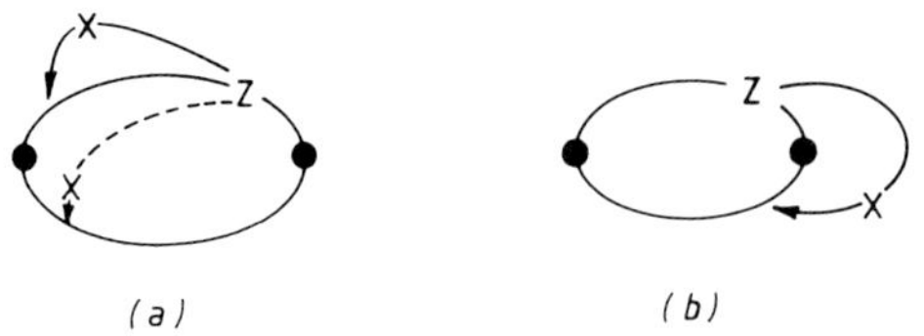

Fig. 3-30. (a) Transannular and diagonal cyclizations resulting from deactivation of the first reaction site, (b) lateral cyclization favored by activation of the reaction site.

Once the various branch chains have been attached, the metalloid templates are eliminated. Depending on the type of macropolycycle formed initially (modes I, II, III), the macrocycles obtained are of various types. I gives rise to two AB macrocycles. II and III both give a very large ABAB macrocycle (Fig. 3-29).

Since the reactions occur via formation of macrobicycles and macrotricycles, this method of macrocyclization is only superior to all others provided that rigidity and directional effects favor formation of the desired product.

The *distannoxane* template, *1*, was formed by reacting dibutyltin oxide with ethylene glycol (Fig. 3-31). Reaction with α,ω-diisocyanates resulted in double cyclization, *2*. Subsequent hydrolysis gave exclusively the dicarbamates, *3* [74].

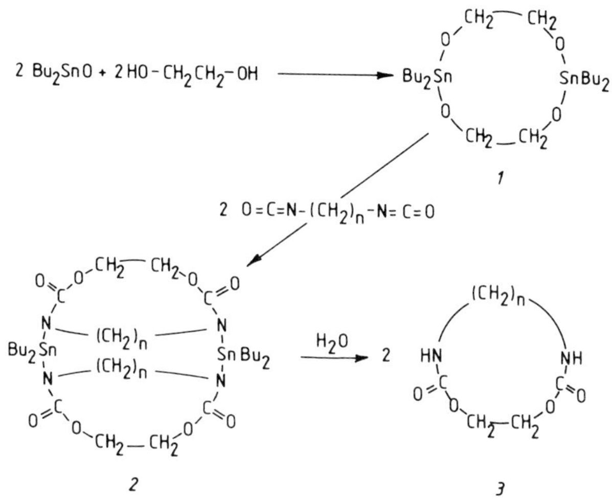

Fig. 3-31. Formation of dicarbamates from a distannoxane and α,ω-diisocyanates.

For n = 6, the yield was 35%, whereas for direct condensation of ethylene glycol with diisocyanate it was less than 5%.

Condensation of distannoxanes with diacyl chlorides (Fig. 3-32) gave tetralactones in quite low yield in preference to dilactones [75].

Monostannoxanes also afforded tetralactones (Fig. 3-33a) [76a]. This reaction took place in two distinct steps (see scheme in Fig. 3-33b). In the first step, the diacyl chloride reacted with two molecules of stannoxane to give derivative *1*. The latter then reacted with a second diacyl chloride molecule to give the tetralactone after cleavage of the metalloid templates from *2* by hydrolysis. The intermediate compound *1* was stable enough to allow introduction of a second diacyl chloride which was different from the first and thus a mixed tetralactone could be obtained.

It should be noted that in this reaction the first insertion into the Sn–O bond deactivated the second. This explains both the intermolecular reaction and the absence of dilactone formation.

N.B. We have kept here the distinction between a di- and a monostannoxane used by Shanzer in his initial work. The actual structure of these compounds in solution is still a subject of some controversy. However, it seems established that stannoxanes have a very marked tendency to associate forming molecular complexes, almost ex-

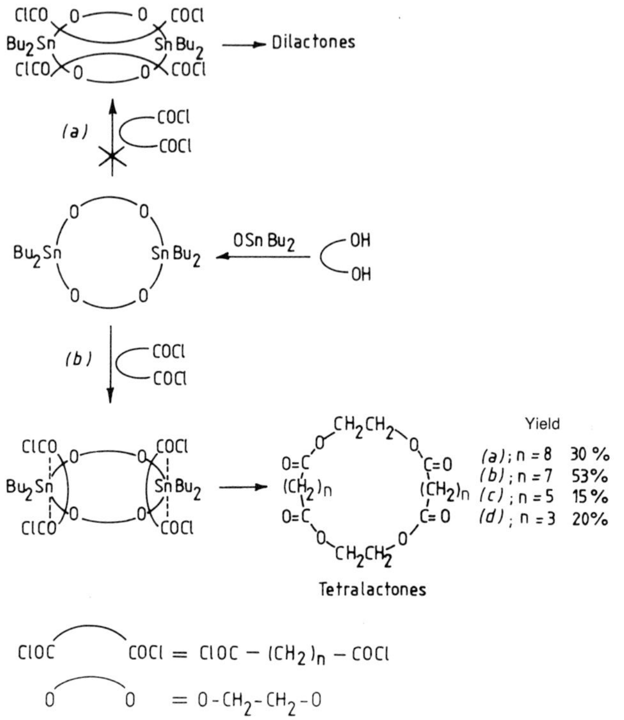

Fig. 3-32. Formation of tetralactones from distannoxanes and diacyl chlorides.

clusively dimers, in solution (Fig. 3-34). The presence of these dimers has been shown by various methods (NMR, IR, Mössbauer studies) [76b]. The nature of the starting diol also plays a part in the phenomenon of association.

These studies do not invalidate all the various results described above (except for certain postulated mechanisms) because distannoxanes can be considered as being a limiting form of *2*. The dimeric structure *2* enables one to better explain, however, some of the results. It shows more clearly the close interdependence of the four Sn–O bonds.

Condensation of a distannoxane with the anhydride of a chiral amino-acid (*2a* or *2b* in Fig. 3-35) afforded chiral tetralactone structures *3a* or *3b* [77a] both having the *syn* configuration. The *anti* compound *4* has also been obtained by treating successively the distannoxane *1* with 1 equivalent of *2a* followed by 1 equivalent of *2b*. The exclusive formation of *3* was explained by the fact that attack by the anhydride molecules takes place on opposite sides of the two stannoxane sites for steric reasons (maximum separation of the bulky groups) [77b]. The same argument could be used to explain the formation of *4* in two steps. The dimeric intermediate *5*, on which was based the interpretation of the relative orientations of the chains, is shown in Fig. 3-35.

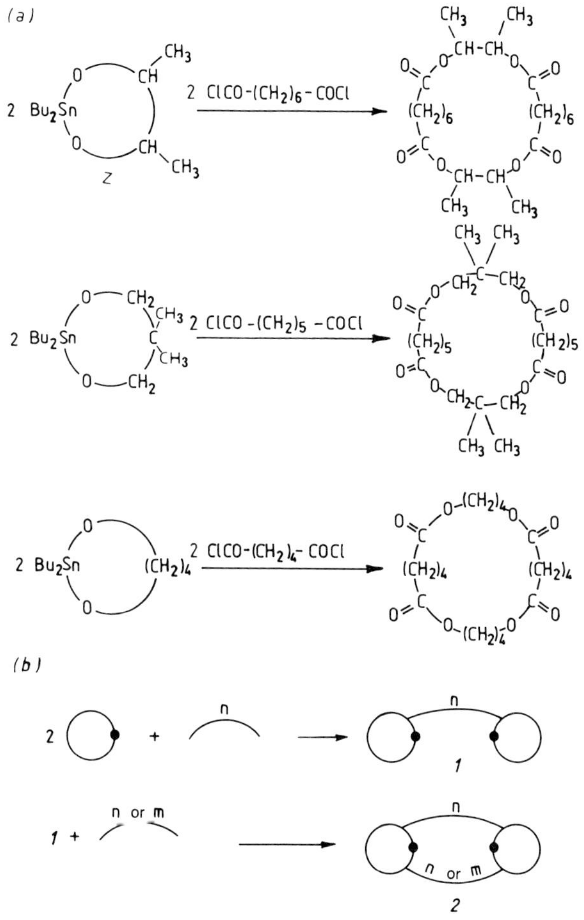

Fig. 3-33. (a) Formation of tetralactones from monostannoxanes, (b) possible mechanism.

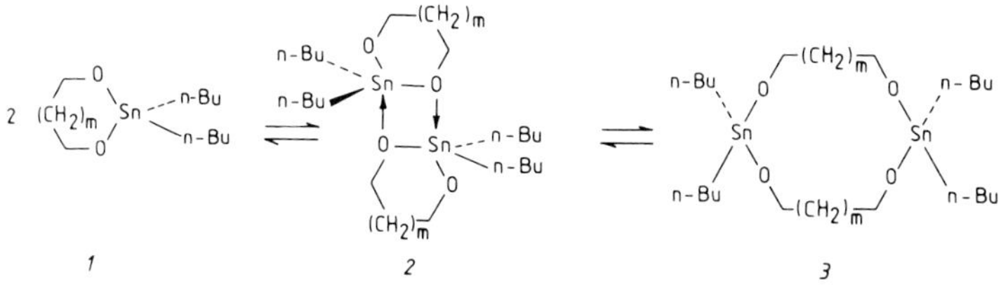

Fig. 3-34. Monomer-dimer interconversion in stannoxanes.

Novel macrocyclic lactones with rotational symmetry axes have been obtained by condensation of the strained lactone, propiolactone, with a distannoxane [78]. The reaction may have occurred by insertion of several propiolactone molecules into the Sn – O bond (Fig. 3-36), followed by expulsion of the macrocyclic product with rearrangement and regeneration of the stannoxane.

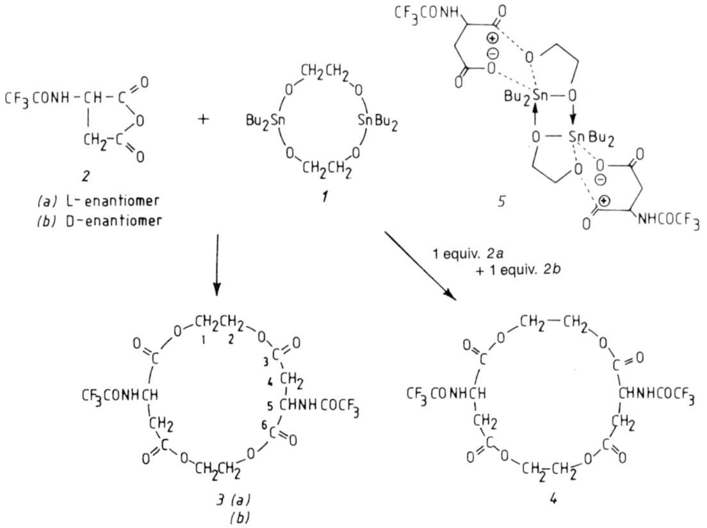

Fig. 3-35. Synthesis of side-chain tetralactones of *syn (3)* and *anti (4)* configuration.

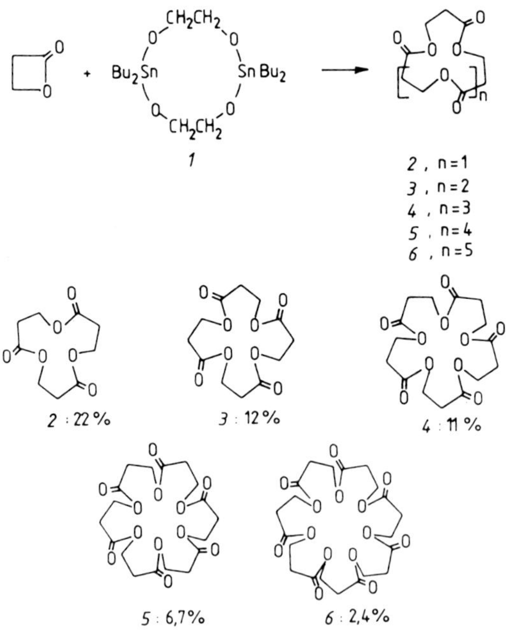

Fig. 3-36. Synthesis of rotolactones from propiolactone and distannoxanes.

This method has led to a new total synthesis of enterobactin, a biological iron carrier [79a] (Fig. 3-37).

Its application in the synthesis of macrocycles is only one of the numerous aspects of stannoxane chemistry [79b].

Fig. 3-37. Synthesis of enterobactin by use of a distannoxane.

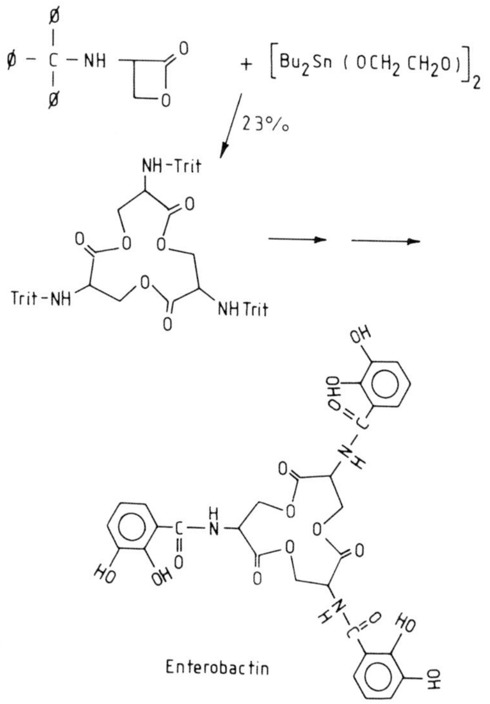

Other Templates. The formation of macrocycles from *silazanes* or *siladithianes* and diacyl chlorides or diisocyanates is very similar to that described using stannoxanes [80, 81] (Fig. 3-38 and 3-39). In certain cases, the silazane can give difunctional macrocycles depending on the nature of the starting diamine and the reactant used [74b]. On the other hand, siladithianes give exclusively tetrafunctional macrocycles.

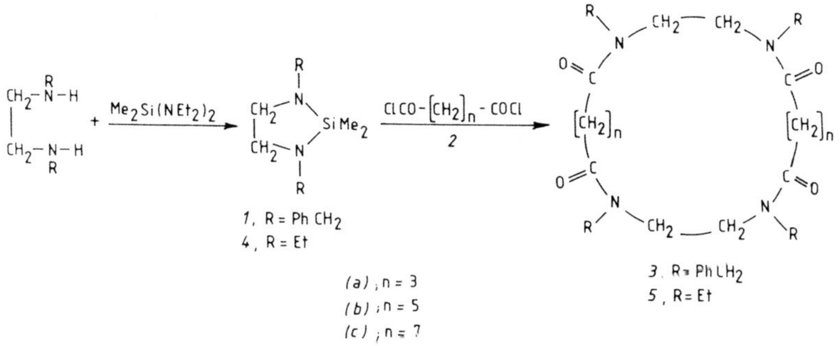

Fig. 3-38. Formation of tetramides using silazanes.

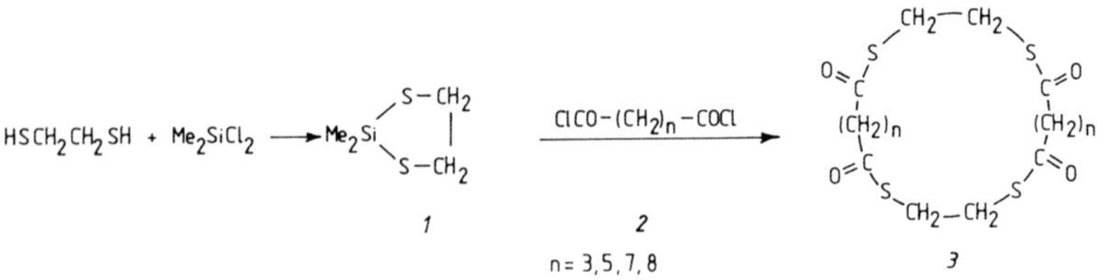

Fig. 3-39. Synthesis of tetrathiolactones from siladithianes.

In the case of *aminoboranes,* the boron acted additionally as an activating agent [82]. This occurred in step (d) in the reaction scheme depicted in Fig. 3-40. This method constitutes a particularly simple route to macrocyclic alkaloid derivatives of spermidine. Compounds *1, 3,* and *4* were obtained with an overall yield of ca. 40%. Alkaloid *2* was formed by irradiation of *1.*

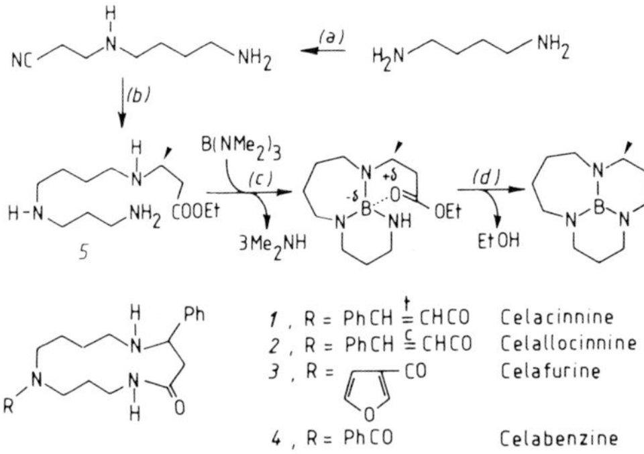

Fig. 3-40. Synthesis of alkaloid spermidine derivatives via aminoborane intermediates. (a) $CH_2=CHCN$, (b) $CH_2=CHCO_2C_2H_5$; H_2/PtO_2. N.B.: The synthetic scheme given is simplified and very general. Compounds *1-4* were obtained starting from the open chain precursor *5* with a phenyl group in the required position (arrowed). Alkylation to introduce R at N−6 was performed on macrocycle *6.*

3.3.4 Cyclization on External Templates: Metal Cations

a) Principle

The synthesis of a macrocyclic ligand in the presence of a metal cation has been used in the formation of numerous macrocycles. The presence of the metal leads to positioning of the reactants such that macrocyclization is very favored, this being the template effect.

This method has both advantages and drawbacks. One of the advantages is the increased yield as a result of partial or total inhibition of the competing reactions, polymerization and formation of non-cyclic products. Another advantage is the possibility of selective cyclization. By judicious choice of the metal ion, formation of a desired macrocycle may be encouraged. However, one of the disadvantages is that the macrocycle is generally coordinated to the metal ion and it is sometimes difficult to separate them. Also, the method is often specific for a certain metal ion and cannot be generalized. However, it has proved very useful in spite of these limitations.

Two types of process can be envisaged:

- The molecules are assembled and become bonded to each other about the cation in a single step (Fig. 3-41 a) e.g., the synthesis of phthalocyanines (Fig. 3-41 b).
- The assembly of one type of unit about the ion followed by bridging by another structural unit (Fig. 3-42). This method offers more possibilities for synthesis than the first since numerous variants are possible depending on the nature of the end groups.

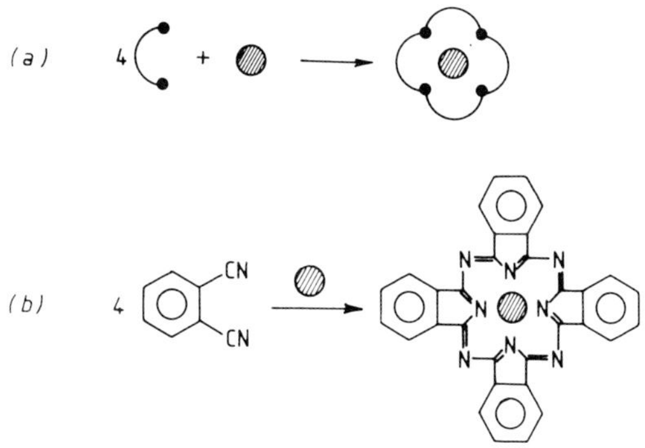

Fig. 3-41. (a) Metal-templated cyclization using a single structural unit in one step. (b) phthalocyanine synthesis.

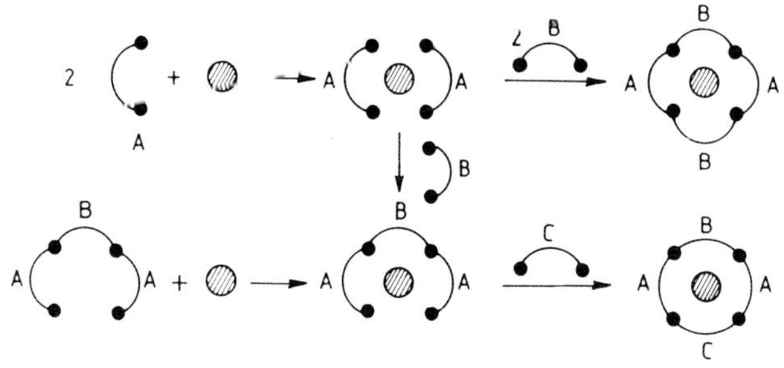

Fig. 3-42. Some examples of the many possibilities for metal-templated cyclization using several structural units.

The type of macrocyclic product obtained depends on:

- the constituent molecules: size, reactivity of end groups, rigidity, and geometry,
- the metal cation: size, favored coordination sites, and coordination geometry.

Great adaptability and variety can therefore be achieved. However, detailed analysis of the factors involved is necessary for controlled use of the method.

In the 1960s, several groups of workers independently synthesized various macrocyclic ligands of the polyamine-imine type by using metal cations as templates.

In the Curtis reaction [83] (Fig. 3-43), the first step is formation of an ethylenediamine-metal complex, and the second step is condensation of a carbonyl compound, such as acetone, with the complex to give a macrocycle. The latter contains two imine and two secondary amine groups.

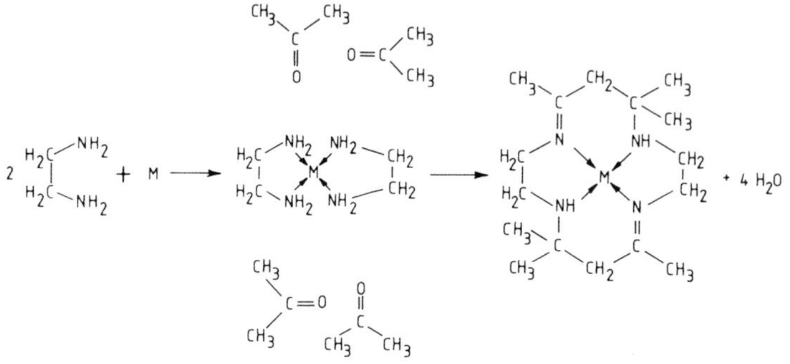

Fig. 3-43. Macrocycle formation by use of the Curtis reaction.

The condensation generally only occurs efficiently when M = Ni(II) or Cu(II).

This *in situ* reaction has been widely used with a large number of diamine-ketone-metal cation combinations [83].

The degree of unsaturation of the macrocycles can be modified, in particular by reduction. With the reduced compound, demetallation using CN^- readily occurs to give the free ligand. Demetallation of non-reduced complexes is more difficult since the free macrocycle, which possesses imine groups, is readily hydrolyzed.

In the reactions described above, the intermediates are generally not isolated. A different approach to the use of metal cations as external templates appeared in 1964. It consisted of first forming an open chelate which was then treated with a reactant capable of giving the macrocyclic complex. The open chelate came either from a previously synthesized ligand or by forming it directly in solution by using the external template effect (see Fig. 3-42).

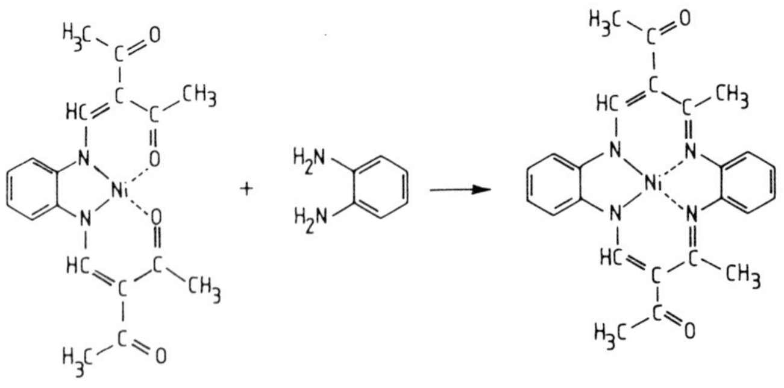

Fig. 3-44. Synthesis of a macrocyclic complex from a nickel chelate complex.

Using an approach of this type, Jäger synthesized a series of macrocyclic complexes of which an example is given in Fig. 3-44 [84a].

The nature of the ligand can be modified (e.g. by replacement of the aromatic nucleus by a 2- or 3-membered hydrocarbon chain). Also, various diamines can be used in the final condensation step.

The presence of carbonyl groups in these macrocyclic complexes was exploited twenty years later by Busch in order to develop the chemistry of cyclidenes *(vide infra).*

Another long-known example [84b] is given in Fig. 3-45a. The intermediate complex, treated with the dibromo derivative, gave the macrocycle by means of S-alkylation reactions.

More recently, complexes of Zn^{2+} and Cd^{2+} have been obtained starting from 2-aminobenzenethiol and 2,6-diacetylpyridine [84c] (Fig. 3-45b).

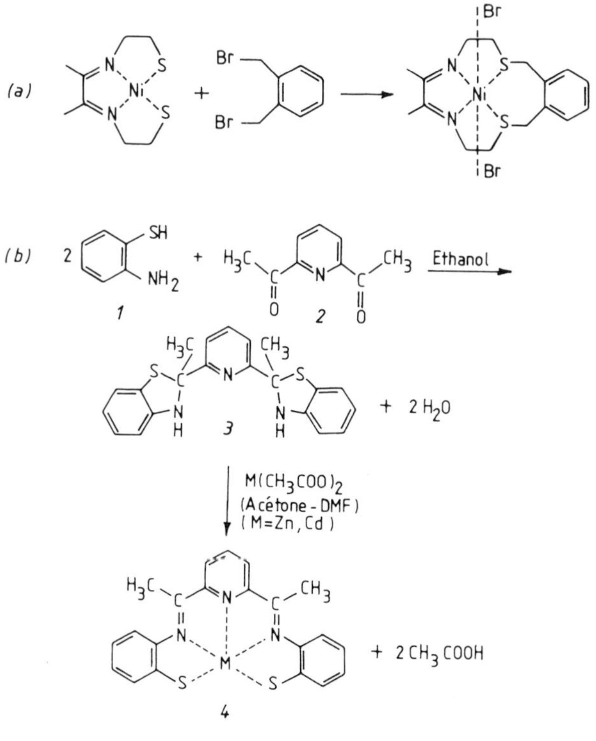

Fig. 3-45. (a) Macrocycle formation by means of double S-alkylation using a dibromo derivative, (b) synthesis of the bis(benzothiazoline) *3* and formation of the complexes *4*.

The ligand was stabilized by complex formation and could not be isolated free of its associated metal ion. The thiolo groups retained sufficient of their nucleophilic character to react with alkyl halides (Fig. 3-46). Addition of a dibromo compound enabled the macrocyclic compound to be obtained.

An insight into the many possibilities of ring closure reactions performed in the presence of a metal cation is given by the following study [85]. In the presence

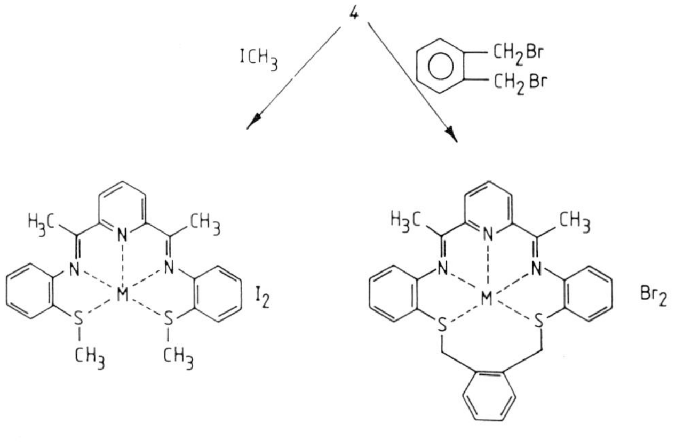

Fig. 3-46. S-alkylation reactions of the complexed thiolo groups.

of the divalent metal cations Ni^{2+}, Cu^{2+} or Co^{2+}, o-phenylenediamines underwent condensation reactions with dicarbonyl compounds to give either the macrocyclic complex *3* or the open chain complex *4* (Fig. 3-47), depending on the relative proportions of the two reactants. The latter afforded the macrocycles *3* or *5* on treatment with the appropriate diamine. On addition of a large excess of ethylenediamine, compound *4* underwent a transamination reaction to give macrocycle *6*. All the above reactions occurred in substantial yield.

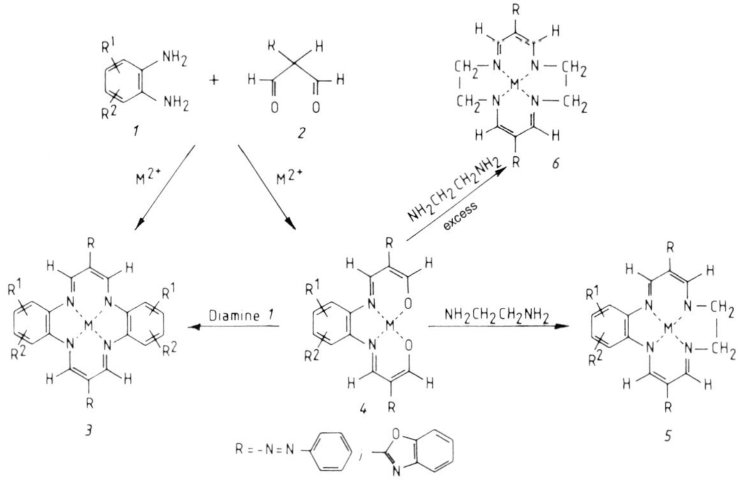

Fig. 3-47. Various synthetic possibilities starting from compounds *1* and *2*.

Complex *3* could be demetallated under anhydrous conditions (hydrochloric acid in absolute ethanol). The free ligand so obtained formed numerous complexes (Fe(II), Ru(II)).

Finally, it should be noted that the R substituents could be groups with potential use in subsequent constructions.

b) Orientation of [1 + 1] and [2 + 2] condensations between diamines and dicarbonyl compounds

In order to study the mechanisms for formation of Schiff's bases and to isolate reaction intermediates, Nelson et al. undertook a very detailed study of a homologous series. The first aspect covered was the orientation of condensation (Fig. 3-48).

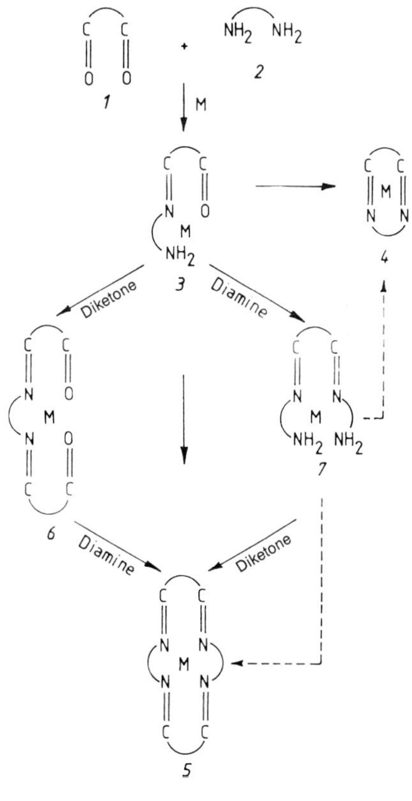

Fig. 3-48. Condensation of a diamine with a dicarbonyl compound, the macrocyclic products formed depending on the relative overall proportion of *1* and *2*, the nature of the reactants (chain length) and the cation used.

Product orientation resulting from condensation either of one molecule of diamine with one of dicarbonyl compound, [1 + 1], or of two molecules of each of the two components, [2 + 2], depended on the nature of the reactants and on the template metal cation [86]. The initial step led in all cases to product *3* resulting from the reaction between one molecule of diamine and one molecule of dicarbonyl compound in the presence of a metal cation.

Possible routes from *3* were the following:

- Intramolecular condensation leading to the [1 + 1] product *4*.
- If the diamine had insufficient chain length to span the two carbonyl groups, or if it were too rigid to fit by folding, the [2 + 2] macrocycle *5* was obtained.

- If the template cations were large in relation to the cavity size of the [1 + 1] ring, then again the [2 + 2] compound was formed.

The coordination type can also play a role in the orientation. A strongly coordinated NH_2 group in intermediate *3* favored [2 + 2] condensation especially when X heteroatoms were present in the diamine chain (Fig. 3-49).

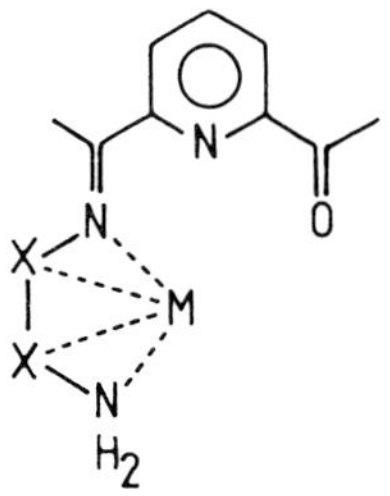

Fig. 3-49. [2 + 2] Condensation favored by the strongly coordinated NH_2 group in intermediate *3* (Fig. 3-48); X = heteroatom (O, N, S).

"Axial" coordination of the cation in intermediate *3* (Fig. 3-50) resulted, on the contrary, in the [1 + 1] product being favored. Additionally, the carbonyl group was activated by coordination to the cation.

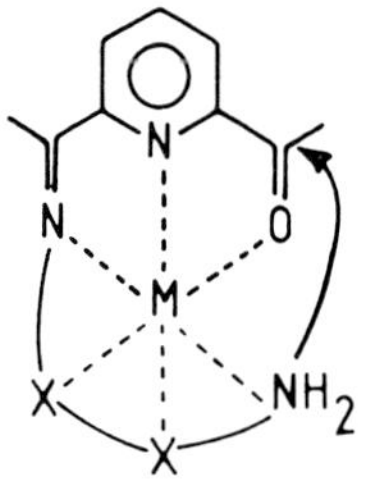

Fig. 3-50. [1 + 1] Condensation favored by axial coordination of the cation in intermediate *3* (Fig. 3-48).

It should be emphasized that the cation effect could play a part in all the various intermediates shown in Fig. 3-48.

Effect of the relative sizes of the cation and the cavity. In the absence of a metal cation, the reaction of 2,6-diacetylpyridine with 3,6-dioxaoctane-1,8-diamine gave exclusively oligomers.

In the presence of various cations (Mg^{2+}, Mn^{2+}, Fe^{2+}, etc.) [1 + 1] condensation occurred and macrocycle formation took place in good yield (Fig. 3-51). The cavity of the 15-membered ring was large enough to accommodate these various small cations [87].

In the presence of Ba^{2+}, Pb^{2+}, and Ag^{+}, [2 + 2] condensation took place leading to a 30-membered ring (compound *4*, Fig. 3-52).

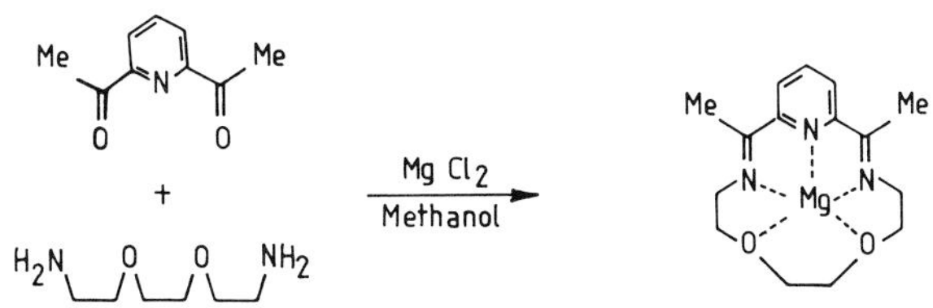

Fig. 3-51. [1 + 1] Condensation in the presence of Mg^{2+} (1 : 1 : 1 reactant mole ratio).

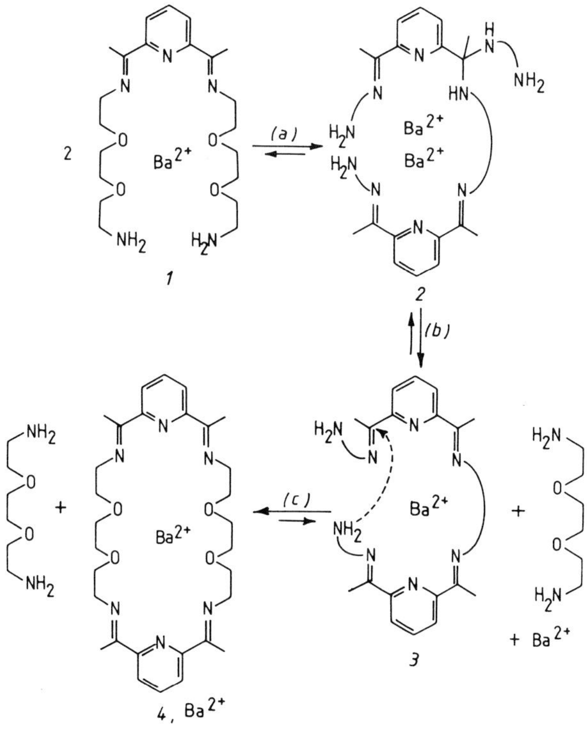

Fig. 3-52. [2 + 2] Condensation in the presence of Ba^{2+}.

The cyclization reaction was especially thoroughly studied in the case of condensation in the presence of Ba^{2+} [86a]. It thus proved possible to show that macrocycle formation occurred *via* an open [2 + 1] intermediate, *1*, which could be isolated and subsequently cyclized on addition of diacetylpyridine.

Moreover, this intermediate *1* could lead to the macrocycle in the absence of any other reactant (diamine or dicarbonyl compound). Thus *1* dissolved in methanol gave *4*. The mechanism is indicated in Fig. 3-52 and consisted of the following steps:

a) bimolecular reaction involving nucleophilic attack of an amine group of one of the two molecules of *1* at a C=N bond of the other,
b) regeneration of the imino-group by elimination of one molecule of diamine,

c) intramolecular nucleophilic attack of an uncoordinated amine group on a neighboring C=N carbon with elimination of a second molecule of diamine thus giving *4*.

The transamination reactions therefore played an essential part in the proposed scheme. The relative proportions of the various reactants determined the nature of the isolated species. The reversibility of transamination has been verified. On treatment of the complex of *4* with Ba^{2+} with two equivalents of 3,6-dioxaoctane-1,8-diamine, the open chain compound *1* was re-formed.

This facile transamination opened wide synthetic possibilities (Fig. 3-53). The binuclear lead complex of ligand *4* on treatment with excess ethylenediamine afforded the 18-membered ring *5*. Treatment of *4* with 3,6-dioxaoctane-1,8-diamine in the presence of Mg^{2+} gave macrocycle *6* [86a].

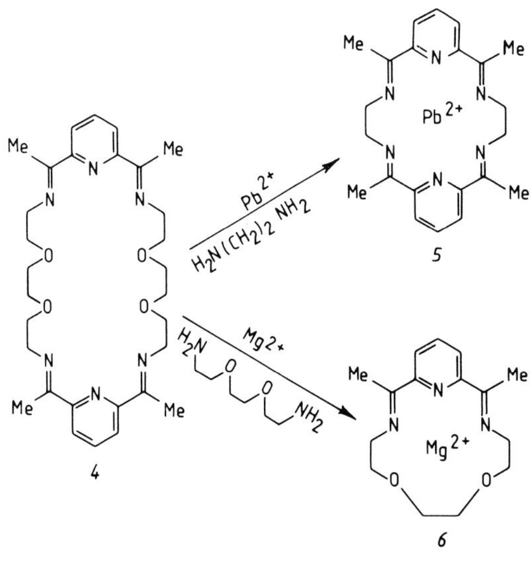

Fig. 3-53. Formation of 18- and 15-membered macrocycles from *4* by transammination in the presence of metal cations.

It should be noted that ligand *4* formed a binuclear complex with Cu^{2+} in which the two cations could bind small substrates such as OH^-, N_3^- or imidazolate [86b, 88]. A heterobinuclear Cu^{2+}/Ni^{2+} complex has also been prepared with this ligand [89].

Effect of the nature of the sites on the type of condensation. The nature of the coordination sites is another factor which determines the type of macrocycle obtained ([1 + 1] or [2 + 2]). This has been shown by studying the condensation of 2,6-diacetylpyridine with various amines [90].

The reaction with 3,6-dioxaoctane-1,8-diamine in the presence of $Pb(SCN)_2$ gave the [2 + 2] macrocycle in the form of the binuclear complex

$[Pb_2(2,2,2\text{-}N_6O_4)\ (SCN)_4]$ in >80% yield (Fig. 3-54a). Conversely, condensation with 3,6-diazaoctane-1,8-diamine or 4,7-diazadecane-1,10-diamine gave the mononuclear [1 + 1] complexes $[Pb(2,2,2\text{-}N_5)\ (NCS)_2]$ and $[Pb(3,2,3\text{-}N_5)\ (NCS)_2]$, respectively (Fig. 3-54b).

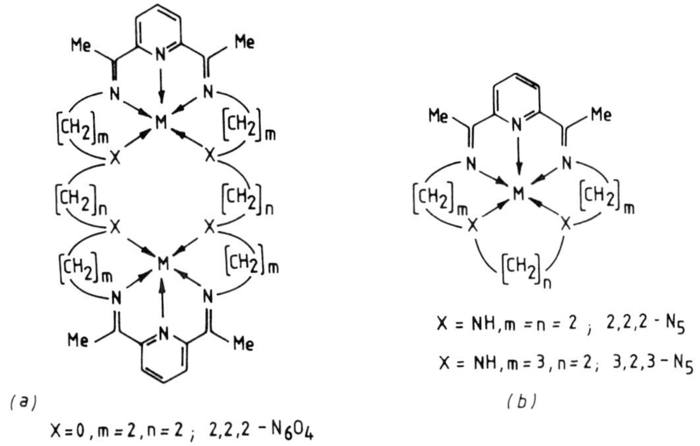

Fig. 3-54. (a) Binuclear complex of a [2 + 2] macrocycle formed from 2,6-diacetylpyridine and 3,6-dioxaoctane-1,8-diamine in the presence of $Pb(SCN)_2$, (b) mononuclear complexes of the [1 + 1] macrocycles formed from 2,6-diacetylpyridine and 3,6-diazaoctane-1,8-diamine or 4,7-diazadecane-1,10-diamine.

Mismatch of the cation and cavity sizes was insufficient to explain the observed differences between the dioxa and diaza diamine derivatives. The structural analysis of the binuclear complex showed that the metal ions lay closer to the nitrogen than to the oxygen donors [90]. This observation has led to a scheme being put forward to explain the different paths of the two reactions (Fig. 3-55).

The ligand conformation in the first-formed intermediate *1* could control the reaction path. When X = NH, strong coordination in the intermediate favored ring closure rather than double ring formation. When X = O, the interaction with the cation was weaker and competed with that between the solvent and the cation. The chain was therefore not contracted thus leading to double ring formation [90].

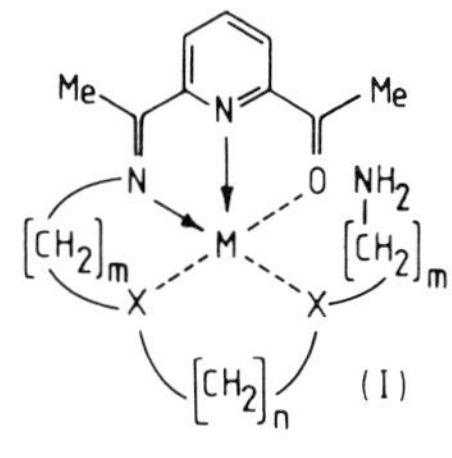

Fig. 3-55. Postulated intermediate in formation of the macrocycles shown in Fig. 3-54. Depending on the nature of X, the reaction is [2 + 2] (X = O), or [1 + 1] (X = NH); M = Pb.

Ring contraction and expansion under cation control. The reaction in methanol between 2,6-diacetylpyridine (1 equivalent) and diethylenetriamine (2 equivalents) in the presence of a salt of an alkaline-earth cation (1 equivalent) gave, at room temperature, the complexes $[ML^1]X_2 \cdot xH_2O$ ($X = ClO_4^-$ or NO_3^-, $x = 0$ or 1) (Fig. 3-56).

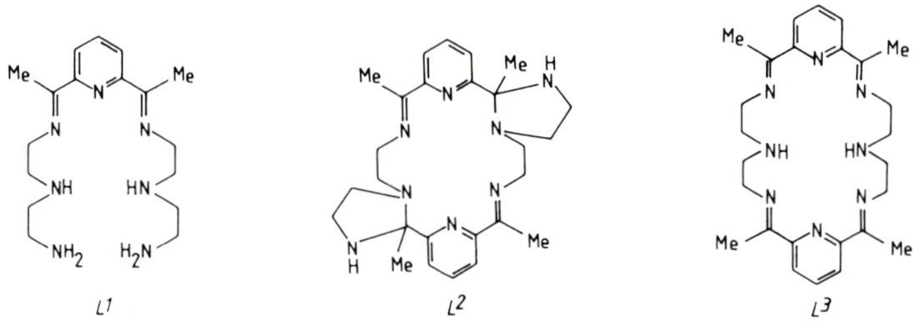

Fig. 3-56. Reaction of 2,6-diacetylpyridine with diethylenetriamine in the presence of a metal cation, the ligands formed depending on the molar ratio of the reactants, the reaction conditions and the nature of the cation.

When this same reaction was carried out with 1:1:1 reactant mole ratio (and the same cations apart from Mg^{2+}) under reflux, complexes of the L^2 macrocycle were obtained [91].

The surprising result was the formation of the 18-membered ring (L^2) rather than the expected 24-membered ring (L^3) (Fig. 3-56). The structure of this L^2 ligand has been confirmed by an X-ray diffraction study of its complex with barium ion.

Various experiments have furnished an explanation for the formation of the L^2 macrocycle and at the same time have enabled the determination of certain of its properties. Heating L^1 complexes (except those of Mg^{2+}) in methanol under reflux afforded the macrocyclic L^2 complexes and this was in the absence of any other reactant. The mechanism of formation is of the same type as that shown in Fig. 3-52 to which are added the steps for the formation of imidazoline rings. The L^1 complexes were thus intermediates in the formation of L^2.

Another series of experiments has shown that the formation of 18-membered macrocycles is strictly controlled by the nature of the cations. Thus, treatment of the complex $[SrL^2(ClO_4)_2]$ with $AgClO_4$ gave the binuclear complex $[Ag_2L^3]$ $[(ClO_4)_2]$ of the 24-membered macrocycle. This latter complex on treatment with $BaBr_2$ re-formed the macrocycle $[BaL^2(ClO_4)_2]$. The reversible reaction $L^2 \rightleftharpoons L^3$ was thus possible depending on the nature of the cation.

The L^2 ring could be considered as resulting from nucleophilic addition of each of the two secondary amine functional groups in L^3 across a C=N bond at each end of the macrocycle. The two imidazolidine rings thus formed were expelled outside the ring which decreased from being 24- to being 18-membered. The ring expansion and contraction mechanisms are shown in Fig. 3-57.

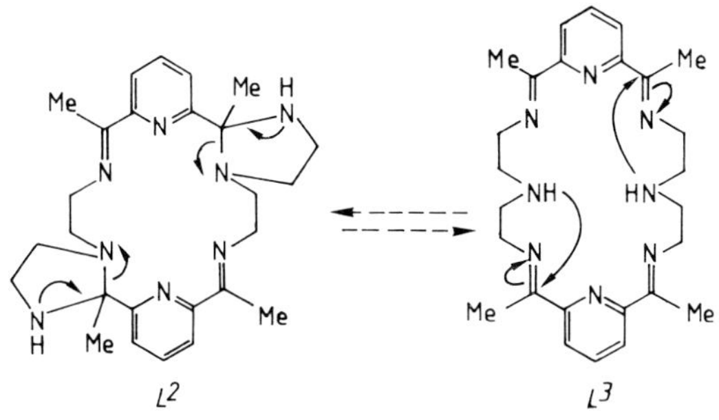

Fig. 3-57. Mechanisms of ring expansion and contraction for L^2 and L^3.

It should be noted that the initial ligand L^1 has been found to be a good complexing agent (heptadentate) for numerous transition metal cations [92].

c) Synthesis of ligands using exo and endo templates

Compounds *3a* and *3c* in Fig. 3-58 were obtained from the same starting materials but by using two totally different methods. In the first case the reaction was carried out in the presence of the metal cations Ni^{2+}, Cu^{2+} or Co^{2+} (exo templates) [93a]. In the second method, the template was an integral part of one of the reactants taking part in the cyclization reaction (endo template) [93b]. The condensation of o,o′-diformyloxalanilide with o-phenylenediamine (and its derivatives) gave the macrocycles *3* under normal conditions of concentration (0.25 mol/l) (Fig. 3-58). The yields were comparable to those obtained with metal cations as templates [93a].

It can be clearly seen in Fig. 3-58 that two hydrogen bonds maintained *1* in the conformation suitable for cyclization. This explains why high dilution conditions were unnecessary.

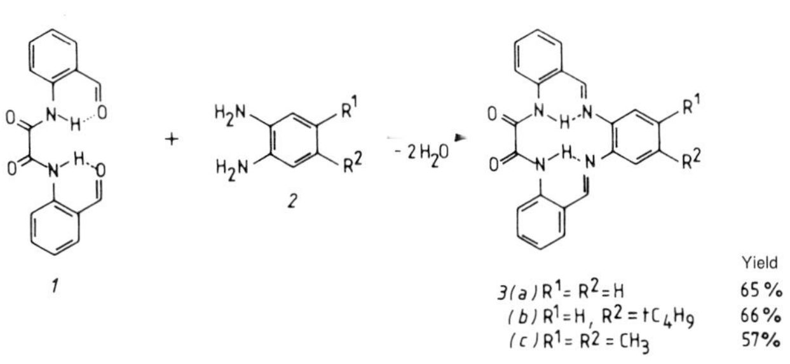

Fig. 3-58. Macrocyclization facilitated by hydrogen bonding which maintains a favorable conformation of *1*.

d) Synthesis of superstructure ligands

We shall see in detail in Part II that macrocycles with sites for interaction (O, N, S, etc.) are complexing agents of choice towards metal cations. This important aspect explains the large number of laboratories in which for the last twenty years efforts have been made to obtain complexing agents of this type. More recently, more intricate structures have been achieved (Fig. 3-59a) in which the metal in the complex could reversibly bind small molecules (O_2, CO, etc.) thus activating them [94, 95].

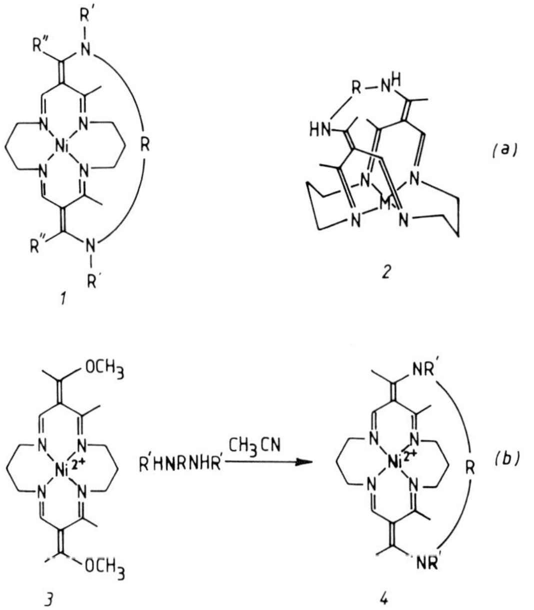

Fig. 3-59. (a) A superstructure ligand, (b) synthesis of this type of compound. The vinyl ether substituents in *3* were obtained by methylation of the precursor dicarbonyl compound which was synthesized using the Jäger method (e.g., see Fig. 3-44). For *4* used in Fig. 3-60, $R = (CH_2)_6$, $R' = CH_3$.

The R bridging group in these bicyclic ligands leads to the formation of a cavity and determines its size and form. This cavity is near to one of the coordination positions of the metal and can be readily seen in structure *2*. Furthermore, the bridging group protects and isolates from external influences the coordination site of the metal leaving it free to bind a substrate. Busch has suggested the term "dry cave" for this type of complex [96].

A nickel complex of this type has been prepared from a different tetracoordinate complex (Fig. 3-59b).

This complex of Ni(II) has been used as the starting material in the preparation of a Co(II) complex (Fig. 3-60) which is of greater bioinorganic interest [95a].

The ligand remained unchanged during this transformation. The Co(II) was tetracoordinate (square planar) and the cavity was well-formed (Fig. 3-61a).

Small ligands could enter the cavity of this complex. Thus, the complex could be oxidized by various reagents, Ce^{4+}, $NOPF_6$ or air, in the presence of $X^- = NCS^-$, N_3^- and NCO^- to give the corresponding Co(III) complexes, $[Co(L)X_2]^+$ (Fig. 3-61b). Various analogous Co(II) complexes can bind molecular oxygen in a manner similar to that of biological ligands [95b].

Fig. 3-60. Transformation of a Ni(II) into a Co(II) complex. The structure of the latter is given in Fig. 3-61.

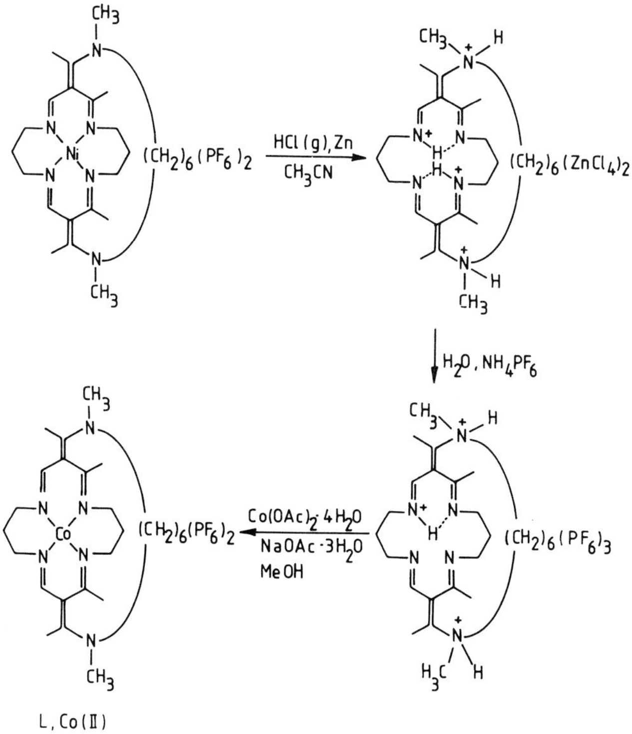

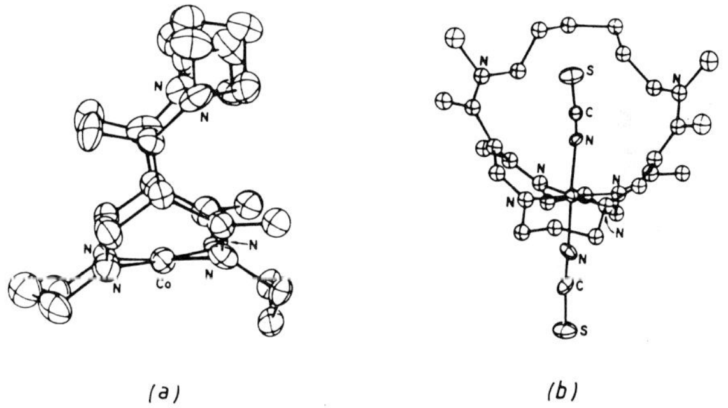

Fig. 3-61. Structures of (a) the L, Co(II) complex and (b) the –NCS inclusion complex, $[Co(III)(L)(NCS)_2]Cl$ (taken from D. H. Busch et al., *J. Am. Chem. Soc.,* **102,** 3283, 1980).

e) Use of metal cations as templates in catenane synthesis

Methods for the synthesis of catenanes are limited and have the drawback that they involve many steps resulting in low yields [97].

Transition metal template effects have recently been applied in the synthesis of these compounds by J.-P. Sauvage et al. [98]. They had observed previously the formation of a very stable pseudotetrahedral complex between Cu(I) and 1,10-

diphenylphenanthroline $[Cu(dpp)_2]^+$ (Fig. 3-62). In this complex, the two ligands "fitted into" each other [98, 99].

This property has been used in the construction of metallo-catenands (or catenates) [98] as shown in the synthetic scheme in Fig. 3-63. The key step in this method was formation of the pre-catenate *3* from *1* and *2* in the presence of Cu(I). It can be seen that the "fitted into" structure of *3* resulted in the two phenol groups being in positions which were relatively favorable for subsequent attack by the diiodo compound *4* to give the copper (I) catenate *5*.

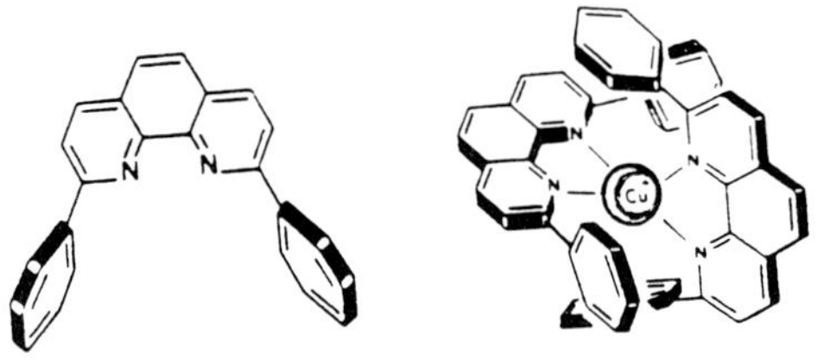

dpp $Cu(dpp)_2^+$

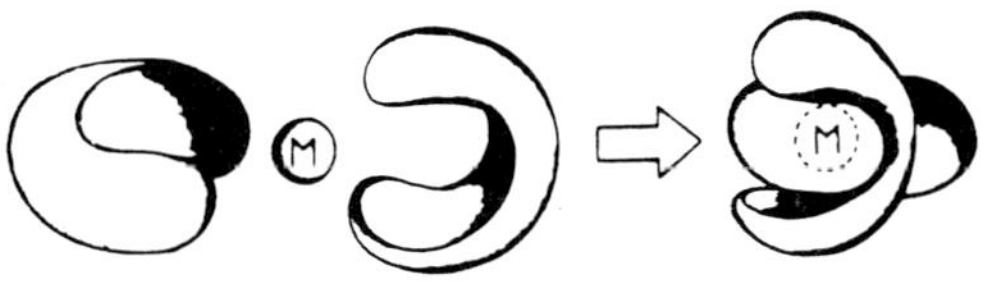

Fig. 3-62. The fitting together of the two dpp ligands around the Cu(I) cation.

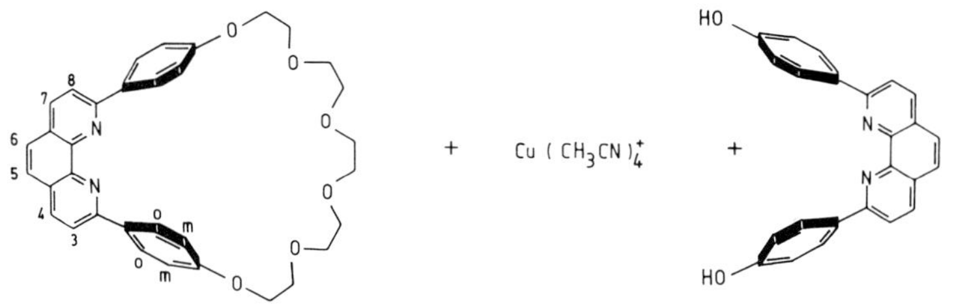

1 *2*

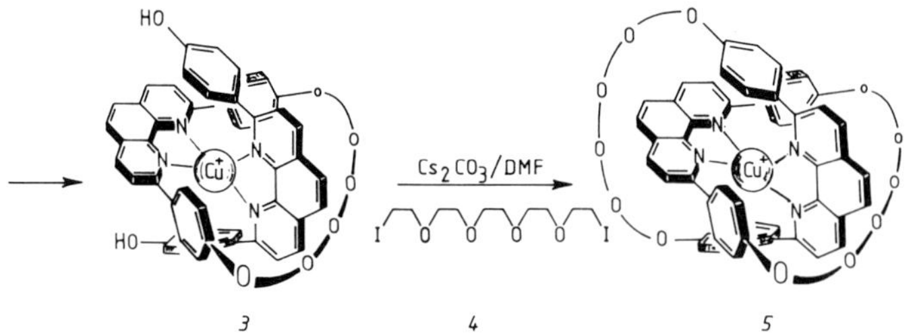

3 *4* *5*

Fig. 3-63. Scheme showing the synthesis of a metallo-catenane.

Although the metallo-catenate and the catenane differed by only one atom (Cu^+), and the bond connectivity of the organic skeleton was identical for both compounds, their respective molecular shapes were totally different as shown by the X-ray study [100a] (Fig. 3-64).

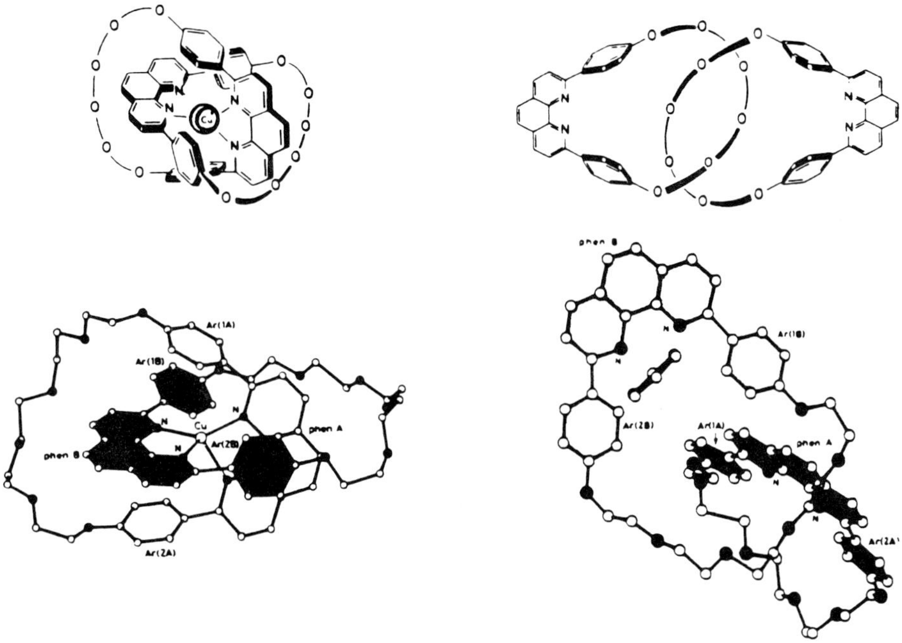

Fig. 3-64. X-ray crystal structures of the Cu(I) catenate and the catenane (taken from C. Pascard et al., *J. Chem. Soc., Chem. Comm.,* 244, 1985).

The use of template synthesis around transition metal cations has been extended to the formation of a trefoil knot. The strategy is based on the formation of a double helix consisting of two molecular threads wrapped around two metal cations [100b].

3.3.5 Template Reagents. Cycloolefin Metathesis

Olefin metathesis using homogeneous or heterogeneous catalysis involves bond reorganization resulting in redistribution of the two alkylidene moieties in each olefin (Fig. 3-65).

Cycloolefin metathesis under high dilution conditions gives macrocycles of various types [101]. Thus, *intermolecular* metathesis reactions result in "double" macrocycles (Fig. 3-66).

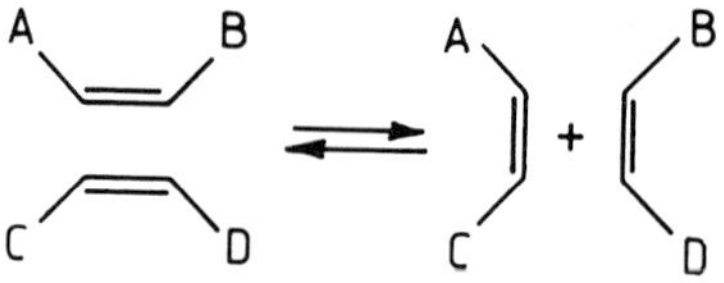

Fig. 3-65. Olefin metathesis reaction.

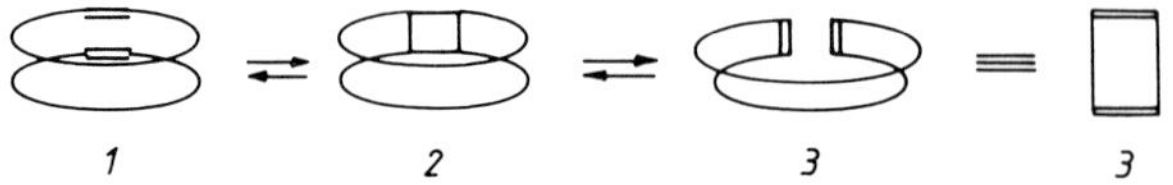

Fig. 3-66. Intermolecular metathesis of cycloolefins.

Once formed, the macrocyclic diene *3* can undergo *intramolecular* metathesis reactions [101, 102]. Because of the large size of the macrocycle, twisting can occur before the intramolecular metathesis reaction takes place (Fig. 3-67).

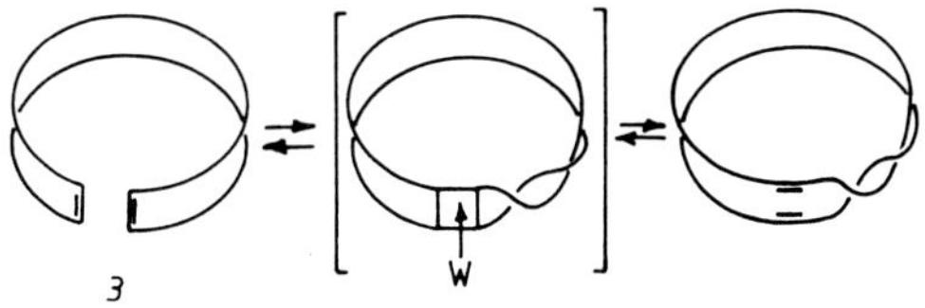

Fig. 3-67. Intramolecular metathesis reaction of a macrocyclic diene having undergone a 180° twist.

Depending on the degree of twisting (180°, 360°, 540°) undergone by molecule *3* (Fig. 3-68) the compounds obtained after metathesis are, respectively, a Möbius strip *4*, catenane *5* and nectinodane *6*.

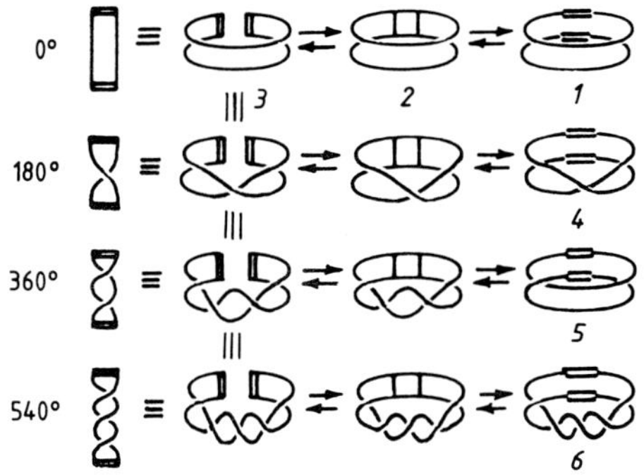

Fig. 3-68. Types of compounds obtained depending on the degree of twisting of ring *3*, which undergoes metathesis: Möbius strip *(4)*, catenane *(5)*, nectinodane *(6)*.

Metathesis reactions of cyclododecene have been performed in n-pentane using a WCl_6-$EtAlCl_2$-EtOH catalyst [102]; C_{24}, C_{36}, C_{48}, C_{60}, C_{72} up to C_{168} oligomers were detected.

A study of the heavier oligomer fractions by mass spectrometry up to 370 °C established that $C_{72} + C_{72}$, $C_{72} + C_{84}$ and $C_{84} + C_{84}$ catenanes had been formed.

3.4 Self-Assembly of Rigid Units

Self-assembly can take place owing to a combination of several factors; rigidity of the units, reaction stereochemistry and conformational folding.

3.4.1 Porphyrin Synthesis

Compounds of this type were first obtained by H. Fischer [103].

Porphyrin synthesis has been the subject of much investigation, and we shall give here only a very brief survey of the matter since numerous studies and reviews have been devoted to this class of compounds.

a) Tetrapyrrole condensation

Treibs and Häberle have examined routes to porphyrin rings starting from pyrroles [104]. The reaction of pyrrole with formic acid (Fig. 3-69) only gave satisfactory results with β,β'-dialkylpyrroles or carboxyalkylpyrroles (route 1). The synthesis of porphin from unsubstituted pyrrole was accomplished by reaction with the aldehyde, followed by oxidation (route II).

These two routes are depicted schematically (without the substituents) in Fig. 3-70. It is accepted that formation of porphin occurs stepwise.

Porphyrin yields vary greatly (Table 3-2) depending on the nature of the aldehyde and the pyrrole substituents [104, 105].

Table 3-2. Yields of porphyrins obtained from pyrroles and aldehydes.

Pyrrole	Aldehyde	Porphyrin yield (%)
Pyrrole	Formaldehyde	<1
”	n-Butyraldehyde	5.5
”	n-Laurylaldehyde	3
”	Benzaldehyde	23.5
”	p-Chlorobenzaldehyde	19.5
”	p-Methoxybenzaldehyde	22.5
”	p-Nitrobenzaldehyde	7
3,4-Dimethylpyrrole	Benzaldehyde	3
3,4-Dimethylpyrrole	Formaldehyde	77.5
3,4-Diphenylpyrrole	Formaldehyde	20

(Taken from A. Treibs et al., *Ann. Chem.* **718**, 183, 1968).

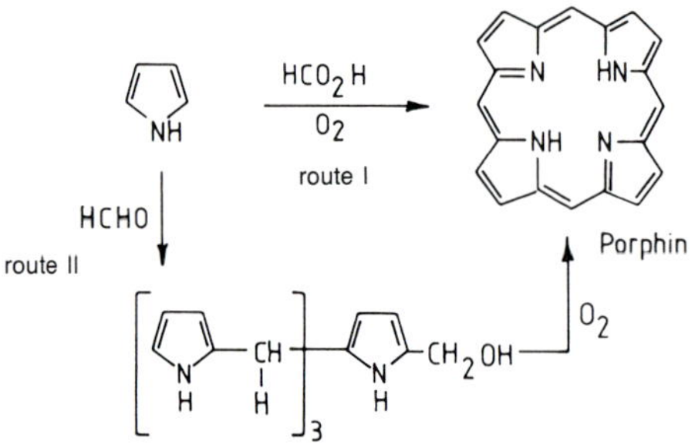

Fig. 3-69. Porphin synthesis using two types of tetrapyrrole condensation.

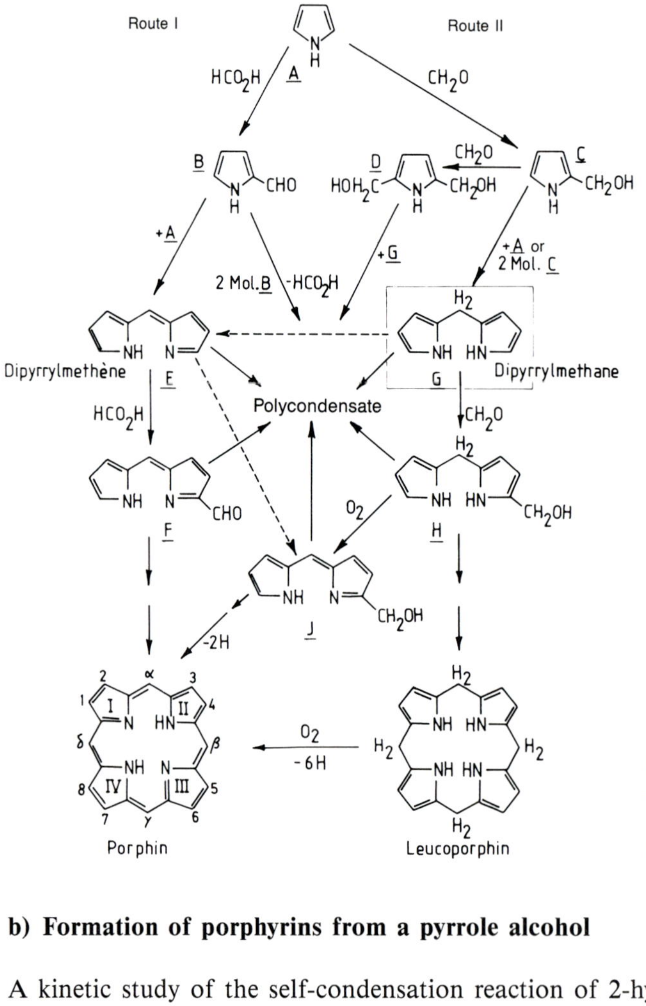

Fig. 3-70. Schematic representation of the steps involved in porphin formation.

b) Formation of porphyrins from a pyrrole alcohol

A kinetic study of the self-condensation reaction of 2-hydroxymethylpyrrole (Fig. 3-71) has been carried out in various solvents at different temperatures in both the presence and absence of acid [106].

The highest yields were obtained in xylenes and ethylbenzene in the absence of catalyst and at moderate temperatures.

Fig. 3-71. Porphin formation through self-condensation of 2-hydroxymethylpyrrole.

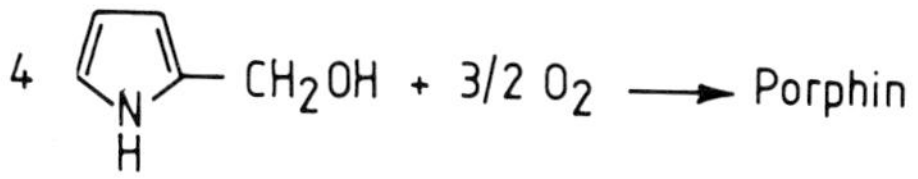

c) Synthesis of hindered porphyrins

Octaalkylporphyrins, including for the first time octaisopropylporphyrin, have been prepared (Fig. 3-72) by (a) anodic oxidation of alkenyl alkyl ethers, followed by (b) ring closure of the coupled product obtained, (c) debenzylation, and (d) condensation with formaldehyde [107].

Fig. 3-72. Synthesis of octaalkylporphyrins from alkenyl alkyl ethers. (a) Anodic oxidation in methanol/$NaClO_4$/2,6-lutidine; (b) $C_6H_5CH_2CO_2NH_2$; (c) H_2/Pd; (d) HCHO/O_2/CH_3COOH/pyridine.

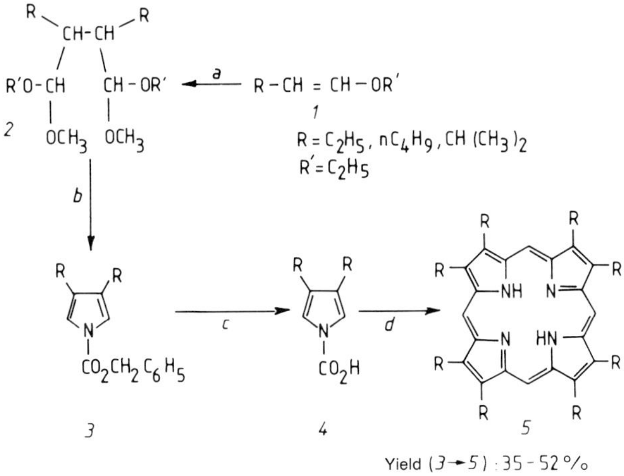

Octasubstituted porphyrins having no substituents in the meso positions could be obtained, also in high yield, by acid catalyzed condensation of formaldehyde with a variety of 3,4-disubstituted pyrroles [108] (Fig. 3-73) (Table 3-3).

Fig. 3-73. Synthesis of octasubstituted porphyrins by condensation of formaldehyde with 3,4-disubstituted pyrroles. R and R′ are given in Table 3-3 along with the nature of the products *1–7*. The alternating R and R′ groups in this figure do not represent the true situation. A mixture of isomers is in fact formed (see text).

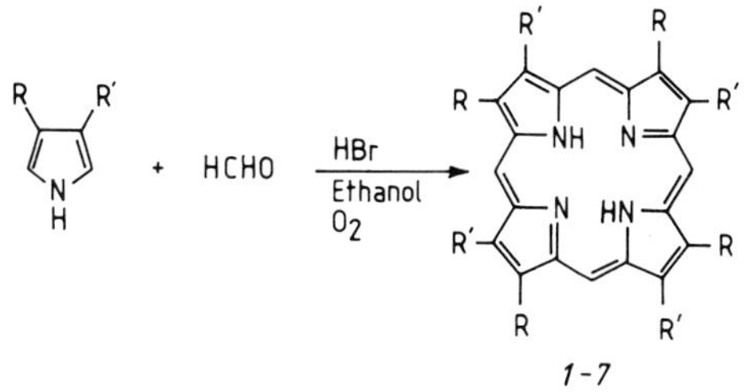

Table 3-3. Octasubstituted porphyrin yield from condensation of formaldehyde with a variety of 3,4-R,R′-pyrroles.

Pyrrole		yield %
1 R = C_2H_5	R′ = CH_3CO	96
2 R = CH_3	R′ = CH_3CO	64
3 R = CH_3	R′ = $CO_2C_2H_5$	92
4 R = CH_3	R′ = $CO_2C_8H_{17}$	52
5 R = Ph	R′ = $CO_2C_2H_5$	86
6 R = R′ = CH_3		76
7 R = R′ = C_2H_5		65

(Taken from E. Legoff et al., *Tetrahedron Letters,* 1469, 1977.)

When R ≠ R′, four isomers are possible and a mixture was in fact obtained. These mixtures could be used for subsequent conversion reactions. Thus, diborane reduction of the isomeric mixture (Fig. 3-74) afforded the octaethylporphyrin in very good yield.

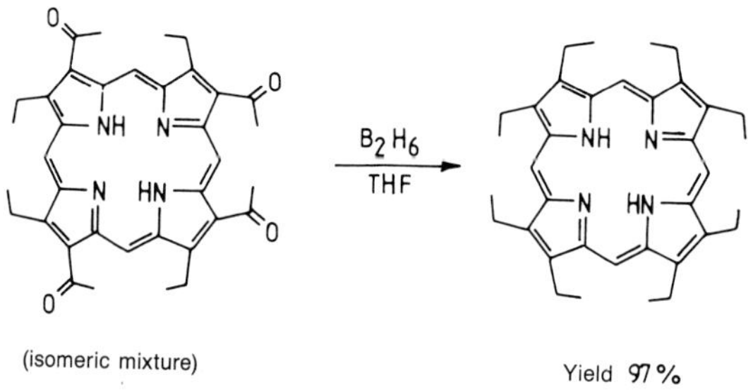

Fig. 3-74. Octaethylporphyrin formation by reduction of an isomeric mixture.

3.4.2 Thiaporphyrins

Porphyrin analogs incorporating thiophene rings have been synthesized. Thus, condensation of diformyl thiophene with tripyrrane diacid derivatives gave the corresponding monothiophene derivatives [109] (Fig. 3-75).

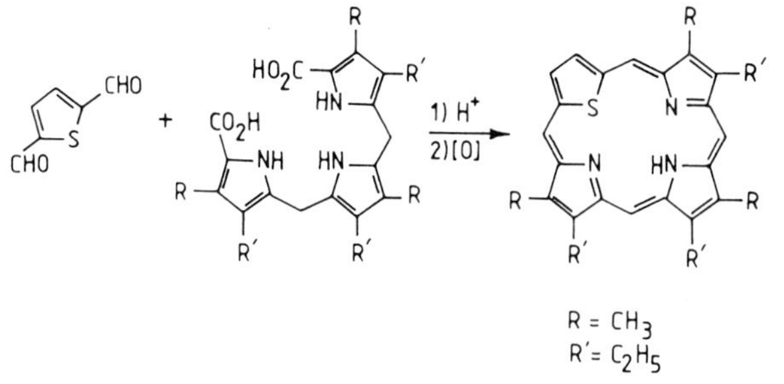

Fig. 3-75. Synthesis of monothiaporphyrins (yield = 12%).

A tetraphenyldithiaporphyrin has been synthesized (Fig. 3-76) by acid-catalyzed reaction of pyrrole with 2,5-bis(phenylhydroxymethyl)thiophene [110].

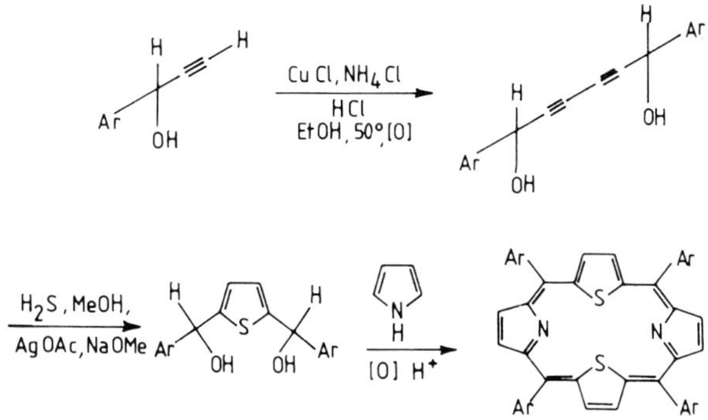

Fig. 3-76. Synthesis of a tetraphenyldithiaporphyrin (yield = 10%).

New macrocyclic aromatic systems related to porphyrins and corroles have been obtained by extrusion of S or SO_2 [111, 112].

The synthesis of 21,24-dioxacorroles by condensation of dipyrromethane-5,5′-dicarboxylic acids with 5,5′-diformyl-2,2′-bifuryl gave the desired product in low yield (7%) [112] (Fig. 3-77). However, the yield increased to 27–30% if bis-(5-formyl-2-furyl) sulfide were used as the starting material. The proposed scheme (Fig. 3-77) involved extrusion of sulfur from a macrocylic intermediate *1* which could not be isolated [112]. The sulfur atom facilitated ring closure and served as a temporary template.

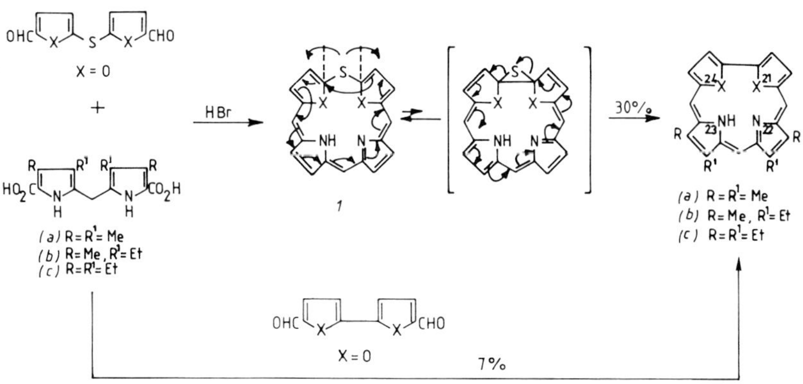

Fig. 3-77. Synthesis of 21,24-dioxacorroles.

3.4.3 Tetraheterocyclic Compounds Similar to Porphyrins

Macrocyclic compounds possessing a variety of subheterocyclic rings have been constructed [113].

a) Pyrrole subunits

The first macrocycle possessing a subheterocyclic ring (pyrrole) was synthesized by A. Baeyer in 1886 by acid-catalyzed condensation of pyrrole with acetone [114] (Fig. 3-78). The yield was 90%.

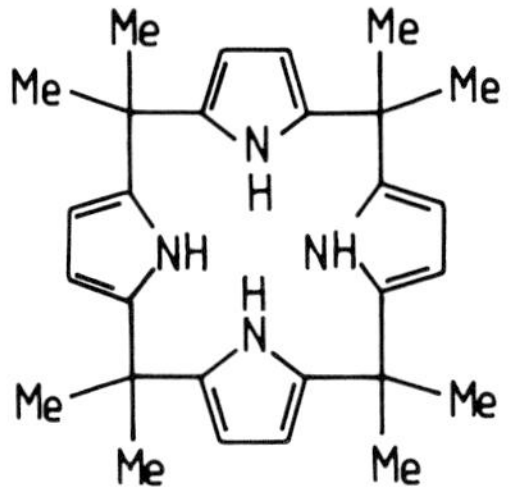

Fig. 3-78. Macrocycle with pyrrole subunits.

Other aldehydes and ketones (e.g., cyclohexanone) have been condensed with pyrrole [115].

Tetrapyrrole compounds can be used as the foundation for constructing macropolycyclic systems if functionalized side chains are introduced into the molecule. This has been achieved in the example given in Fig. 3-79 with a yield of 50% [116a].

Many substitution reactions have been carried out on compound *1* replacing X = Cl by other groups [116b]. Thus, the octacyano derivative, the X-ray structure of which is shown in Fig. 3-79b, has been obtained [116c].

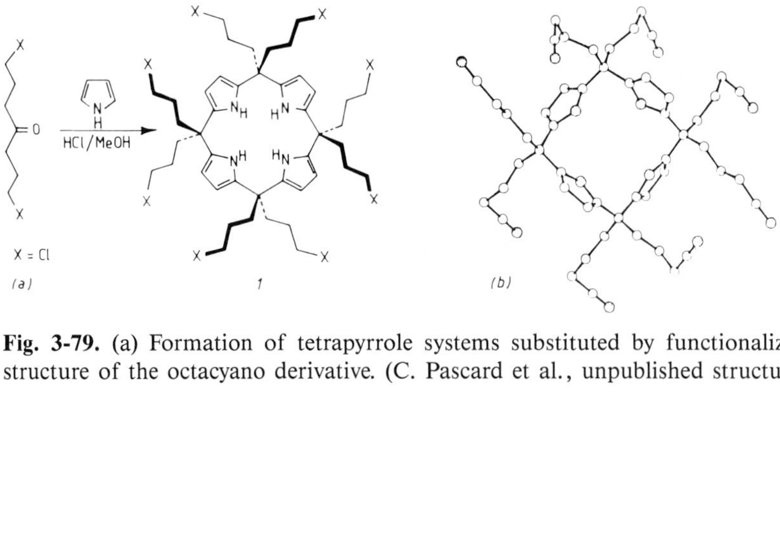

Fig. 3-79. (a) Formation of tetrapyrrole systems substituted by functionalized chains, (b) structure of the octacyano derivative. (C. Pascard et al., unpublished structure).

b) Furan subunits

Tetraoxaquaterenes constitute the class with the largest number of macrocyclic compounds that contain furan rings. The 16-membered compound has been synthesized by acid-catalyzed condensation of furan with a dialkyl ketone (Fig. 3-80). The yields were generally low as a result of predominant formation of polymers [117]. They have since been improved by use of metal ions which apparently acted as external templates *(vide infra)*. Many intermediates have been isolated, and in certain cases could be converted to the macrocyclic system when subjected to additional acidic condensation conditions [113].

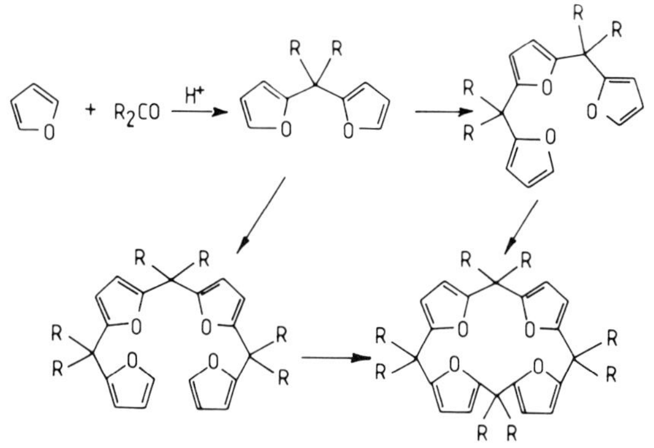

Fig. 3-80. Synthesis of tetraoxaquaterenes via intermediates which could be isolated.

The increased yield of cyclic product resulting from the addition of cations such as Li^+, Ca^{2+}, Zn^{2+} and Mg^{2+} [118] or those of transition metals (Mn^{2+}, Ni^{2+}, Cu^{2+}) [119] seemed to be mainly due to a pH effect. The arguments in favor of a pH effect rather than an external template effect (even though the latter could not be completely ruled out in certain cases) have been developed by de Souza Healy and Rest [120], based on various observations. If the coordination (external template effect) were to play a dominant part, the wide variety of cations, varying both in size and also in their affinity for the oxygen sites, should have resulted in the following:

- Formation of macrocycles of different sizes (which was not in fact observed).
- Absence of effects with certain cations such as Mn^{2+}, Ni^{2+} and Cu^{2+} since these have little affinity for the oxygen sites (according to the HSAB principle). In fact, these cations gave excellent results.

Furthermore, in the absence of metal cations, increase in the concentration of hydrochloric acid led to a marked increase in yield. Finally, it has been shown that addition of neutral salts (LiCl, NaCl, NaBr, etc.) to solutions of strong acids results in increased acidity of the solution. Consequently, the cation template effect was of minor importance and the dominant effect was one due to pH [120]. It is advisable in this case to be prudent in interpreting the effect of the cation.

c) Thiophene subunits

A large number of macrocycles of this type have been prepared.

The formation of tetrathiaquaterenes necessitated more vigorous reaction conditions (72% sulfuric acid) for the reaction of the heterocycle with acetone (Fig. 3-81) than in the cases of the pyrrole and furan analogs. The yields were low [121].

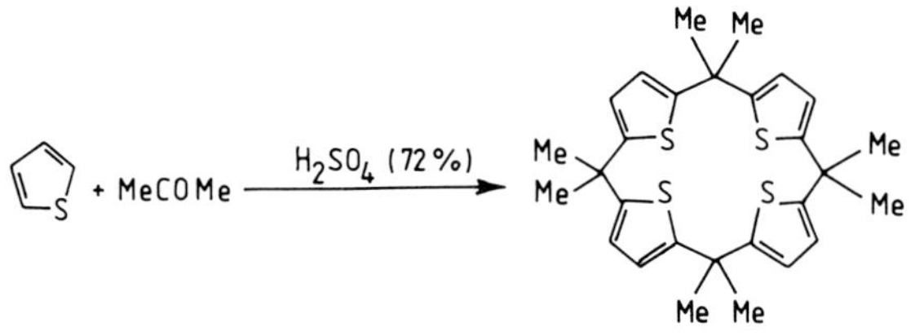

Fig. 3-81. Synthesis of a tetrathiaquaterene.

Other compounds in the same series have been obtained (Fig. 3-82) by condensing 2,2,-bis-(5′-lithio-2′-thienyl)propane with 2,2-bis-(5′-formyl-2′-thienyl)propane or 2,2-bis-(5′-acetyl-2′-thienyl)propane [122]. The yields were, however, still very low (~5%).

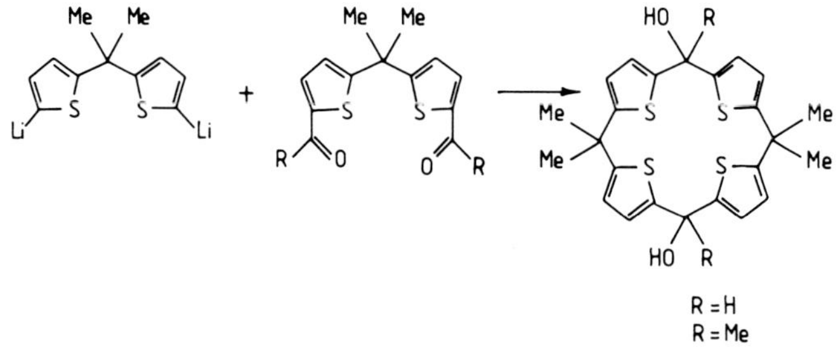

Fig. 3-82. Synthesis of tetrathiaquaterenes by condensation between a dilithio and a diacyl derivative.

The synthesis of a thiophene-containing macrocycle with bridging silicon atoms (Fig. 3-83) has been achieved by condensation of 2,2-bis-(5′-lithio-2′-thienyl)-2-silapropane with dichlorodimethylsilane in THF at 0 °C [123].

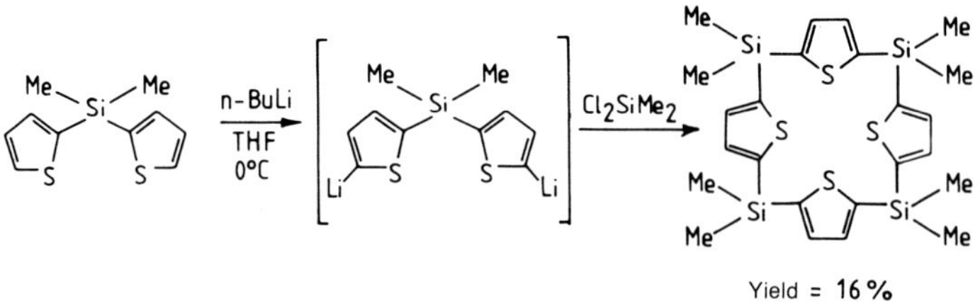

Fig. 3-83. Synthesis of a silathiophene macrocycle.

d) Combination of different subunits

There is a large number of macrocycles containing at least two different heterocyclic subunits such as pyrrole and furan.

The synthesis of tetraoxaquaterenes involved, as we have seen, numerous intermediates which could be isolated and characterized. When the latter were treated with pyrrole and acetone under acid conditions, they gave a series of "mixed quaterenes" [124].

3.4.4 Phenol-Aldehyde Macrocycles

The acid-catalyzed condensation of a phenol and an aldehyde generally results in a complex mixture of high molecular weight products. Nevertheless, it has long been known that some phenols such as resorcinol react with aldehydes such as benzaldehyde or salicylaldehyde to give crystalline compounds the structures of which have only recently been established.

a) Condensation of resorcinol with benzaldehyde

This acid-catalyzed reaction giving macrocycle *1* has been studied by Högberg who showed that two stereoisomeric octaphenol tetramers were formed [125] (Fig. 3-84).

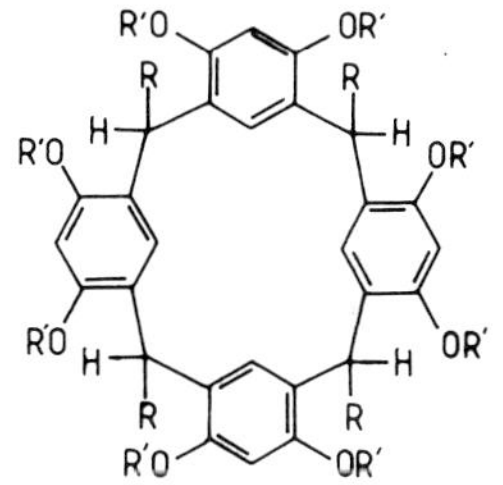

1

$R = C_6H_5$, $R' = H$

Fig. 3-84. Macrocycle obtained by condensation of resorcinol with benzaldehyde.

The kinetic product *(1a)*, which was the predominant isomer at the start of the reaction, was transformed into the thermodynamic product *(1b)*. The latter was obtained in greater than 80% yield if the reaction time was extended (see the yield plots in Fig. 3-85).

An NMR study of octabutyrate derivatives ($R' = COC_3H_7$, Fig. 3-84) enabled the stereochemistry of the two isomers to be assigned. Isomer *1a* possessed a chair-like conformation whereas isomer *1b* had a boat-like conformation (Fig. 3-86). In isomer *1a* (chair type), the Ar phenyl groups are axial in pairs on either side of the plane of the macrocycle (C_{2h} symmetry). In isomer *1b* (boat type), the four Ar phenyl groups are axial on the same side of the plane of the macrocycle (C_{2v} symmetry).

The stereoselectivity of the reaction has been attributed to a combination of three factors: the intramolecular steric interactions within the triphenylmethane units, the reversibility of the cyclooligomerization, and the difference in solubility of the two macrocyclic compounds (*1 b* is the less soluble).

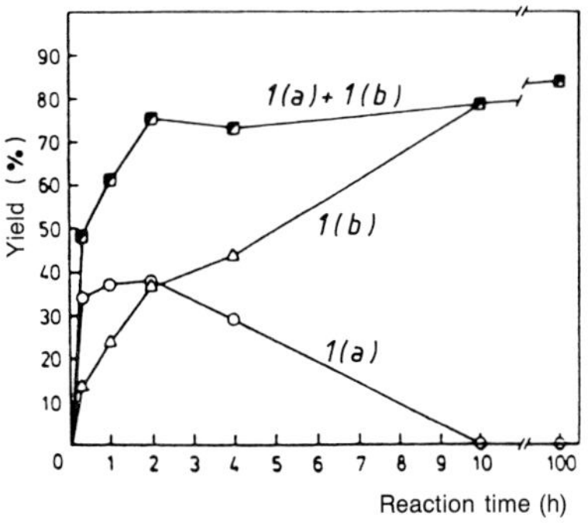

Fig. 3-85. Plot of yields of macrocyclic oligomers *1 (a)* and *1 (b)* vs. reaction time.

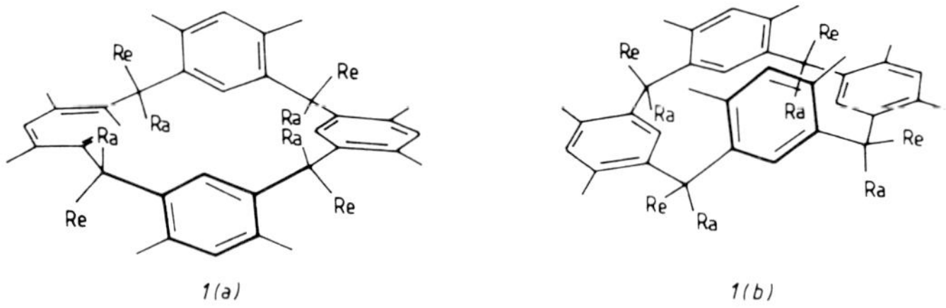

Fig. 3-86. Macrocyclic stereoisomers: (a) kinetic and (b) thermodynamic product. Ra = Ar, Re = H. For reasons of clarity, the OR′ substituents are denoted by a line.

b) Condensation reaction of phenol with formaldehyde

This reaction gives cyclic oligomers known since long and obtained by several different procedures from phenol and formaldehyde in the presence of base [126].

Some recent studies [127a] have been devoted to macrocyclic tertiary butyl phenol derivatives, known as calixarenes, which exhibit various interesting properties. Some examples of calixarenes are given in Fig. 3-87. They have cavities of adequate dimensions to allow formation of complexes with certain organic molecules or with cations. The presence of functional groups allows modification of solubility, complexing ability and polycyclic structure. They are models for cyclodextrins.

The nature of the macrocyclic products obtained depended on the reaction conditions. The cyclic octamer, *p*-tert-butylcalix[8]arene, was isolated in yields which reached 64% in a one step reaction in the presence of a trace of base. When a higher concentration of base was used, *p*-tert-butylcalix[6]arene became the major product. The yield of this latter product was 74% in the case of rubidium hydroxide. The cyclic tetramer was obtained in a yield of 20 to 25% in a three step process.

An X-ray study has confirmed the structure of the octamer [128a] (Fig. 3-88).

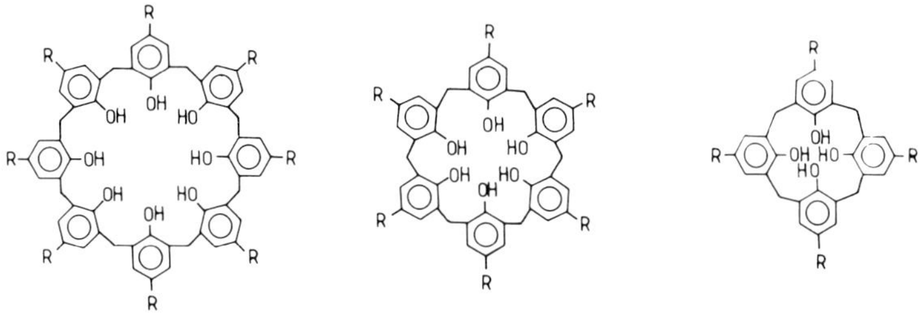

Fig. 3-87. Examples of calixarenes; R = t-Bu. The figure enclosed by brackets, [ ], indicates the number of phenol units per macrocycle.

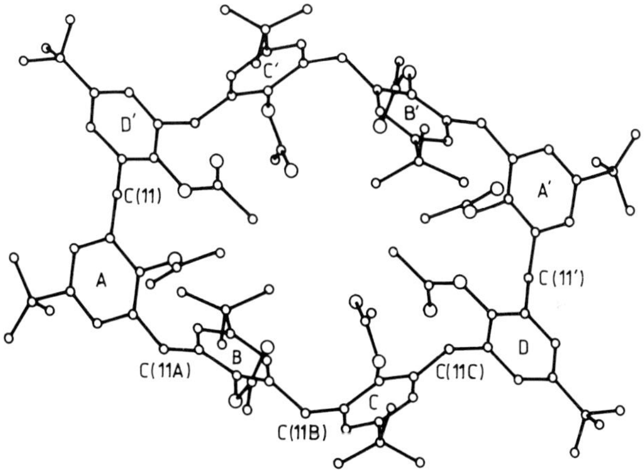

Fig. 3-88. X-ray crystal structure of *p*-tert-butylcalix[8]arene (taken from G. D. Andreetti et al., *J. Chem. Soc., Chem. Comm.*, 533, 1981. N. B.: The structure is that of the octa-acetate.)

Recrystallization of the tetrameric calixarene in toluene gave an inclusion complex with one molecule of solvent. Its X-ray crystal structure has been determined [128b] (Fig. 3-89). The tertiary butyl groups play an important part in binding the guest species.

Calixarenes are in certain ways analogous to cyclodextrins, e.g., the torus shape of the molecules and the existence of an intramolecular cavity. Important differences exist, however, between the two classes of compounds. Cyclodextrins are quite soluble in water and their cavities are lipophilic, whereas calixarenes are insoluble in water and the presence of phenol groups results in their cavities being polar. From the point of view of complexing properties, in water cyclodextrins form inclusion complexes of high stability with organic substrate molecules (almost exclusively aromatic), whereas for calixarenes such complexes are almost only ever observed in the solid state.

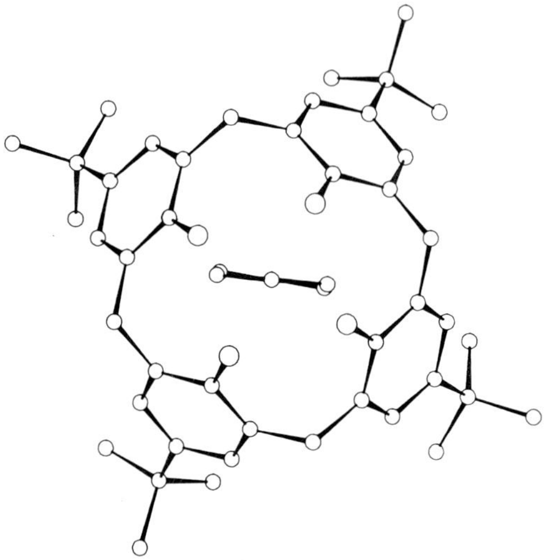

Fig. 3-89. X-ray crystal structure of *p*-tert-butylcalix[4]arene: toluene clathrate (taken from [128b]). We thank Professor Andreetti for supplying this figure.

Metal cations (alkali and alkaline earth) are transported through chloroform membranes by *p*-tert-butylcalixarenes of the [4], [6] and [8] types, but only if the source phase is very basic (ionization of one or more of the phenol functional groups) [127b].

The insolubility of calixarenes in water being a serious impediment to their use, several types of derivatives possessing greater solubility in polar solvents have been synthesized [127b and c]. *p*-Sulfonatocalix[5]arene and *p*-Sulfonatocalix[6]arene, synthesized by Shinkai, complex very strongly and very selectively the uranyl cation, UO_2^{2+}, in water (the stability constants are ca. 10^{20}) [127b].

c) Condensation reaction of veratrole with formaldehyde

This reaction was described as early as 1915, but with an excusable error of structure attribution for the product obtained [129a].

The acid-catalyzed condensation reaction gave a mixture of the trimer (major product) and the tetramer (Fig. 3-90). Condensation of 3,4-dimethoxybenzyl alcohol gave the same products [129b, 130a].

An NMR study over a wide temperature range has shown that cyclotriveratrylene (Fig. 3-90a) has a rigid structure and that the molecule undergoes neither inversion nor rotation. This rigidity has been confirmed by the resolution of enantiomers of certain cyclotriveratrylene derivatives [130c].

On the other hand, the tetramer is more flexible and pseudorotation is possible about the methylene groups [130b] (Fig. 3-91).

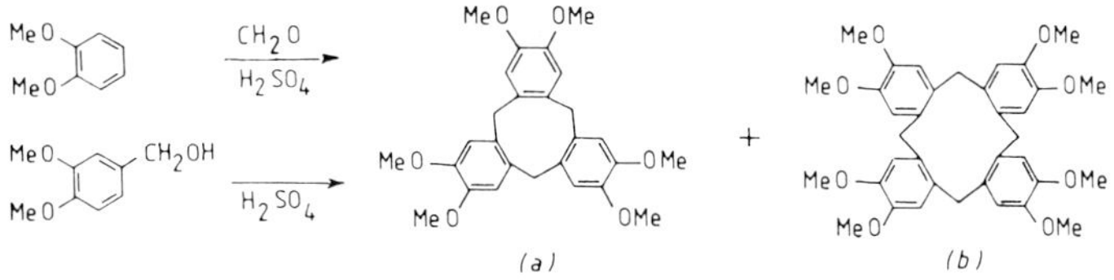

Fig. 3-90. (a) Trimer and (b) tetramer obtained by condensation of veratrole with formaldehyde or condensation of 3,4-dimethoxybenzyl alcohol.

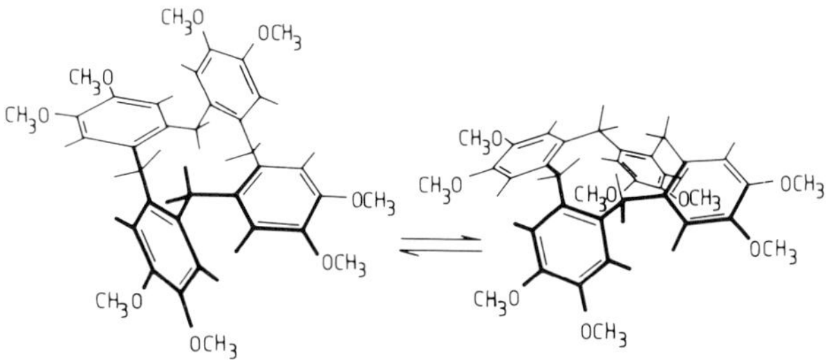

Fig. 3-91. Conformational mobility of cyclotetraveratrylene.

d) The replication of a cyclotriveratrylene derivative

Cyclotriveratrylene forms crystalline solvates with many solvents, both aromatic and other. By making use of both this property and also the possibility of obtaining chiral cyclotriveratrylenes, Collet et al. have elaborated a new family of compounds capable of complexing organic molecules efficiently [131 a].

A macrocage, (D_3)-bis(cyclotriveratrylenyl) *4*, has been synthesized [131 b]. The chiral D_3 symmetry originates in the spatial arrangement of the six equivalent achiral aromatic residues (Fig. 3-92 b).

Acid catalyzed self-condensation of *1 a* gave the trimer (C_3)(±)-*2 a* which was transformed into the compound (C_3)(±)-*2 e* via *2 b, c, d* (Fig. 3-92 a).

The reaction of (C_3)(±)-*2 e* with the iodide *1 b* in hexamethylphosphoric triamide in the presence of aqueous sodium hydroxide afforded the derivative (C_3)(±)-*3* with a cyclotriveratrylene-like crown conformation. When treated with formic acid at 90 °C for 30 min, it was cyclized to give a single racemic product (D_3)(±)-*4* (Fig. 3-92 b, Exp. A). The stereochemistry of this reaction was established by using an optically active precursor *2 e* (Exp. B). The transformation of the triphenol (+)-*2 e* into optically active (−)-*3* and then into optically active (D_3)(+)-*4* showed that intramolecular ring closure *3* → *4* was a stereospecific replication, the new ring (D_3) having the same chirality as the ring (C_3) formed initially. The macrocage *4* formed a very stable complex with chloroform.

Compound *4* was the first representative of a new family known as the *cryptophanes.* Other molecules of this type synthesized more recently complex dichloromethane, dibromomethane, chloroform and bromoform [132].

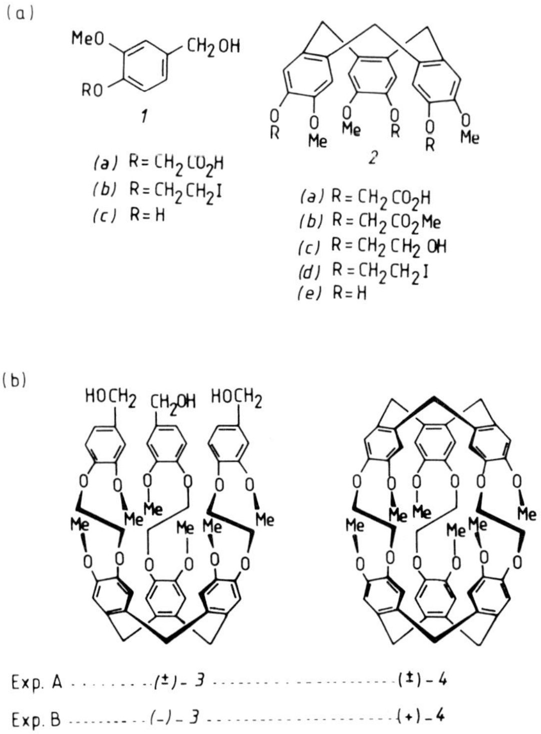

Fig. 3-92. Formation of (D_3)-bis(cyclotriveratrylenyl), *4*.

e) Cavitands

The class name cavitand was proposed by Cram for synthetic organic compounds that contain *enforced cavities* large enough to complex the smaller molecules or ions. In these compounds, the cavity is preexistent before binding the substrate species, which is not at all the case for crown ethers nor to a certain extent for cryptands. In these latter two cases, in the uncomplexed state, the ligand more or less fills its own cavity. The cavity is only fully formed by conformational reorganization once binding to the substrate species occurs.

The compound resulting from the duplication of cyclotriveratrylene which we have just discussed, and the family of spherands mentioned below, are both models for cavitands since their cavities are formed during synthesis rather than during the binding process. The calixarenes are not classified as cavitands since their conformational mobility is too great. However, certain of them serve as starting materials for new types of cavitands synthesized by Cram et al. [133].

Thus, cavitand *3* was obtained by treating compound *1*, described by Högberg [125], with base and $BrCH_2Cl$ (Fig. 3-93). Compound *4* was synthesized in an analogous way from *2,* and it could be readily transformed into *5*.

Fig. 3-93. Formation of cavitands *3–5* from *1* and *2*.

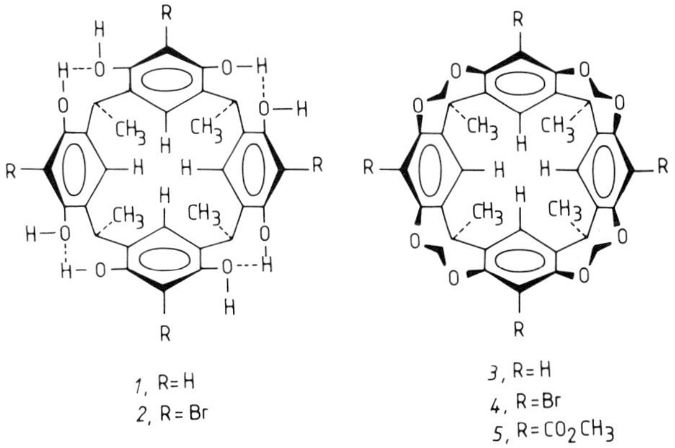

Molecular models and NMR spectra showed that conformational constraints impart to cavitands *3–5* the shape of a bowl supported on a base of four methyl groups [134].

On treatment with 4 equivalents of 2,3-dichloro-1,4-diazanaphthalene and excess potassium hydroxide in DMF, compound *1* afforded the vase-shaped cavitand *6* (Fig. 3-94) which could exist in either the "compact" structure *6a* or the "extended" structure *6b*. This new cavitand *6* was exclusively in the form *6a* at 25 °C and above, and exclusively in the form *6b* at −100 °C and below [133, 134].

The molecular model of *6a* showed that the cavity was of sufficient size to contain one molecule of [2.2]paracyclophane or twelve molecules of water.

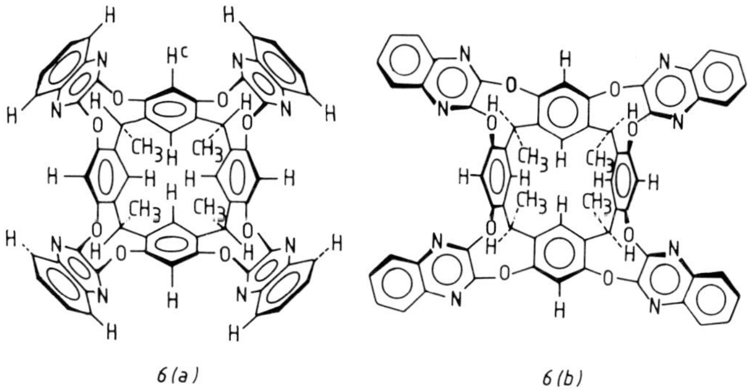

Fig. 3-94. (a) Compact and (b) extended structures of a cavitand.

Another series of cavitands has been synthesized. These were macrocyclic oligomers (Fig. 3-95) composed of four or six diethyldibenzofuran units [135].

Compound *1* is saddle-shaped and molecular models showed it to contain 2 cavities ca. 12 Å long, 4.3 Å wide, and 3.4 Å deep.

The four oxygen atoms occupy the same plane, and they line a hole which can contain various species such as CH_4, NH_4^+ or Cs^+.

Macrocycle *2* was a secondary product of the reaction and it contains a large cavity of dimensions ca. 11 × 7 × 7 Å. The six oxygens lie in one plane. Synchronized rotations of the aryl groups about their Ar-Ar bonds were possible in the model, but resulted in aryl-aryl bond deformations. When of maximum size, the cavity could contain seven benzene rings; one in the plane of the oxygens, and the other six stacked in groups of three on each side of a plane perpendicular to the first plane.

Unlike other cavitands, macrocycles *1* and *2* were more soluble in benzene, toluene or the xylenes than in CH_2Cl_2 or $CHCl_3$. This was explained by the fact that the benzene rings "fitted into" the cavity. When crystallized from benzene, compound *2* formed heavily solvated crystals that released benzene too readily for the X-ray crystal structure to be determined.

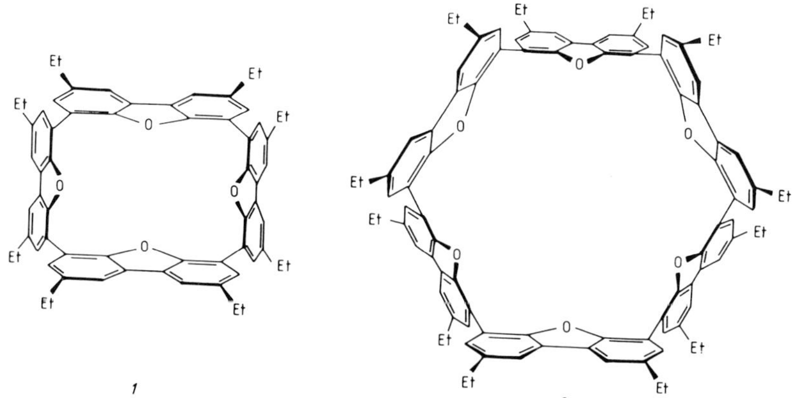

1 *2*

Fig. 3-95. Macrocyclic oligomers of diethyldibenzofuran.

f) Spherands

Although spherands are not obtained by self-assembly, it is appropriate to discuss them here.

Historically, spherands were the first representatives of the class of cavitands.

Spherand *1* was synthesized (Fig. 3-96) in six steps from p-cresol which was initially oxidatively coupled to give oligomer *2*. Subsequent bromination followed by methylation afforded *4* which on treatment with butyl lithium gave the aryl lithium. The latter was oxidatively coupled using $Fe(acac)_3$ to form *5* which was converted into *6* which in turn was decomplexed to give *1* [136].

The crystal structures of *1* and *6* (Fig. 3-97) are very similar, which shows that complexation resulted in almost no conformational change.

The six oxygens define an octahedral cavity formed during the synthesis and imposed by the rigidity of the benzene rings as well as steric hindrance resulting from the methoxyl groups. Three of the latter groups point up and the other three point down with respect to the median plane of the macrocycle thus ensuring very

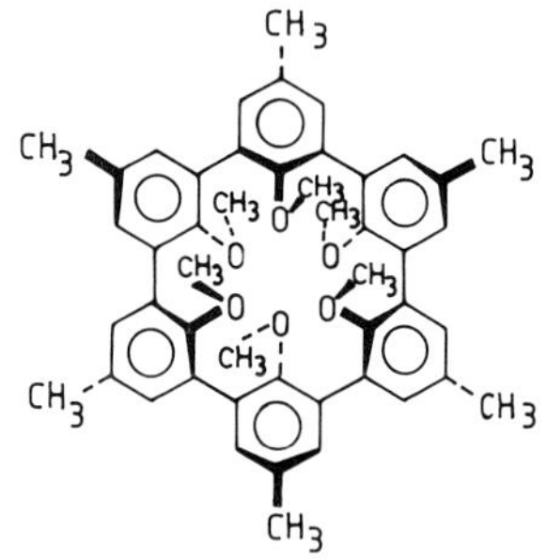

1

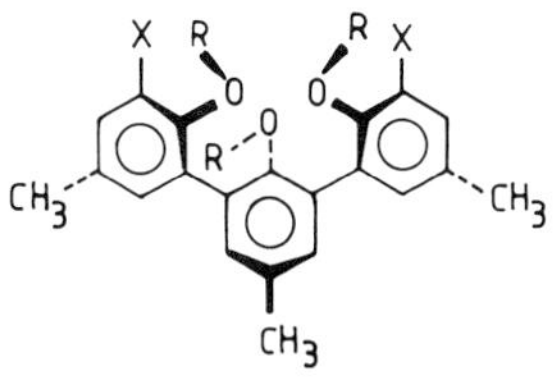

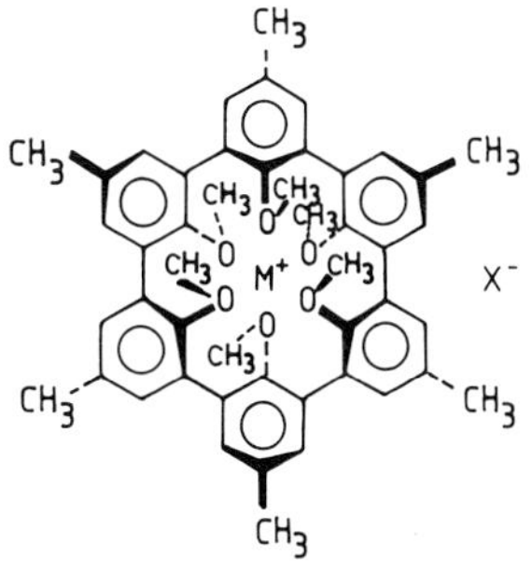

2, R = X = H
3, R = H, X = Br
4, R = CH_3, X = Br

5, M = Li, X = $FeCl_4$
6, M = Li, X = Cl

Fig. 3-96. Synthesis of spherand *1* (overall yield from *p*-cresol ~6%).

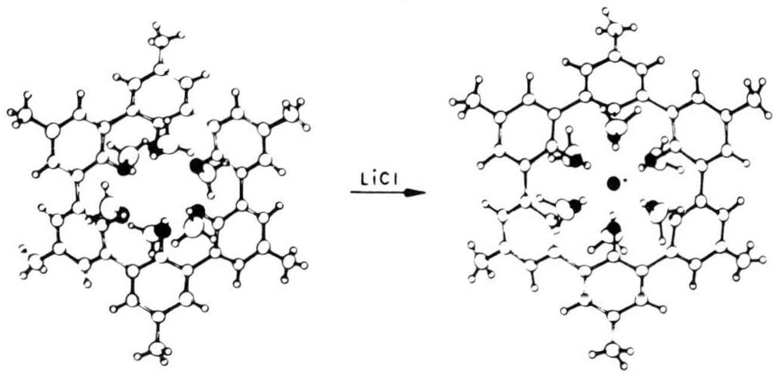

Fig. 3-97. X-ray crystal structures of spherand *1* and of its complex with Li^+ cation (taken from D. Cram et al., *J. Am. Chem. Soc.,* **103,** 5594, 1981).

good protection of the cavity interior. This spherand is the best binder known for Li^+ and Na^+ ions. It is also very selective for these two ions since K^+, Rb^+, Cs^+, Mg^{2+} and Ca^{2+} are not at all complexed.

The chemistry of spherands has become widely diversified (Fig. 3-98) through both the introduction of anisyl and cyclic urea units *1* into the macrocyclic framework and also the synthesis of hemispherands *2* and *3* [137].

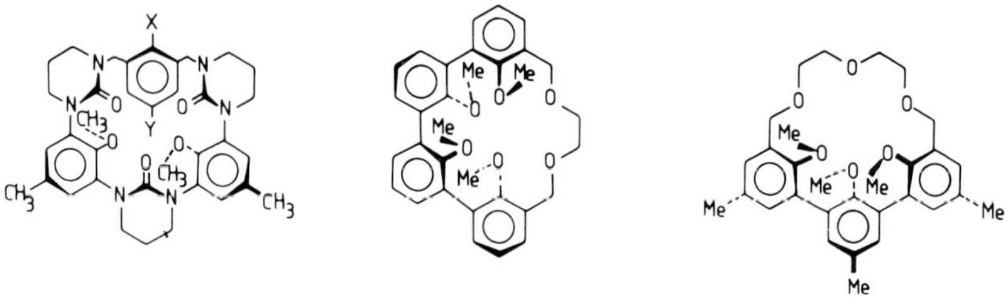

1 2 3

Fig. 3-98. Spherands *(1)* and hemispherands *(2, 3)*.

3.5 Crown Ether Macrocyclic Compounds

3.5.1 General Synthetic Methods

Although a few examples of cyclic polyethers had been known in the literature before the work of Pedersen, it was he who synthesized numerous compounds of this type and showed their ability to complex cations, both alkali metal and other [138a].

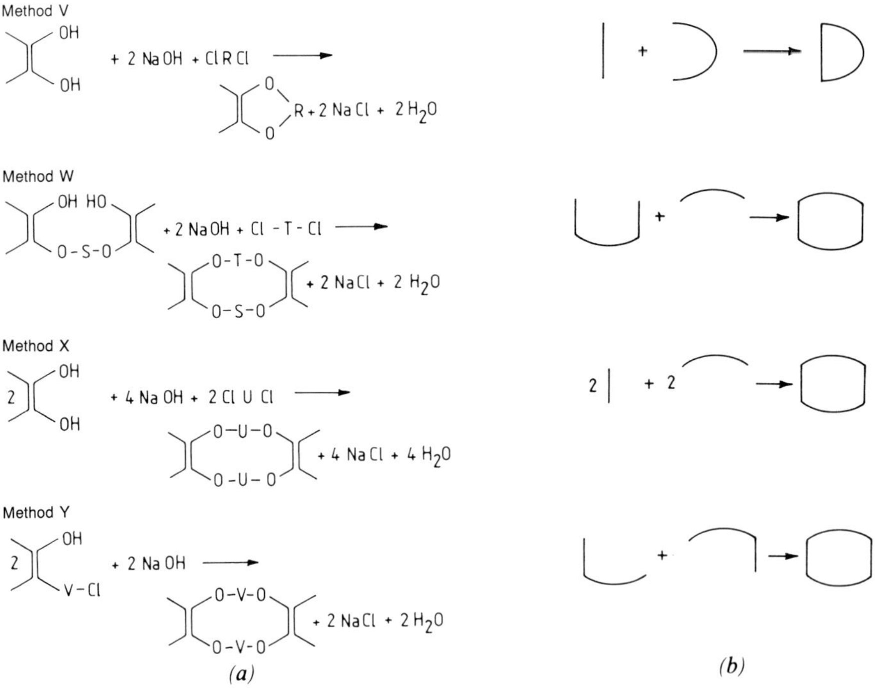

Fig. 3-99. Methods of crown ether synthesis.

As a result of the pioneering role played by C. J. Pedersen in the discovery of crown ethers, in 1987 the Nobel committee conferred on him (jointly with D. J. Cram and J.-M. Lehn) the highest distinction, the Nobel prize for chemistry. His unusual career (exclusively industrial) has been described by one of his colleagues [138 b].

Since these investigations, hundreds of macrocyclic polyethers have been synthesized. A similar number of publications has been devoted to them in the last twenty years or so and many reviews have appeared [139]. Some methods of synthesis are given in Fig. 3-99 a but the schematic representations (Fig. 3-99 b) illustrate better the different synthetic routes.

The first compound described by Pedersen was dibenzo-18-crown-6 obtained in ca. 45% yield from catechol and bis(2-chloroethyl) ether in butanol in the presence of sodium hydroxide (Fig. 3-100). Its hydrogenation product was a mixture of two isomers of dicyclohexyl-18-crown-6.

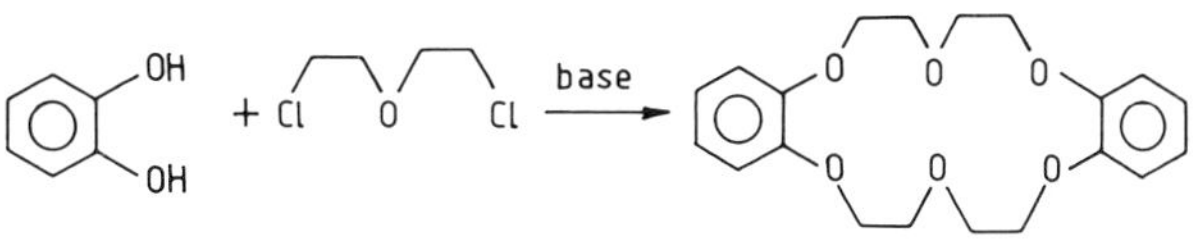

Fig. 3-100. Synthesis of dibenzo-18-crown-6.

Heteroatoms other than oxygen have been incorporated into the macromolecular framework since macrocycles are used almost exclusively for binding cations. In view of the vast extent of the literature, we shall indicate here only the major types with a few references. In Part II, more details on the synthesis and properties will be given for certain of them.

a) Polyaza- and other polyheteroatomic macrocycles

Polyazamacrocycles constitute a huge collection of efficient complexing agents for a large number of transition metal cations [55, 140]. Some examples together with the nomenclature of these compounds are given in Fig. 3-101.

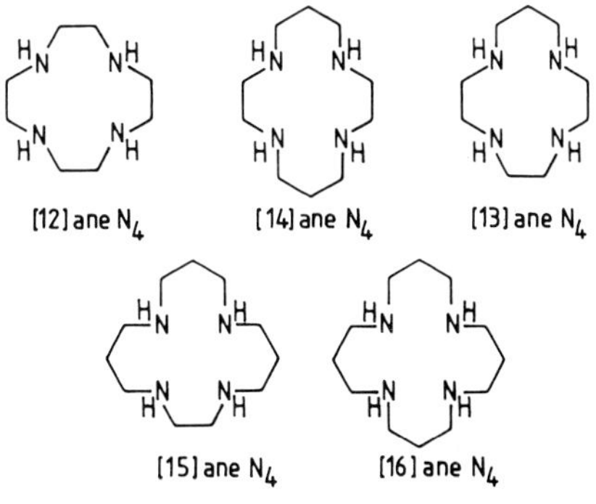

Fig. 3-101. Structures and nomenclature for tetraaza macrocycles (following nomenclature proposed by D. H. Busch, *Inorg. Chem.,* **11,** 1979, 1972).

More recently, protonated forms of very large macrocycles (24- to 38-membered rings) (Fig. 3-102) have been found to be anion complexones [141].

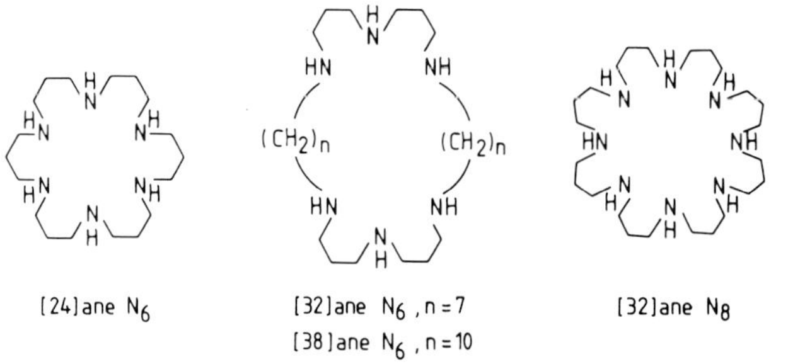

Fig. 3-102. Structures and nomenclature of hexa- and octaazamacrocycles.

Numerous *sulfur*-containing macrocycles such as those in Fig. 3-103 have been synthesized [139, 142].

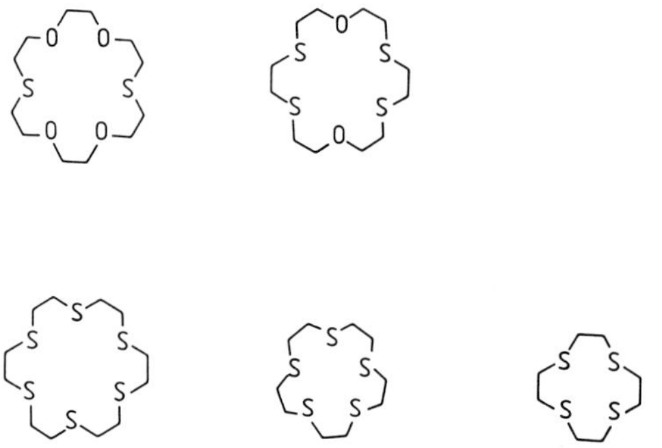

Fig. 3-103. Sulfur-containing macrocycles.

An important application of certain sulfur-containing macrocycles discovered by Boekelheide is worth mentioning. This is the use of sulfur as a temporary template facilitating cyclization in cyclophane syntheses (Fig. 3-104).

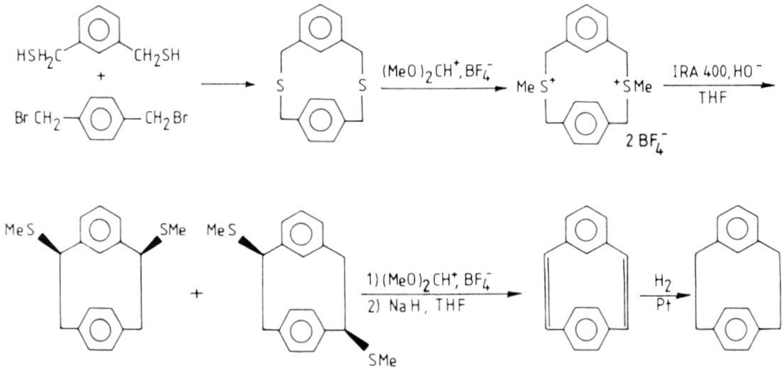

Fig. 3-104. Use of sulfur as a temporary template in cyclophane synthesis.

Transformation of C–S–C bonds into double bonds via sulfur extrusion was effected by a Stevens rearrangement followed by a Hofmann elimination [143].

Other heteroatoms (*phosphorus* [144], *arsenic* [145], *tin* [146]) have also been incorporated into macrocycles (Fig. 3-105).

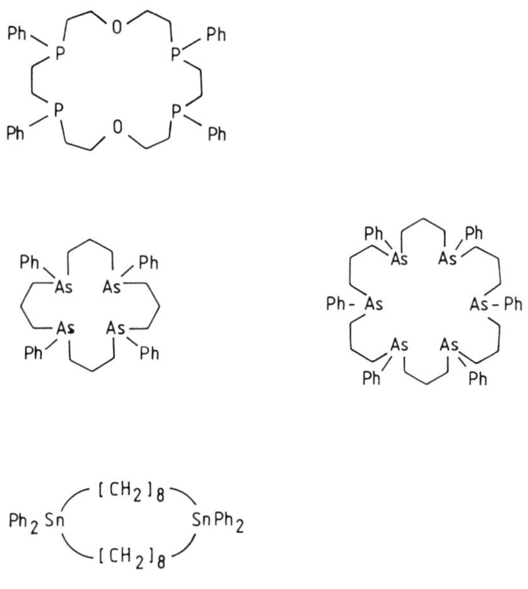

Fig. 3-105. Macrocycles containing various heteroatoms.

b) Macrocycles with functionalized side arms

This group of compounds was developed more recently and is expanding fast. These macrocycles are characterized by the appendage of side arms which possess binding sites (Fig. 3-106).

The arms are attached in particular to nitrogen or carbon atoms in the macrocycle. The name *lariat* has been proposed for this type of ligand [147a].

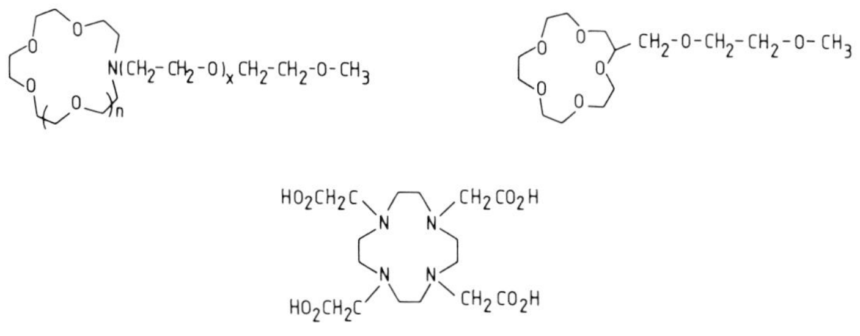

Fig. 3-106. Macrocycles with ligating arms.

The field of macrocyclic systems with pendant chains has been extensively developed. The incorporation of reactive functional groups into the side chains gives rise to systems possessing both a complexation site (the macrocycle cavity) and a reactive site in the lateral chain. The proximity effects thus exhibited provide good enzyme models [147b]. Methods have been developed for the attachment of macrocycles, *via* pendant chains, to antibodies. Complexation of radioisotopes (e. g. ^{64}Cu and ^{67}Cu) by the macrocycles affords radiolabelled antibodies [147c].

3.5.2 Introduction of Cyclic Subunits

Numerous crown ethers containing a wide variety of heterocyclic units have been synthesized [113].

Thus, bis-2,5-hydroxymethylfuran, obtained from sucrose, on treatment with tetraethyleneglycol ditosylate in the presence of potassium *tert*-butoxide in THF gave a furan-containing polyether [148] (Fig. 3-107).

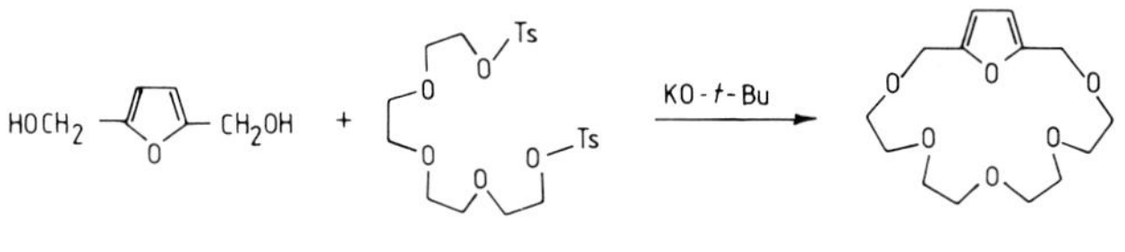

Fig. 3-107. Synthesis of a furan-containing macrocyclic polyether.

The pyrimidino subunit has been incorporated into crown ethers (Fig. 3-108) by treating 4,6-dichloropyrimidine with the disodium salt of diethyleneglycol. The same reaction carried out with the disodium salt of triethyleneglycol gave the [1 + 1] *(5)* and [2 + 2] *(6)* products [149]. However, the yield was low in each case.

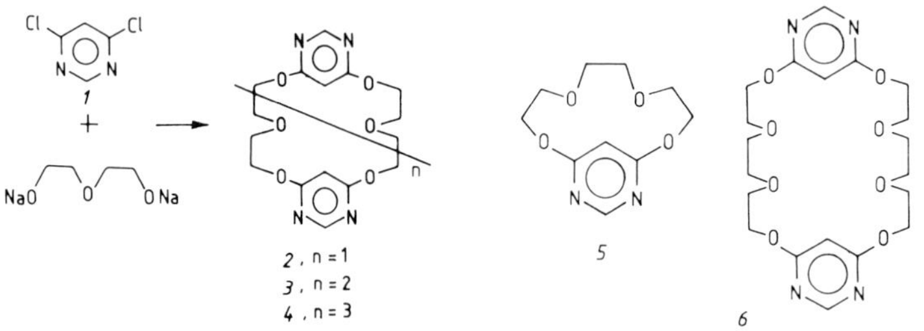

Fig. 3-108. Crown ethers containing the pyrimidino subunit.

Many other cyclic subunits, such as pyridine, dipyridine and thiophene, have been incorporated into macrocyclic frameworks [113, 150].

3.5.3 Introduction of Chiral Subunits

Chiral recognition by a macrocyclic ligand requires the presence of a suitable cavity to ensure good complexation, and also chiral barriers. For there to be chiral discrimination, only one of the two enantiomers in the racemic mixture should be complexed strongly by the macrocycle.

The first syntheses of chiral macrocycles (Fig. 3-109) were carried out by Wudl starting from *L*-proline or *D*-ψ-ephedrine [151].

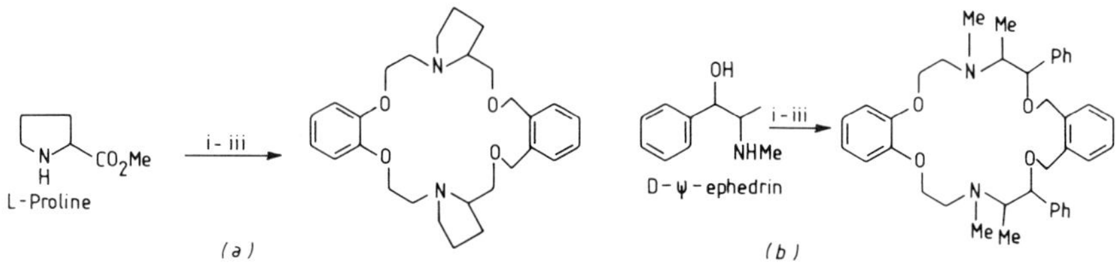

Fig. 3-109. Synthesis of chiral macrocycles from (a) *L*-proline or (b) D-ψ-ephedrine. (a) i: *o*-$C_6H_4(OCH_2COCl)_2$: ii: LAH; iii: *o*-$C_6H_4(CH_2Br)_2$/NaH/DMSO. (b) i: *o*-$C_6H_4(OCH_2COON_p)_2(N_p = p\text{-nitrophenyl})$; ii: LAH, iii: *o*-$C_6H_4(CH_2Br)_2$/NaH/DMSO.

Various laboratories have elaborated a whole class of chiral compounds which are derivatives of [18-0-6] crown ether. This crown ether, as we shall see in Part II, is particularly efficient at complexing the ions NH_4^+ and RNH_3^+. The construction of the macrocycle is carried out using synthons which are quite easily accessible from carbohydrates. We shall only mention a few examples, of increasing complexity, from this vast field.

A chiral diethyleneglycol unit in the form of the fragment 2,5-anhydro-D-mannitol, D-*1*, has been introduced into a crown ether [152]. The synthesis of compound D-*2* (Fig. 3-110) was performed by condensation of D-*5* (NaH/DMSO) with tetraethyleneglycol ditosylate (19% yield), the compound D-*5* being obtained in several steps from D-*1*.

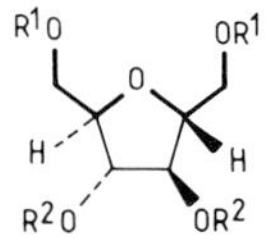

	R^1	R^2
D-*1*	H	H
D-*3*	$SiMe_2Bu^t$	H
D-*4*	$SiMe_2Bu^t$	Me
D-*5*	H	Me

D-*2*

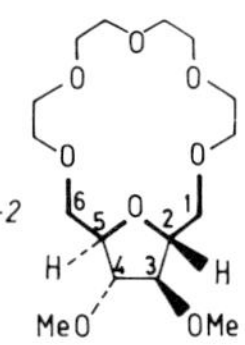

Fig. 3-110. Synthesis of a crown ether containing a chiral diethyleneglycol unit.

The tetrahydrofuran ring was chosen for its complexing properties and its ability to ensure that macrocyclic structures were preserved. Indeed, it was found that the crown ether formed (in $CDCl_3$) extremely stable complexes with alkali metal cations and also NH_4^+ and RNH_3^+ ions.

Condensation reactions of the diols α-D-*7* and α-D-*9* with their derived ditosylates α-D-*8* and α-D-*10* (Fig. 3-111 a) gave respectively the bisglucosides 2,3 : 2′, 3′-αα-DD-*3* and 2,3 : 3′,2′-αα-DD-*4*, and the bisgalactosides 2,3 : 2′,3′-αα-DD-*5* and 2,3 : 3′,2′-αα-DD-*6* [153] (Fig. 3-111 b).

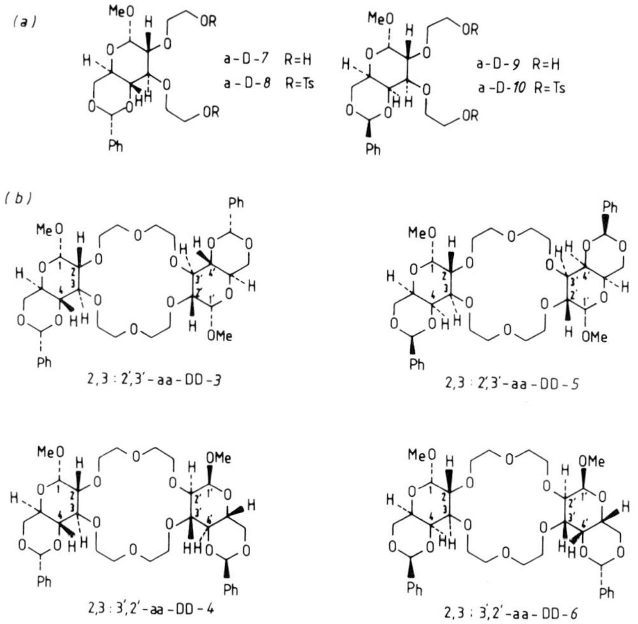

Fig. 3-111. Macrocyclic bisglucosides (DD-*3* and DD-*4*) and bisgalactosides (DD-*5* and DD-*6*). (a) Starting materials, (b) macrocycles formed.

Another method for the synthesis of the isomers αα-DD-*4* (a) and αα-DD-*3* (b) from benzylidene-glucopyranoside *1* and *2b* has been described [154] (Fig. 3-112).

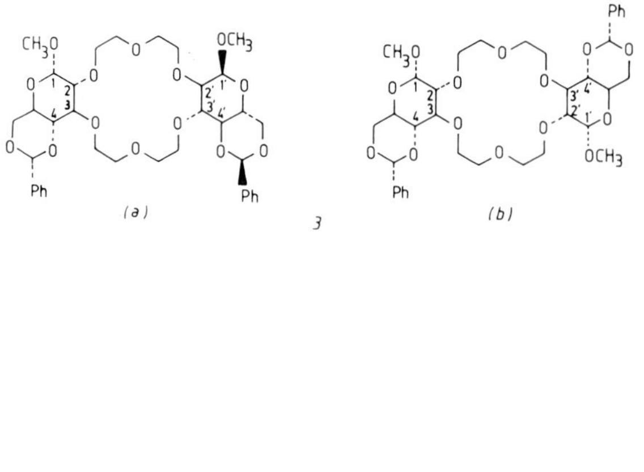

Fig. 3-112. Synthesis of the macrocycles DD-*4* (a) and DD-*3* (b) from benzylidene-glucopyranoside *1* and diethylene glycol ditosylate (*2b*).

Deprotection by removal of the benzylidene groups from the isomers of *3* enabled numerous derivatives to be synthesized by using the alcohol groups thus liberated [154b].

Stoddart's previous studies led him to believe that chiral recognition of the *R* or *S* enantiomers of a racemic salt *(RS)*-$PhCHMeNH_3^+ClO_4^-$ (*(RS)-(1)*$HClO_4$) might be improved by including in the ligand several residues of different sizes with the aim of ensuring a better steric match with one of the isomers than with the other. Fig. 3-113a shows the arrangement of the large (L), medium (M) and small (S) substituents in the *R* and *S* forms. In order to test the validity of his hypotheses, Stoddart synthesized the macrocycle DDD-*5* containing two D-*3* residues and one D-*6* residue [155] (Fig. 3-113b). The circles L′, M′ and S′ symbolize the sizes of the cavities available on the face of the ligand. As a result, the *S* isomer was the matching enantiomer. The macrocycle DDD-*5* in fact exhibited modest chiral recognition in favor of the *S-(1)*$HClO_4$ isomer (see Fig. 3-113a).

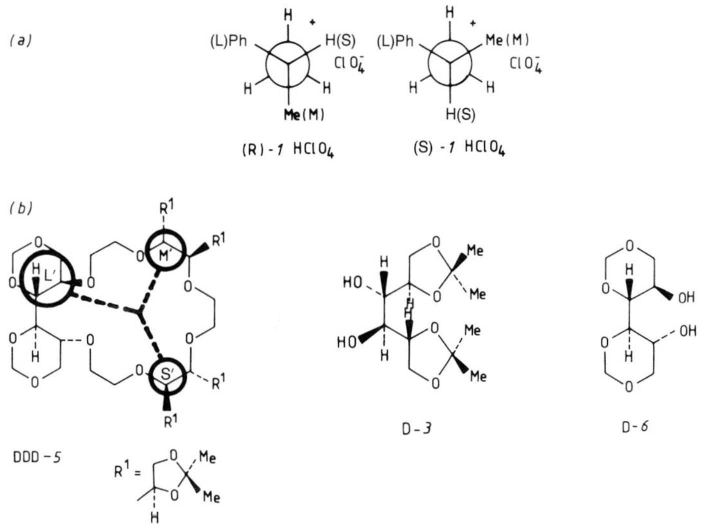

Fig. 3-113. (a) Arrangement of the substituents in the *R* and *S* isomers of $PhCHMeNH_3^+$, (b) macrocycle DDD-*5* and starting materials D-*3* and D-*6*.

Stoddart concluded from these results that chiral crown ethers containing three carbohydrate residues disposed trigonally have much more potential to discriminate between enantiomeric salts than do chiral crown ethers containing two carbohydrate residues disposed diagonally.

A chiral receptor suitable for primary ammonium cations should have the following properties:

- It should be a macrocyclic polyether of the 18-C-6 type.

- It should have functional groups which are directly attached to the ring in order to permit construction of specific receptors.
- It should be a chiral macrocycle of known absolute configuration [156].

Lehn et al. synthesized a number of functionalized and chiral macrocyclic polyethers by using a modified Williamson synthesis based on that described by Seebach [157]. This involved condensation of the dithallium alcoholate of *(R,R)*-(+)-tartaric acid derivatives with α, ω-dihalides.

Different synthetic routes are possible involving either one step (route A, Fig. 3-114a) or two steps (routes B and C). Route A is the one most used. Depending on the nature of the dihalide and the route used, the macrocycle formed is the monomer (n = 1), the dimer (n = 2) or in certain cases the trimer (n = 3). The reactants and the macrocycles formed are given in Fig. 3-114b and c, respectively.

Reaction of *1c* with *2a* gave a mixture containing mostly the macrocycle *3a* (dimer) and a little of the compound *7a* (trimer).

When the dihalo reactant is a long chain, formation of the monomer is possible. Thus, the reaction between *1c* and the diiodide *2c* led to a mixture of *8a* (monomer) and *9a* (dimer).

Reaction between *1c* and *2d* gave macrocycle *6* with two pyridine rings. Finally, reaction of *1c* with *2e* gave the macrocycles *11, 12,* and *13.*

Similarly, reaction between *1e* and *2a* gave *3b* which could be converted by acid hydrolysis into bis-imide *5.*

Compounds *1f* and *1g* were the starting materials for synthesis by the two-step routes (Fig. 3-114a, routes B and C).

An extremely important macrocycle is the tetracarboxylic acid *3c* which is obtained by acid hydrolysis of *3a*. This tetraacid can be converted into the tetraacyl chloride *3d* and the dianhydride *3e*. The tetraacid *3c* was the starting material for synthesis of the derivatives *4b* to *4e*. Compound *4a* was obtained by reduction of *3a*.

The yields varied greatly depending on the case:

[18]-O_6 $(CONMe_2)_4$, *3a*:	20%
[27]-O_9 $(CONMe_2)_6$, *7a*:	1 to 2%
[15]-O_5 $(CONMe_2)_2$, *8a*:	13%
[30]-$O_{10}$$(CONMe_2)_4$, *9a*:	13%

Compounds of the same type, prepared by the above method, have been converted to crown ether monoamides with an acid functional group and also bearing a long chain alkyl group on the amide group [158] (Fig. 3-115).

Fig. 3-114. (a) Synthetic paths for synthesis of functionalized and chiral macrocyclic polyethers, (b) starting materials, (c) macrocycles obtained.

(a)

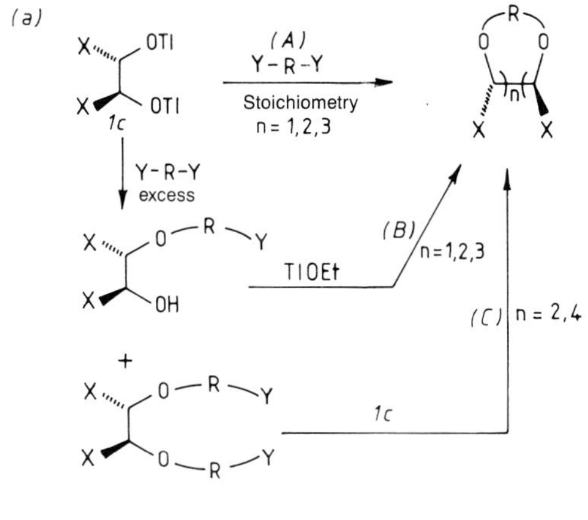

X = CONMe$_2$

(b)

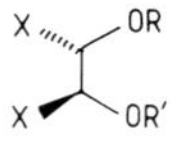

1 X = COOH, R = R' = H

1(a) X = COOEt, R = R' = H

1(b) X = CONMe$_2$, R = R' = H

1(c) X = CONMe$_2$, R = R' = Tl

1(d) X = CONHMe, R = R' = H

1(e) X = CONHMe, R = R' = Tl

1(f) X = CONMe$_2$, R = H, R' = (CH$_2$)$_2$O(CH$_2$)$_2$I

1(g) X = CONMe$_2$, R = R' = (CH$_2$)$_2$O(CH$_2$)$_2$I

2(a) Y = I

2(b) Y = OTs

2(c) Y = OCH$_2$CH$_2$I

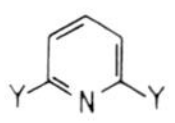

2(d) Y = CH$_2$Br

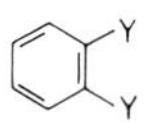

2(e) Y = CH$_2$Br

(c)

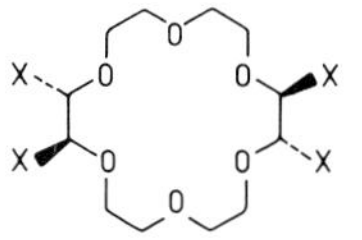

3(a) X = CONMe$_2$

3(b) X = CONHMe

3(c) X = COOH

3(d) X = COCl

3(f) X = CON(n-Bu)$_2$

3(g) X = H

4(a) X = CH$_2$NMe$_2$

4(b) X = CH$_2$OH

4(c) X = CH$_2$OTs

4(d) X = CH$_2$CN

4(e) X = CH$_2$COOH

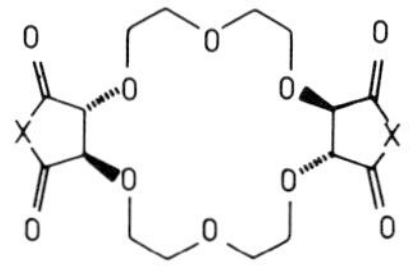

3(e) X = O

5 X = NMe

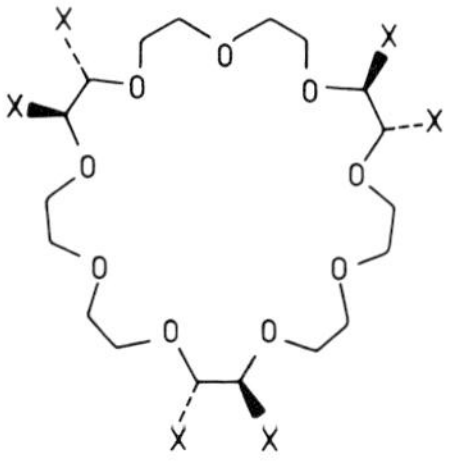

7(a) X = CONMe$_2$

7(b) X = COOH

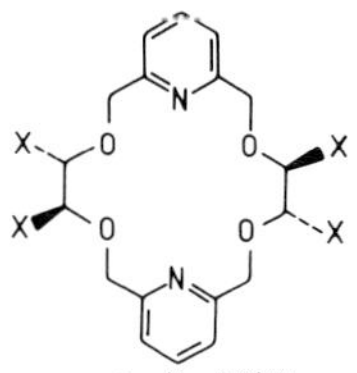

6 X = CONMe$_2$

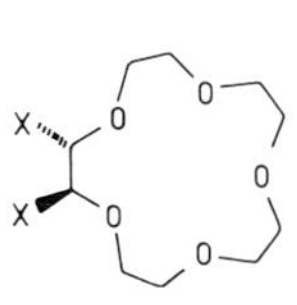

8(a) X = CONMe$_2$

8(b) X = COOH

8(c) X = COOMe

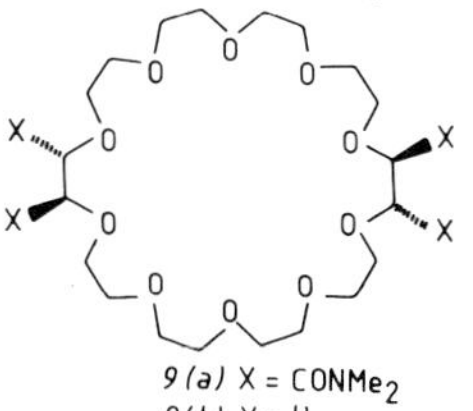

9(a) X = CONMe$_2$

9(b) X = H

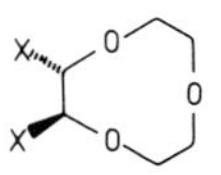

10 X = CONMe$_2$

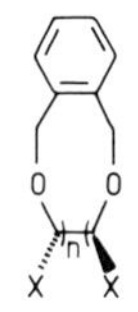

11 n = 1 X = CONMe$_2$

12 n = 2 " "

13 n = 3 " "

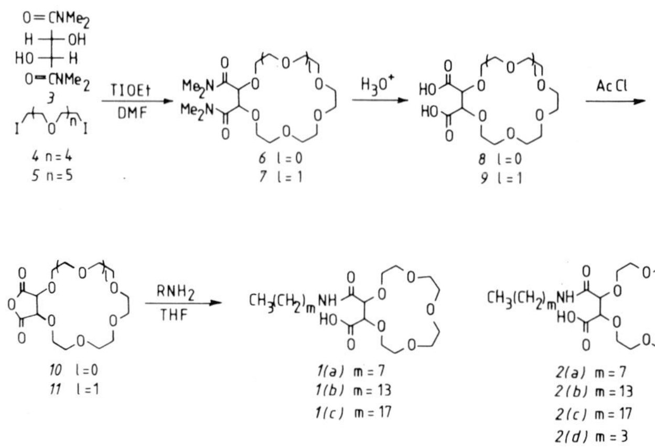

Fig. 3-115. Synthesis of functionalized and lipophilic macrocycles.

3.5.4 Cationic Cyclooligomerization of Ethylene Oxide

a) In the presence of a Lewis acid

Since the repeat unit in macrocyclic polyethers is $-CH_2-CH_2-O-$, a logical route to them should be the cyclooligomerization of ethylene oxide. Dale et al. have studied this route. Treatment of ethylene oxide with 1% BF_3/HF (1:1) in dioxan gave a mixture of the cyclic products, cyclo$(CH_2-CH_2-O-)_n$, but the yields were quite low [159a] (Table 3-4).

Table 3-4. Distribution of cyclic products formed by oligomerization of ethylene oxide with BF_3/HF.

n	2	3	4	5	6	7	8	9	10	11
%	40	1	15	5	4	3	2	2	1	1

The mechanism of this reaction has been studied using labelled products [159b]. Thus, cyclooligomerization of ethylene oxide-d_4 with BF_3 in CH_2Cl_2 has been carried out in the presence of successively 1,4-dioxan, 9-crown-3, 12-crown-4, 15-crown-5 and 18-crown-6 (all these cyclic oligomers were non-deuteriated). Each experiment was conducted in the presence of a single one of the cyclic oligomers. The various non-deuteriated macrocycles were incorporated (via conversion into oxonium ions) into the growing chains. The nature of the products obtained, that is their degree of deuteriation, enabled the course of the reaction to be followed precisely.

Rigorous analysis of the evolution of all these reactions as a function of time led to the following conclusions (Fig. 3-116):

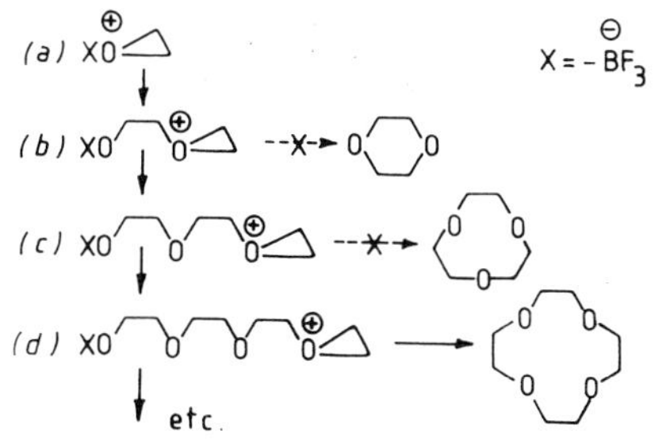

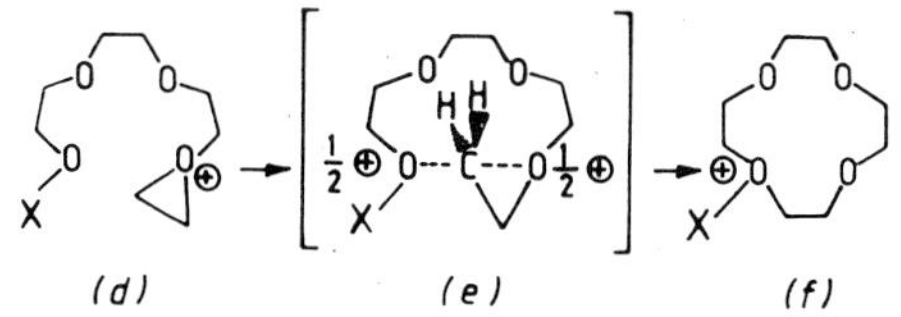

Fig. 3-116. Oligomerization of ethylene oxide. Chain growth (a, b, c, d) and length necessary for direct cyclization (d → e → f).

- Only the BF_3 adduct *a* of the strained 3-membered ring could bring about oligomerization.
- The intermediate oxonium ions *b* and *c* reacted preferentially with the monomer and did not give 1,4-dioxan and 9-crown-3 cyclic products.
- The tetramer was the smallest ring which could be formed directly from ethylene oxide. The linear S_N2-type transition state *e* (Fig. 3-116) required a minimum chain length which corresponded to formation of tetramer *f*.
- A growing chain (Fig. 3-117) had at any one time only two possibilities. Either, in the presence of monomer, it could continue growing or, in the absence of monomer, it could cyclize.

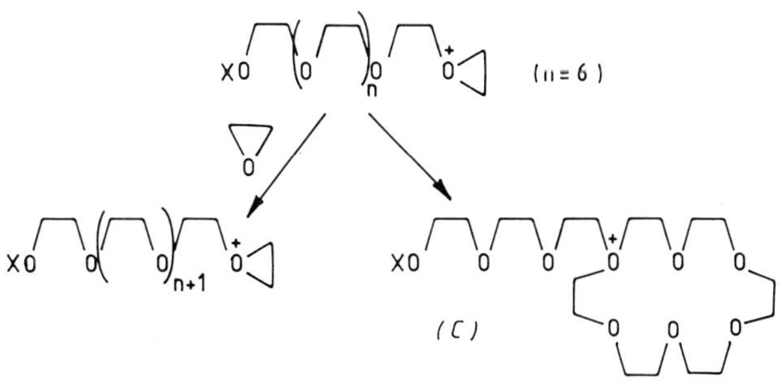

Fig. 3-117. Chain growth vs. cyclization depending on the quantity of monomer present.

- The ring formed could evolve in many ways (Fig. 3-118). It can be seen that routes 1 and 11 give the trimer, and routes 5, 7, 8, 10 and 11 afford dioxan. Thus, the formation of these two products was explained. As pointed out above, they were not obtained directly. Dioxan, which is thermodynamically the most stable product, was formed in abundance since it was the end product of ring degradation. Finally, rings of different size were formed.

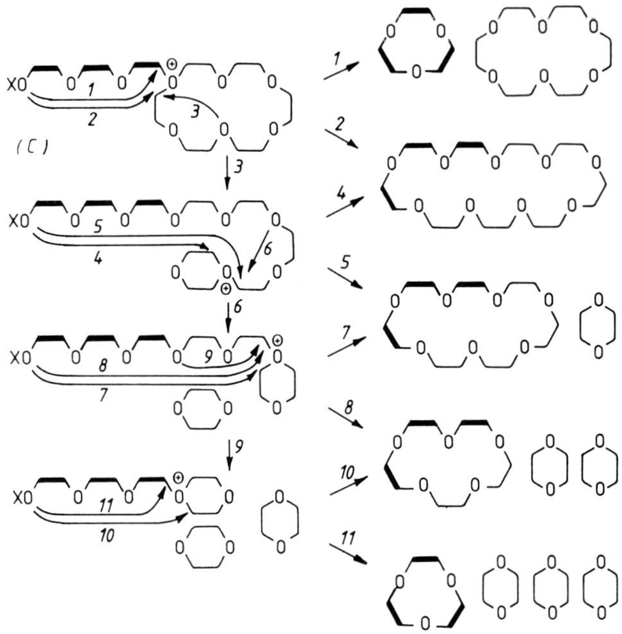

Fig. 3-118. Possibilities for evolution of system *C* (see Fig. 3-117).

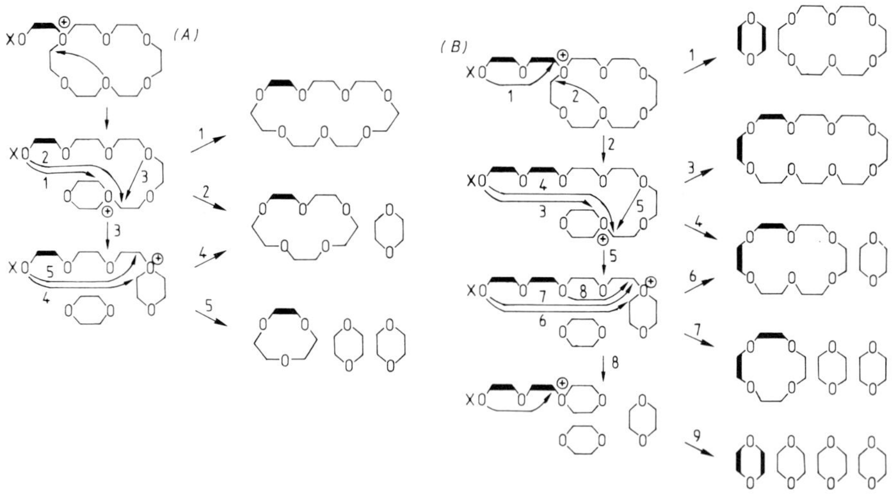

Fig. 3-119. Oligomerization of ethylene oxide-d_4 in the presence of [18-O_6]. Heavy lines symbolize $-CD_2-CD_2-$ units.

Figures 3-118 and 3-119 indicate how the mechanisms of formation were determined using the experiments described above, that is the oligomerization reaction of ethylene oxide-d_4 in the presence of 18-crown-6 (or other non-deuteriated macrocycles).

The addition of one, two or three molecules of ethylene oxide-d_4 to the macrocycle gave products *A, B* (Fig. 3-119) and *C* (Fig. 3-118), respectively. Identification of all the degradation products of *A, B* and *C* by mass spectrometry led to the mechanisms shown in Fig. 3-116 being postulated [159b].

b) In the presence of metal salts

The yields of 12-, 15- and 18-membered rings were substantially improved when the oligomerization by BF_3 was conducted in the presence of metal salts [159c] (Table 3-5).

It should be noted that only these three macrocycles were formed. The cations seem to play a double role. First, the cation template effect promotes the appropriate folding of the growing chain thus favoring cyclization. Second, the positive charge on the metal protects the rings which are formed by preventing formation of the oxonium salt needed to start degradation to dioxan.

Table 3-5. Oligomerization of ethylene oxide with BF_3 in the presence of metal salts. Distribution of cyclic oligomers formed.

Salt	% Tetramer[a]	% Pentamer[b]	% Hexamer[c]
–	15	5	4
$LiBF_4$	30	70	–
$NaBF_4$	25	50	25
KBF_4	–	50	50
KPF_6	20	40	40
$KSbF_6$	40	20	40
$RbBF_4$	–	–	100
$CsBF_4$	–	–	100
$Ca(BF_4)_2$	50	50	–
$Sr(BF_4)_2$	10	45	45
$Ba(BF_4)_2$	10	30	60
$AgBF_4$	35	30	35
$Hg(BF_4)_2$	20	70	10
$Ni(BF_4)_2$	20	80	–
$Cu(BF_4)_2$	5	90	5
$Zn(BF_4)_2$	5	90	5

(a) 1,4,7,10-Tetraoxacyclododecane (b) 1,4,7,10,13-Pentaoxacyclopentadecane (c) 1,4,7,10,13,16-Hexaoxacyclooctadecane. (Taken from J. Dale et al., *J. Chem. Soc., Chem. Comm.,* 295, 1976.)

3.5.5 Cyclization Using Templates

a) Kinetic study of the template effect

Many studies have shown the influence of alkali and alkaline earth metal cations on the course of numerous organic reactions. Thus, formation of crown ethers is notably facilitated by the presence of these cations which act as templates. Mandolini et al. have studied quantitatively this template effect by determining the rates for the cyclization of phenoxides to give benzo[18 – C – 6] in the presence of a variety of cations in water and also in methanol [160, 161].

The acceleration observed in water in the presence of certain cations (Na^+, K^+, Sr^{2+}, Ba^{2+}) has been attributed to greater reactivity of A^-M^+ than A^- solvated by water (Fig. 3-120). This hypothesis could not be verified since the reactivity of A^-M^+ could not be measured. However, it was confirmed by the kinetic study in methanol in which the cation-ligand interactions are much stronger than in water.

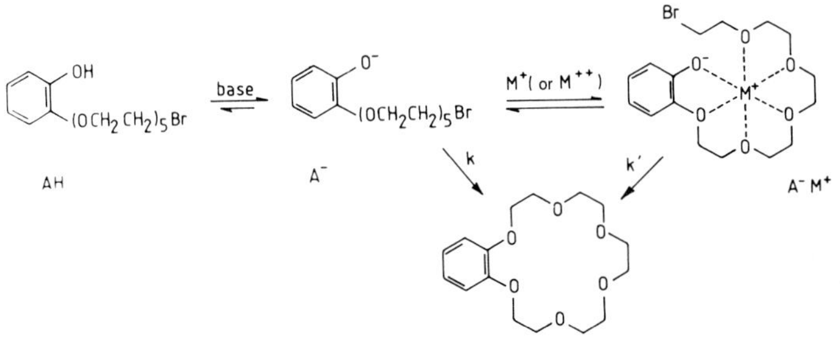

Fig. 3-120. Synthesis of benzo-18-crown-6 in the presence of a metal cation.

The cyclization kinetics were studied by following the disappearance of A^- spectrophotometrically. A^- was formed by addition of excess quaternary ammonium base ($Et_4N^{+\,-}OMe$).

The reaction was strongly and specifically accelerated by added alkali (Na^+, K^+, Rb^+ and Cs^+) and alkaline earth (Ca^{2+}, Sr^{2+} and Ba^{2+}) bromides, the observed rate enhancements ranging from 13.2 for Cs^+ to 540 for Sr^{2+}. A simple model involving contributions from both unassociated and associated forms of phenoxide rationalized the results,

$$A^- + M^+ \overset{K}{\rightleftharpoons} A^-M^+ \qquad A^- \xrightarrow{k} \text{B-18-C-6} \xleftarrow{k'} A^-M^+$$

where k and k' are the rate constants for the anions in the free and associated forms, respectively. The kinetics were measured in the presence of the cation Et_4N^+ either alone, or together with the alkali or alkaline earth cation being studied. Since the interaction of the anion with Et_4N^+ was negligible, the reaction carried out in the presence of Et_4N^+ alone allowed the rate constant k for the free A^- ion to be determined. In the presence of a metal cation acting as a template, the measured rate constant could be considered as being equal to k' since the concentration of free ion was very low (large K). The value of the ratio $k'/k = k_{rel}$ (Fig. 3-121) was a measure of the efficiency of the template effect due to the cation.

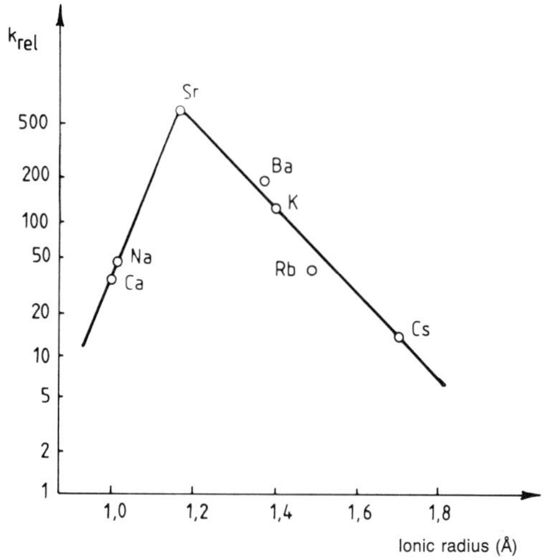

Fig. 3-121. Change in the relative rate constant for the phenoxide anion associated to a cation (A^-M^+), relative to that of the free anion A^-, as a function of the nature of M^+ (taken from L. Mandolini et al., *J. Am. Chem. Soc.,* **103,** 2780, 1981).

b) Interpretation of the metal template effect

This interpretation is based on a combination of chemical and proximity effects.

The nucleophilicity of the phenoxide ion should be reduced by interaction with the cation but this effect could not be measured. However, it was less marked than for a pair of simple ions since the cation M^+ interacted not only with the anion but also with the oxygen atoms in the chain (see Fig. 3-120).

The complexation led to increased statistical proximity of the chain ends, and a loss of conformational entropy. Consequently, in order to attain the transition state for cyclization, the loss in entropy was much less great for the associated than for the unassociated species.

These two effects, chemical and proximity, acted in opposite directions and both were large. Nevertheless, a quite distinct template effect resulted (Fig. 3-121), but one which did not precisely follow the complexation constants which increase in the order: $Li^+ < Na^+ < K^+ < Sr^{2+} < Ba^{2+}$.

c) Examples of template effects in the synthesis of macrocycles [15-O_5], [12-O_4] and [24-O_8]

The crown ether [15-O_5] has been synthesized from triethylene glycol and bis(2-chloroethyl) ether in the presence of various alkali metal hydroxides (Fig. 3-122a). The cyclization yield (Table 3-6) indicated a template effect for cations of sufficient size, this being an effect related to the relative stability of the complexes [162].

Table 3-6. Yield of 15-crown-5 depending on the nature of the hydroxide.

Hydroxide	Yield (%)
LiOH	4
NaOH	39
KOH	20
RbOH	15
CsOH	12
TlOH	19
$N(nBu)_4OH$	<1

(Taken from A. J. Rest et al., *Inorg. Chim. Acta,* **45,** L 5, 1980.)

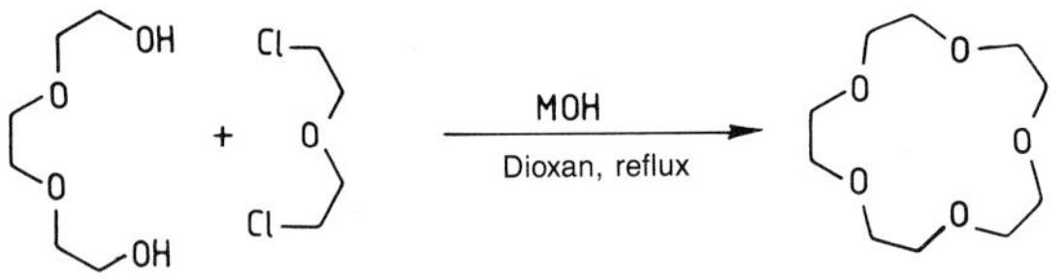

(a)

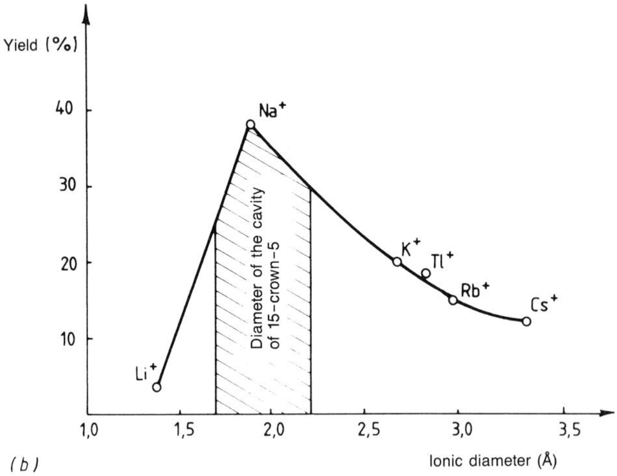

(b)

Fig. 3-122. (a) Synthesis of the crown ether [15-O_5] in the presence of hydroxides of alkali metal cations, (b) effect of the metal cation on the reaction yield (taken from A. J. Rest et al., *Inorg. Chim. Acta,* **45,** L 5, 1980).

The asymmetry in the plot (yield vs. ionic diameter, Fig. 3-122b) indicated that the reaction took place in reasonable yield even when the cation was larger than the diameter of the cavity. As in the previous example, formation of an intermediate complex ion pair was postulated (Fig. 3-123).

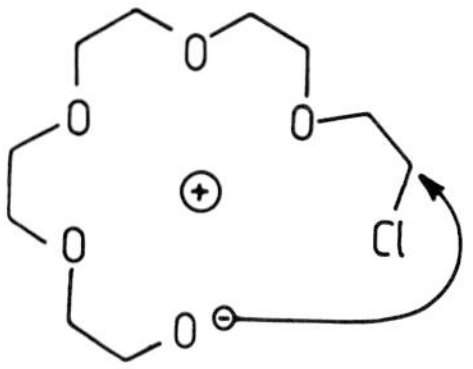

Fig. 3-123. Intermediate ion pair in the formation of the crown ether [15-O_5] (see Fig. 3-122).

Reaction between diethylene glycol and bis(2-chloroethyl) ether gave a mixture of [12-O_4] and [24-O_8] (Fig. 3-124) [163].

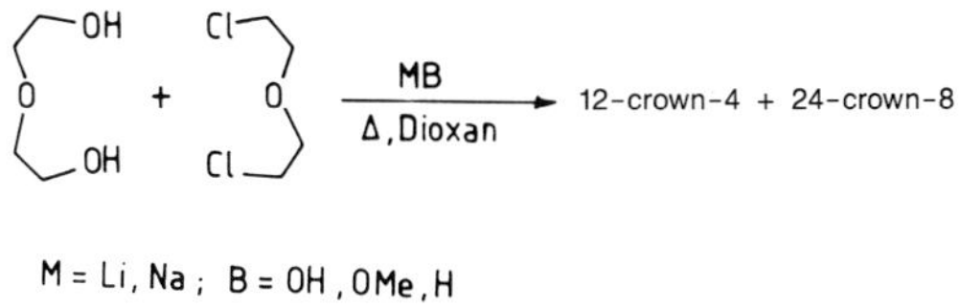

Fig. 3-124. Synthesis of [12-O_4] and [24-O_8] from diethylene glycol and bis(2-chloroethyl) ether.

The results (Table 3-7) showed that in this case too the nature of the cation was important. Use of Na^+ favored the 24-membered compound because this ion was too large for the cavity in [12-O_4]. Furthermore, the nature of the base also had an effect on the reaction product distribution.

Table 3-7. Effect of the cation and basicity on the formation of the crown ethers [12-O-4] and [24-O-8].

Base	Yield of 12-crown-4 (%)	Yield of 24-crown-8 (%)
LiOH	0	0
LiOMe	8	< 2
LiH	13	< 2
NaOH	< 1	4
NaOMe	4	20
NaH	6	9

(Taken from A. J. Rest et al., *Inorg. Chim Acta* **53**, L175, 1981.)

d) Stabilization by complexation

Photocyclization of the bis-anthracene derivative *1* (Fig. 3-125) has been performed in the presence of Li^+. Under these conditions, the crown ether formed was stable whereas in the absence of Li^+ it underwent spontaneous ring-opening. The half-life for this ring-opening reaction, which was the reverse of cyclization, was ca. 3 min. The crown ether was thus stabilized by complexation [164].

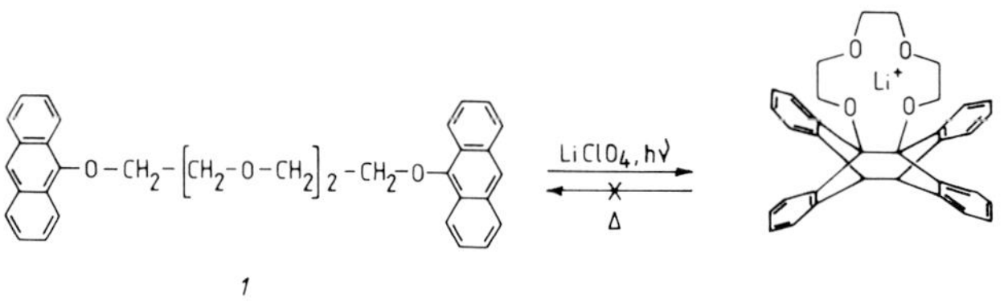

Fig. 3-125. Photocyclization of a bisanthracene derivative in the presence of Li^+.

3.6 Macrocyclic Lactones – Macrolides

The term macrolide is used to describe a macrocyclic system with one or more lactone groups in the ring. The development of this chemistry has accelerated over the past twenty years or so chiefly because of the large number of natural products possessing a macrocyclic lactone structure. For instance, two examples in the realm of perfumes are exaltolide and ambrettolide (Fig. 3-126).

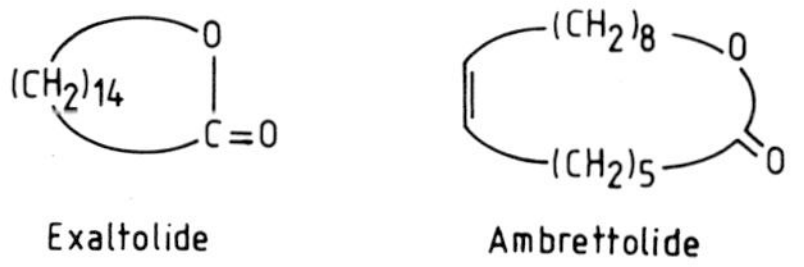

Fig. 3-126. Natural macrocyclic lactones.

However, the most important such products are the lactone and polylactone macrocyclic antibiotics: picromycin (the first to be isolated), erythromycin, magnamycin, nonactin (tetralactone), valinomycin (cyclodepsipeptide), carpaine (dilactone).

Because of their biological interest, numerous synthetic methods have been developed. We shall not comment here on the difficult construction of the acylic skeleton but rather solely concern ourselves with the various methods of macrocyclization.

3.6.1 General Methods of Macrolactonization

The starting products are mainly ω-halogeno or ω-hydroxy acids which afford lactones by means of various types of reaction. The direct reactions (Fig. 3-127) have been known since long. Thus, reaction (a) in the presence of silver oxide or potassium carbonate or potash, and reaction (b) in the presence of benzene- or toluene-sulfonic acid both give macrocyclic lactones in very variable yield. Amongst these long-established reactions, Carothers transesterification should be recalled (see Figures 3-5 to 3-7).

Fig. 3-127. Lactone synthesis from ω-halogeno or ω-hydroxy acids.

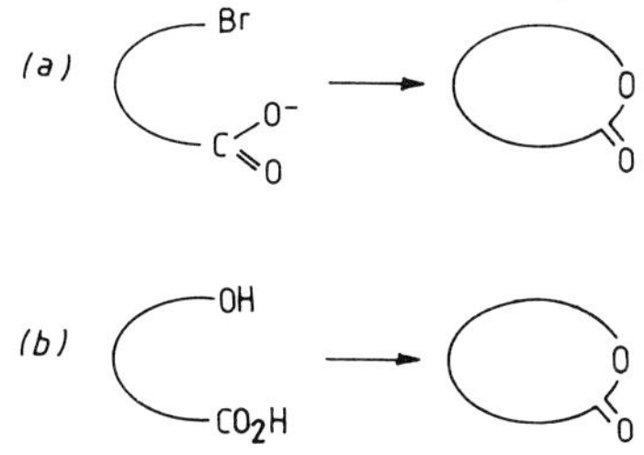

However, these methods have serious limitations. The development of activating groups for acid and alcohol functional groups has led to great advances in macrolide chemistry. Many such activating groups for acids (the most common) (Fig. 3-128) or alcohols or both functional groups together have been used to improve macrocyclization: mixed anhydrides *1*, N-acylimidazolides *2*, thiol esters *3*, 2-acyloxypyridinium *4* or 2-acyloxybenzothiazolium *5* salts. In addition, diazoester "inverse" activation and other methods described below are also employed.

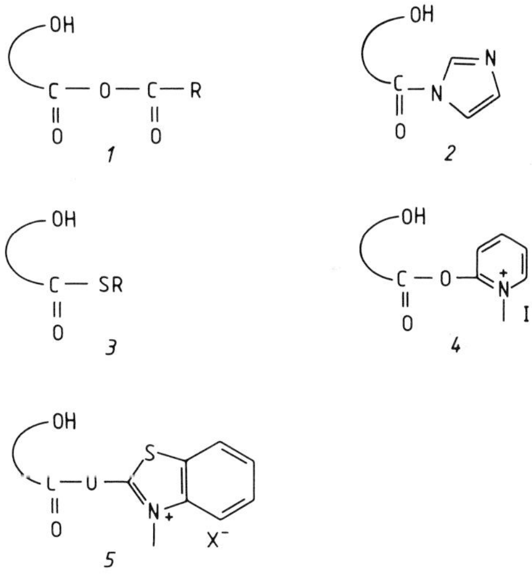

Fig. 3-128. Activating groups for acids.

3.6.2 ω-Halocarboxylate Cyclization

a) Kinetic study

We have already noted the work of G. Illuminati et al. (cf. Section 1.1) who carried out a quantitative investigation of the lactonization of ω-bromoalkanoic acids with strong bases in 99% aqueous dimethylsulfoxide DMSO) [9]. They measured the intramolecular rate constants, k_{intra}, for ring sizes n = 3–16, 18 and 24 as well as the intermolecular rate constant k_{inter} for the reaction between potassium butanoate and butyl bromide. The effective molarities EM, k_{intra}/k_{inter}, which represent an absolute measure of the ring-closure tendency of bifunctional substrates, could thus be obtained (see Fig. 1.8 and Table 1-2).

b) Double lactonization

The [32-O_8] macrocycle *4* has been synthesized by reaction between the dibromoester *2* and the diacid potassium salt *3* in DMF (Fig. 3-129). This macrocycle is a model for complexing tetralactone antibiotics, that is the nactins [165a].

Other nactin analogues have also been prepared and their ability to transport cations across membranes studied [165b and c].

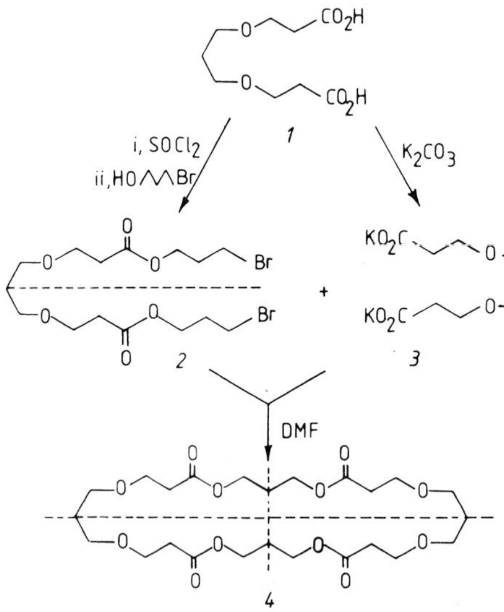

Fig. 3-129. Synthesis of a nactin model. 30% cyclization yield.

c) Cesium Effect

Cyclization of cesium ω-halocarboxylates (Fig. 3-130) has been carried out in DMF by treatment of ω-halo-acids of various chain lengths with dry cesium carbonate [166]. The reaction gave a mixture of the monomer (macrolide *2*) and the dimer (dilide *3*).

The monomer/dimer ratio varied with chain length (Table 3-8).

The highest monomer yield was found for 16- and 17-membered macrocycles (85%) which is comparable to that obtained using the method of Corey et al.

(vide infra), except in the case of small rings for which the latter method gave better results. The 14- and 20-membered dimers (n = 5 and 8) were those formed in highest yield.

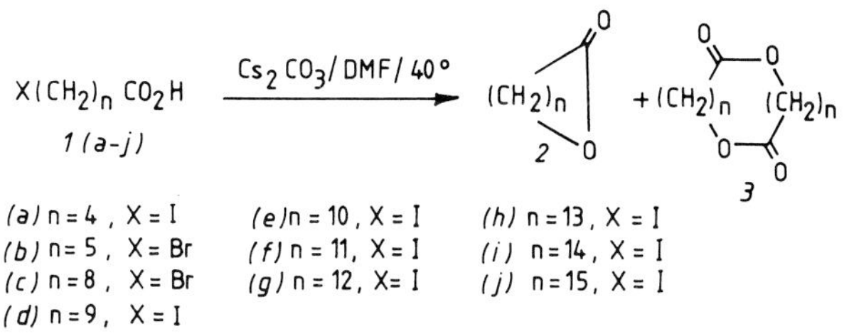

Fig. 3-130. Macrolide and dilide formation by cyclization of cesium ω-halocarboxylates.

Table 3-8. Macrolide, *2*, and dilide, *3*, yields from ω-halogenoaliphatic acids and Cs_2CO_3 in DMF.

ω-Halogeno acid	yield in %	
	2	*3*
1a	70	4
1b	–	88
1c	–	95
1d	23	55
1e	33	54
1f	62	30
1g	77	18
1h	72	13
1i	83	17
1j	85	15

(Taken from W. H. Kruizinga and R. M. Kellogg, *J. Am. Chem. Soc.,* **103,** 5183, 1981.)

Comparison between the effects of cesium and other cations has been made using 15-iodopentadecanoic acid *(1i)*. The superiority of cesium which gave the cleanest and most complete reactions should be noted. DMF was the best of the solvents studied (Table 3-9).

Various proposals, in particular increased reactivity of the carboxylate anion when associated to the large Cs^+ cation (decrease in cation-anion interactions), have been put forward to explain the particularly high efficiency of cesium salts. The postulated formation of a triple ion is the most attractive explanation (Fig. 3-131).

The Cs^+ cation is both attached to the carboxylate group and also provides electrophilic assistance to the leaving group. In this hypothesis the reacting groups are brought into proximity thus accounting for the high yield of macrocycles. The cesium effect has been extensively discussed in several articles [8, 166, 167].

Table 3-9. 15-Iodopentadecanoic acid cyclization. Effect of carbonate metal cation (in the case of DMF solvent) and solvent (in the case of cesium carbonate) on macrolide *(2i)* and dilide *(3i)* yield.

		yield (%)	
		2i	*3i*
Carbonate:	Li_2CO_3	–	–
	Na_2CO_3	54	10
	K_2CO_3	67	9
	Rb_2CO_3	68	12
	Cs_2CO_3	80	12
	Ag_2CO_3	53	6
	Tl_2CO_3	64	8
	$MgCO_3$	–	–
	$SrCO_3$	–	–
	$BaCO_3$	–	–
Solvent (in the presence of Cs_2CO_3)	DMF	72	18
	DME	47	28
	THF	30	8
	DMSO	35	40
	NMP*	29	15
	HMPTA	11	2

* N-methylpyrrolidone
(Taken from W. H. Kruizinga and R. M. Kellogg, *J. Am. Chem. Soc.*, **103**, 5183, 1981.)

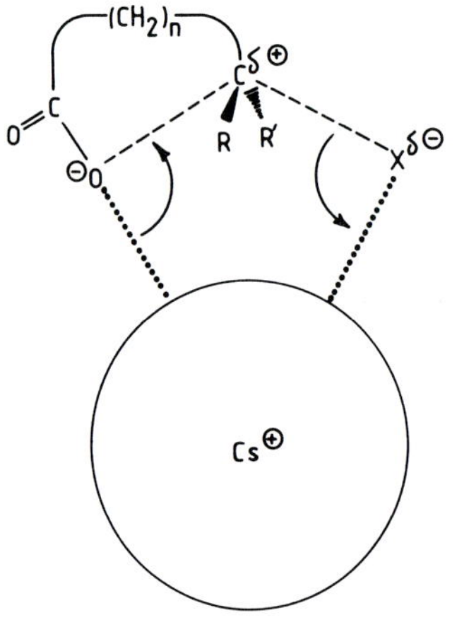

Fig. 3-131. Triple ion postulated to explain the reactivity of cesium ω-halocarboxylates.

d) Application of the Cesium Effect

This lactonization method has been applied to several natural products.

The optically pure *S*-lactone *4* of *R*-ricinelaidic acid (Fig. 3-132) has been obtained in 80% yield with inversion of configuration as expected for an S_N2 reaction [166].

S-Zearalenone has been converted into its *R* enantiomer by using cesium ion (Fig. 3-133). Following protection of the phenol and ketone groups of *S*-zearalenone

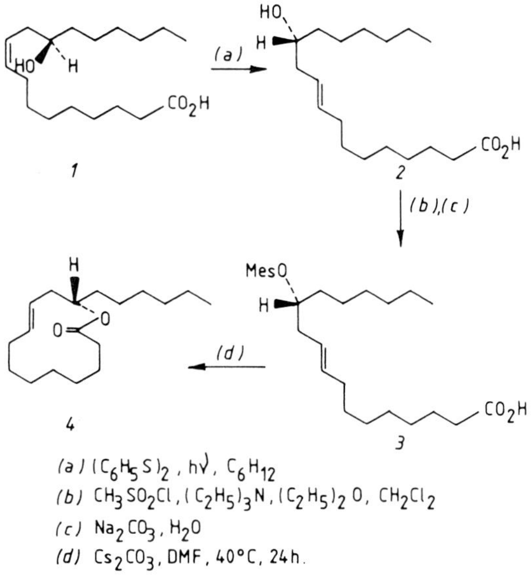

Fig. 3-132. Application of the cesium effect in the formation of the *S*-lactone of *R*-ricinelaidic acid.

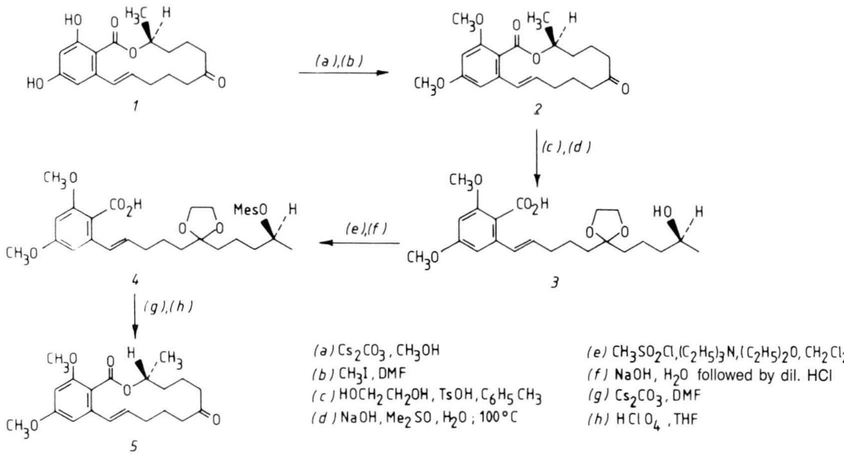

Fig. 3-133. Conversion of *S*-zearalenone into its *R* enantiomer.

1 (steps (a), (b), (c)), the macrocycle was opened (step (d)) to give *3*. This latter compound was mesylated to give *4* (steps (e), (f)). Cyclization of *4* in the presence of cesium carbonate (step (g)) gave, after ketone deprotection (step (h)), the (−)-*R*-zearalenone derivative, the yield and enantiomeric excess both being 80% [166].

e) Tetraalkylammonium carboxylates

Compound *2*, obtained by electroreduction of 2-pyrrolidone in the presence of a quaternary ammonium salt, is a very efficient base. It led to formation of macrocyclic lactones in high yield [168] (Fig. 3-134).

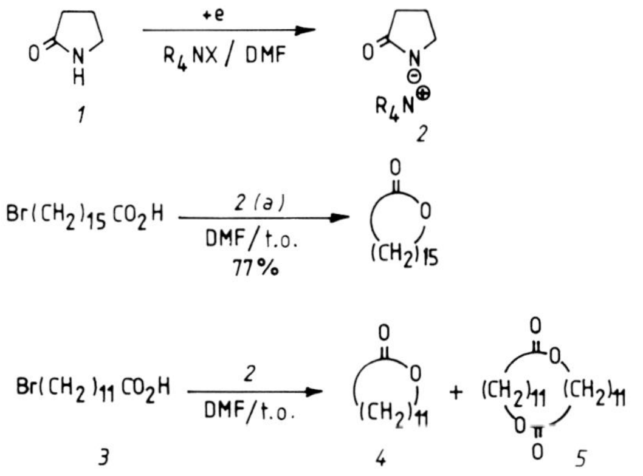

	Yield (%) 4	5	Yield (% 4+5)
2(a), R = Et	75	25	64
2(b), R = Bu	93	7	72
2(c), R = Oct	100	0	66

Fig. 3-134. Lactonization by use of the tetraalkylammonium carboxylate formed by treatment of an ω-halo acid with the electrogenerated base *2*.

It can be seen that alkylammonium cation size had a considerable influence on the nature of the products obtained. Bulky substituents favored the macrocyclic lactone at the expense of the dilactone with little effect on the overall yield. The efficiency of this novel method was attributed to the steric bulkiness of the ammonium carboxylate which made the intramolecular reaction more favorable than the intermolecular one.

3.6.3 Cyclization of ω-Hydroxy Acids

a) Double Activation

Certain reactions giving esters, peptides, etc. and carried out under mild conditions are categorized as being "oxidation-reduction condensations". A particularly convenient reagent for this type of condensation involving elimination of water is the mixture triphenylphosphine/2,2′-dipyridyl disulfide (Mukaiyama reagent). Triphenylphosphine by removing the oxygen atom from water is oxidized to the phosphine oxide Ph_3PO. 2,2′-Dipyridyl disulfide through picking up the two hydrogen atoms is reduced to pyridinethione [169a]. Application of this method to the formation of peptide links is illustrated in Fig. 3-135 [169b].

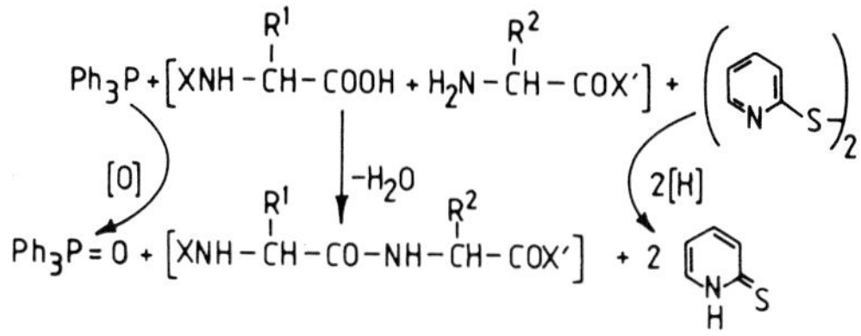

Fig. 3-135. Formation of a peptide link by use of the Mukaiyama reagent.

Corey and Nicolaou, as a result of the following considerations, were the first to apply the Mukaiyama reagent to lactonization [170]. Since lactone formation is relatively slow for large rings, high temperatures or extremely slow addition of hydroxy acid (high dilution conditions) are necessary unless the reacting groups can be activated.

One way of simultaneously activating both the carboxyl and hydroxyl groups is to use a carboxylic derivative such as the 2-pyridinethiol ester of an ω-hydroxy acid *1* which favors proton transfer from hydroxyl to carboxylic oxygen so giving *2*. The latter dipolar intermediate *2* should undergo facile electrostatically driven cyclization thus affording the desired macrocyclic lactone *4* (Fig. 3-136). It can be seen that protonation activates the carboxyl group and deprotonation activates the hydroxyl, and hence this process constitutes *double activation*.

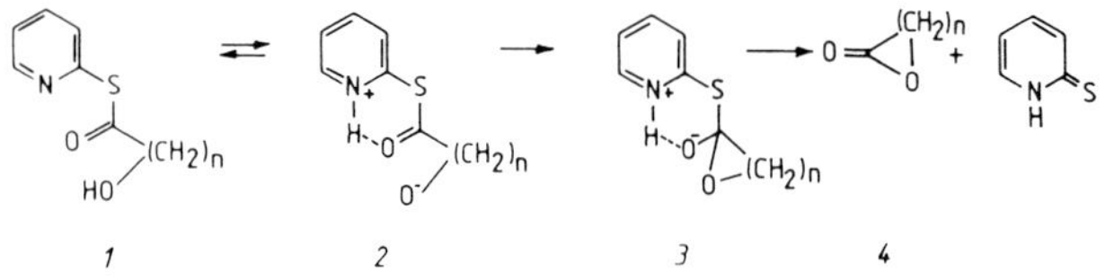

Fig. 3-136. Lactonization using double activation (carboxyl and hydroxyl groups).

A series of ω-hydroxy acids $HO(CH_2)_nCO_2H$ with n = 5, 7, 10, 11, 12, or 14 has been cyclized in this way after conversion into 2-pyridinethiol esters obtained by treatment of the hydroxy acids with the mixture triphenylphosphine/2,2′-dipyridyl disulfide. The results were most satisfactory.

The yields (Table 3-10) were equal or superior to those obtained using cesium salts. However, the thiol ester had to be prepared and triphenylphosphine was consumed in the reaction.

This method has been used to cyclize *monensin*, which belongs to the class of non-macrocyclic antibiotics. It is a polyether with an acid functional group at one end and two alcohol groups at the other.

Table 3-10. Lactone and diolide formation by cyclization of 2-pyridinethiol esters of ω-hydroxy carboxylic acids *1*.

n	Solvent	Lactone			Diolide	
		Ring size	GLC yield (%)	Isolated yield (%)	Ring size	Isolated yield (%)
5	Benzene	7	87	71	14	7
7	Xylenes	9	25	8	18	41
10	"	12	64	47	24	30
11	"	13	76	66	26	7
12	"	14	79	68	28	6
14	"	16	88	80	32	5

(Taken from E. J. Corey et al., *J. Am. Chem. Soc.* **96**, 5614, 1974.)

Treatment of monensin, *1*, with Ph_3P and 2,2′-dipyridyl disulfide (DPDS) [171] gave the lactone derivative *2* in 95% yield (Fig. 3-137). It would be interesting to study the complexing properties of this latter compound.

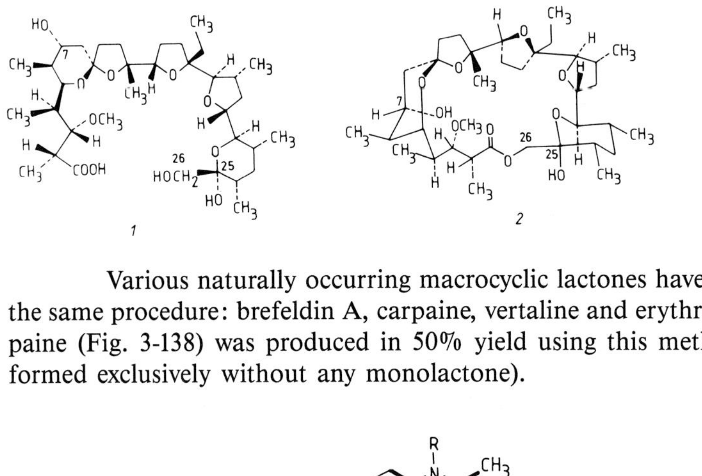

Fig. 3-137. Monensin and its lactone derivative.

Various naturally occurring macrocyclic lactones have been synthesized by the same procedure: brefeldin A, carpaine, vertaline and erythronolide B [172]. Carpaine (Fig. 3-138) was produced in 50% yield using this method (the diolide was formed exclusively without any monolactone).

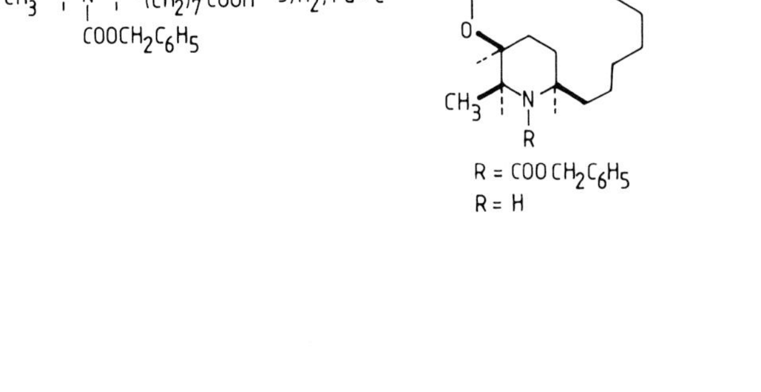

Fig. 3-138. Synthesis of carpaine by lactonization using the Mukaiyama reagent. (Carpaine, R = H, was obtained by hydrogenation of the protected derivative, R = $COOCH_2C_6H_5$.)

b) Activation of the Mukaiyama reagent by complexation with Ag^+

In the presence of silver salts, the Mukaiyama reaction is considerably accelerated [173]. Since the kinetic aspect is very important in macrocyclization reactions, this variant has been used in several syntheses.

The macrolide antibiotic nonactin *1* (Fig. 3-139), containing four units of a C_{10} hydroxy-acid, has been synthesized from the two derivatives *3* and *4* of nonactic acid *2* (Fig. 3-140) [174, 175].

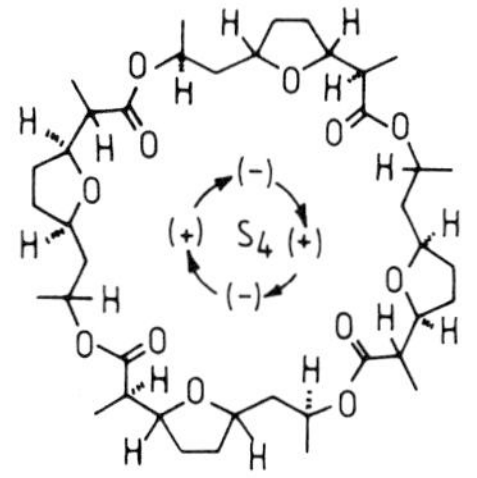

1

Fig. 3-139. Structure of nonactin.

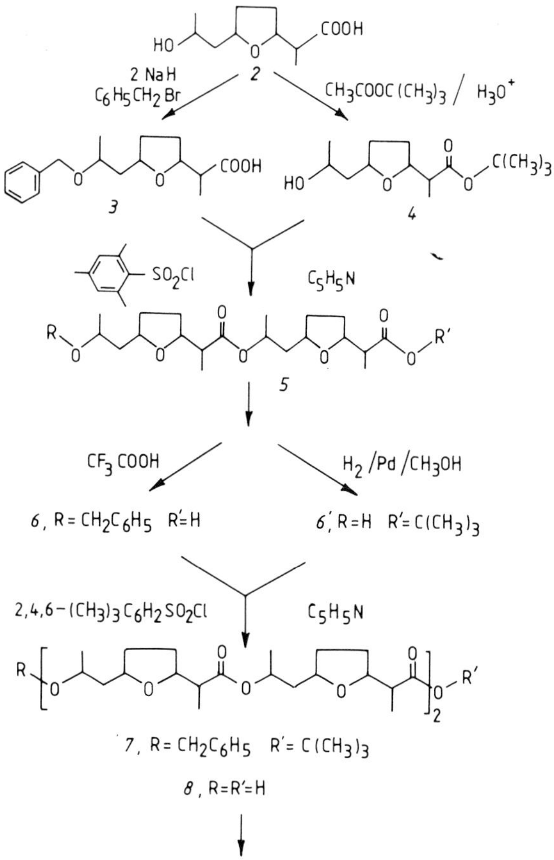

Fig. 3-140. Synthesis of nonactin.

Condensation of *3* and *4* via a mixed anhydride afforded the dimer *5* which on appropriate deprotection gave the compounds *6* and *6'*. The latter were condensed as above to give *7* which was deprotected to afford the tetrameric precursor *8*.

The Mukaiyama reagent in the presence of silver perchlorate was used in thc final cyclization step to give the macrolide, the silver ion exerting its metal template effect on S-(2-pyridyl) ω-hydroxy carbothioate (Fig. 3-141 a) as illustrated in Fig. 3-141 b.

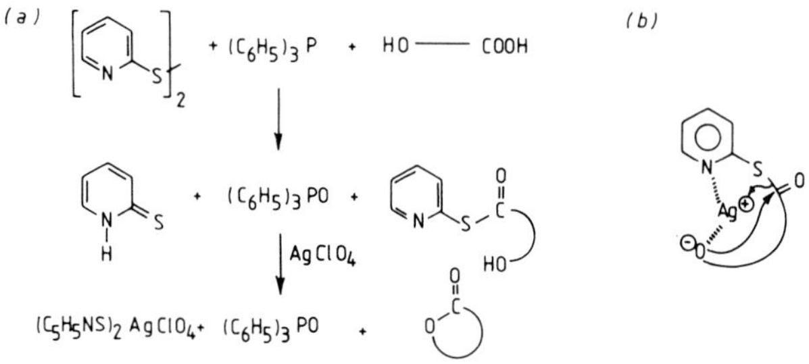

Fig. 3-141. (a) Lactonization of an ω-hydroxyacid by the Mukaiyama reagent in the presence of Ag^+, (b) the role of Ag^+ in the cyclization step.

A mixture of nonactin diastereoisomers has been obtained from racemic nonactic acid [175].

The stereoselective total synthesis of nonactin has been carried out starting from both enantiomers of nonactic acid [176–178].

Recifeiolide, which is a mould metabolite, has been obtained by an analogous method [179].

c) Other activating groups for lactonization

A survey has been made of various heterocyclic disulfides (Fig. 3-142 a) which in principle are capable of reacting in the same way as 2,2'-dipyridyl disulfide [180]. Compounds *1* to *6* were less effective than dipyridyl disulfide. In the case of the disulfide *7*, cyclization of 16-hydroxyhexadecanoic acid was very fast but the yield did not exceed 37% because of simultaneous formation of the N- and S-acyl derivatives *10* and *11* (Fig. 3-142 b), of which only the latter could serve as a lactone precursor.

With the compounds *8* and *9*, which are both substituted in the 4 position by a *tert*-butyl group, the lactone yields were improved over those using 2,2'-dipyridyl disulfide, and lower reaction temperature could be employed. In both cases, N-acylation was difficult for steric reasons.

Another very efficient method, in which the *tert*-butyl thioate activating group is used, has been described by Masamune [181]. It was applied in the synthesis of zearalenone dimethyl ether (Fig. 3-143), the yield being 90%.

The Hg(II) ion plays a fundamental role in the intermediate step (Fig. 3-144).

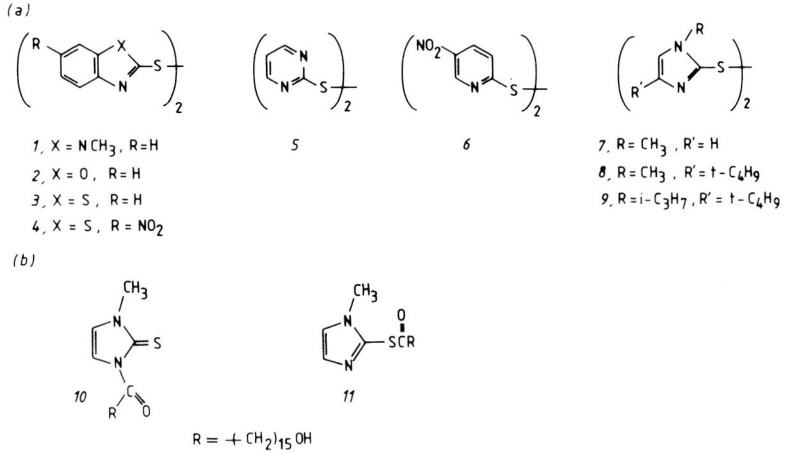

Fig. 3-142. (a) Heterocyclic disulfides acting as activating groups in lactonization, (b) N- and S-acyl products formed by treating 16-hydroxyhexadecanoic acid with *7.*

Fig. 3-143. Synthesis of zearalenone dimethyl ether using the *tert*-butyl thioate activating group.

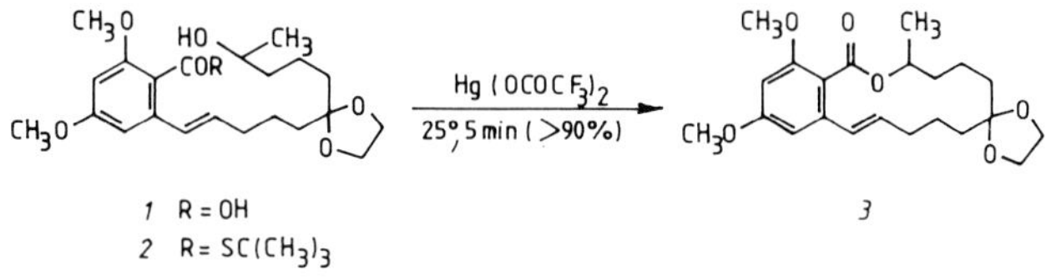

Fig. 3-144. Role of the Hg(II) ion in the intermediate step in the synthesis of zearalenone dimethyl ether.

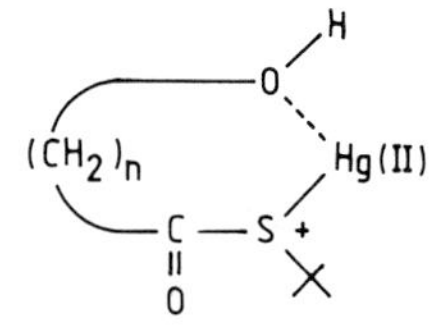

The syntheses of other macrolides, e. g. methymycin [182], have been carried out using this process.

A systematic study has shown that other thiol esters can be used, as well as other thiophilic cations such as Ag(I), Cu(I) and Cu(II) [183].

d) Activation and ring closure using onium salts

Ring closure of the hydroxy acids $HO(CH_2)_nCO_2H$ (where n = 5–7, 10, 11, 14) has been brought about using 2-chloro-1-methylpyridinium iodide in the presence of triethylamine (Fig. 3-145). The method is based on having a good leaving group with stabilization of the final step transition state [184]. These reactions take place with concomitant formation of dilactones. There is exclusive formation of the lactone for n = 5, and of the dilactone for n = 6. For n > 7, the yield of lactone (Table 3-11) increases with ring size, whereas that of dilactone decreases.

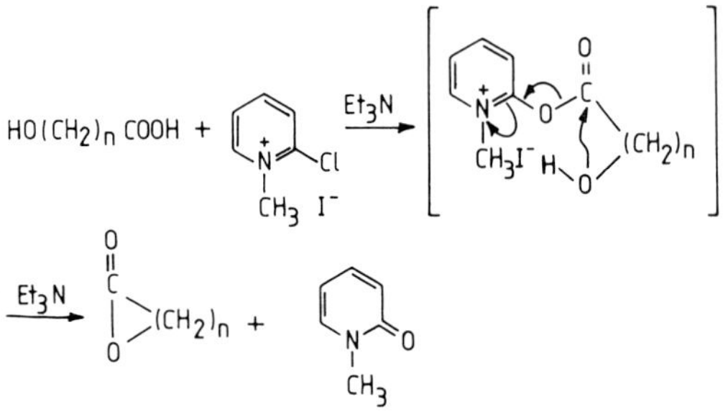

Fig. 3-145. Cyclization of hydroxy acids by 2-chloro-1-methylpyridinium iodide in the presence of triethylamine.

Table 3-11. Lactonization of ω-hydroxy acids using 2-chloro-1-methylpyridinium iodide.

			Lactone		Dilactone	
n	Solvent	Reaction Time (h)	Ring Size	yield (%)	Ring size	yield (%)
5	CH_2Cl_2	7.5	7	89	14	0
6	CH_3CN	7.5	8	0	16	93
7	"	8.0	9	13	18	34
10	"	9.0	12	61	24	24
11	"	8.0	13	69	26	14
14	"	8.5	16	84	32	3

(Taken from T. Mukaiyama et al., *Chem. Letters,* 49, 1976.)

A more efficient cyclization method has been developed by Mukaiyama [185]. The long chain hydroxy acids *1* (n = 10, 11, 14) were converted into 6-phenyl-2-pyridyl esters by treatment with 6-phenyl-2-pyridone and 2-chloro-1-methylpyridinium iodide in the presence of triethylamine (Fig. 3-146). These esters *2* were cyclized to the lactones *3* and *4* using *p*-toluenesulfonic acid. The yields were relatively high (Table 3-12), the ring closure step being favored by protonation of the intermediate [185].

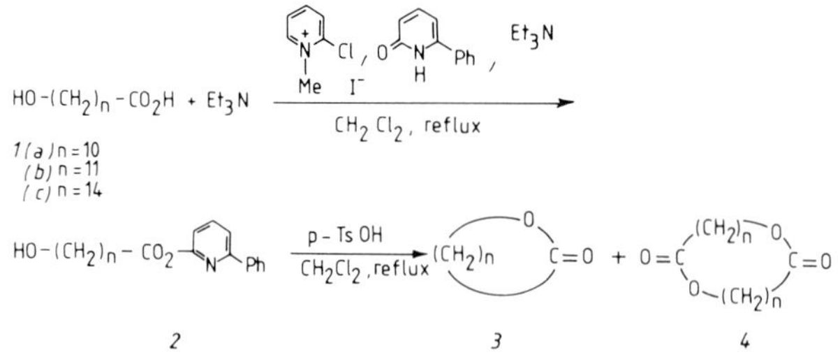

Fig. 3-146. Lactonization of ω-hydroxy acids via the 6-phenyl-2-pyridyl esters, *2*.

Table 3-12. Yields of macrocyclic lactones obtained by treating long chain (n = 10, 11, 14) 6-phenyl-2-pyridyl esters of ω-hydroxy acids with *p*-toluenesulfonic acid.

	yield (%)	
n	Lactone *3*	Dilactone *4*
10	73 (32)*	25 (23)*
11	99 (72)*	<1 (16)*
14	quant. (92)*	0 (trace)*

( )* Yields obtained at room temperature
(Taken from T. Mukaiyama et al., *Chem. Letters,* 441, 1977.)

However, the difficulties arising from the use of *p*-toluenesulfonic acid, and also the decomposition of the 2-chloropyridinium salts through attack by refluxing triethylamine, have led to the quest for single step cyclization methods using stable onium salts. Thus, in the presence of 2-chloro-6-methyl-1,3-diphenylpyridinium tetrafluoroborate *(1)*, 2,4,6-triphenylpyridine and benzyltriethylammonium chloride (both the first two being very sterically hindered), the lactonization step occurred in high yield (Fig. 3-147) [186].

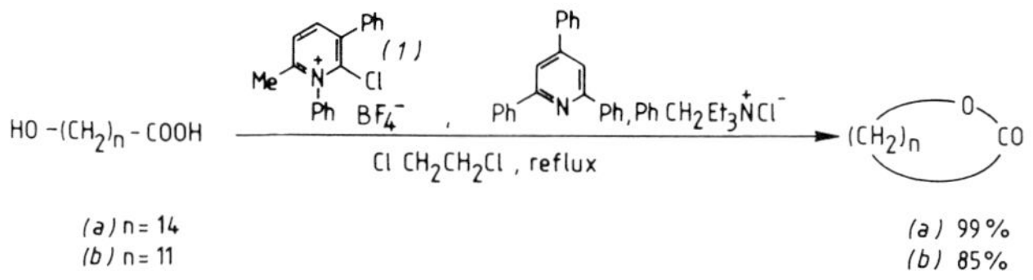

Fig. 3-147. Lactonization using the system 2-chloro-6-methyl-1,3-diphenylpyridinium tetrafluoroborate/2,4,6-triphenylpyridine/benzyltriethylammonium chloride.

Lactonization of *(R)*-(+)-ricinelaidic acid (Fig. 3-148) using this method led to the *(R)*-(+)-lactone without epimerization.

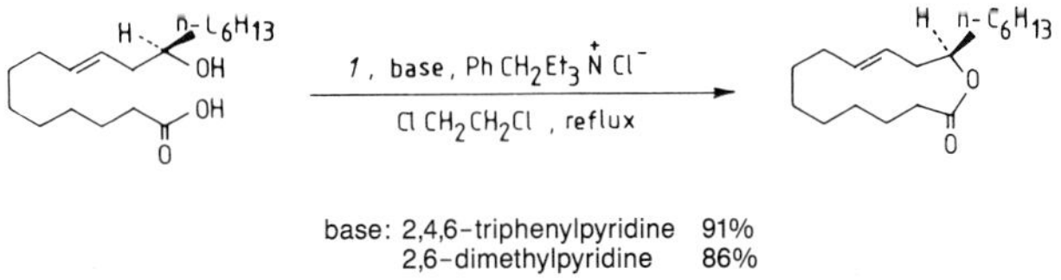

Fig. 3-148. Conversion of *(R)*-(+)-ricinelaidic acid to the *(R)*-(+)-lactone using reagent *1* in Fig. 3-147.

More complex hydroxy acids have been treated with the reagent *1* (Fig. 3-149). The cyclization of prostaglandin $F_{2\alpha}$-9,11-bis(tetrahydropyranyl) derivative, *2*, gave the 1,15-lactone *4*, deprotection of which resulted in *5*. The lactonization of prostaglandin $F_{2\alpha}$, *3*, afforded the 1,9-lactone, *6*, along with a small amount of the 1,15-lactone *5* [186].

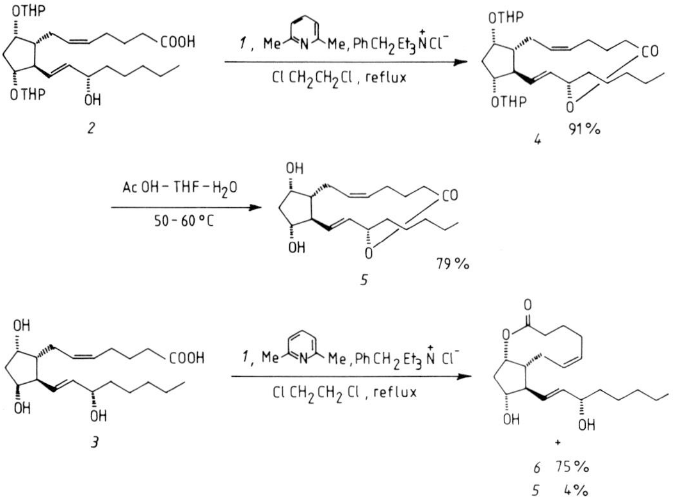

Fig. 3-149. Lactonization of prostaglandin $F_{2\alpha}$ *(3)* and its bis(tetrahydropyranyl) derivative *(2)*.

The effect of $PhCH_2NEt_3^+Cl^-$, the presence of which was necessary for the reaction to occur, is due to formation of the ion pair $HO(CH_2)_nCO_2^-, Et_3N^+CH_2Ph$ thus facilitating the formation of the reactive intermediate (Fig. 3-150).

An extensive review of onium salt chemistry has been published [187].

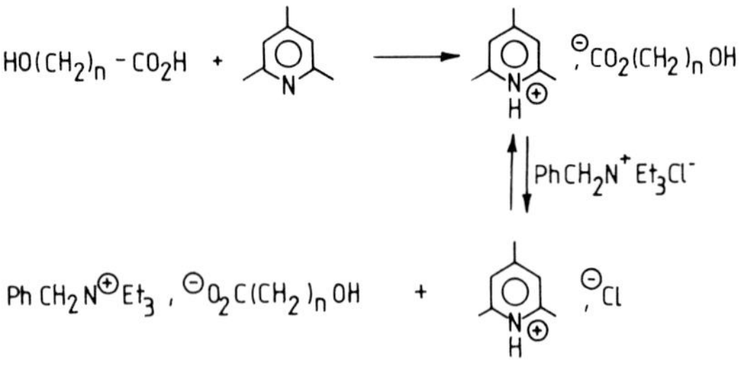

Fig. 3-150. Formation of the intermediate ion pair.

e) Activation and cyclization using dibutyltin oxide

Lactones and lactams can be prepared from ω-hydroxy and ω-amino carboxylic acids in the presence of catalytic amounts of organotin oxides, e. g. $n\text{-}Bu_2SnO$ (Fig. 3-151) [188, 79b].

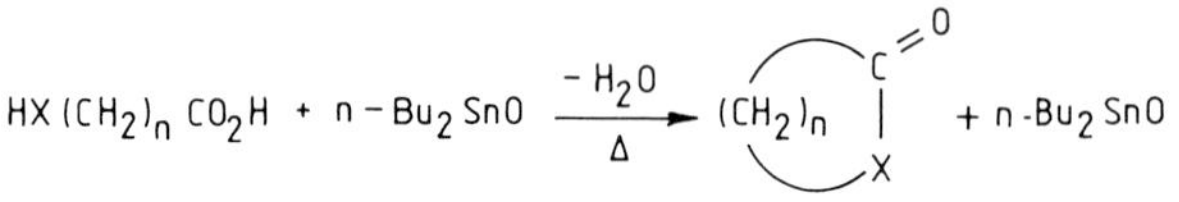

Fig. 3-151. Preparation of lactones and lactams by treating ω-hydroxy (X = O) and ω-amino (X = NH) acids with dibutyltin oxide.

15-Hydroxypentadecanoic and 16-hydroxyhexadecanoic acid each gave the corresponding monolactone in yields of ca. 60% together with small amounts of the dilactone (Table 3-13). However, the 9-membered lactone could not be isolated.

Table 3-13. Yields of macrocyclic lactones and lactams obtained by treating ω-hydroxy (X = O) and ω-amino (X = NH) acids with dibutyltin oxide.

X	n	Solvent	Reaction Time (h)	Lactone or Lactam (%)	Dilactone (%)
O	7	Mesitylene	19	0	20
O	14	”	23	63	8
O	15	”	17	60	15
HN	3	Xylene	12	>95	
HN	4	”	12	>95	
HN	5	”	20	>95	
HN	10	Mesitylene	24	22	
HN	11	”	24	25	

(Taken from S. Hanessian et al., *J. Am. Chem. Soc.* **102**, 7578, 1980.)

Before the above study was carried out, this type of reaction was unknown, although the reverse reaction (cleavage of the ester bond by tin oxides) had been reported.

A comparative study showed that bis(tributyltin) oxide, which led to the intermediate (a) (Fig. 3-152), gave unsatisfactory results. However, dibutyltin oxide afforded macrocyclic lactones in high yields. Structures (b) and (c) were postulated as being intermediates. The ring closure step giving (d) is accompanied by release of R_2SnO, and hence the reaction is catalytic.

This method has been applied in the synthesis of numerous antibiotics, e.g., zearalenone, ingramycin, and nodusmicin [189].

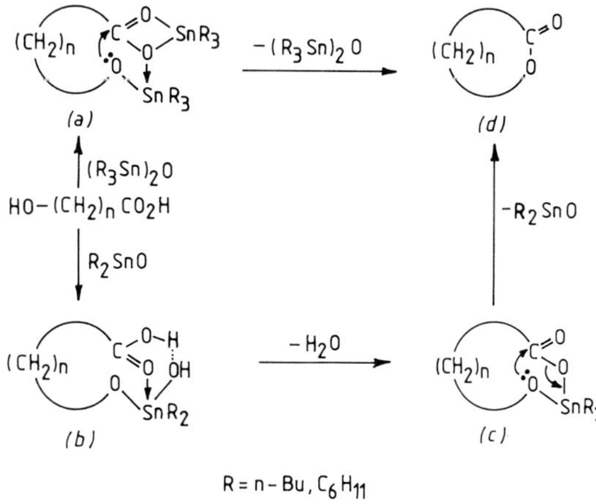

Fig. 3-152. Lactonization in the presence of bis(tributyltin) oxide or dibutyltin oxide.

f) Activation using 1-phenyl-2-tetrazoline-5-thione

This tetrazoline derivative *1* gave a mixture of the thioformimidate *2a* and the formamidine *2b* on treatment with *tert*-butyl isocyanide. Subsequent treatment of the resulting mixture with the hydroxy acid (R = ω-hydroxyalkyl) gave a mixture of the N-and S-acyl derivatives via the intermediates *3a* and *3b* (Fig. 3-153).

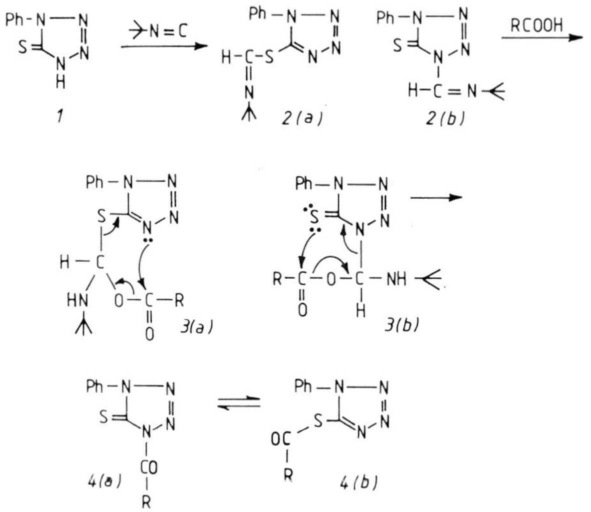

Fig. 3-153. Formation of the N- and S-acyl derivatives of 1-phenyl-2-tetrazoline-5-thione. Compounds *4(a)* and *4(b)* are the reactive species giving the lactones (R = $(CH_2)_nOH$).

These derivatives *4a* and *4b* proved to be very readily lactonized as a result of their strong acylating ability: 16-, 17-, 18- and 20-membered lactones were formed in 90% yield, e.g., exaltolide (Fig. 3-154) [190].

The 13-membered ring was obtained in only 55% yield probably because of the need for prolonged heating of the intermediates *4* in the case of medium rings.

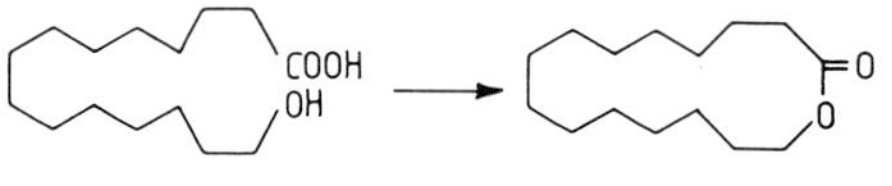

Fig. 3-154. Synthesis of exaltolide using the method shown in Fig. 3-153.

g) Activation using N,N,N′,N′-tetramethylchloroformamidinium chloride

Another activating reagent for carboxylic functional groups is tetramethylchloroformamidinium chloride [191]. This compound *(1)* is obtained very readily from tetramethylurea by reaction with oxalyl chloride (Fig. 3-155).

Slow addition of an ω-hydroxy acid to a solution of this salt gave the corresponding macrocyclic lactone in acceptable yield; 54%, 68% and 90% for n = 11, 12 and 14, respectively.

Fig. 3-155. Lactonization using tetramethylchloroformamidinium chloride.

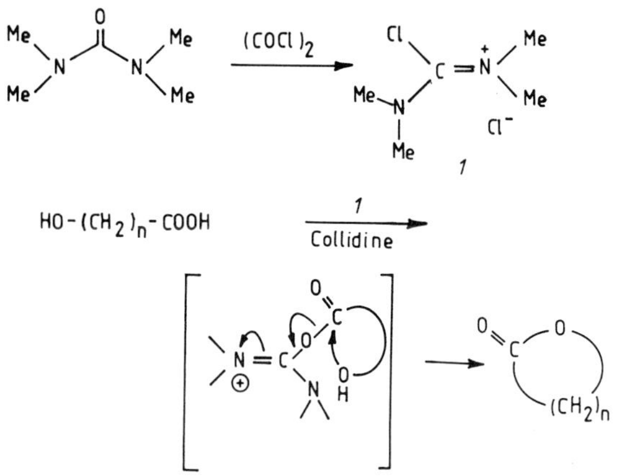

h) Macrolactonization using control of proton-transfer steps

This recent method is based on Steglich esterification [192] for which it was shown that the yield of the esterification reaction using dicyclohexylcarbodiimide is greatly increased on addition of 4-dimethylaminopyridine (DMAP) (Fig. 3-156).

Fig. 3-156. Esterification using dicyclohexylcarbodiimide (DCC) in the presence of 4-di-methyl-aminopyridine (DMAP).

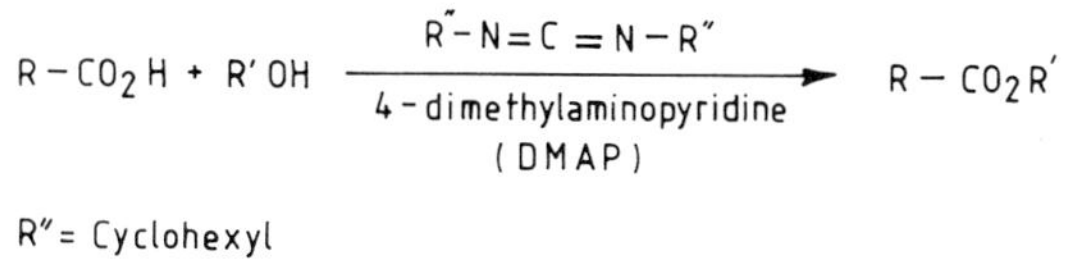

However, in the case of ω-hydroxy acids under these reaction conditions, the N-acylurea derivative is formed almost exclusively (Fig. 3-157a) with hardly any lactones [193]. The reason for these low yields appeared to lie in the proton-transfer steps (Fig. 3-157b).

By operating in the presence of a proton source (2 equivalents of DMAP-HCl) the lactonization occurred in excellent yield (95% for n = 14 and 15) [193].

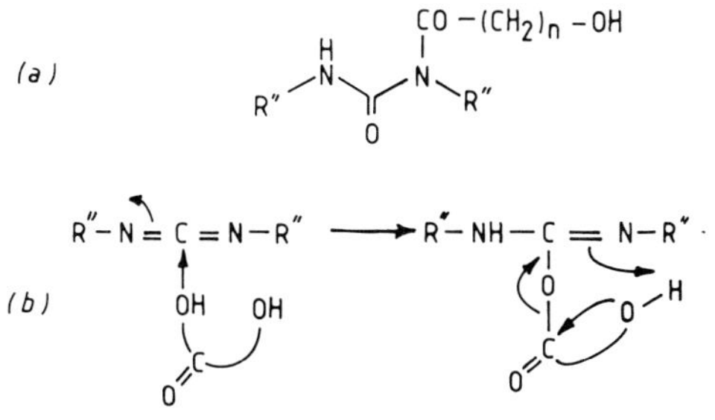

Fig. 3-157. (a) Product of the reaction of an ω-hydroxy acid with DCC/DMAP (b) lactonization using DCC: proton-transfer steps.

i) Reverse activation

In this case, contrary to that for the preceding examples, the hydroxyl functional group is activated and the carboxylate group behaves as a nucleophile. The room temperature reaction of the hydroxy acid with diethyl azodicarboxylate and triphenylphosphine gives rise to ring closure via an alkoxyphosphonium carboxylate intermediate (Fig. 3-158), this being an example of *reverse activation* [194].

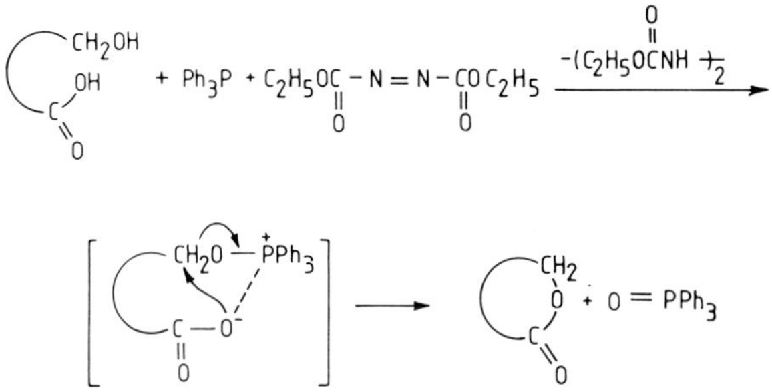

Fig. 3-158. Lactonization of an ω-hydroxy acid via an alkoxyphosphonium carboxylate intermediate (reverse activation).

This method has been used in the synthesis of (±)-vermiculine (Fig. 3-159). Compound *2*, which was synthesized from Hagemann's ester *1*, was oxidatively cleaved and then hydrolyzed to give the hydroxy acid *3*. Lactonization of the latter gave (±)-vermiculine in 15% yield [195].

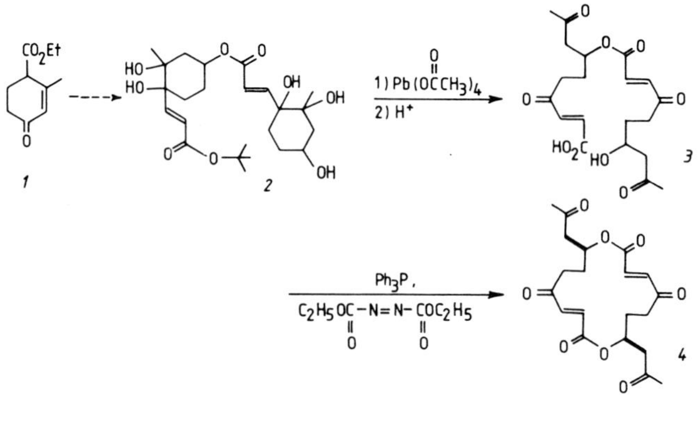

Fig. 3-159. Synthesis of (±)-vermiculine using reverse activation.

j) Lactonization of doubly protected ω-hydroxy acids

Because of its simplicity, this process is extremely elegant. The ω-hydroxyacid is doubly protected with trimethylsilyl groups (Fig. 3-160), and the resulting product is then treated with a Lewis acid (e.g., dipropylboryl triflate). The yields obtained using the method are remarkably high [196].

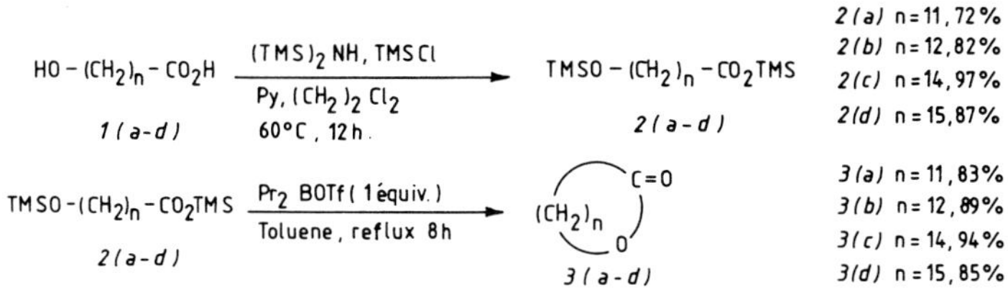

Fig. 3-160. Lactonization of doubly protected ω-hydroxy acids.

k) Macrolide closure of ω-hydroxy acids via conversion into potassium alkoxides. Use of crown ethers as internal templates

As we have repeatedly seen above, high yields for macrolide closure are observed when the two chain ends are drawn together and held in proximity. The reaction can be template-driven by hydrogen bonding [170], Ag(I) [173], Hg(II) [181, 183], or to a certain extent Cs^+ [167]. A very original method has been described by Rastetter et al. initially for one example [197] and then for a whole series of compounds.

The crown ether ω-hydroxy thioester derivatives *1, 2, 3,* and *4* (Fig. 3-161) have been synthesized [198] and used as reagents for macrolide closure. The results were compared to those obtained using the control ethyl thioester *5* without an internal template [199]. It should be noted that the thiol appendages of cyclopentano crown thioesters *3* and *4* converge and diverge, respectively, from the crown ether.

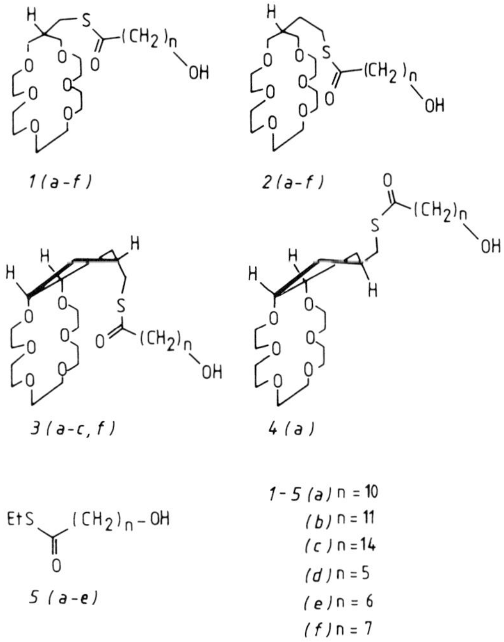

Fig. 3-161. ω-Hydroxy thioester derivatives of crown ethers.

As an example, Figure 3-162 shows the successive steps leading to lactone formation in the case of *3*. The ring closure step takes place *via* an intermediate in which proximate binding of the ω-alkoxide and thioester is achieved by the ionic anchoring of alkoxide against the potassium-crown template.

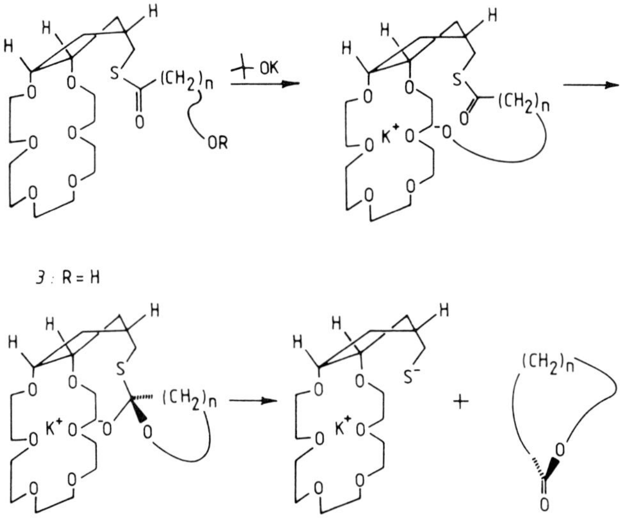

Fig. 3-162. Lactone formation from ω-alkoxy thioesters.

All the experiments were performed under conditions of both medium (MD) and high dilution (HD). Under HD, the yields were clearly the highest (Table 3-14), but the crown ether's role of internal template was most evident under MD conditions. This effect was particularly apparent for the series of experiments with *2* and *3* (favorable orientation of the thioester functional group). It was least evident for the series *1* and *4* (difficult access to the reactive center or incorrect orientation). However, in almost every case, the macrolide yields were superior to those obtained for the control series *5*.

The formation of the lactone was accompanied by that of the dilactone (Table 3-14) and of higher polymers as a result of competing reactions of the alkoxide (Fig. 3-163).

The formation of smaller rings (n = 5–7) occurred in low yield (ca. 20% in the best of cases). Two explanations were put forward. Firstly, formation of these macrolides was kinetically too slow with respect to the competing formation of the dilactones (the latter were indeed observed). Secondly, the small rings obtained were unstable under the reaction conditions.

Table 3-14. Ring closure yields[a].

	11-undecanolide (n = 10)		12-dodecanolide (n = 11)		15-pentadecanolide (n = 14)	
	M.D.[d]	H.D.[d]	M.D.	H.D.	M.D.	H.D.
1	9–10		4–6		38–45	
2	44	72 (68 plus 14 diolide)	37–47	64–74 (71–74 plus 6 diolide)	57–58	73–76 (73–74 plus 6 diolide)
3	60–63	68–77 (75 plus 2 diolide)	58		61–69	72–76 (81 plus trace diolide)
4	11–12	54 (52 plus 11 diolide)				
5[b]	5		9		40–43	
	~1[c]	57[c]				

a) Yields (in percent) determined by GLC; isolated yields (in percent) appear in parentheses.
b) Yield with added 19-crown-6.
c) Yield without added 19-crown-6.
d) M.D.: medium dilution
H.D.: high dilution
(Taken from W. H. Rastetter et al., *J. Org. Chem.* **46**, 3209, 1981.)

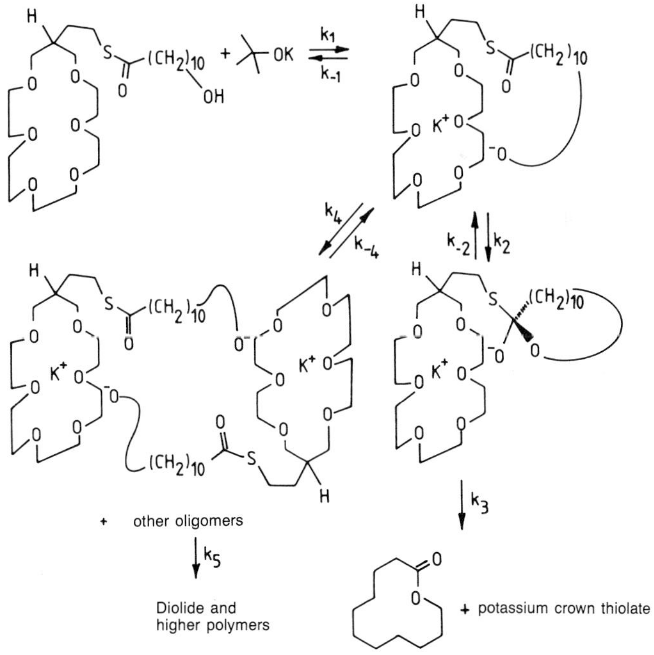

Fig. 3-163. Competing formation of the dilactone and polymers in the ring closure of an ω-alkoxy thioester.

3.6.4 Various Methods of Macrolactonization

a) Translactonization

The reversibility of the translactonization reaction (Fig. 3-164), under either acid or base catalysis, implies that this process will normally lead to the thermodynamically controlled products. Thus, it can also be expected to provide quite precise information concerning the relative stabilities of isomeric lactones of different ring sizes. Depending on the case, it leads to ring expansion or contraction [200].

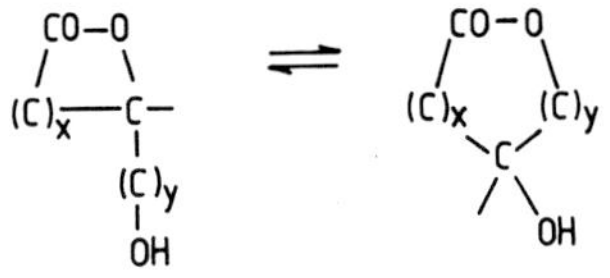

(a) (b)

Fig. 3-164. Translactonization reaction.

The lactone acid *1* was reduced to *2* which underwent translactonization on treatment with *p*-toluenesulfonic acid in dichloromethane at 23–25 °C affording the product *3* in 97% yield (Fig. 3-165).

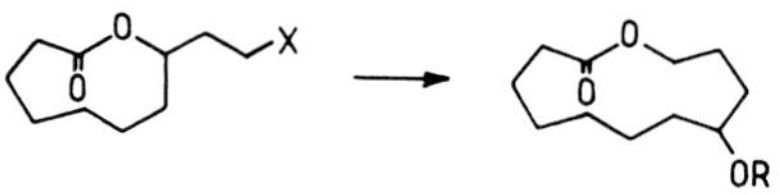

1, X = COOH
2, X = CH_2OH
3, R = H

Fig. 3-165. Ring expansion (9 → 12) via translactonization.

The acid *1* also served as a starting point for generation of a 15-membered lactone (Fig. 3-166). The yield from the alcohol *5* was 90% after 36 h at 23 °C. In this case, a six carbon ring expansion occurred.

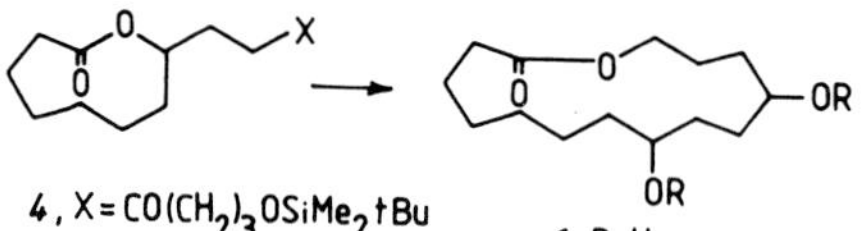

4, X = $CO(CH_2)_3OSiMe_2tBu$
5, X = $CH(OH)(CH_2)_3OH$
6, R = H

Fig. 3-166. Ring expansion (9 → 15) via translactonization.

These two ring expansion reactions are driven mainly by the relative instability of the 9-membered cyclic system. A number of other experiments provided useful information concerning the possibilities of this type of reaction for the generation of macrocyclic lactones.

The 8-membered hydroxy lactone *7* underwent ring expansion to form *8* (Fig. 3-167), but more slowly and in lower yield (69%) than for the larger 9-membered homolog *2* (see Fig. 3-165).

On the other hand, the 7-membered hydroxy lactone *9* did not undergo ring expansion to the 10-membered lactone *10* probably because *9* is more stable than is *10* (Fig. 3-167).

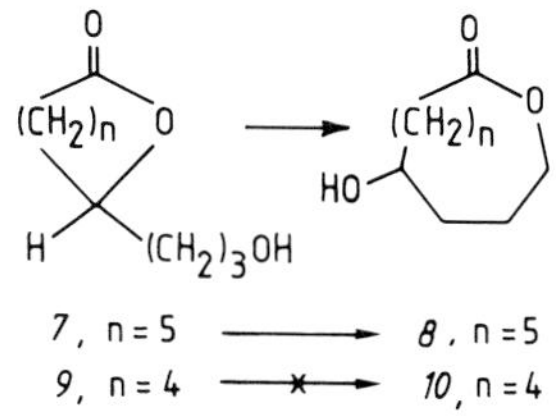

Fig. 3-167. Ring expansion (8 → 11) via translactonization. This reaction does not occur in the 7 → 10 case.

Thy hydroxy lactone *11* either in chloroform solution (without addition of acid) or when treated with *p*-toluenesulfonic acid gave in both cases a 35/65 mixture of *11* and *12* (Fig. 3-168).

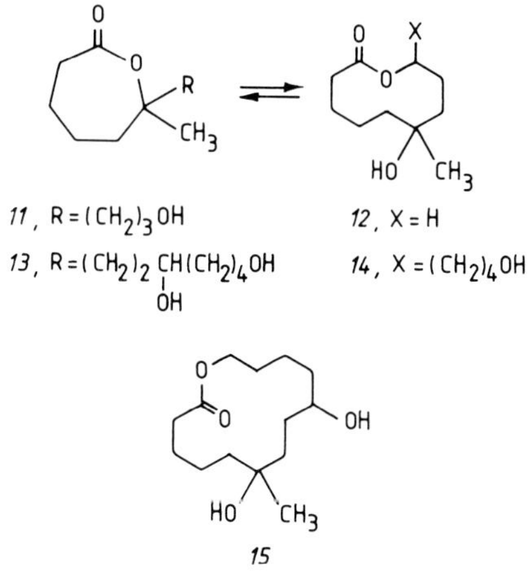

Fig. 3-168. Translactonization of *11* and *13* gave, respectively, a 35/65 mixture of *11* and *12* and a 50/50 mixture of *13* and *14*. Translactonization of *14* → *15* did not occur.

The lactone *13* afforded an equimolar mixture of *13* and *14* (Fig. 3-168). The latter compound *14* did not give the expected translactonization product *15*, it being more difficult to attain the 8-membered than the 7-membered cyclic transition state. It therefore seems that the scheme in Fig. 3-164 applies only for y = 1–3 and not for y = 4.

The following ring expansions can be expected to be most favorable based on both the relative stability of each lactone and the length of the side chain:

8 → 11 → 13 or 14;   9 → 12 → 14 or 15;   10 → 13.

b) Oligomerization

We have already mentioned several methods for the generation of macrocycles by means of controlled polymerization (Shanzer [78] and Dale [159]). Another method has also been described using a bicyclic oxalactone [201], the latter being either the racemic monomer or the optically active oxalactone *3* (Fig. 3-169).

The preparation of (+)-(1*R*,5*R*)-6,8-dioxabicyclo[3.2.1]octan-7-one, *3*, is shown in Fig. 3-169 involving the use of dehydroabietylamine (DAA) for the optical resolution of 3,4-dihydro-2*H*-pyran-2-carboxylic acid to give *2*. The latter diastereoisomer was treated with Na_2CO_3 solution (to eliminate the amine) followed by hydrochloric acid thus affording the free acid which on distillation gave the oxalactone *3*.

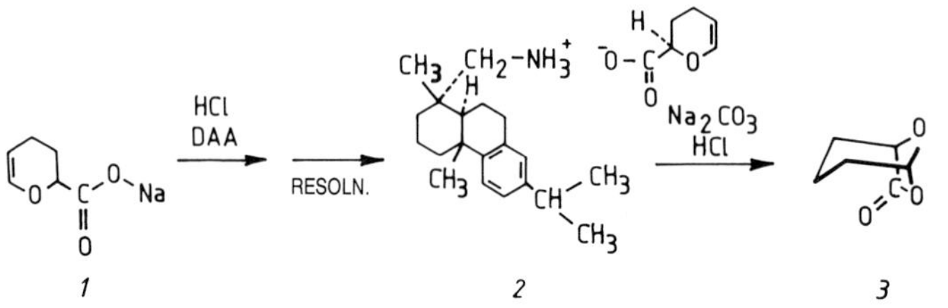

Fig. 3-169. Synthesis of the (+)-oxalactone *3*.

Oligomerization of *3* was carried out using $BF_3 \cdot OEt_2$ at −40 °C in a variety of solvents. The macrocycles which could be formed are shown in Fig. 3-170.

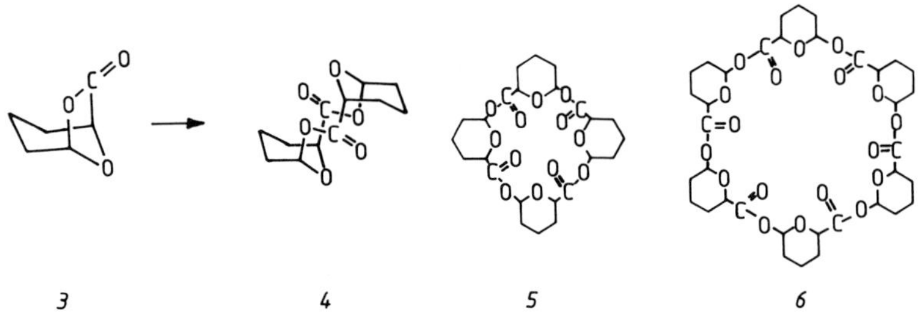

Fig. 3-170. Macrocycles which can be formed by oxalactone oligomerization.

It can be seen from Table 3-15 that the cyclic hexamer was preferentially formed and that the selectivity was particularly high in 1-nitropropane probably owing to the low solubility of the macrocycle *6* in this solvent. Precipitation of *6* thus prevented formation of higher oligomers.

The absence of cyclic dimer should be noted, whereas this compound was obtained when starting from the racemic monomer. In the latter case, the dimer was formed from a pair of different enantiomeric units (Fig. 3-171 a). This was not possible in the case of the optically active monomer (Fig. 3-171 b).

Table 3-15. Oligomerization of (+)-(1 *R*, 5 *R*)-6,8-dioxabicyclo[3.2.1]octan-7-one.

Monomer g	Solvent ml	Initiator mol %	Time h	Conversion (%) dimer *4*	tetramer *5*	hexamer *6*	other	total
0.5	MC 0.5	1	48	0	5	63	7	75
1.0	AN 1.0	1	24	0	25	59	0	84
1.0	CF 1.0	5	24	0	0	46	0	46
1.0	NP 1.0	1	24	0	0	81	0	81

MC, methylene chloride; AN, acetonitrile; CF, chloroform; NP, l-nitropropane.
(Taken from M. Okada et al., *J. Am. Chem. Soc.* **101**, 4013, 1979.)

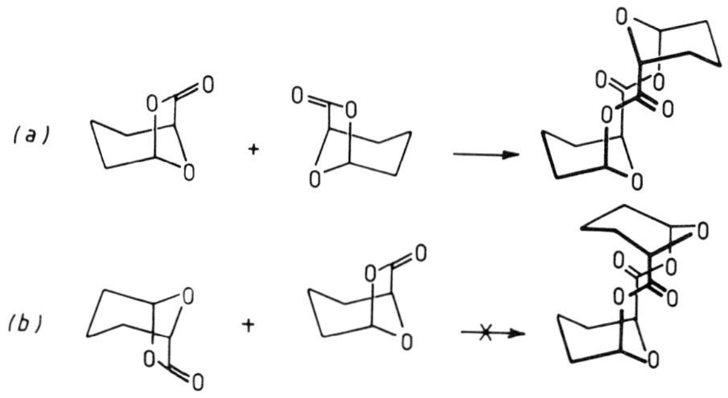

Fig. 3-171. (a) Dimer formation by oligomerization of racemic oxalactone. (b) This dimerization does not occur with (+)-oxalactone *3* (see Table 3-15).

c) Triphase catalysis

This relatively recent technique is clearly different from classical phase transfer catalysis, in spite of similarities between the two. In the case of the latter method, the catalyst (quaternary ammonium or phosphonium salt, crown ether, cryptand, etc.) *extracts* the reagent (the associated anion) into the organic phase containing the substrate. In triphase catalysis, the catalyst is *immobilized* on a polymer (solid phase) which is in contact with the aqueous and organic phases containing the reactant and substrate, respectively. This method has the great advantage of resolving the problem of catalyst and product separation at the end of the reaction [202 a].

Macrolide synthesis using a triphase system has been carried out as shown in Fig. 3-172. The hydroxy acid was first activated by mesylation of its alcohol functional group and then added to a toluene/aqueous potassium bicarbonate/phosphonium mesylate resin three phase system [202 b]. Lactonization occurred on the resin which at the end of the reaction was once again in the mesylate form.

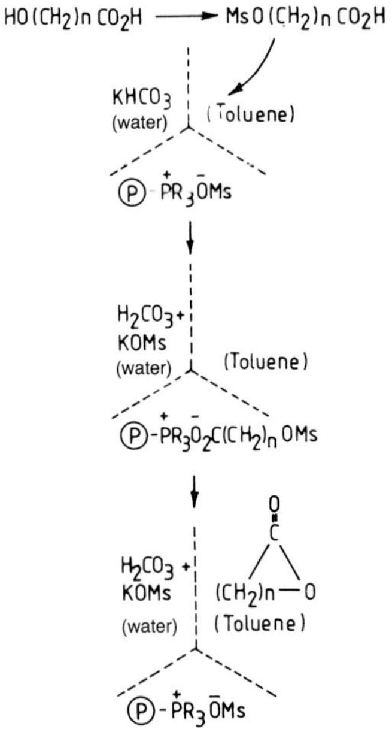

Fig. 3-172. Scheme for macrolide synthesis using triphase catalysis.

The results were comparable to those obtained using other methods (Table 3-16), except in the case of the medium ring (n = 7). They were explained by intramolecular nucleophilic displacement of "kinetically isolated" pendant groups favoring formation of macrolides over that of oligomers or polymers. When n = 7, the internal cyclization reaction was too slow for "kinetic isolation" to be achieved.

The efficiency of the method can be seen in an example involving a secondary hydroxyl group (Fig. 3-173). By means of the reactions shown in Fig. 3-173, ricinelaidic acid afforded the corresponding lactone in 40% yield. The relatively high

Table 3-16. Macrolide synthesis results obtained using triphase catalysis, compared to those obtained using the Corey-Nicolaou and Kruizinga-Kellogg methods, starting from $X-(CH_2)_n-CO-R$ (in %).

n	ring size	Triphase catalysis (R = OH, X = OMs)	Corey-Nicolaou[a] R = 2-thiopyridine X = OH	Kruizinga-Kellogg[b] R = OH, X = I
7	9	0	25	
10	12	47	64	33
11	13	66	76	62
12	14	76	79	77
14	16	72	88	83

a) see [170]; b) see [166].
(Taken from S. L. Regen and Y. Kimura, *J. Am. Chem. Soc.*, **104**, 2064, 1982.)

optical purity of the product indicated that the displacement had proceeded with predominant inversion of configuration [202b].

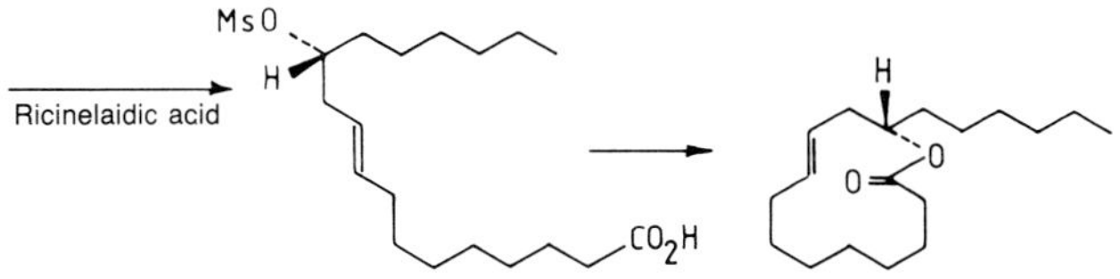

Fig. 3-173. Lactonization of ricinelaidic acid.

d) Solid-liquid phase transfer catalysis

In this type of phase transfer catalysis, the insoluble salt incorporating the active anionic species is in suspension in the organic solvent containing the phase transfer catalyst and the substrate [203a]. This method has shown promising results in the field of macrolides [203b].

The catalytic cycle for lactone formation is illustrated in Fig. 3-174. It should be noted that this reaction is a special case of solid-liquid catalysis since the anionic species, solubilized by the catalyst, is both reactant and substrate.

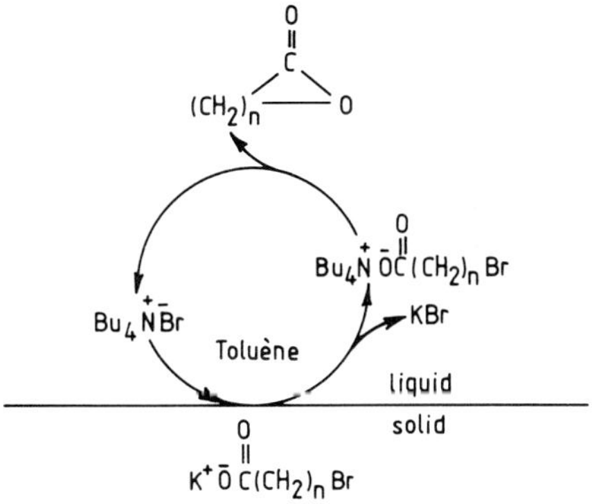

Fig. 3-174. Lactone formation using solid-liquid phase transfer catalysis.

Because the tetrabutylammonium salt is used in catalytic quantities, the concentration of the active species (tetrabutylammonium carboxylate) is extremely low throughout the reaction (high dilution conditions). It can be seen in Table 3-17 that the macrocycle yields are high (except in the case of the 9-membered ring). It can also be seen that increase in catalyst concentration brings about a decrease in yield of cyclic product (increase in intermolecular reaction). From the point of view of carboxylate reactivity, this reaction should be compared to the method discussed above involving the use of tetraalkylammonium salts (cf. Section 3.6.2e and [168]).

Table 3-17. Cyclization of ω-bromocarboxylic acids[a] using solid-liquid phase transfer catalysis.

n	ring size	$Bu_4N^+Br^-$	Time h	yield %
5	7	2.5×10^{-3} mmole	3	92
7	9	1.5 "	24	26
11	13	2.5 "	3	95
		10.0 "	3	94
		58.0 "	3	85
		100.0 "	3	73
14	16	2.5 "	3	92
15	17	2.5 "	3	94

(a) 0.1 mmol of potassium salt per ml of solvent, T = 90 °C
(Taken from Y. Kimura and S. L. Regen, *J. Org. Chem.*, **48**, 1533, 1983.)

e) Cyclization on a polymer support

We have already mentioned (cf. Section 3.1.2) the use of polymer supports to obtain macrocycles. The method has been applied in the synthesis of macrolides [204], but the yields are low in this case (at best 13% for a dodecanolide).

3.6.5 Macrolactonization via Carbon-Carbon Bond Formation

The numerous examples given up to this point show that macrocyclic lactones are most usually obtained by reaction of acid and alcohol functional groups (or their derivatives). However, there are also some methods involving carbon-carbon bond formation. This methodology, which is little used, presents many advantages as pointed out by Trost [205]. Firstly, there is a greater degree of flexibility in synthetic design since the synthons bearing the alcohol and acid functional groups can be elaborated separately. These are then coupled to give the ester, and it is only after this step that cyclization is carried out through formation of a C–C bond. Another advantage is that this method provides access to cycloalkanes.

a) Cyclization of allylic dibromides using nickel tetracarbonyl

First described for the formation of macromolecular carbocycles [206], this method has been successfully applied to the synthesis of macrolides [207]. An example of coupling leading to a macrocyclic lactone is given in Fig. 3-175.

A review concerning this type of reaction has appeared [208].

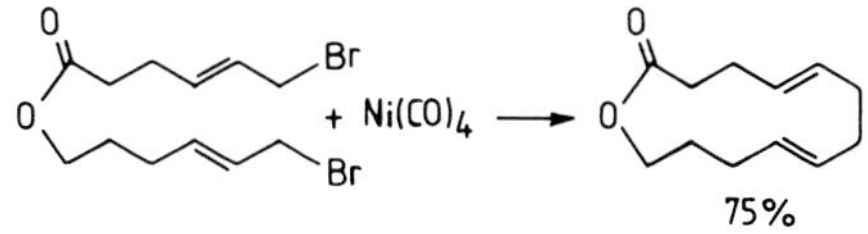

Fig. 3-175. Synthesis of a macrocyclic lactone from an allylic dibromide in the presence of $Ni(CO)_4$.

b) Cyclization using organopalladium intermediates

The prototype reaction for this type of cyclization is alkylation of the palladium complex formed with an allylic acetate [209] (Fig. 3-176).

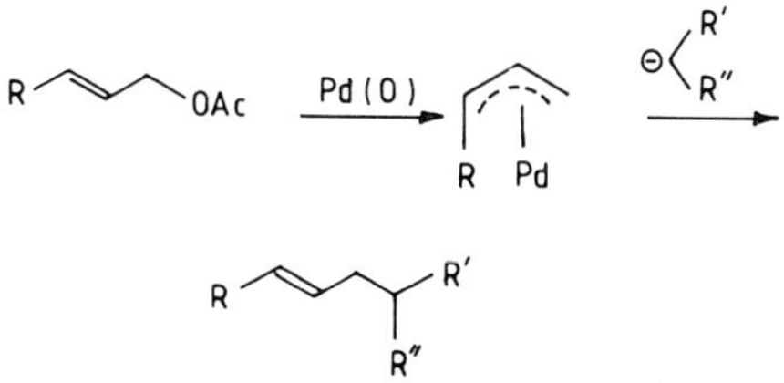

Fig. 3-176. Alkylation of the complex formed by palladium with an allylic acetate.

This reaction has been used in the synthesis of recifeiolide (Fig. 3-177). Cyclization of compound *1* (the synthesis of which we leave undescribed) was carried out by conversion to the carbanion (NaH in THF) to which was slowly added the $(Ph_3P)_4Pd$ catalyst. The macrocycle *2* thus formed in high yield (78%) gave recifeiolide after decarbomethoxylation followed by desulfonylation [205].

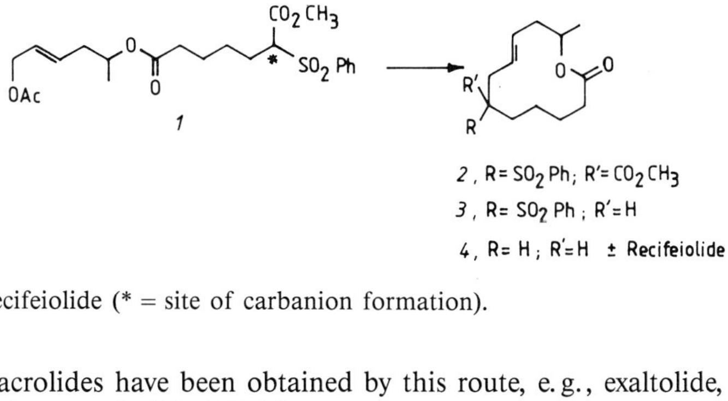

Fig. 3-177. Synthesis of recifeiolide (* = site of carbanion formation).

Many other macrolides have been obtained by this route, e. g., exaltolide, phoracantholide I and phoracantholide J [205].

An important characteristic of this synthetic procedure affording macrolactones should be noted, that is, its low sensitivity to the size of the ring formed. In other words, medium rings (n = 8–11) are obtained as readily as large rings. The minima in intramolecular reaction rates and yields found for all other direct cyclization methods described so far, are not observed in this case. The reasons for these somewhat surprising results are extensively discussed by Trost. One important factor seems to be the presence of the ester functional group which strongly decreases intramolecular repulsions in the transition state (heteroatom effect, cf. Section 1.2.1) [205].

Other palladium complexes formed, in this case with vinyl epoxides, also lead to very efficient macrocyclizations [210]. The principle of the reaction (a) and an example of cyclization (b) are given in Fig. 3-178.

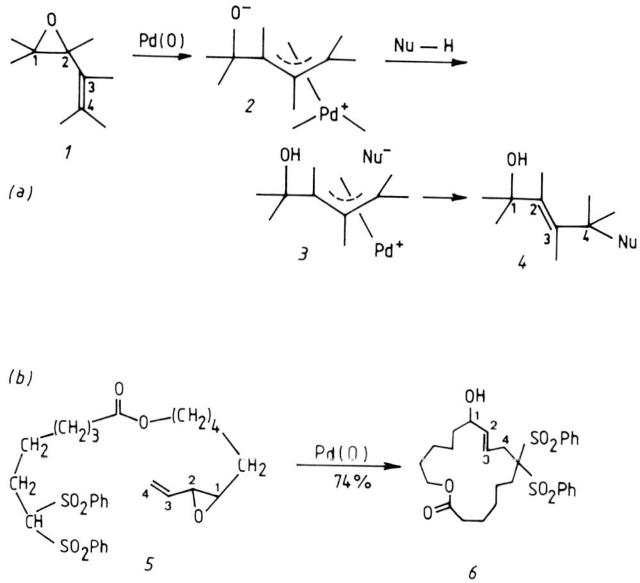

Fig. 3-178. Macrocyclization via a vinyl epoxide palladium complex. (a) Principle, (b) application.

It should be noted that this reaction takes place without prior nucleophile formation and that addition of base is unnecessary. The intermediate species *2* plays the part of the base thus affording the nucleophile. Experimentally, the ring closure reaction was found to occur very readily on treatment of a refluxing solution of *5* (0.2 mol/l in THF) with the polystyrene-bound catalyst (5 mol %).

All these reactions involving palladium complexes are highly regio- and stereoselective.

c) Photochemical synthesis

Irradiation of *1* (Fig. 3-179) gave intramolecular formation of the oxetane *2* (cf. Section 2.3.2). The latter afforded the macrocyclic lactone *3* on treatment with silica gel [211].

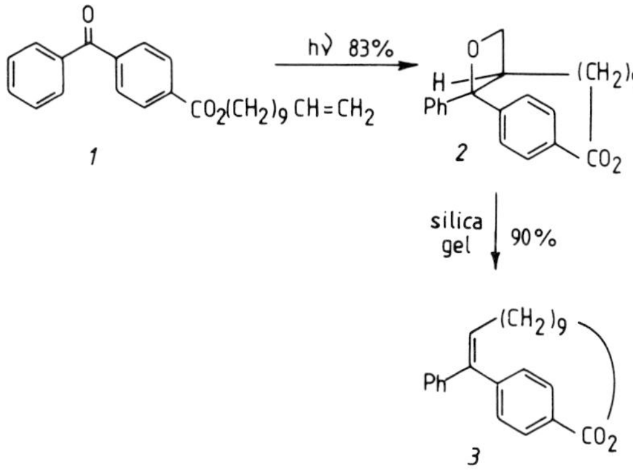

Fig. 3-179. Photochemical synthesis of macrocyclic lactones. The conversion of *2* into *3* occurred with loss of formaldehyde.

d) Oxidative coupling of diacetylenic esters

This route to macrolides has been employed with a variety of acetylenic compounds (65–80% yields) [212]. In like fashion, exaltolide was obtained in 88% yield (Fig. 3-180) [213]. The diacetylenic compound *1* was treated with copper(II) acetate in pyridine giving *2* which on catalytic hydrogenation afforded exaltolide.

Other examples of oxidative coupling are given below.

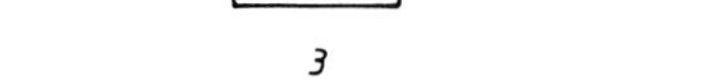

Fig. 3-180. Lactone synthesis by oxidative coupling of diacetylenic esters.

Some excellent reviews give a good overall survey of macrolide synthesis [214, 215].

3.6.6 Lactonization of very Complex Systems. Thiolactones

Synthetic routes to other systems which are highly complex, such as the *trichothecene* family [216] (Fig. 3-181), involve the extensive use of certain of the methods discussed above.

Thus, for example, the crucial cyclization steps in the syntheses of *verrucarin A* and *verrucarin B* (Fig. 3-182) are carried out by using the reagent triphenylphosphine-diethyl azodicarboxylate in the first case, and via a mixed anhydride intermediate in the second case [217].

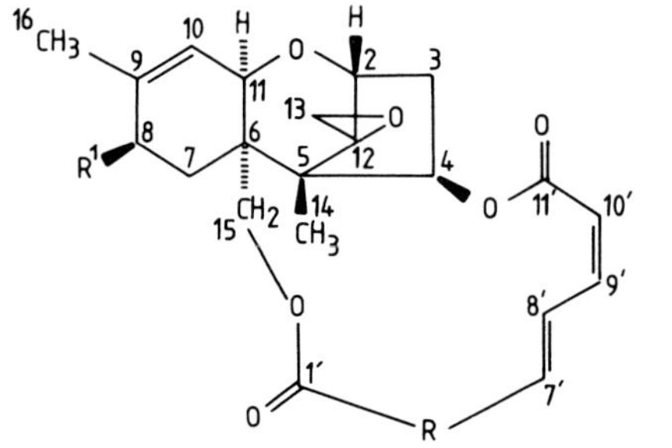

Fig. 3-181. Examples of trichothecenes.

	R	R¹
Verrucarin A	2′ 3′ 12′ 4′ 5′ 6′ $-CH\,OH\,CH\,CH_3CH_2CH_2OC(=O)-$ *S* *R*	H
Verrucarin J	$-CH{=}CCH_3CH_2CH_2OC(=O)$ *E*	H
Roridin A	2′ 3′ 12′ 4′ 5′ 6′ 13′ 14′ $-CHOH\,CH\,CH_3CH_2CH_2OCH\,CHOH\,CH_3$ *S* *R* *R* *R*	H

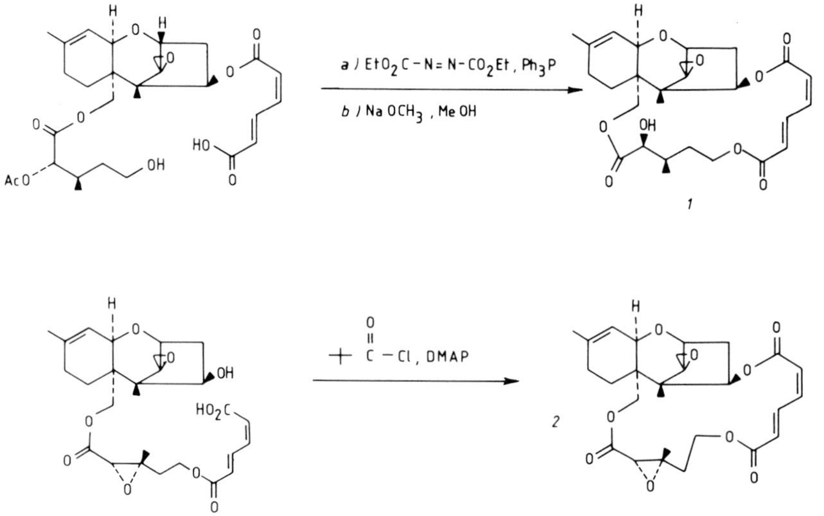

Fig. 3-182. Synthesis of verrucarin A and verrucarin B.

The total synthesis of *polyenic macrolides* is also a current topic of investigation [218], amphotericin B being an important example (Fig. 3-183).

Fig. 3-183. Structure of amphotericin B.

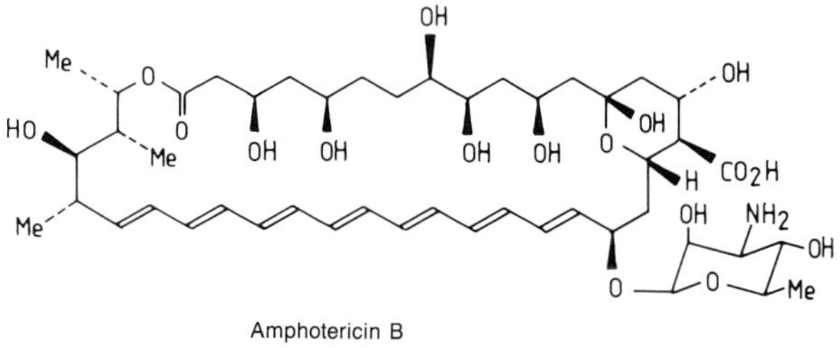

To close this section on lactones, we describe the synthesis, using a stannoxathiane, of a new family of macrocyclic compounds containing both lactone and thiolactone groups (Fig. 3-184). Treatment of the stannoxathiane *1* with a series of diacyl dichlorides (n = 4–7) provided the macrocycles *2–5*. The macrocyclic dimers *3* were in every case the major products (12–47% yield depending on the value of n). The other macrocycles *2, 4* and *5* were formed in low yields (generally ~10%).

These substances are of potential use for complexing heavy cations [219].

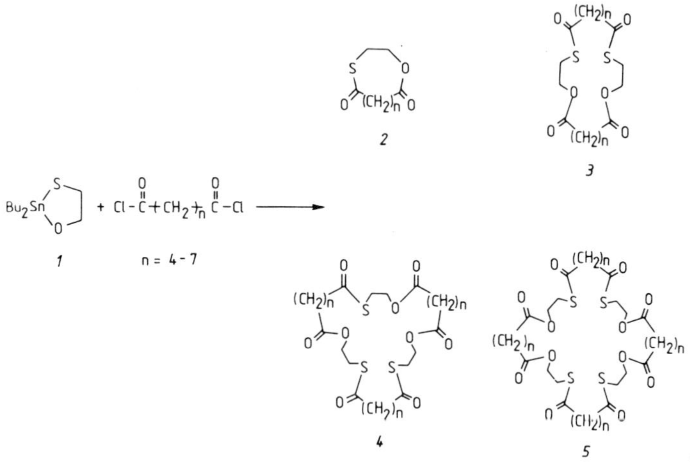

Fig. 3-184. Synthesis of lactone-thiolactone macrocycles from stannoxathianes.

3.7 Macrocyclic Lactams

The lactam functional group is commonly encountered in numerous macrocyclic natural products.

3.7.1 Macrocyclic Spermidine Alkaloids

A first important group is that of the alkaloids containing biogenic polyamines (spermidine: $NH_2(CH_2)_3NH(CH_2)_4NH_2$; spermine: $NH_2(CH_2)_3NH(CH_2)_4NH(CH_2)_3NH_2$) in their ring systems [220].

Numerous syntheses have been described [221]. As is the case for macrolides, it is important that the final cyclization step should occur in high yield if the linear chain has had to be elaborated, given the cost of the starting material. Some examples will illustrate the various methods used. Only the macrocyclization steps will be discussed.

a) Intramolecular reaction between acyl chloride and amine functional groups

Macrocyclization was carried out by slow addition of the precursor *1* into a large volume of benzene containing triethylamine (Fig. 3-185). The amine functional group thus released reacted with the acyl chloride to form the cyclic product *2*. The yield of the cyclization step is unknown since only the overall yield (30%) for the four reaction steps was reported. In a series of additional steps *oncinotin 3* was obtained [222].

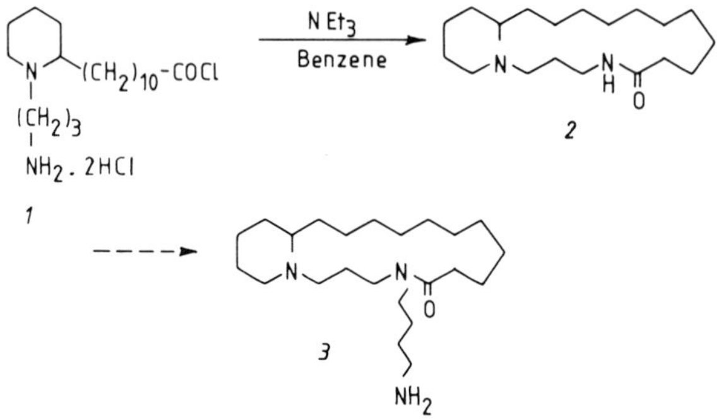

Fig. 3-185. Synthesis of oncinotin *3*.

b) Intramolecular reaction of an aminoester in the presence of imidazole

Compound *3, dihydropalustrine,* was synthesized from aminoester *1* in a melt with imidazole (activating group for carboxylate) (Fig. 3-186). The experimental method used may explain the low yield obtained (~10%) [223].

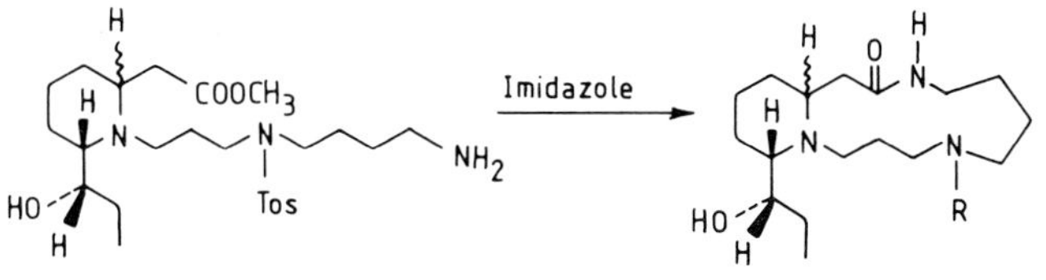

1

2, R = Tos
3, R = H

Fig. 3-186. Synthesis of dihydropalustrin *3*.

c) Activation by hydroxypiperidine

Tetrahydrocodonocarpine 2 has been prepared by direct reaction of spermidine with *1* (Fig. 3-187). The disadvantage of this relatively rapid method is the formation of two isomers *2* and *3* as a result of the two possible orientations for spermidine in the condensation step. After purification, the desired product *2* was isolated in 17% yield [224].

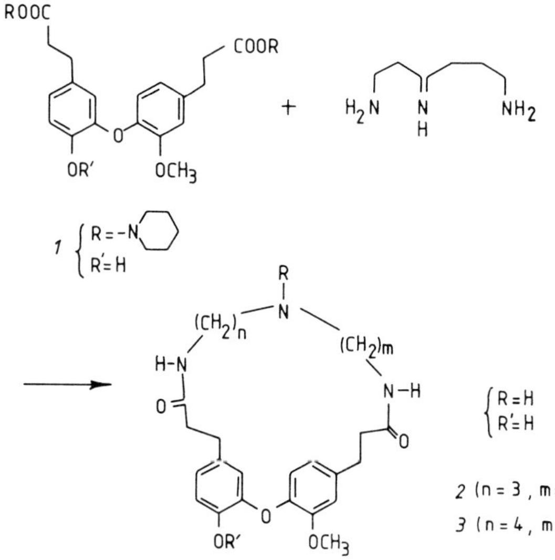

1

2 (n = 3, m = 4)
3 (n = 4, m = 3)

Fig. 3-187. Synthesis of tetrahydrocodonocarpine *2*.

d) Activation by thiophenol

A more lengthy method afforded *codonocarpine 4* specifically [225]. The acid functional group of the open precursor *1* (Fig. 3-188) was activated by conversion into the thioester *2*. Following amine deprotection, the primary amine reacted under basic conditions with the thioester group to give macrocycle *3*. The latter afforded *codonocarpine 4* on removal of the benzyl group (60% cyclization yield).

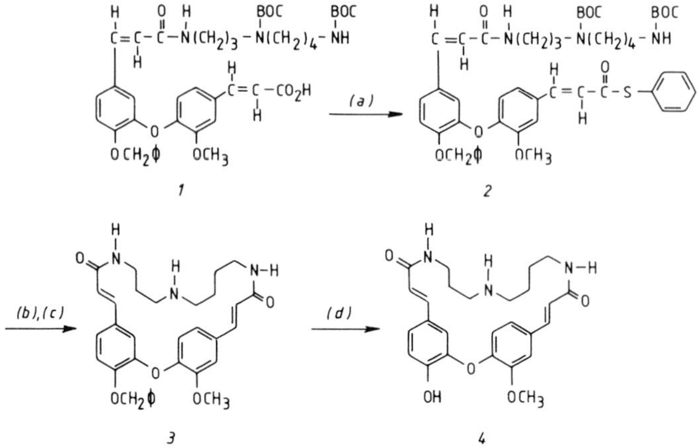

Fig. 3-188. Synthesis of codonocarpine *4* (TFA, trifluoroacetic acid; TMSI, trimethylsilyl iodide).

e) Activation by thiazolidine-2-thione

Treatment of a carboxylic acid or the derived acyl chloride with thiazolidine-2-thione or its salt gives 3-acylthiazolidine-2-thione (Fig. 3-189). The latter is very reactive towards amines giving the amides *3* in high yield [226].

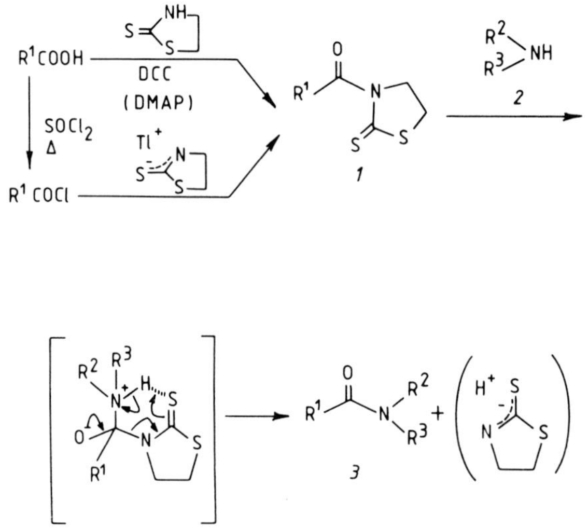

Fig. 3-189. Amide synthesis via activation of a carboxylic acid or the derived acyl chloride by thiazolidine-2-thione (DCC, dicyclohexylcarbodiimide; DMAP, 4-dimethylaminopyridine).

By application of this method to diacids, the difunctional derivatives are readily obtained. The latter react very efficiently with diamines (under high dilution conditions) to give macrolactams in yields of 35–90% [227]. The synthesis of *codonocarpine* using this procedure is shown in Fig. 3-190 [228].

Numerous other macrocyclic alkaloids of the type described above (lunaridine, lunarine, etc.) have been obtained in this way [228].

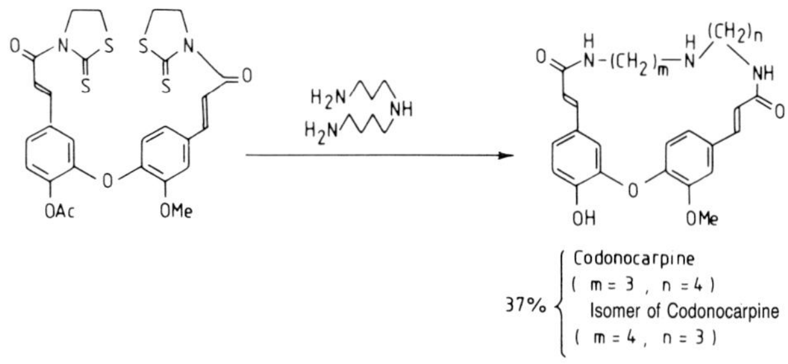

Fig. 3-190. Synthesis of macrolactams using activated diacids.

f) Cyclization via the potassium tosylamide salt

A different type of cyclization was employed in this series of compounds [229]. The reaction took place (Fig. 3-191) via attack of the potassium tosylamide salt on the carbon bearing the OMs leaving group.

We shall see in Part II that the application of this extremely simple coupling method is very extensive.

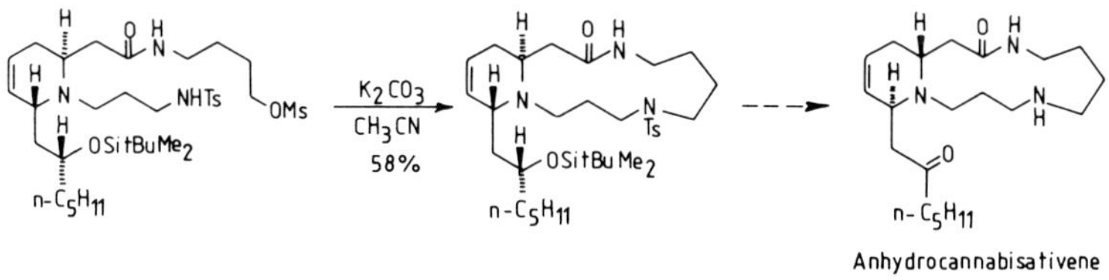

Fig. 3-191. Synthesis of anhydrocannabisativene.

3.7.2 Formation of the Lactam Group in Rifamycin

Another group of compounds incorporating the lactam group is the very large one consisting of the *ansamycins* [230]. Within this, one of the best known sub-groups is that of the *rifamycins* (Fig. 3-192).

Chemists in quest of difficult syntheses have been attracted to these highly complex systems. It is quite evident that the macrocyclization step is only one small part of the synthesis since the construction of the linear precursor is of extraordinary difficulty in view of the number of asymmetric centers. However, the final cyclization step has a great effect on the overall yield and can present difficulties. Thus, in the

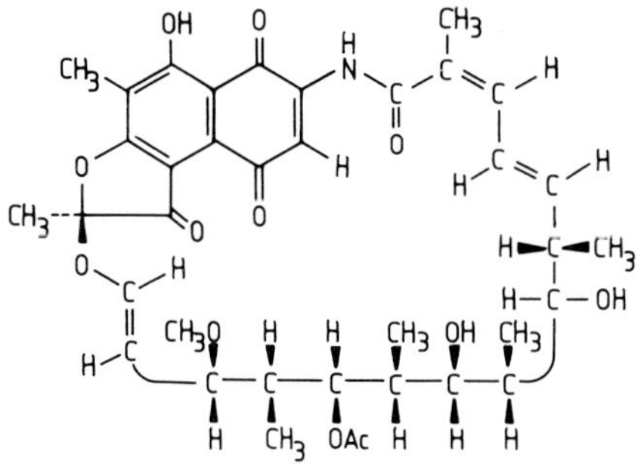

Fig. 3-192. Structure of rifamycin S.

total synthesis of rifamycin S, the cyclization reaction necessitated numerous attempts (Fig. 3-193). A wide variety of activating groups for the acid functional group were tried unsuccessfully [231]; the quinone system deactivates the amine group. This difficulty was overcome by reducing the quinone to the hydroquinone which was then re-oxidized to the quinone after coupling. Several reaction sequences have been described [231, 232] (Fig. 3-193). The two methods (b and c) using mixed anhydrides gave the best results.

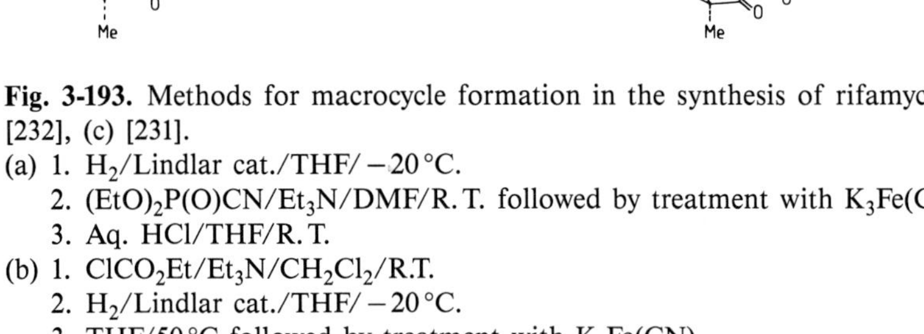

Fig. 3-193. Methods for macrocycle formation in the synthesis of rifamycin S; (a) and (b) [232], (c) [231].

(a) 1. H_2/Lindlar cat./THF/-20 °C.
2. $(EtO)_2P(O)CN/Et_3N$/DMF/R.T. followed by treatment with $K_3Fe(CN)_6$.
3. Aq. HCl/THF/R.T.

(b) 1. $ClCO_2Et/Et_3N/CH_2Cl_2$/R.T.
2. H_2/Lindlar cat./THF/-20 °C.
3. THF/50 °C followed by treatment with $K_3Fe(CN)_6$
4. Aq. HCl/THF/R.T.

(c) 1. $ClCO_2$ i-Bu/Et_3N/Et_2O/23 °C.
2. $H_2/Pd\text{-}CaCO_3$/THF/-40 °C.
3. THF/50 °C followed by treatment with $K_3Fe(CN)_6$.
4. Aq. $HClO_4$/THF/25 °C.

3.7.3 Lactam Synthesis Using Diazasilolidines

This method, which has already been mentioned (cf. Section 3.3.3), has been studied in more detail [233]. The interesting observation was made that either the dilactam or the tetralactam was exclusively formed depending on (a) the number of methylene groups in the chain of the diamine derivative, (b) the nature of R, and (c) the nature of the diacyl dichloride (Fig. 3-194, Table 3-18).

The selectivity of these reactions was attributed to a template effect exerted by the silyl group undergoing noncovalent interactions with the carbonyl oxygens.

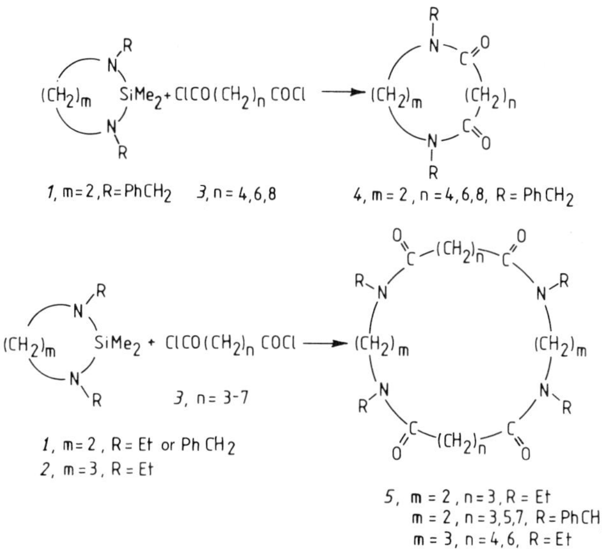

Fig. 3-194. Synthesis of macrocyclic polylactams from diazasilolidines.

Table 3-18. Yields of dilactams *4* and tetralactams *5*.

Starting materials	yield (%) Dilactams *4*	yield (%) Tetralactams *5*
1 (m = 2, R = CH_2CH_3) *3* (n = 3)	–	24
1 (m = 2, R = CH_2Ph) *3* (n = 3)	–	31
1 (m = 2, R = CH_2Ph) *3* (n = 4)	15	–
1 (m = 2, R = CH_2Ph) *3* (n = 5)	–	12
1 (m = 2, R = CH_2Ph) *3* (n = 6)	17.5	–
1 (m = 2, R = CH_2Ph) *3* (n = 7)	–	40.5
1 (m = 2, R = CH_2Ph) *3* (n = 8)	24.5	–
2 (m = 3, R = CH_2CH_3) *3* (n = 4)	–	21
2 (m = 3, R = CH_2CH_3) *3* (n = 6)	–	15

(Taken from A. Shanzer et al., *J. Org. Chem.* **50**, 5469, 1985.)

These interactions stabilize specific conformations during the reaction and direct subsequent ring formation towards either the di- or the tetralactam.

It can be seen (Table 3-18) that the type of macrocycle obtained depended on the nature of the chain (m or n even or odd). For m = 2, diacyl dichlorides with even values of n (4, 6, 8) gave dilactams, and those with odd values of n (3, 5, 7) afforded tetralactams. The latter were also obtained for m = 3 with even values of n (4, 6).

3.7.4 Cyclopeptides

The biological importance of these compounds is widely known (oxytocin, vasopressin, cyclosporin, etc.) [234]. Furthermore, because of the limited number of conformations for synthetic cyclic peptides, these could be studied in detail using various techniques (X-ray diffraction, NMR, and CD) [235]. It proved possible to apply the results to the conformational study of more complex natural proteins thus contributing to the understanding of structure/activity relationships for these substances [236].

The synthesis of cyclopeptides poses no particular fundamental problems. The general principles governing macrocyclization, which have been extensively covered in previous chapters, can be applied in this case. The methodology (Fig. 3-195) consists of constructing the linear chain step by step to obtain the linear precursor. At this point the acid moiety is activated using classical reagents for peptide synthesis whilst keeping the amine moiety protected (by protonation) after removal of Y. The cyclization reaction takes place on slow addition of the activated product to a basic medium (e.g., pyridine).

A large part of a recent work is devoted to cyclopeptide formation [237].

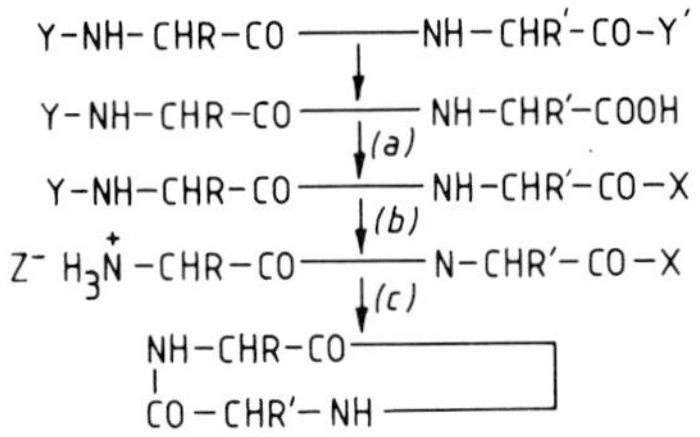

Fig. 3-195. Scheme for cyclopeptide synthesis: Y, Y', protecting groups; X, activating group. (a) activation of carboxyl group, (b) acid deprotection of the N-terminal amine, (c) cyclization in basic medium (formation of the free amine group).

3.8 Macrocarbocycles

Amongst compounds containing only carbon atoms in the macrocyclic framework, we mention first the *annulenes* [238a]. An important method of synthesis is the oxidative coupling of diacetylenes using copper acetate. Thus, the coupling of 1,5-hexadiyne (Fig. 3-196a) gave the cyclic trimer *1* which on rearrangement and partial hydrogenation afforded *3*. This method was used to obtain [14]-, [16]-, [18]-, [22]- and [24]annulene. Annulenes containing an annelated benzene ring [238 b] have also been synthesized using the same procedure, as have annulenones [238 c] (Fig. 3-196 b).

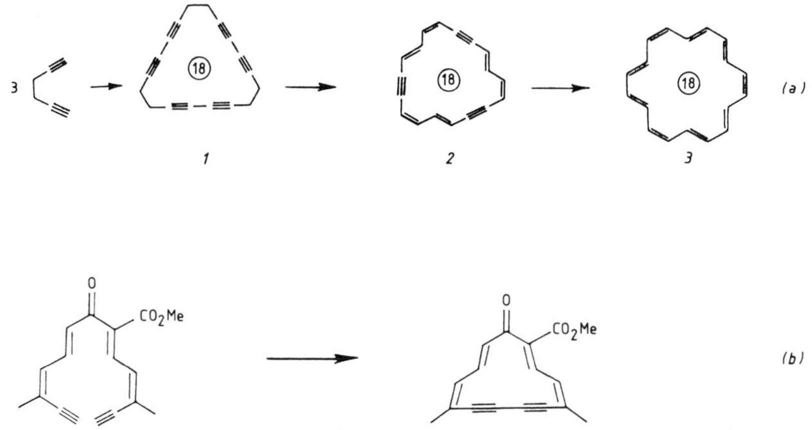

Fig. 3-196. Formation of macrocycles by oxidative coupling of diacetylenes; (a) [18]annulene, (b) bisdehydro[13]annulenone.

In a photochemical route (Fig. 3-197), the starting material was cyclooctatetraene which on irradiation afforded [16]annulene in 60% yield [239].

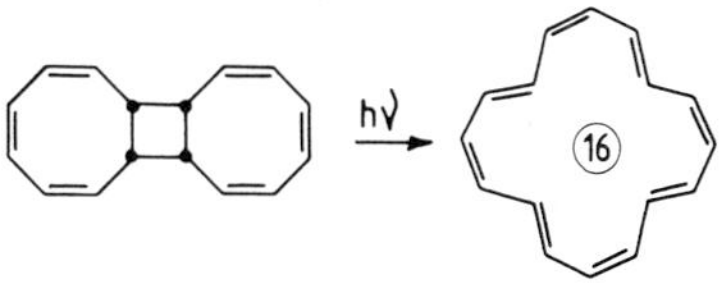

Fig. 3-197. Annulene synthesis using a photochemical route.

A recently discovered and increasingly important group of natural substances is that of the *cembranoids*. The framework in these compounds is based on the diterpene, cembrene, containing a 14-membered macromolecular carbocycle (Fig. 3-198).

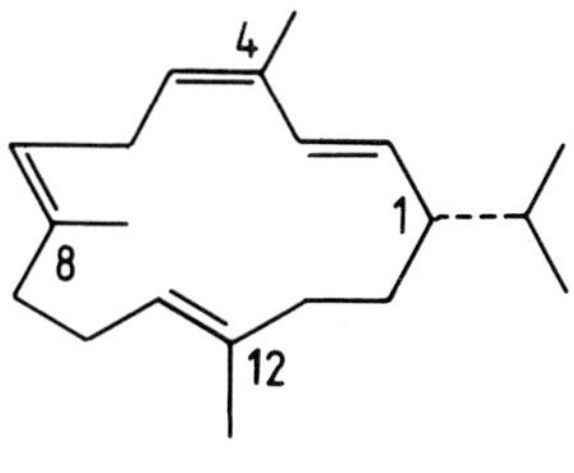

Fig. 3-198. Structure of cembrene.

Cembranoids are found in numerous terrestrial plants and in various marine organisms (sponges and coral) [240]. They are chiefly interesting for the biological activity which a large number of them possess [241]. Macrocyclization through formation of a carbon-carbon bond, which was mentioned above as a possibility in macrolide synthesis, becomes crucially important in this case. Great progress has been achieved in recent years. One approach involves intramolecular alkylation of carbanions. This is the method most often employed and we shall illustrate it with some examples (Fig. 3-199).

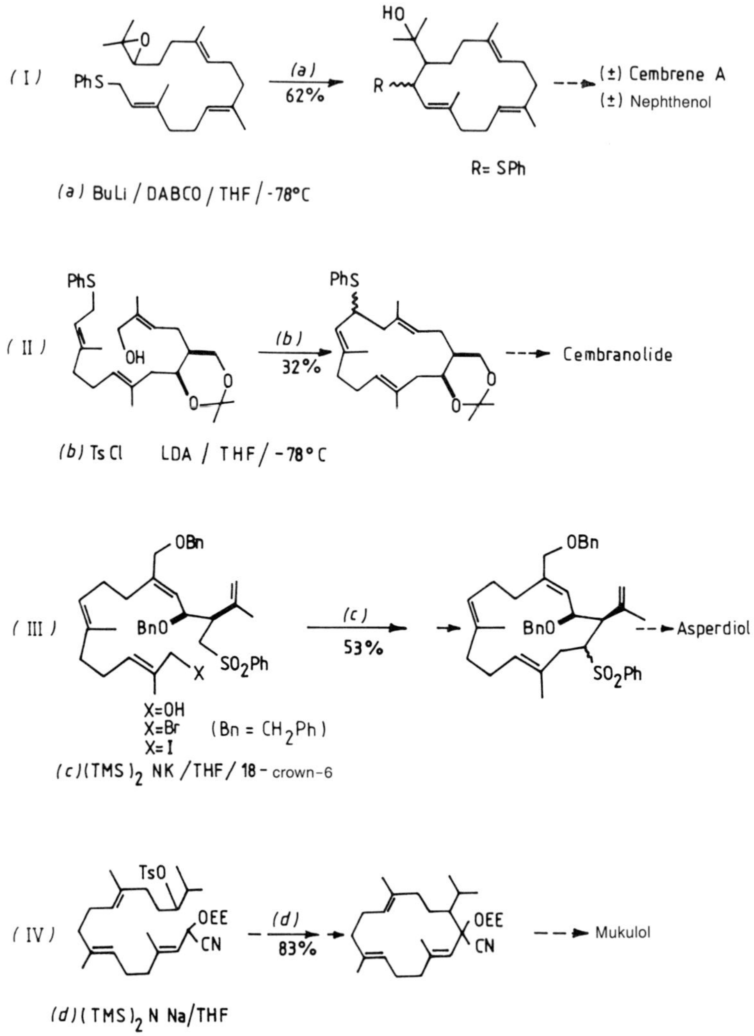

Fig. 3-199. Cembranoid synthesis via carbon-carbon bond formation.

It should be noted that carbanions formed α to sulfur are extensively used: I [242], II [243], III [244]. Procedure IV, in which a different type of carbanion is generated, is particularly efficient [245].

In very many coupling methods the aldehyde group is used for the formation of the carbon-carbon bond. One example is cyclization using the Horner-Wadsworth-Emmons reaction under the conditions described by Masamune and Roush (LiCl, DBU, CH_3CN) [246, 247] (Fig. 3-200).

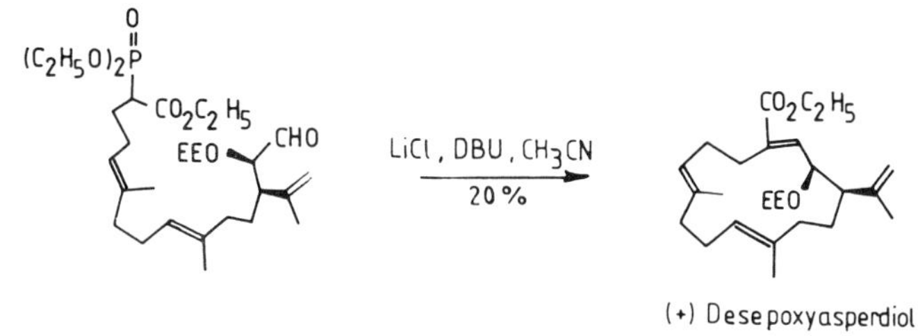

Fig. 3-200. Macrocyclization using the Horner-Wadsworth-Emmons method (DBU; 1,8-diazabicyclo[5.4.0]undec-7-ene).

Another very elegant method led to cembranolides via reaction of α-alkoxyallylstannanes with aldehydes [248]. This involved the adaptation for macrocyclization of a procedure developed for acyclic systems, the reaction scheme for which is shown [249] (Fig. 3-201). The yield of cyclic product was remarkably high.

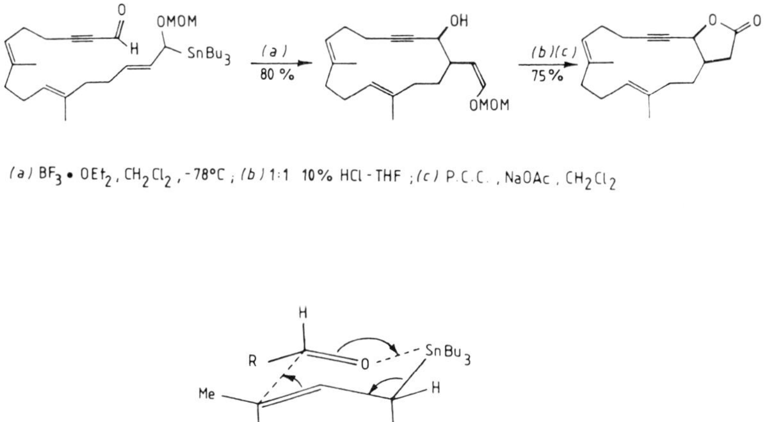

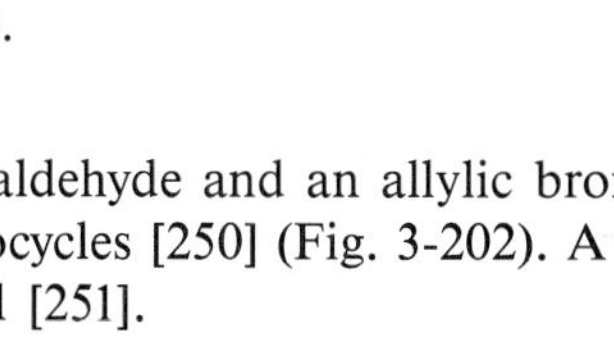

Fig. 3-201. Synthesis of a cembranolide from an α-alkoxyallylstannane and an aldehyde (PCC; pyridinium chlorochromate).

Coupling between an aldehyde and an allylic bromide in the presence of Cr(II) is another route to macrocycles [250] (Fig. 3-202). A possible mechanism for this reaction has been proposed [251].

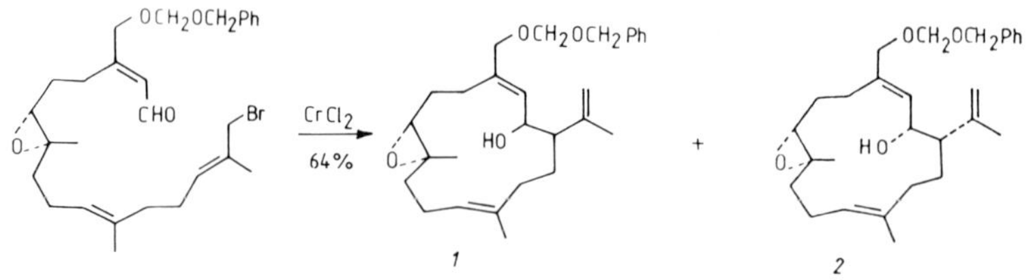

Fig. 3-202. Cyclization via coupling between an aldehyde and an allylic bromide in the presence of Cr(II). The product distribution was *1*:*2* = 4:1, *1* being the desired product (asperdiol).

Many other methods for cembranoid synthesis have been described involving the following:

- Coupling between an acyl chloride and a double bond in the presence of $SnCl_4$ [252].
- Macroexpansion methods [253].
- Ring contraction reactions [254].

Other types of macrocycles have been obtained, e.g., by coupling a linear precursor possessing a carbonyl at each end [255] (Fig. 3-203), or by intramolecular Wittig reaction [256] (Fig. 3-204).

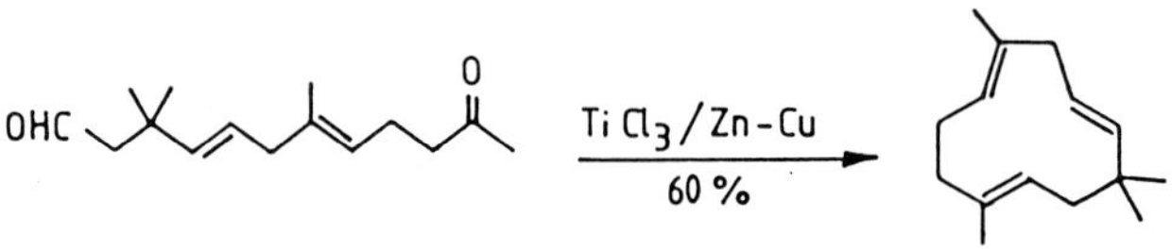

Fig. 3-203. Macrocycle formation by coupling of a dicarbonyl compound.

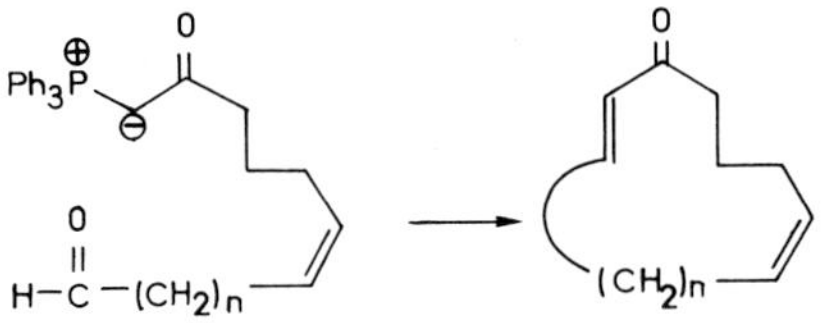

Fig. 3-204. Macrocycle formation by intramolecular Wittig reaction.

References for Part I

1. K. Ziegler, *Methoden der Organischen Chemie* (Houben Weyl), Georg Thieme Verlag, Stuttgart, 4/2, 1955, p. 729.
2. a) E. L. Eliel. *Stereochemistry of Carbon Compounds,* McGraw-Hill, New York, 1962, p. 188. b) A. Greenberg, J. F. Liebman. *Strained Organic Molecules,* Academic Press, New York 1978; J. F. Liebman, A. Greenberg, *Chem. Rev.,* **76,** 311, 1976.
3. L. Ruzicka, W. Brugger, M. Pfeiffer, H. Schinz, M. Stoll, *Helv. Chim. Acta,* **9,** 499, 1926; L. Ruzicka, M. Stoll, H.W. Huyser, H.A. Boekenoogen, *ibid,* **13,** 1152, 1930.
4. K. Ziegler, R. Aurnhammer, *Ann.,* **513,** 43, 1934; K. Ziegler, W. Hechelhammer, *Ann.,* **528,** 114, 1937.
5. G. Illuminati, L. Mandolini, *Acc. Chem. Res.,* **14,** 95, 1981; L. Mandolini, *Adv. Phys. Org. Chem.* **22,** 1, 1986.
6. M. Stoll, A. Rouvé, G. Stoll-Comte, *Helv. Chim. Acta,* **17,** 1289, 1934; M. Stoll, A. Rouvé, *ibid.,* **18,** 1087, 1935.
7. A. J. Kirby, *Advances in Physical Organic Chemistry,* **17,** 183, 1980, Academic Press.
8. C. Galli, L. Mandolini, *J. Chem. Soc., Chem. Comm.,* 251, 1982.
9. C. Galli, G. Illuminati, L. Mandolini, P. Tamborra, *J. Am. Chem. Soc.,* **99,** 2591, 1977.
10. J. Dale, *Tetrahedron,* **30,** 1683, 1974; *Israel J. Chem.,* **20,** 3, 1980.
11. K. Ziegler, H. Holl, *Ann.,* **528,** 143, 1937.
12. H. Stetter, J. Marx, *Ann.,* **607,** 59, 1957.
13 H.E. Simmons, C.H. Park, *J. Am. Chem. Soc.,* **90,** 2428, 1968; C.H. Park, H.E. Simmons, *Ibid,* 2429, 2431, 1968.
14. B. Dietrich, J.M. Lehn, J.P. Sauvage, *Tetrahedron Letters,* 2885, 1969.
15. B. Dietrich, J.M. Lehn, J.P. Sauvage, J. Blanzat, *Tetrahedron,* **29,** 1629, 1973.
16. W. Baker, J.F.W. McOmie, W.D. Ollis, *J. Chem. Soc.,* 200, 1951.
17. G. Illuminati, L. Mandolini, B. Masci, *J. Am. Chem. Soc.,* **97,** 4960, 1975.
18. G. Illuminati, L. Mandolini, B. Masci, *J. Am. Chem. Soc.,* **96,** 1422, 1974.
19. a) L. Mandolini, B. Masci, S. Roelens, *J. Org. Chem.,* **42,** 3733, 1977.
    b) P. Deslongchamps, *Aldrichimica Acta,* **17,** 59, 1984.
20. a) G. Borgen, J. Dale, *Acta Chem. Scand.,* **26,** 952, 1972. b) C. Galli, G. Giovannelli, G. Illuminati, L. Mandolini, *J. Org. Chem.,* **44,** 1258, 1979.
21. a) N. Kawamura, M. Miki, I. Ikeda, M. Okahara, *Tetrahedron Letters,* 535, 1979;
    b) P.J. Hammond, P.D. Beer, C.D. Hall, *J. Chem. Soc., Chem. Comm.,* 1161, 1983.
22. M.A. Winnik, *Acc. Chem. Res.,* **10,** 173, 1977.
23. M.A. Winnik, *Chem. Rev.,* **81,** 491, 1981.
24. J.F. Freimanis, J.J. Dregeris, L.E. Berzina, I.V. Turovsky, S.J. Bitte, L.M. Blynov, V.G. Rumyancev, *Adv. Mol. Relaxation Processes,* **5,** 33, 1973.
25. H.A.H. Craenen, J.W. Verhoeven, Th.J. de Boer, *Tetrahedron,* **27,** 2561, 1971.
26. F. Hirayama, *J. Chem. Phys.,* **42,** 3163, 1965.
27. a) K. Zachariasse, W. Kühnle, *Z. Phys. Chem. Neue Folge,* **101,** 267, 1976; b) F.C. de Schryver, P. Collart, J. Vandendriessche, R. Goedeweek, A. Swinnen, M. van der Auweraer, *Acc. Chem. Res.,* **20,** 159, 1987.
28. A.M. Halpern, M.W. Legenza, B.R. Ramachandran, *J. Am. Chem. Soc.,* **101,** 5736, 1979.
29. J.H. Borkent, A.W.J. De Jong, J.W. Verhoeven, Th.J. De Boer, *Chem. Phys. Letters,* **57,** 530, 1978.

30. K. Shimada, M. Szwarc, *J. Am. Chem. Soc.,* **97,** 3313, 3321, 1975; Y. Shimozato, K. Shimada, M. Szwarc, *ibid.,* **97,** 5831, 5834, 1975; M. Sisido, K. Shimada, *ibid.,* **99,** 7785, 1977.
31. M.A. Winnik, C.K. Hsiao, *Chem. Phys. Letters,* **33,** 518, 1975.
32. A. Mar, M.A. Winnik, *Chem. Phys. Letters,* **77,** 73, 1981.
33. A. Mar, S. Fraser, M.A. Winnik, *J. Am. Chem. Soc.,* **103,** 4941, 1981.
34. P.J. Wagner, *Acc. Chem. Res.,* **16,** 461, 1983.
35. G.J. Kavarnos, N.J. Turro, *Chem. Rev.,* **86,** 401, 1986.
36. M.A. Winnik, *Acc. Chem. Res.,* **18,** 73, 1985; J. A. Semlyen, *Cyclic Polymers,* Elsevier Applied Science Publ. London, 1986.
37. a) C.R. Eck, D.J. Hunter, T. Money, *J. Chem. Soc., Chem. Comm.,* 865, 1974; b) M.A. Winnik, D.S. Saunders, *ibid.,* 156, 1976.
38. G.K. Eigendorf, C.L. Ma, T. Money, *J. Chem. Soc., Chem. Comm.,* 561, 1976.
39. R. Breslow, *Chem. Soc. Rev.,* **1,** 553, 1972.
40. R. Breslow, M.A. Winnik, *J. Am. Chem. Soc.,* **91,** 3083, 1969; R. Breslow, J. Rothbard, F. Herman, M.L. Rodriquez, *J. Am. Chem. Soc.,* **100,** 1213, 1978.
41. R. Breslow, P. Kalicky, *J. Am. Chem. Soc.,* **93,** 3540, 1971.
42. R. Breslow, S. Baldwin, T. Flechtner, P. Kalicky, S. Liu, W. Washburn, *J. Am. Chem. Soc.,* **95,** 3251, 1973.
43. R. Breslow, *Acc. Chem. Res.,* **13,** 170, 1980.
44. R. Breslow, S.W. Baldwin, *J. Am. Chem. Soc.,* **92,** 732, 1970.
45. G. A. Russell, *J. Am. Chem. Soc.,* **80,** 4987, 1958; D.D. Tanner, P.B. Van Bostelen, *J. Org. Chem.,* **32,** 1517, 1967.
46. R. Breslow, R. J. Corcoran, J.A. Dale, S. Liu, P. Kalicky, *J. Am. Chem. Soc.,* **96,** 1973, 1974.
47. R. Breslow, R. J. Corcoran, B. B. Snider, R. J. Doll, P. L. Khanna, R. Kaleya, *J. Am. Chem. Soc.,* **99,** 905, 1977.
48. R. Breslow, R. J. Corcoran, B.B. Snider, *J. Am. Chem. Soc.,* **96,** 6791, 1974.
49. R. Breslow, B. B. Snider, R. J. Corcoran, *J. Am. Chem. Soc.,* **96,** 6792, 1974.
50. R. Breslow, R. L. Wife, D. Prezant, *Tetrahedron Letters,* 1925, 1976.
51. M. F. Czarniecki, R. Breslow, *J. Am. Chem. Soc.,* **101,** 3675, 1979.
52. R. Breslow, R. Rajagopalan, J. Schwarz, *J. Am. Chem. Soc.,* **103,** 2905, 1981.
53. P. Ruggli, *Ann.,* **392,** 92, 1912; **399,** 174, 1913; **412,** 1, 1917.
54. G. Salomon, *Helv. Chim. Acta,* **17,** 851, 1934; **19,** 743, 1936.
55. a) L. Rossa, F. Vögtle, *Topics in Current Chemistry,* **113,** 1, 1983; b) P. Knops, N. Sendhoff, H.-B. Mekelburger, F. Vögtle, *ibid.* **161,** 1, 1991.
56. V. Prelog, L. Frenkiel, M. Kobelt, P. Barman, *Helv. Chim. Acta,* **30,** 1741, 1947.
57. M. Stoll, A. Rouvé, *Helv. Chim. Acta,* **30,** 1822, 1947.
58. A. Patchornik, M. A. Kraus, *Pure Appl. Chem.,* **43,** 503, 1975.
59. J. I. Crowley, H. Rapoport, *Acc. Chem. Res.,* **9,** 135, 1976.
60. a) C. C. Leznoff, *Acc. Chem. Res.,* **11,** 327, 1978; *Chem. Soc. Rev.,* **3,** 65, 1974; b) C. Birr, *Aspects of the Merrifield Peptide Synthesis,* Springer-Verlag, Berlin 1978, p. 3.
61. J. W. Hill, W. H. Carothers, *J. Am. Chem. Soc.,* **55,** 5023, 5031; 5039; 1933; W. H. Carothers, J. W. Hill, *J. Am. Chem. Soc.,* **55,** 5043, 1933.
62. E. W. Spanagel, W. H. Carothers, *J. Am. Chem. Soc.,* **57,** 929, 1935.
63. U. Kramer, A. Guggisberg, M. Hesse, H. Schmidt, *Angew. Chem. Int. Ed. Eng.,* **16,** 861, 1977; A. Guggisberg, B. Dabrowski, U. Kramer, C. Heidelberger, M. Hesse, H. Schmid, *Helv. Chim. Acta,* **61,** 1039, 1978.

64. U. Kramer, H. Schmid, A. Guggisberg, M. Hesse, *Helv. Chim. Acta,* **62,** 811, 1979.
65. a) U. Kramer, A. Guggisberg, M. Hesse, *Helv. Chim. Acta,* **62,** 2317, 1979; b) U. Kramer, A. Guggisberg, M. Hesse, H. Schmid, *Helv. Chim. Acta,* **61,** 1342, 1978.
66. C. Jenny, M. Hesse, *Helv. Chim. Acta,* **64,** 1807, 1981.
67. N. J. Leonard, S. Swann Jr., J. Figueras Jr., *J. Am. Chem. Soc.,* **74,** 4620, 1952; N. J. Leonard, S. Swann Jr., E. H. Mottus, *ibid,* 6251.
68. H. Stetter, H. Spangenberger, *Ber.* **91,** 1982, 1958.
69. a) V. Bhat, R. C. Cookson, *J. Chem. Soc., Chem. Comm.,* 1123, 1981; b) R. C. Cookson, P. S. Ray, *Tetrahedron Letters,* **23,** 3521, 1982.
70. J. R. Mahajan, H. C. Araujo, *Synthesis,* 54, 1975 and 111, 1976.
71. J. R. Mahajan, G. A. L. Ferreira, H. C. Araujo, B. J. Nunes, *Synthesis,* 112, 1976.
72. Y. Ito, T. Sugaya, M. Nakatsuka, T. Saegusa, *J. Am. Chem. Soc.,* **99,** 8366, 1977.
73. K. Rühlmann, *Synthesis,* 236, 1971.
74. a) A. Shanzer, N. Shochet, D. Rabinovich, F. Frolow, *Angew. Chem. Int. Ed. Engl.,* **19,** 326, 1980; b) A. Shanzer, J. Libman, F. Frolow, *Acc. Chem. Res.,* **16,** 60, 1983.
75. A. Shanzer, N. Mayer-Shochet, *J. Chem. Soc., Chem. Comm.* 176, 1980.
76. a) A. Shanzer, N. Mayer-Shochet, F. Frolow, D. Rabinovich, *J. Org. Chem.,* **46,** 4662, 1981; b) J. C. Pommier, J. Valade, *J. Organometal. Chem.,* **12,** 433, 1968; P. J. Smith, R. F. M. White, L. Smith, *ibid.,* **40,** 341, 1972; R. H. Herber, A. Shanzer, J. Libman, *Organometallics,* **3,** 586,. 1984.
77. a) A. Shanzer, E. Berman, *J. Chem. Soc., Chem. Comm.,* 259, 1980; b) S. Lifson, C. E. Felder, A. Shanzer, J. Libman, *Synthesis of Macrocycles,* John Wiley and Sons, New York, 1987, p. 241.
78. A. Shanzer, J. Libman, F. Frolow, *J. Am. Chem. Soc.,* **103,** 7339, 1981.
79. a) A. Shanzer, J. Libman, *J. Chem. Soc., Chem. Comm.,* 846, 1983; b) S. David, S. Hanessian, *Tetrahedron,* **41,** 643, 1985.
80. E. Schwartz, A. Shanzer, *J. Chem. Soc., Chem. Comm.,* 634, 1981.
81. A. Shanzer, E. Schwartz, *Tetrahedron Letters,* 5019, 1979.
82. H. Yamamoto, K. Maruoka, *J. Am. Chem. Soc.,* **103,** 6133, 1981.
83. N. F. Curtis, *Coordin. Chem. Rev.,* **3,** 3, 1968.
84. a) E. G. Jäger, *Z. Chem.,* **4,** 437, 1964; *ibid.,* **8,** 30, 392, 470, 1968; *Z. Anorg. Allg. Chem.,* **364,** 177, 1969; b) M. C. Thompson, D. H. Busch, *J. Am. Chem. Soc.,* **86,** 3651, 1964; c) L. F. Lindoy, D. H. Busch, *Inorg. Chem.,* **13,** 2494, 1974; D. H. Busch, *Helv. Chim. Acta, Fasc. Extraord.* A. Werner, 174, 1967; *Coordination Chemistry of macrocyclic compounds.* Ed. G. A. Melson, Plenum Press, 1979.
85. F. A. L'Eplattenier, A. Pugin, *Helv. Chim. Acta,* **58,** 917, 1975.
86. a) S. M. Nelson, C. V. Knox, M. MacCann., M. G. B. Drew, *J. Chem. Soc., Dalton,* 1669, 1981; b) S. M. Nelson, *Inorg. Chim. Acta,* **62,** 39, 1982.
87. D. H. Cook, D. E. Fenton, M. G. B. Drew, S. G. McFall, S. M. Nelson, *J. Chem. Soc., Dalton,* 446, 1977; M. G. B. Drew, A. H. B. Othman, S. G. McFall, P. D. A. McIlroy, S. M. Nelson, *ibid.,* 1173, 1977.
88. M. G. B. Drew, M. McCann, S. M. Nelson, *J. Chem. Soc., Chem. Comm.,* 481, 1979.
89. M. G. B. Drew, M. McCann, S. M. Nelson, *J. Chem. Soc., Dalton,* 1868, 1981.
90. M. G. B. Drew, A. Rodgers, M. McCann, S. M. Nelson, *J. Chem. Soc., Chem. Comm.,* 415, 1978.
91. M. G. B. Drew, J. Nelson, S. M. Nelson, *J. Chem. Soc., Dalton,* 1678, 1981.
92. M. G. B. Drew, J. Nelson, S. M. Nelson, *J. Chem. Soc., Dalton,* 1685, 1981.
93. a) D. St. C. Black, C. H. B. Vanderzalm, A. J. Hartshorn, *Inorg. Nucl. Chem. Letters,* **12,** 657, 1976; b) C. Skötsch, E. Breitmaier, Synthesis, 680, 1978.

94. W. P. Schammel, K. S. B. Mertes, G. G. Christoph, D. H. Busch, *J. Am. Chem. Soc.,* **101,** 1622, 1979; W. P. Schammel, L. L. Zimmer, D. H. Busch, *Inorg. Chem.,* **19,** 3159, 1980.
95. a) J. C. Stevens, P. J. Jackson, W. P. Schammel, G. G. Christoph, D. H. Busch, *J. Am. Chem. Soc.,* **102,** 3283, 1980; b) J. C. Stevens, D. H. Busch, *J. Am. Chem. Soc.,* **102,** 3285, 1980.
96. D. H. Busch, *Pure and Appl. Chem.,* **52,** 2477, 1980; D. H. Busch, C. Cairns, *Synthesis of Macrocycles, Progress in Macrocyclic Chemistry,* vol. 3, J. Wiley and Sons, New York, 1987, p. 1.
97 G. Schill, *Catenanes, Rotaxanes and Knots,* Academic Press, New York, 1971.
98. C. O. Dietrich-Buchecker, J. P. Sauvage, J. P. Kintzinger, *Tetrahedron Letters,* **24,** 5095, 1983; J. P. Sauvage, *Nouv. J. Chimie,* **9,** 299, 1985; C. O. Dietrich-Buchecker, J. P. Sauvage, *Chem. Rev.,* **87,** 795, 1987.
99. C. O. Dietrich-Buchecker, P. A. Marnot, J. P. Sauvage, *Tetrahedron Letters,* **23,** 5291, 1982.
100. a) M. Cesario, C. O. Dietrich-Buchecker, J. Guilhem, C. Pascard, J. P. Sauvage, *J. Chem. Soc., Chem. Comm.,* 244, 1985; b) J.-P. Sauvage, *Acc. Chem. Res.,* **23,** 319, 1990; C. O. Dietrich-Buchecker, J.-P. Sauvage, *Bioorganic Chemistry Frontiers,* vol. 2, Springer-Verlag, Berlin, 1991, p. 195.
101. N. Calderon, *Acc. Chem. Res.,* **5,** 127, 1972.
102. R. Wolovsky, *J. Am. Chem. Soc.,* **92,** 2132, 1970; D. A. Ben-Efraim, C. Batich, E. Wasserman, *ibid,* 2133.
103. H. Fischer, K. Zeile, *Ann. Chem.,* **468,** 98, 1929.
104. A. Treibs, N. Häberle, *Ann. Chem.,* **718,** 183, 1968.
105. A. D. Adler, F. R. Longo, J. D. Finarelli, J. Goldmacher, J. Assour, L. Korsakoff, *J. Org. Chem.,* **32,** 476, 1967.
106. F. R. Longo, E. J. Thorney, A. D. Adler, S. Dym, *J. Heterocycl. Chem.,* 1305, 1975.
107. H. J. Callot, A. Louati, M. Gross, *Angew. Chem. Int. Ed. Engl.,* **21,** 285, 1982; *Bull. Soc. Chim. Fr.* II, 317, 1983.
108. D. O. Cheng, E. LeGoff, *Tetrahedron Letters,* 1469, 1977.
109. M. J. Broadhurst, R. Grigg, A. W. Johnson, *J. Chem. Soc., C.* 3681, 1971.
110. A. Ulman, J. Manassen, *J. Am. Chem. Soc.,* **97,** 6540, 1975.
111. M. J. Broadhurst, R. Grigg, A. W. Johnson, *J. Chem. Soc., Chem. Comm.,* 1480, 1969; 807, 1970; 23, 1969.
112. M. J. Broadhurst, R. Grigg, A. W. Johnson, *J. Chem. Soc., Perkin I,* 1124, 1972.
113. G. R. Newkome, J. D. Sauer, J. M. Roper, D. C. Hager, *Chem. Rev.,* **77,** 513, 1977.
114. A. Baeyer, *Ber.,* **19,** 2184, 1886.
115. W. H. Brown, B. J. Hutchinson, M. H. MacKinnon, *Canad. J. Chem.,* **49,** 4017, 1971.
116. a) J. C. Lalloz, J. M. Lehn; b) J. C. Lalloz, J. M. Lehn, A. K. Willard; c) M. Cesario, C. Pascard. Unpublished results.
117. W. H. Brown, W. N. French, *Canad. J. Chem.,* **36,** 537, 1958.
118. M. Chastrette, F. Chastrette, *J. Chem. Soc., Chem. Comm.,* 534, 1973.
119. A. J. Rest, S. A. Smith, I. D. Tyler, *Inorg. Chim. Acta,* **16,** L 1, 1976.
120. M. de Sousa Healy, A. J. Rest, *J. Chem. Soc., Chem. Comm.,* 149, 1981; *J. Chem. Soc., Perkin,* I, 973, 1985.
121. M. Ahmed, O. Meth-Cohn, *Tetrahedron Letters,* 1493, 1969.
122. M. Ahmed, O. Meth-Cohn, *J. Chem. Soc.,* **C,** 2104, 1971.
123. T. Kauffmann, H. H. Kniese, *Tetrahedron Letters,* 4043, 1973.
124. W. H. Brown, W. N. French, *Canad. J. Chem.,* **36,** 371, 1958.
125. A. G. S. Högberg, *J. Am. Chem. Soc.,* **102,** 6046, 1980.

126. A. Zinke, R. Kretz, E. Leggewie, K. Hössinger, *Monatsh. Chem.,* **83,** 1213, 1952; J. W. Cornforth, P. D'Arcy Hart, G. A. Nicholls, R. J. W. Rees, J. A. Stock, *Brit. J. Pharmacol.,* **10,** 73, 1955; B. T. Hayes, R. F. Hunter, *Chem. Ind.,* 193, 1956.
127. a) C. D. Gutsche, B. Dhawan, K. H. No, R. Muthukrishnan, *J. Am. Chem. Soc.,* **103,** 3782, 1981; C. D. Gutsche, *Acc. Chem. Res.,* **16,** 161, 1983; *Topics in Current Chemistry,* **123,** 1, 1984; b) C. D. Gutsche, *Calixarenes,* Ed. J. F. Stoddart, Royal Society of Chemistry 1989;
c) M.-J. Schwing, F. Arnaud, E. Marques, *Pure and Appl. Chem.* **61,** 1597, 1989.
128. a) G. D. Andreetti, R. Ungaro, A. Pochini, *J. Chem. Soc., Chem. Comm.,* 533, 1981; b) G. D. Andreetti, R. Ungaro, A. Pochini, *J. Chem. Soc., Chem. Comm.,* 1005, 1979; C. Rizzoli, G. D. Andreetti, R. Ungaro, A. Pochini, *J. Mol. Struct.,* **82,** 133, 1982.
129. a) G. M. Robinson, *J. Chem. Soc.,* **107,** 267, 1915; b) H. Erdtman, F. Haglid, R. Ryhage, *Acta Chem. Scand.,* **18,** 1249, 1964.
130. a) A. S. Lindsey, *J. Chem. Soc.,* 1685, 1965; b) J. D. White, B. D. Gesner, *Tetrahedron Letters,* 1591, 1968; c) A. Collet, J. Jacques, *ibid,* 1265, 1978.
131. a) A. Collet, *Inclusion Compounds,* vol. 2, Eds. J. L. Atwood, J. E. D. Davies, D. D. MacNicol, Academic Press, 1984, p. 97. b) J. Gabard, A. Collet, *J. Chem. Soc., Chem. Comm.,* 1137, 1981.
132. J. Canceill, M. Cesario, A. Collet, J. Guilhem, C. Pascard, *J. Chem. Soc., Chem. Comm.,* 361, 1985; J. Canceill, M. Cesario, A. Collet, J. Guilhem, C. Riche, C. Pascard, *J. Chem. Soc., Chem. Comm.,* 339, 1986; J. Canceill, L. Lacombe, A. Collet, *J. Am. Chem. Soc.,* **108,** 4230, 1986.
133. D. J. Cram, *Science,* **219,** 1177, 1983.
134. J. R. Moran, S. Karbach, D. J. Cram, *J. Am. Chem. Soc.,* **104,** 5826, 1982.
135. R. C. Helgeson, M. Lauer, D. J. Cram, *J. Chem. Soc., Chem. Comm.,* 101, 1983.
136. D. J. Cram, T. Kaneda, R. C. Helgeson, G. M. Lein, *J. Am. Chem. Soc.,* **101,** 6752, 1979; K. N. Trueblood, C. B. Knobler, E. Maverick, R. C. Helgeson, S. B. Brown, D. J. Cram, *J. Am. Chem. Soc.,* **103,** 5594, 1981; D. J. Cram, K. N. Trueblood, *Topics in Current Chemistry,* **98,** 43, 1981.
137. D. J. Cram, I. B. Dicker, M. Lauer, C. B. Knobler, K. N. Trueblood, *J. Am. Chem. Soc.,* **106,** 7150, 1984; K. D. Stewart, M. Miesch, C. B. Knobler, E. F. Maverick, D. J. Cram, *J. Org. Chem.,* **51,** 4327, 1986; S. P. Artz, D. J. Cram, *J. Am. Chem. Soc.,* **106,** 2160, 1984; D. J. Cram, M. de Grandpre, C. B. Knobler, K. N. Trueblood, *J. Am. Chem. Soc.,* **106,** 3286, 1984; D. N. Reinhoudt, P. J. Dijkstra, *Pure and Appl. Chem.,* **60,** 477, 1988.
138. a) C. J. Pedersen, *J. Am. Chem. Soc.,* **89,** 2495, 7017, 1967; b) H. E. Schroeder (and C. J. Pedersen), *Pure and Appl. Chem.,* **60,** 445, 1988.
139. J. S. Bradshaw, P. E. Stott, *Tetrahedron,* **36,** 461, 1980; J. J. Christensen, D. J. Eatough, R. M. Izatt, *Chem. Rev.,* **74,** 351, 1974; R. M. Izatt, J. J. Christensen, (Eds.). *Synthetic Multidentate Macrocyclic Compounds,* Academic Press, New York, 1978; G. W. Gokel, S. H. Korzeniowski. *Macrocyclic Polyether Synthesis,* Springer Verlag, Berlin, 1982.
140. D. H. Busch, *Acc. Chem. Res.,* **11,** 392, 1978; R. D. Hancock, *Acc. Chem. Res.,* **23,** 253, 1990.
141. B. Dietrich, *Inclusion Compounds,* vol. 2, Eds. J. L. Atwood, J. E. D. Davies, D. D. MacNicol, Academic Press, 1984, p. 337.; E. Kimura, *Topics in Current Chemistry,* **128,** 113, 1985; F. Vögtle, H. Sieger, W. M. Müller. *Host Guest Complex Chemistry Macrocycles,* Eds. F. Vögtle, E. Weber, Springer Verlag, Berlin 1985, p. 319; J. L. Pierre, P. Baret, *Bull. Soc. Chim. Fr.,* 367, 1983; A. Bianchi, M. Micheloni, P. Paoletti, *Pure and Appl. Chem.,* **60,** 525, 1988.
142. F. Vögtle, E. Weber, *Kontakte Merck,* 11, 1977.

143. R. H. Mitchell, V. Boekelheide, *J. Am. Chem. Soc.,* **96,** 1547, 1974; V. Boekelheide, P. H. Anderson, T. A. Hylton, *ibid,* 1558; S. A. Sherrod, R. L. da Costa, R. A. Barnes, V. Boekelheide, *ibid,* 1565; V. Boekelheide, *Topics in Current Chemistry.* **113,** 87, 1983; E. Block, *Aldrichimica Acta,* **11,** 51, 1978.
144. H. Christol, H. J. Cristau, F. Fallouh, P. Hullot, *Tetrahedron Letters,* 2591, 1979; H. J. Cristau, L. Chiche, F. Fallouh, P. Hullot, G. Renard, H. Christol, *Nouv. J. Chimie,* **8,** 191, 1984; M. Ciampolini, N. Nardi, P. L. Orioli, S. Mangani, F. Zanobini, *J. Chem. Soc., Dalton,* 1425, 1985.
145. J. Ennen, T. Kauffmann, *Angew. Chem. Int. Ed. Engl.,* **20,** 118, 1981; *Tetrahedron Letters,* 5035, 1981.
146. M. Newcomb, M. T. Blanda, Y. Azuma, T. J. Delord, *J. Chem. Soc., Chem. Comm.,* 1159, 1984.
147 a) G. W. Gokel, D. M. Dishong, C. J. Diamond, *J. Chem. Soc., Chem. Comm.,* 1053, 1980; G. W. Gokel, K. A. Arnold, M. Delgado, L. Echeverria, V. J. Gatto, D. A. Gustowski, J. Hernandez, A. Kaifer, S. R. Miller, L. Echegoyen, *Pure and Appl. Chem.* **60,** 461, 1988; b) K. Koga, S. Sasaki, *Pure and Appl. Chem.* **60,** 539, 1988; T. A. Kaden, D. Tschudin, M. Studer, U. Brunner, *ibid.* **61,** 879, 1989; c) D. Parker, J. R. Morphy, K. Jankowski, J. Cox, *Pure and Appl. Chem.,* **61,** 1637, 1989; D. Parker, *Chem. Soc. Rev.* **19,** 271, 1990.
148. J. M. Timko, D. J. Cram, *J. Am. Chem. Soc.,* **96,** 7159, 1974; J. M. Timko, R. C. Helgeson, M. Newcomb, G. W. Gokel, D. J. Cram, *J. Am. Chem. Soc.,* **96,** 7097, 1974.
149. G. R. Newkome, A. Nayak, M. G. Sorci, W. H. Benton, *J. Org. Chem.,* **44,** 3812, 1979.
150. V. K. Majestic, G. R. Newkome, *Topics in Current Chemistry,* **106,** 79, 1982.
151. F. Wudl, F. Gaeta, *J. Chem. Soc., Chem. Comm.,* 107, 1972.
152. J. A. Haslegrave, J. F. Stoddart, D. J. Thompson, *Tetrahedron Letters,* 2279, 1979; J. F. Stoddart, *Chem. Soc. Rev.,* **8,** 85, 1979.
153. D. A. Laidler, J. F. Stoddart, J. B. Wolstenholme, *Tetrahedron Letters,* 465, 1979; J. F. Stoddart, *Bioenergetics and Thermodynamics: Model Systems,* A. Braibanti Ed. D. Reidel, 1980, p. 43.
154. a) W. Hain, R. Lehnert, H. Röttele, G. Schröder, *Tetrahedron Letters,* 625, 1978; L. Töke, L. Fenichel, P. Bako, J. Szejtli, *Acta Chim. Acad. Sci. Hung.* **98,** 357, 1978; b) P. Bako, L. Fenichel, L. Töke, M. Czugler, *Ann.,* 1163, 1981.
155. D. G. Andrews, P. R. Ashton, D. A. Laidler, J. F. Stoddart, J. B. Wolstenholme, *Tetrahedron Letters,* 2629, 1979.
156. J. M. Girodeau, J. M. Lehn, J. P. Sauvage, *Angew. Chem. Int. Ed. Engl.,* **14,** 764, 1975; J. P. Behr, J. M. Girodeau, R. C. Hayward, J. M. Lehn, J. P. Sauvage, *Helv. Chim. Acta,* **63,** 2096, 1980.
157. H. O. Kalinowski, D. Seebach, G. Crass, *Angew. Chem. Int. Ed. Engl.,* **14,** 762, 1975.
158. L. A. Frederick, T. M. Fyles, N. P. Gurprasad, D. M. Whitfield, *Canad. J. Chem.,* **59,** 1724, 1981; L. M. Dulyea, T. M. Fyles, D. W. Whitfield, *Canad. J. Chem.,* **62,** 498, 1984; T. M. Fyles, D. M. Whitfield, *Canad. J. Chem.,* **62,** 507, 1984.
159. a) J. Dale, G. Borgen, K. Daasvatn, *Acta Chem. Scand.,* **B28,** 378, 1974; b) J. Dale, K. Daasvatn, *Acta Chem. Scand.,* **B34,** 327, 1980; c) J. Dale, K. Daasvatn, *J. Chem. Soc., Chem. Comm.,* 295, 1976.
160. L. Mandolini, B. Masci, *J. Am. Chem. Soc.,* **99,** 7709, 1977.
161. G. Ercolani, L. Mandolini, B. Masci, *J. Am. Chem. Soc.,* **103,** 2780, 1981.
162. B. R. Bowsher, A. J. Rest, *Inorg. Chim. Acta,* **45,** L 5, 1980.
163. B. R. Bowsher, A. J. Rest, *Inorg. Chim. Acta,* **53,** L 175, 1981.
164. J. P. Desvergne, H. Bouas-Laurent, *J. Chem. Soc., Chem. Comm.,* 403, 1978.

165. a) A. Samat, J. Elguero, J. Metzger, *J. Chem. Soc., Chem. Comm.,* 1182, 1979; b) A. Samat, M.E.M. Bibout, M. Chanon, *Nouv. J. Chimie,* **6,** 483, 1982; c) A. Samat, M.E.M. Bibout, J. Elguero, *J. Chem. Soc., Perkin I,* 1717, 1985.
166. W.H. Kruizinga, R.M. Kellogg, *J. Chem. Soc., Chem. Comm.,* 286, 1979; *J. Am. Chem. Soc.,* **103,** 5183, 1981.
167. B. Klieser, L. Rossa, F. Vögtle, *Kontakte Merck, 3,* 1984; A. Ostrowicki, E. Koepp, F. Vögtle, *Topics in Current Chemistry,* **161,** 37, 1991.
168. T. Shono, O. Ishige, H. Uyama, S. Kashimura, *J. Org. Chem.,* **51,** 546, 1986.
169. a) T. Mukaiyama, *Angew. Chem., Intern. Ed. Engl.,* **15,** 94, 1976; b) T. Mukaiyama, R. Matsueda, H. Maruyama, *Bull. Chem. Soc. Japan,* **43,** 1271, 1970.
170. E.J. Corey, K.C. Nicolaou, *J. Am. Chem. Soc.,* **96,** 5614, 1974.
171. E.J. Corey, K.C. Nicolaou, L.S. Melvin Jr., *J. Am. Chem. Soc.,* **97,** 653, 1975.
172. E.J. Corey, K.C. Nicolaou, L.S. Melvin Jr., *J. Am. Chem. Soc.,* **97,** 654, 1975.
173. H. Gerlach, A. Thalmann, *Helv. Chim. Acta,* **57,** 2661, 1974.
174. H. Gerlach, H. Wetter, *Helv. Chim. Acta,* **57,** 2306, 1974.
175. H. Gerlach, K. Oertle, A. Thalmann, S. Servi, *Helv. Chim. Acta,* **58,** 2036, 1975.
176. J. Gombos, E. Haslinger, H. Zak, U. Schmidt, *Tetrahedron Letters,* 3391, 1975.
177. U. Schmidt, J. Gombos, E. Haslinger, H. Zak, *Chem. Ber.,* **109,** 2628, 1976.
178. J. Gombos, E. Haslinger, U. Schmidt, *Chem. Ber.,* **109,** 2645, 1976.
179. H. Gerlach, K. Oertle, A. Thalmann, *Helv. Chim. Acta,* **59,** 755, 1976.
180. E.J. Corey, D.J. Brunelle, *Tetrahedron Letters,* 3409, 1976.
181. S. Masamune, S. Kamata, W. Schilling, *J. Am. Chem. Soc.,* **97,** 3515, 1975.
182. S. Masamune, C.U. Kim, K.E. Wilson, G.O. Spessard, P.E. Georghiou, G.S. Bates, *J. Am. Chem. Soc.,* **97,** 3512, 1975; S. Masamune, H. Yamamoto, S. Kamata, A. Fukuzawa, *J. Am. Chem. Soc.,* **97,** 3513, 1975.
183. S. Masamune, Y. Hayase, W. Schilling, W.K. Chan, G.S. Bates, *J. Am. Chem. Soc.,* **99,** 6756, 1977.
184. T. Mukaiyama, M. Usui, K. Saigo, *Chem. Letters,* 49, 1976.
185. T. Mukaiyama, K. Narasaka, K. Kikuchi, *Chem. Letters,* 441, 1977.
186. K. Narasaka, K. Maruyama, T. Maukaiyma, *Chem. Letters,* 885, 1978.
187. T. Mukaiyama, *Angew. Chem. Int. Ed. Engl.,* **18,** 707, 1979.
188. K. Steliou, A. Szczygielska-Nowosielska, A. Favre, M.A. Poupart, S. Hanessian, *J. Am. Chem. Soc.,* **102,** 7578, 1980.
189. K. Steliou, M.A. Poupart, *J. Am. Chem. Soc.,* **105,** 7130, 1983.
190. U. Schmidt, M. Dietsche, *Angew. Chem. Int. Ed. Engl.,* **20,** 771, 1981.
191. T. Fujisawa, T. Mori, K. Fukumoto, T. Sato, *Chem. Letters,* 1891, 1982.
192. B. Neises, W. Steglich, *Angew. Chem., Intern. Ed. Engl.,* **17,** 522, 1978.
193. E.P. Boden, G.E. Keck, *J. Org. Chem.,* **50,** 2394, 1985.
194. T. Kurihara, Y. Nakajima, O. Mitsunobu, *Tetrahedron Letters,* 2455, 1976.
195. Y. Fukuyama, C.L. Kirkemo, J.D. White, *J. Am. Chem. Soc.,* **99,** 646, 1977.
196. N. Taniguchi, H. Kinoshita, K. Inomata, H. Kotake, *Chem. Letters,* 1347, 1984.
197. W.H. Rastetter, D.P. Phillion, *Tetrahedron Letters,* 1469, 1979.
198. W.H. Rastetter, D.P. Phillion, *J. Org. Chem.,* **46,** 3204, 1981.
199. W.H. Rastetter, D.P. Phillion, *J. Org. Chem.,* **46,** 3209, 1981.
200. E.J. Corey, D.J. Brunelle, K.C. Nicolaou, *J. Am. Chem. Soc.,* **99,** 7359, 1977.
201. M. Okada, H. Sumitomo, I. Tajima, *J. Am. Chem. Soc.,* **101,** 4013, 1979; M. Okada, H. Sumitomo, M. Atsumi, *J. Am. Chem. Soc.,* **106,** 2101, 1984.
202. a) S.L. Regen, *Angew. Chem. Int. Ed. Engl.,* **18,** 421, 1979; *Nouv. J. Chimie* **6,** 629, 1982; b) S.L. Regen, Y. Kimura, *J. Am. Chem. Soc.,* **104,** 2064, 1982.

203. a) R. J. Gallo, M. Makosza, H. J. M. Dou, P. Hassanaly, *Adv. Heterocycl. Chem.,* **36,** 175, 1984; b) Y. Kimura, S. L. Regen, *J. Org. Chem.* **48,** 1533, 1983.
204. S. Mohanraj, W. T. Ford, *J. Org. Chem.,* **50,** 1616, 1985.
205. B. M. Trost, T. R. Verhoeven, *J. Am. Chem. Soc.,* **102,** 4743, 1980.
206. E. J. Corey, E. K. W. Wat, *J. Am. Chem. Soc.,* **89,** 2757, 1967; E. J. Corey, P. Helquist, *Tetrahedron Letters,* 4091, 1975.
207. E. J. Corey, H. A. Kirst, *J. Am. Chem. Soc.,* **94,** 667, 1972.
208. D. C. Billington, *Chem. Soc. Rev.,* **14,** 93, 1985.
209. B. M. Trost, T. R. Verhoeven, *J. Am. Chem. Soc.,* **99,** 3867, 1977; B. M. Trost, *Pure and Appl. Chem.,* **51,** 787, 1979; B. M. Trost, *Tetrahedron,* **33,** 2615, 1977.
210. B. M. Trost, G. A. Molander, *J. Am. Chem. Soc.,* **103,** 5969, 1981; J. Tsuji, H. Kataoka, Y. Kobayashi, *Tetrahedron Letters,* 2575, 1981; B. M. Trost, R. W. Warner, *J. Am. Chem. Soc.,* **105,** 5940, 1983; B. M. Trost, *Aldrichimica Acta,* **14,** 43, 1981; B. M. Trost, *Angew. Chem. Int. Ed. Engl.,* **28,** 1173, 1989.
211. D. Bichan, M. Winnik, *Tetrahedron Letters,* 3857, 1974.
212. L. D. Bergelson, J. G. Molotkovsky, M. M. Shemyakin, *Chem. and Ind.,* 558, 1960.
213. J. Carnduff, G. Eglinton, W. McCrae, R. A. Raphael, *Chem. and Ind.,* 559, 1960.
214. K. C. Nicolaou, *Tetrahedron,* **33,** 683, 1977; S. Masamune, *Aldrichimica Acta,* **11,** 23, 1978.
215. T. G. Back, *Tetrahedron,* **33,** 3041, 1977; I. Paterson, M. M. Mansuri, *Tetrahedron* **41,** 3569, 1985; Q. Meng, M. Hesse, *Topics in Current Chemistry,* **161,** 108, 1991.
216. B. B. Jarvis, E. P. Mazzola, *Acc. Chem. Res.,* **15,** 388, 1982.
217. P. G. McDougal, N. R. Schmuff, *Progress in the Chemistry of Organic Natural Products,* **47,** 153, 1985.
218. B. H. Lipshutz, H. Kotsuki, W. Lew, *Tetrahedron Letters,* **27,** 4825, 1986; B. M. Trost, P. Metz, J. T. Hane, *Tetrahedron Letters,* **27,** 5691, 5695, 1986.
219. Y. Tor, J. Libman, F. Frolow, H. E. Gottlieb, R. Lazar, A. Shanzer, *J. Org. Chem.,* **50,** 5476, 1985.
220. M. M. Badawi, K. Bernauer, P. van den Broek, D. Gröger, A. Guggisberg, S. Johne, I. Kompis, F. Schneider, H. J. Veith, M. Hesse, H. Schmid, *Pure and Applied Chem.,* **33,** 81, 1973.
221. B. Ganem, *Acc. Chem. Res.,* **15,** 290, 1982.
222. F. Schneider, K. Bernauer, A. Guggisberg, P. van den Broek, M. Hesse, H. Schmid, *Helv. Chim. Acta,* **57,** 434, 1974.
223. E. Wälchli-Schaer, C. H. Eugster, *Helv. Chim. Acta,* **61,** 928, 1978.
224. C. Poupat, *Tetrahedron Letters,* 1669, 1976.
225. M. J. Humora, D. E. Seitz, J. Quick, *Tetrahedron Letters,* 3971, 1980.
226. Y. Nagao, K. Seno, K. Kawabata, T. Miyasaka, S. Takao, E. Fujita, *Tetrahedron Letters,* 841, 1980.
227. Y. Nagao, K. Seno, T. Miyasaka, E. Fujita, *Chem. Letters,* 159, 1980.
228. Y. Nagao, K. Seno, E. Fujita, *Tetrahedron Letters,* 4931, 1980; Y. Nagao, S. Takao, T. Miyasaka, E. Fujita, *J. Chem. Soc., Chem. Comm.,* 286, 1981; E. Fujita, *Pure and Appl. Chem.,* **53,** 1141, 1981.
229. S. M. Weinreb, *Acc. Chem. Res.,* **18,** 16, 1985.
230. K. L. Rinehart Jr., *Acc. Chem. Res.,* **5,** 57, 1972; K. L. Rinehart Jr., L. S. Shield, *Progress in the Chemistry of Organic Natural Products,* **33,** 231, 1976; W. Wehrli, *Topics in Current Chemistry,* **72,** 21, 1977.
231 E. J. Corey, D. A. Clark, *Tetrahedron Letters,* 2045, 1980.
232. Y. Kishi, *Pure and Appl. Chem.,* **53,** 1163, 1981.

233. E. Schwartz, H.E. Gottlieb, F. Frolow, A. Shanzer, *J. Org. Chem.,* **50,** 5469, 1985.
234. G. Schmidt, *Topics in Current Chemistry,* **136,** 109, 1986.
235. C.M. Deber, V. Madison, E.R. Blout, *Acc. Chem. Res.,* **9,** 106, 1976.
236. H. Kessler, *Angew. Chem. Int. Ed. Engl.,* **21,** 512, 1982.
237. M. Bodanszky, *Principles of Peptide Synthesis,* Springer-Verlag, Berlin, 1984, p. 217.
238. a) F. Sondheimer, *Acc. Chem. Res.,* **5,** 81, 1972; b) N. Darby, T.M. Cresp, F. Sondheimer, *J. Org. Chem.,* **42,** 1960, 1977; c) V.K. Sharma, H. Shahriari-Zavareh, P.J. Garratt, F. Sondheimer, *J. Org. Chem.,* **48,** 2379, 1983.
239. G. Schröder, J.F.M. Oth, *Tetrahedron Letters,* 4083, 1966.
240. A.J. Weinheimer, C.W.J. Chang, J.A. Matson, *Progress in the Chemistry of Organic Natural Products,* **36,** 285, 1979.
241. Y. Kashman, A. Groweiss, S. Carmely, Z. Kinamoni, D. Czarkie, M. Rotem, *Pure and Appl. Chem.,* **54,** 1995, 1982; H.C. Krebs, *Progress in the Chemistry of Organic Natural Products,* **49,** 151, 1986.
242. M. Kodama, Y. Matsuki, S. Itô, *Tetrahedron Letters,* 3065, 1975; K. Shimada, M. Kodama, S. Itô, *Tetrahedron Letters,* 4275, 1981.
243. W.G. Dauben, R.K. Saugier, I. Fleischhauer, *J. Org. Chem.,* **50,** 3767, 1985.
244. J.A. Marshall, D.G. Cleary, *J. Org. Chem.,* **51,** 858, 1986.
245. T. Takahashi, H. Nemeto, J. Tsuji, I. Miura, *Tetrahedron Letters,* 3485, 1983.
246. M.A. Tius, A.H. Fauq, *J. Am. Chem. Soc.,* **108,** 1035, 1986.
247. M.A. Blanchette, W. Choy, J.T. Davis, A.P. Essenfeld, S. Masamune, W.R. Roush, T. Sakai, *Tetrahedron Letters,* 2183, 1984.
248. J.A. Marshall, B.S. DeHoff, S.L. Crooks, *Tetrahedron Letters,* 527, 1987.
249. A.J. Pratt, E.J. Thomas, *J. Chem. Soc., Chem. Comm.,* 1115, 1982.
250. W.C. Still, D. Mobilio, *J. Org. Chem.,* **48,** 4785, 1983.
251. C.T. Buse, C.H. Heathcock, *Tetrahedron Letters,* 1685, 1978.
252. T. Kato, M. Suzuki, T. Kobayashi, B.P. Moore, *J. Org. Chem.,* **45,** 1126, 1980.
253. P.A. Wender, D.A. Holt, *J. Am. Chem. Soc.,* **107,** 7771, 1985.
254. J.A. Marshall, T.M. Jenson, B.S. DeHoff, *J. Org. Chem.,* **51,** 4316, 1986.
255. J.E. McMurry, J.R. Matz, *Tetrahedron Letters,* 2723, 1982.
256. H.J. Bestmann, H. Lütke, *Tetrahedron Letters,* 1707, 1984.

Part II

Macrocyclic Complexes Cryptates

Introduction

Selective complexation by natural or synthetic macrocycles involves the recognition of spherical substrates (alkali or alkali earth metal cations) by macrocyclic and macrobicyclic molecular receptors. Two aspects, in particular, of this theme will be developed:

- Analysis of the chemistry of receptors for the simplest substrates, in other words analysis of *spherical recognition.*
- A discussion of the coordination chemistry of *alkali metal cations* and its applications.

In addition, a wide variety of fundamental chemical processes and species will be examined, such as:

- Desolvation and coordination of ions.
- Thermodynamics and kinetics of complexation.
- Stability and selectivity of complexation.
- Ligand exchange.
- Protection and modification of the cation properties etc.

The chemistry of macrocyclic complexes of alkali metal cations has developed enormously since the end of the 1960s because of the discovery of, on the one hand, the ability of macrocyclic antibiotics to transport these ions [1], and, on the other, the strong complexing properties shown by the "crown"-type macrocyclic polyethers [2].

The alkali metal cations are spherical species bearing a positive charge. The recognition of these species involves the selection of a sphere from a collection of spheres of different radii (Fig. 1-1).

A cation is characterized by its ionic radius, charge and charge density, polarizability, and solvation energy, and also the ionization potential of the corresponding metal.

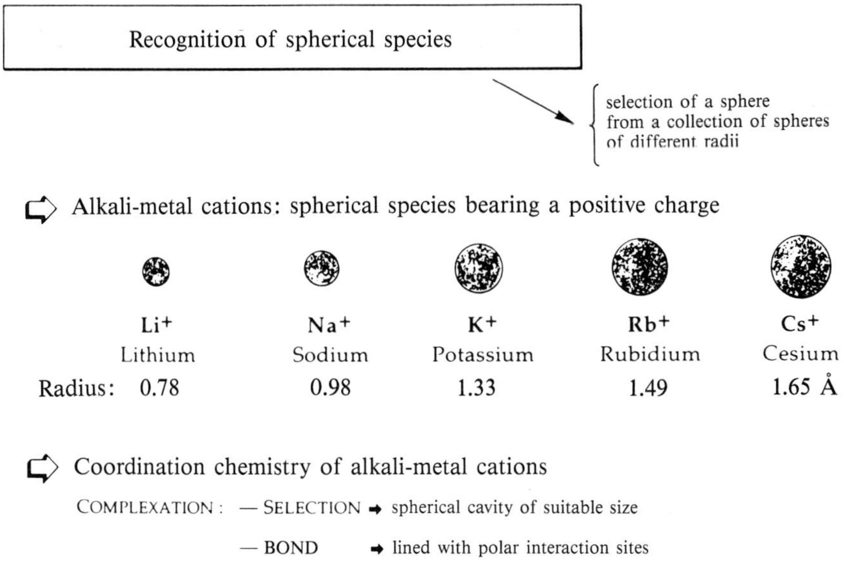

Fig. 1-1. Recognition of alkali-metal cations.

Chapter 1

Synthetic Macrocyclic Complexes

Cation complexation is governed by coordination chemistry. The selection of the cation is ensured by a cavity of adequate size, and the bond to the macrocycle arises through the presence of polar interaction sites lining the cavity.

1.1 Complementarity of the Cavity and Cation

The complementarity of the macrocycle cavity (shape, size, number of binding sites) and the cationic substrate determines the type of complex.

1.1.1 Conformation of the Macrocycle

Two extreme situations can be envisaged.

a) "Key" and "lock" type complementarity

In this case, the size of the cation and that of the preformed cavity in the macrocycle are very similar. The cation sits in the center of the cavity. The cation and cavity correspond like a key and lock, according to the image proposed by Emil Fischer. However, the cation remains accessible from both sides of the macrocycle thus leading to formation of an ion pair. This is so for the complex of dibenzo-18-crown-6 with rubidium thiocyanate (Fig. 1-2). The thiocyanate ion is bound to Rb^+ by its most highly charged site, this being the nitrogen atom [3].

The crystalline dibenzo-18-crown-6/potassium iodide/thiourea ternary complex has a polymeric structure [4] (Fig. 1-3). In this "super-complex", the K^+I^- ion pair is conserved, and the I^- anion acts as a bridge between the K^+ cation and the chain of thiourea molecules.

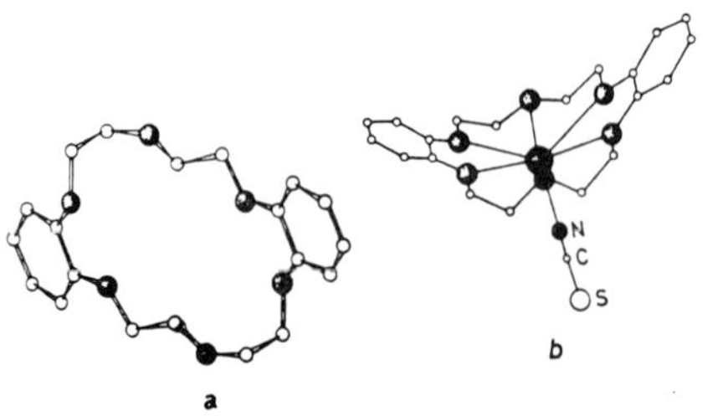

Fig. 1-2. Crystal structure and conformation of (a) uncomplexed dibenzo-18-crown-6, (b) its complex with rubidium thiocyanate (taken from D. Bright and M. R. Truter, *J. Chem. Soc. (B)*, 1544, 1970).

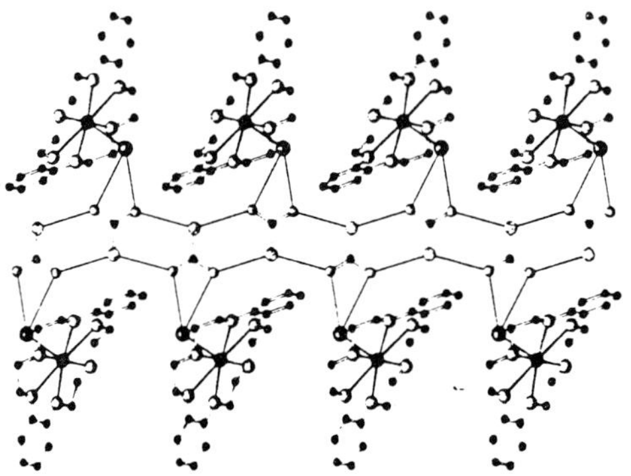

Fig. 1-3. Structure of the ternary "super-complex" between dibenzo-18-crown-6, potassium iodide and thiourea (taken from W. Saenger et al., *Angew. Chem., Int. Ed. Engl.*, **20**, 1045, 1981).

b) Induced conformational adjustment

In this case, the macrocycle, which is too large for the cation, is extensively reorganized in the latter's presence with resulting formation of an induced cavity. Thus, in the complex of dibenzo-30-crown-10 with potassium iodide, the ligand wraps itself round the cation so forming a three dimensional cavity (Fig. 1-4b). The conformation is therefore completely different from that for the free ligand (Fig. 1-4a). No ion pair is formed [5].

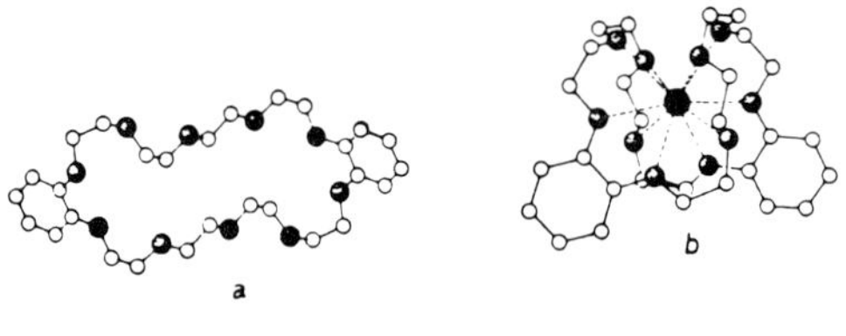

Fig. 1-4. Crystal structure and conformation of (a) dibenzo-30-crown-10, (b) its complex with potassium iodide (taken from M. R. Truter et al., *J. Chem. Soc., Perkin II*, 345, 1972).

The same conformational modifications were observed for the complexes of this ligand with RbSCN [6] and KSCN [7].

The conformational change has an effect on the stability and selectivity of complexation:

The crown ether 18-O-6 has a center of symmetry and, in the nomenclature of Klyne and Prelog [8a], the ring conformation is approximately described in terms of (ap,ap,ap), (ap,+sc,ap) and (ap,−sc,+sc) for the three non-equivalent $-OCH_2CH_2O-$ units [8b] (Fig. 1-5, and 1-6c).

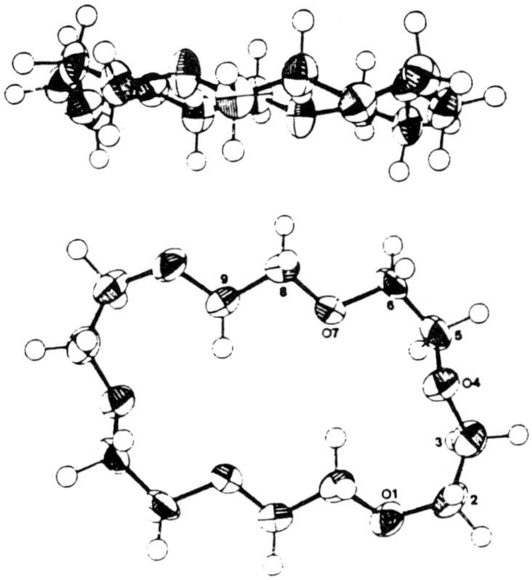

Fig. 1-5. Crystal structure of 18-crown-6; viewed both along and normal to the mean plane (taken from J.D. Dunitz et al., *Acta Cryst.,* **B30**, 2739, 1974).

The complex of this hexaether with KSCN has virtual D_{3d} symmetry [9] (Fig. 1-6a), whereas in the molecular complex with benzenesulfonamide, the hexaether has the centrosymmetric biangular conformation (Fig. 1-6b) [10, 11].

These conformational modifications arising from complexation result in decreased stability and also either increased or decreased selectivity.

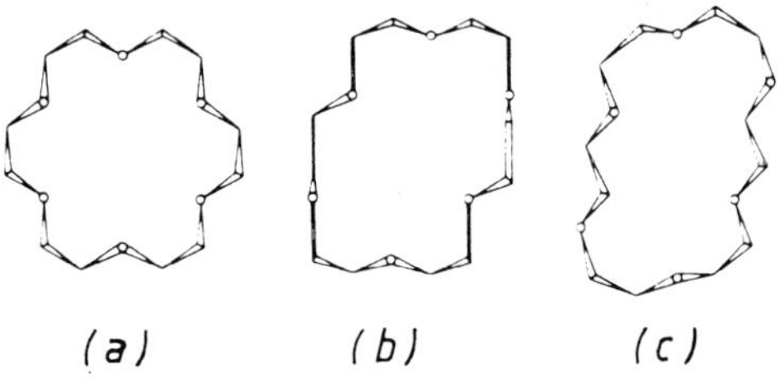

Fig. 1-6. Crystal structures of (a) the complex formed by 18-crown-6 with rubidium thiocyanate, (b) the complex of 18-crown-6 with benzenesulfonamide, and (c) the free ligand (taken from P. Groth, *Acta Chem. Scand.,* **A36**, 109, 1982. N.B.: The substrates are not shown in (a) and (b)).

1.1.2 Relative Sizes of the Cation and the Cavity

a) Cation > cavity

Two types of complex have been observed with the cyclic pentaether benzo-15-O-5 depending on cation size:

- The Na^+ cation fits the cavity giving rise to formation of a 1/1 complex which crystallizes with one molecule of water. The cation, sitting slightly above the mean plane of the cavity, is coordinated to the five macrocycle oxygen atoms and one molecule of water (W) (Fig. 1-7). The water molecule forms a hydrogen bond with the I^- anion [12].

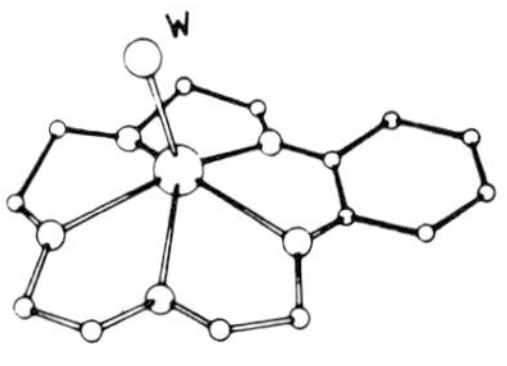

Fig. 1-7. Structure of the complex benzo-15-crown-5/Na^+/H_2O showing the pentagonal pyramid arrangement of the oxygen atoms around the sodium cation (taken from M. R. Truter et al., *J. Chem. Soc., Perkin II,* 341, 1972).

- The K^+ cation is too large to fit the cavity and so a 2/1 sandwich-type complex is formed [13] (Fig. 1-8). There is no interaction between the I^- anion and the complexed cation.

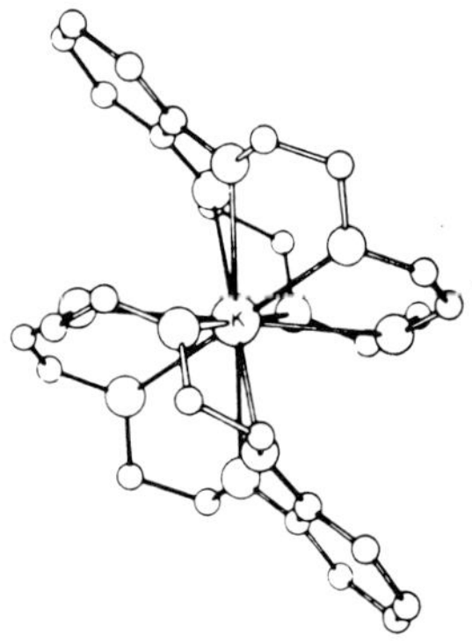

Fig. 1-8. Structure of the 2/1 complex formed between benzo-15-crown-5 and potassium iodide (taken from M. R. Truter et al., *J. Chem. Soc., Perkin II,* 1818, 1972).

The complex of Na^+ with 12-O-4 crown ether is also of the sandwich type. The cation is too large to fit the cavity and so two ligands replace the solvation sphere (Fig. 1-9) thus leading to octacoordination of the cation [14]. Regardless of the nature of the counterion (Cl^- [14a], OH^- [14b], ClO_4^- [14c], no interaction between the cation and the anion is observed.

The cation Cs^+ has been found to form a "sandwich dimer" complex of 1/1 stoichiometry with dibenzo-18-O-6, in which the external cations are bridged by the anions [15] (Fig. 1-10).

The proximity of ligands in a dimer or a polymer favours the formation of sandwich-type species, resulting in specific complexation and selectivity effects. Thus, the study of the complexation of cations (introduced as picrates) by various bis-macrocyclic polyethers (Fig. 1-11) showed that the latter are much more effective complexing agents for K^+ and NH_4^+ than are the corresponding monocyclic ligands [16]. The system shown in Fig. 1-11a with 5 atoms between the two ester functions

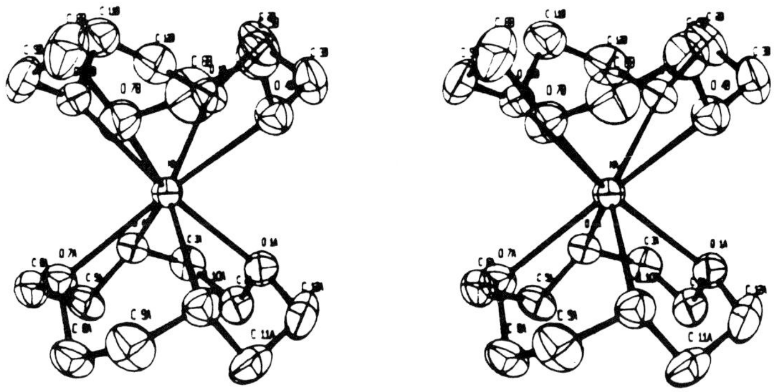

Fig. 1-9. Crystal structure of the 2/1 complex formed between 12-crown-4 and Na^+ (taken from F.P. van Remoortere et al., *Inorg. Chem.,* **13,** 2826, 1974; stereopair view).

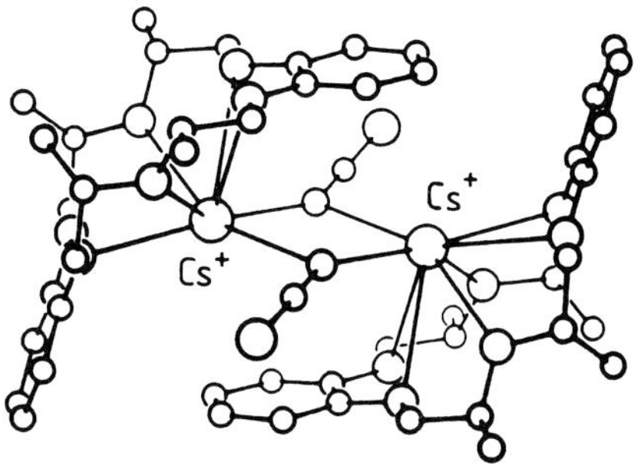

Fig. 1-10. Dimeric structure of the complex formed between dibenzo-18-crown-6 and cesium thiocyanate (taken from M.R. Truter, *Structure and Bonding,* **16,** 71, 1973).

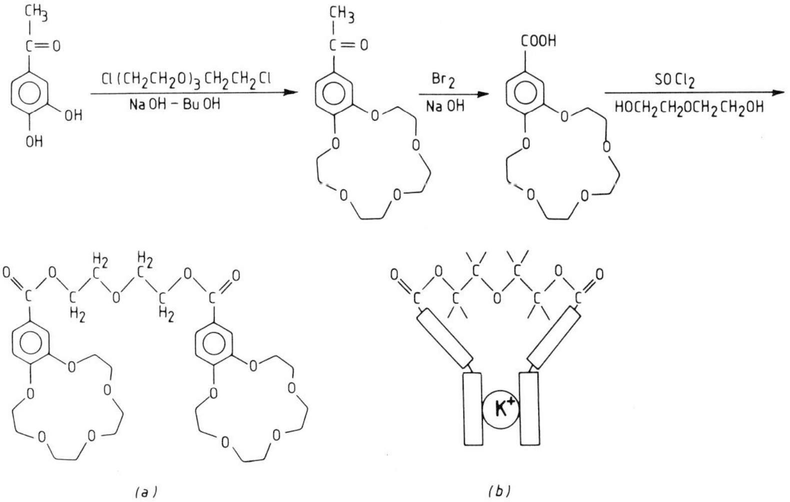

Fig. 1-11. (a) Synthesis of a bis-benzo-15-crown-5 derivative in which the two crown ethers are connected by a chain, (b) schematic representation of the complex with potassium.

gave the most stable complexes with these two cations. A schematic representation of this type of complex is shown in Fig. 1-11 b.

Another system with two [12]-macrocycles (Fig. 1-12 a) was found to form "sandwich" complexes with lithium, sodium and potassium cations. The Na^+ cation was the most strongly complexed [17 a]. The structures of the free ligand and of its Li^+ complex (Fig. 1-12 b) have been determined [17 b].

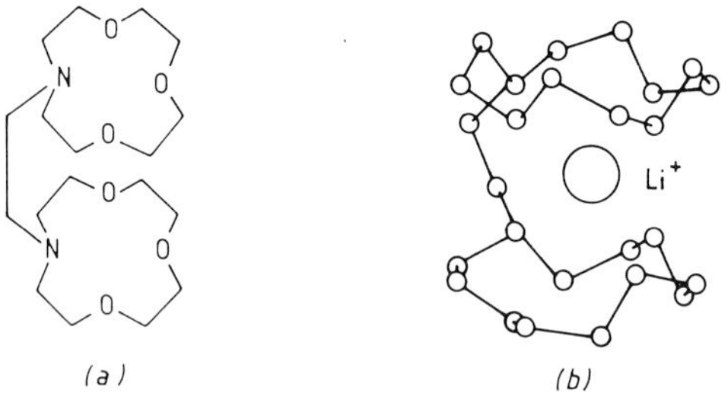

Fig. 1-12. (a) Bis-12-crown-4, (b) its complex with lithium cation (taken from P. Groth, *Acta. Chem. Scand.*, **A 38,** 337, 1984).

b) Cation < cavity

Whereas Cs^+ is too large for the 18-O-6 polyether cavity thus giving a "sandwich" dimer complex [18] (Fig. 1-13 a), Na^+ is too small. In the latter case, ring deformation occurs giving rise to an apical oxygen atom. Furthermore, Na^+ completes its coordination sphere with a water molecule (W) the latter forming a bridge with the SCN^- anion [19] (Fig. 1-13 b and c).

However, the cation can also complete its coordination sphere with other external moieties. Thus, in the 18-O-6/THF/$Na^+P(CN)_2^-$ ternary complex the macrocycle was found to retain its regular D_{3d} symmetry as was the case for its complexes with K^+, Rb^+ and Cs^+. The sodium cation interacts either with THF (Fig. 1-14 a) or with the $P(CN)_2^-$ anion (Fig. 1-14 b) [20]. The (a) and (b) units are positively and negatively charged, respectively.

There are two ways of accommodating a cavity which is too large:

- By formation of an induced cavity as in the case of KI or KSCN with DB-30-O-10 [21] (Fig. 1-15). Other examples have been given above [5–7].
- Formation of a dinuclear complex, e. g., as with sodium *o*-nitrophenolate and DB 24-O-8 where the anion plays an important part by providing additional interaction sites [22]. Each Na^+ cation interacts with three macrocycle oxygen atoms (note the $Na^+\cdots O$ distance in Fig. 1-16 a). The other sites necessary for cation hexacoordination are provided by the two phenolate anions (Fig. 1-16 b).

This ligand also forms a dinuclear complex with potassium cations [23].

The macrocycle shown in Fig. 1-17 a has been found to form a dinuclear complex with sodium cations [24]. The structure of this complex is remarkable; note the two aromatic residues stacked in the center thus creating two cavities each capable of complexing a cation.

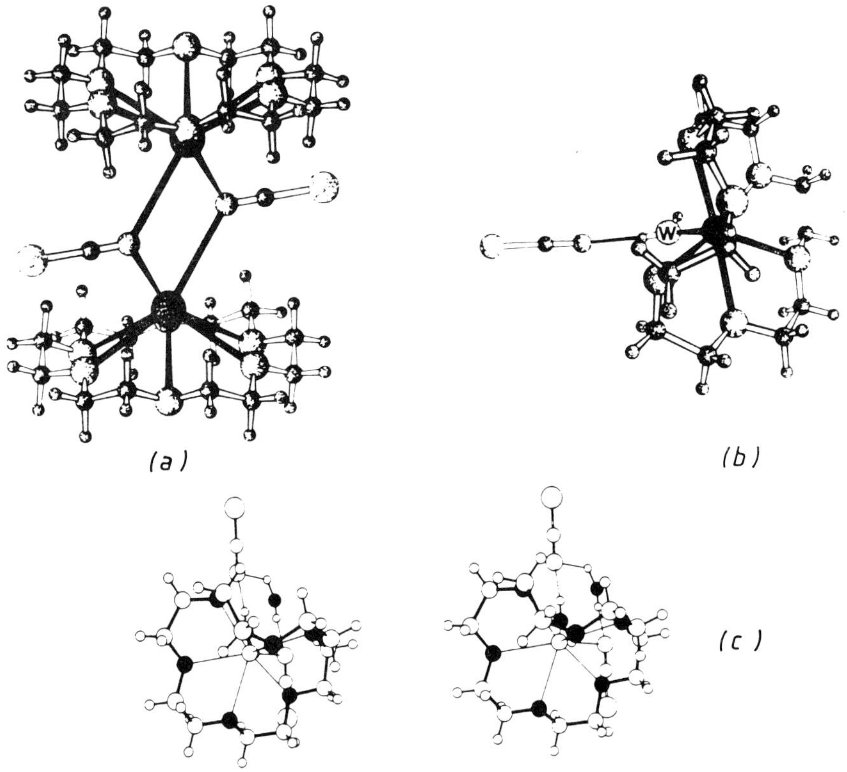

Fig. 1-13. Structures of the complexes of 18-crown-6 with (a) cesium thiocyanate (2/2) and (b and c) sodium thiocyanate (1/1) (W represents a water molecule) ((c): stereopair view) (taken from M. Dobler et al., *Acta Crys.* **B 30**, 2748, 1974; and J. D. Dunitz et al., *Acta Cryst.*, **B 30**, 2741, 1974).

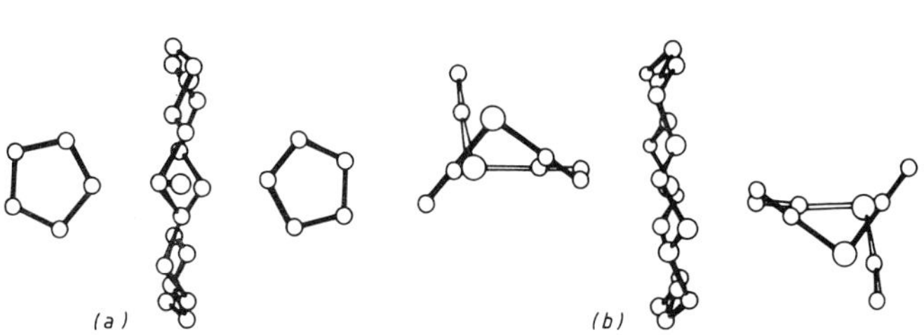

Fig. 1-14. Structure of the ternary complex [18-O-6]/THF/ $Na^+P(CN)_2^-$. Views of (a) the $\{[18\text{-}O\text{-}6]/Na/(THF)_2\}^+$, and (b) the $\{[18\text{-}O\text{-}6]/Na/[P(CN)_2]_2\}^-$ units perpendicular (top) and parallel (bottom) to the macrocycle plane (taken from A. Schmidpeter et al., *Angew. Chem., Int. Ed. Engl.,* **18**, 934, 1979).

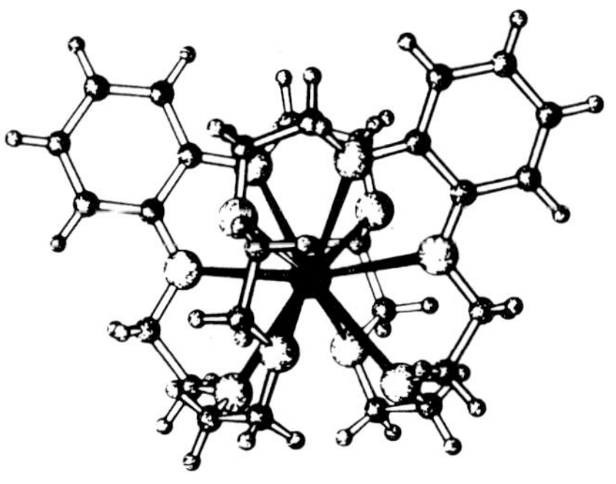

Fig. 1-15. Structure of the 1/1 complex of dibenzo-30-crown-10 with potassium iodide (taken from R. Hilgenfeld et al., *Topics in Current Chem.,* **101**, 1, 1982).

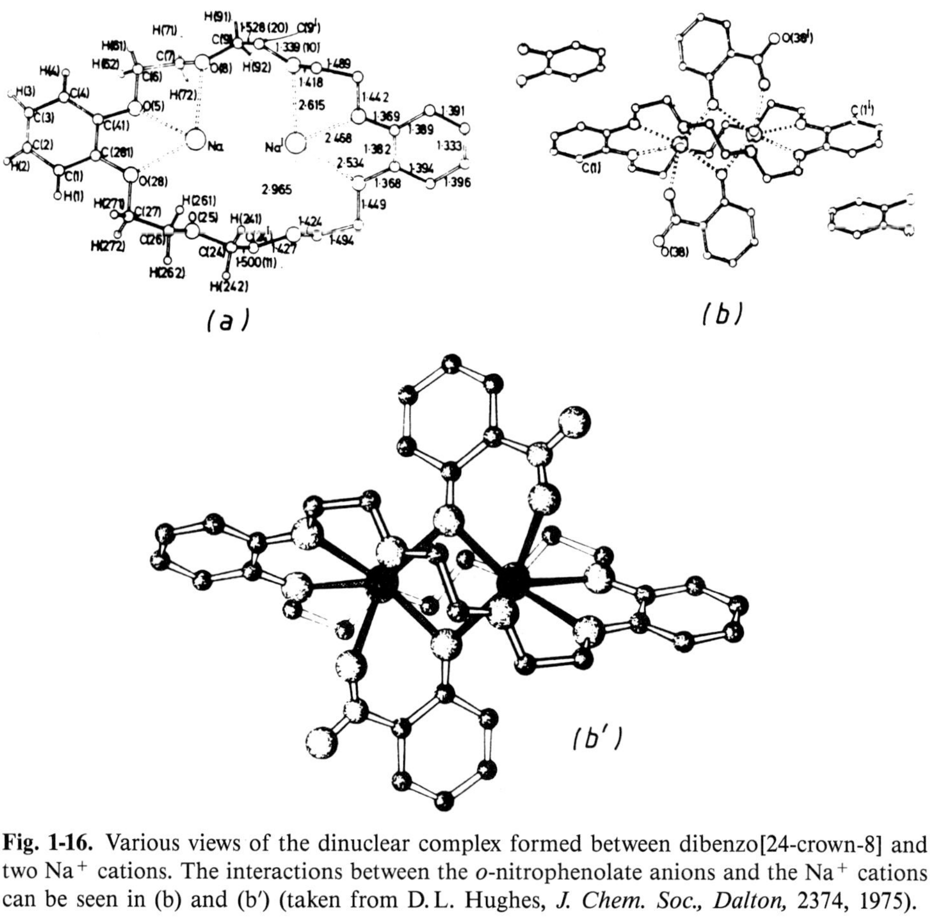

Fig. 1-16. Various views of the dinuclear complex formed between dibenzo[24-crown-8] and two Na^+ cations. The interactions between the *o*-nitrophenolate anions and the Na^+ cations can be seen in (b) and (b') (taken from D. L. Hughes, *J. Chem. Soc., Dalton,* 2374, 1975).

A systematic study of the Na^+ and K^+ complexes formed with the ligands B-15-O-5, DB-18-O-6, DB-24-O-8, and DB-30-O-10 (isolation of the complex, microanalysis) enabled the stoichiometries of the various complexes obtained to be determined [25].

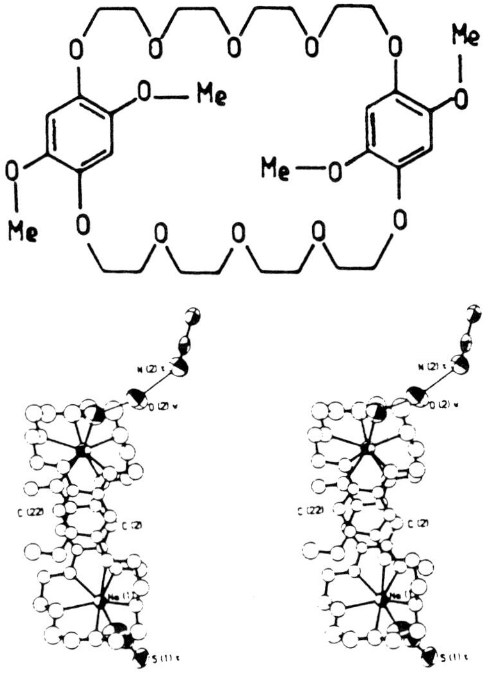

(a) (b)

Fig. 1-17. (a) The ligand and (b) stereoview of its dinuclear complex with two sodium cations (taken from J. D. Owen, *Acta Cryst.*, **C40**, 951, 1984).

1.2 Effect of the Anion

The anion can greatly influence the type and structure of the complex, depending on its ability to coordinate to the cation:

- Formation of an *ion pair or a sandwich molecule.*
- *Ligand distortion:* e. g., in 18-O-6/Na^+ complexes, the macrocycle is found to be distorted when in the presence of one SCN^- anion (19) whereas it has D_{3d} symmetry when in the presence of two $P(CN)_2^-$ anions or of two THF molecules [20].
- *Cation position:* The cation can be pulled out of the cavity by a strong interaction with a strongly chelating anion such as that of ethylacetoacetate. Thus, K^+ was found to be displaced by 0.9 Å out of the mean plane of the ring in the case shown in Fig. 1-18 [26]. The picrate anion was found to have the same effect in the complex of sodium with benzo-[15-O-5] [27].

- Formation of *dinuclear complexes* such as those with *o*-nitrophenolate anions which act as bridges in the complex $2Na^+[24\text{-}O\text{-}8]$ [22].

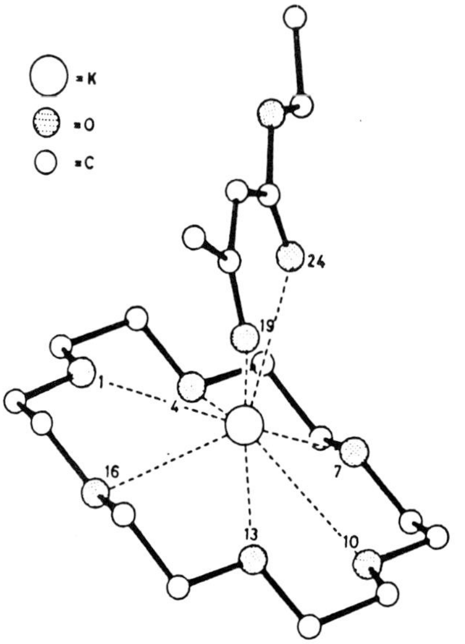

Fig. 1-18. Structure of the complex formed between 18-crown-6 and potassium ethyl acetoacetate enolate (taken from C. Riche et al., *J. Chem. Soc., Chem. Comm.,* 183, 1977).

The influence of the anion on the structure of the complex is particularly marked in the case of the alkaline earth metal cations. The structures of the complexes of benzo-15-crown-5 with $Mg(NCS)_2$, $Ca(NCS)_2 \cdot MeOH$, and $Ca(NCS)_2 \cdot H_2O$ have been determined [28]. The Mg^{2+} cation is small enough to fit inside the ring and it is coordinated to seven atoms, these being the five cyclic ether oxygens and the two nitrogens from the anions (Fig. 1-19a). The coordination polyhedron was found to be a pentagonal bipyramid.

The larger Ca^{2+} cation sits outside the cavity and it is coordinated to eight atoms consisting of the five cyclic ether oxygens, the two nitrogens from the anions, and an oxygen from the solvent (MeOH or H_2O) (Fig. 1-19b). The structure of this Ca^{2+} complex differs from that found with calcium 3,5-dinitrobenzoate [29]. In the latter case, the cation was coordinated to nine oxygens (five ligand oxygens and four from the carboxylate groups of the two benzoate molecules). As a result, the cation is displaced further from the median plane of the macrocycle oxygen atoms (1.38 Å) than in the case of the NCS anions (1.22 Å). It should be noted that, because of their relatively high charge density, the alkaline earth metal cations always form ion pairs with two anions.

The structure of the complex of benzo-15-crown-5 with barium 3,5-dinitrobenzoate shows that the Ba^{2+} cation sits well outside the cavity (1.877 Å out of the plane) (Fig. 1-19c). The fundamental unit is a dimer consisting of two ligands, two Ba^{2+} and four dinitrobenzoate anions. The Ba^{2+} cation is involved in ninefold coordination through the five ring oxygen atoms and four carboxylate anions (each of the latter is coordinated to two cations) [30].

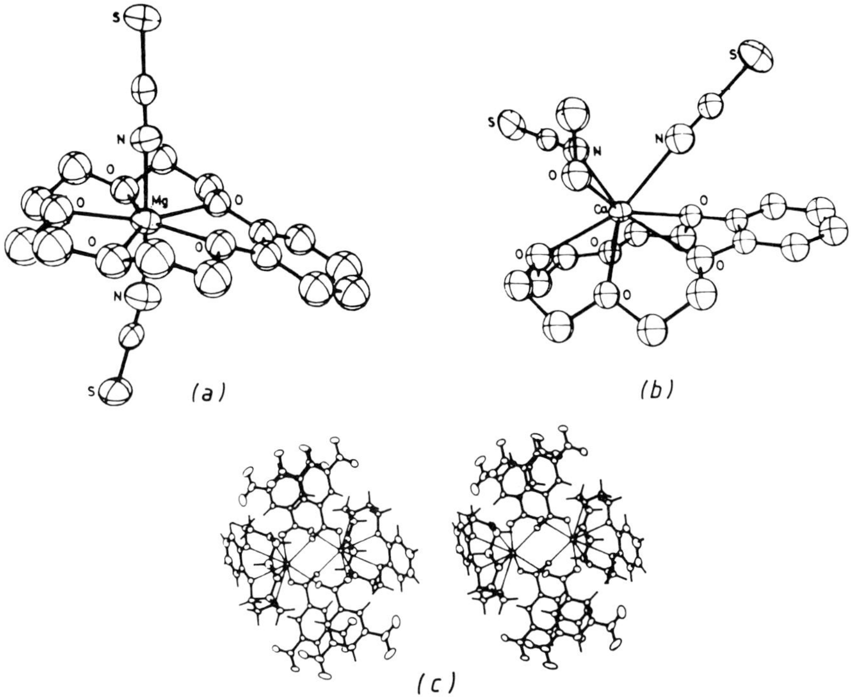

Fig. 1-19. Crystal structures of complexes of benzo-15-crown-5 with (a) $Mg(NCS)_2$, (b) $Ca(NCS)_2 \cdot MeOH$, and (c) Ba(3,5-dinitrobenzoate)$_2$ ((c): three-dimensional stereopair view). ((a) and (b) taken from J. D. Owen, *J. Chem. Soc, Dalton,* 1418, 1978; (c) taken from D. L. Ward et al., *Acta Cryst.,* **C40,** 1183, 1984).

A spectacular anion effect has been observed for the complex $[Cs_2(18\text{-crown-}6)]^{2+}[Al_3Me_9SO_4]^{2-}$ [31]. A cesium-(18-crown-6)-cesium(sulfate/aluminium)-cesium chain is formed (Fig. 1-20). Each cesium ion can be considered to

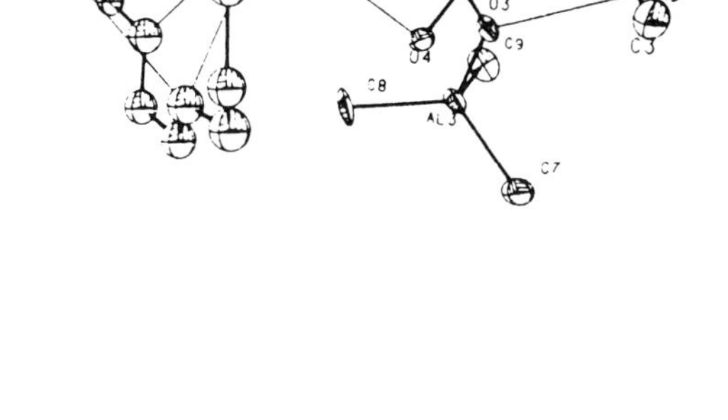

Fig. 1-20. Structure of the complex $(Cs_2[18\text{-C-}6]^{2+}\ (Al_3Me_9SO_4)^{2-}$ (taken from J. L. Atwood et al., *J. Am. Chem. Soc.,* **106,** 7627, 1984).

be nine-coordinate interacting with the six crown ether oxygens, two sulfate oxygens and the other Cs^+ ion. The $Cs^+ \cdots Cs^+$ distance was found to be only 3.92 Å whereas the sum of the ionic radii is 3.56 Å.

The two Cs^+ cations are well outside the macrocycle plane (1.79 and 2.37 Å) whereas for the complex in Fig. 1-13a this distance is only 1.44 Å.

1.3 Participation of Water Molecules. Simultaneous Complexation of a Cation and Water

The role of water molecules is of particular interest since the progressive loss of the solvation sphere occurring on complexation corresponds to the steps of complex formation.

In the complex between benzo-15-crown-5 and calcium picrate, the cation more or less fits the cavity. However, both the strong anion-cation (picrate-Ca^{2+}) interaction in the ion pair and also coordination of the remaining sites with three water molecules give rise to external sevenfold coordination of Ca^{2+} (Fig. 1-21). This structure is also stabilized by hydrogen-bonding between the two water molecules and the crown ether [32].

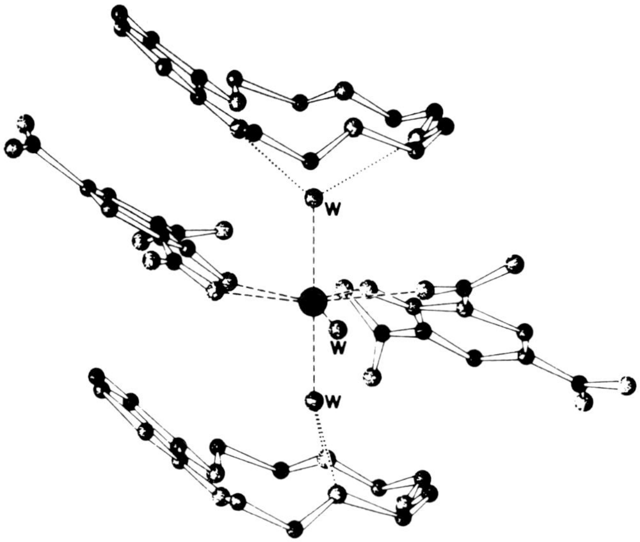

Fig. 1-21. Structure of the complex of benzo-15-crown-5 with calcium picrate and three molecules of water (W) (taken from R. Hilgenfeld and W. Saenger, *Topics in Current Chem.,* **101,** 1, 1982).

An analogous complex is formed by Li^+ picrate [32].

In these two examples the crown ether does not interact directly with the cations, and its only role is to stabilize the crystal structure.

The Li^+ cation is much too small for the cavity in 18-O-6 crown ether. However, when two [19-O-6] crown ether units are joined in a spiro derivative (Fig. 1-22a), each of the two rings can contain a cation and a water molecule. This internal solvation ensures partial filling of the cavity [33].

The structure (Fig. 1-22b and b′) shows that in each ring, the Li^+ cation is pentacoordinate interacting with three ring oxygens, the internal water molecule (which is hydrogen bonded to two ring oxygens) and one external water molecule.

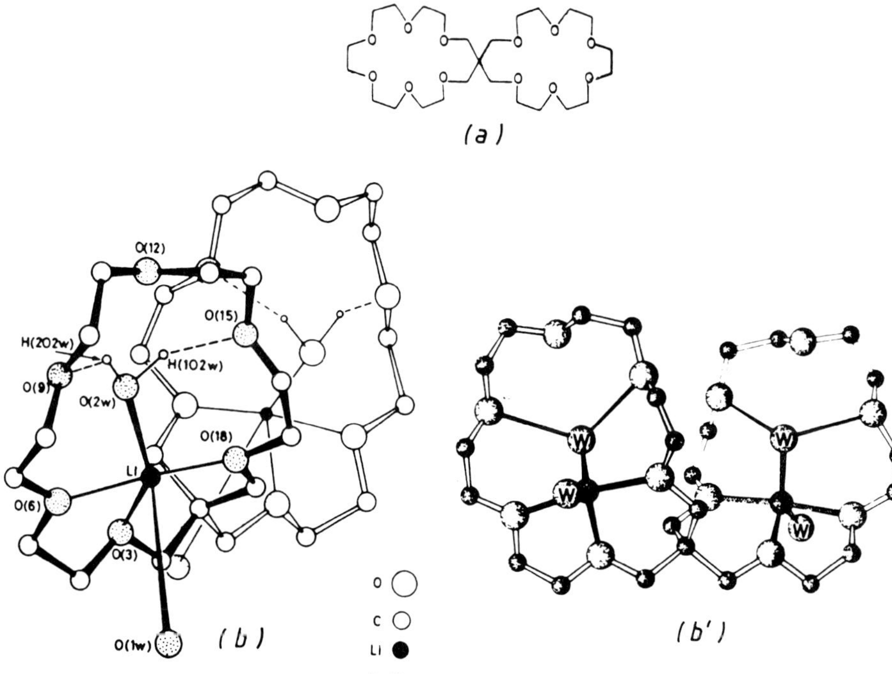

Fig. 1-22. Structure of the 1/2 complex of spiro-bis[19-crown-6] with lithium iodide and water molecules (W). ((a) and (b) taken from E. Weber, et al., *J. Chem. Soc., Chem. Comm.*, 472, 1981; (b′) taken from R. Hilgenfeld and W. Saenger, *Topics in Current Chem.*, **101**, 1, 1982).

The same type of phenomenon is observed with Li^+ and 18-O-6 crown ether. The cavity is partially filled by water molecules bonded on the one hand to the cation and on the other hydrogen-bonded to the crown ether oxygens and the anion [11].

In the case of the (barium picrate)-(dibenzo-24-O-8) complex (Fig. 1-23 a), only five crown ether oxygens are coordinated to barium, and two of the three other ether oxygens are hydrogen-bonded to a water molecule. The latter water molecule, which is symbolized by O(1), interacts with the Ba^{2+} cation, which in turn is coordinated to another water molecule represented by O(2) as well as two phenolate oxygens and a nitro-group oxygen thus giving 10-coordination [34 a].

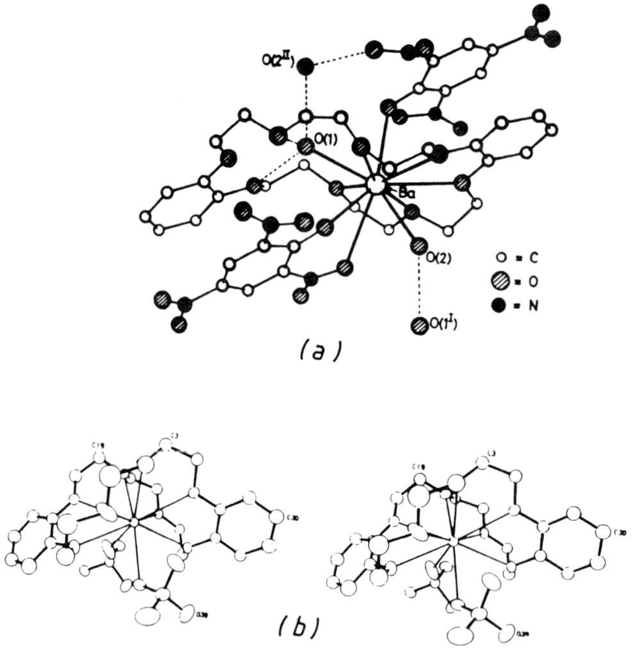

Fig. 1-23. Structures of the complexes of dibenzo-24-crown-8 with (a) barium picrate and two water molecules, and (b) barium perchlorate ((b) three-dimensional stereopair view); ((a) taken from D. L. Hughes et al., *J. Chem. Soc., Chem. Comm.,* 804, 1977; (b) taken from M. R. Truter et al., *Acta Cryst.,* **B 34,** 800, 1978).

The important part played by the picrate anion in this structure should be noted, since in the case of perchlorate anion, the Ba^{2+} cation is bound to all the surrounding crown ether oxygens as well as to the two ClO_4^- anions [34 b] (Fig. 1-23 b).

The cavity in the ligand 12-O-4 is too small for the Mg^{2+} cation. Furthermore, the latter is strongly solvated and forms hexahydrates in which the cation is octahedrally coordinated by the water oxygens. In the complex (Fig. 1-24), the Mg^{2+} was found to lie externally and to have retained its hydration sphere, the water molecules being hydrogen-bonded to the crown ether which occupies a second coordination sphere [35].

In conclusion, the type of complex formed depends on the fit of the cation to the cavity, the nature of the anion, and inclusion and participation of other species, in particular water.

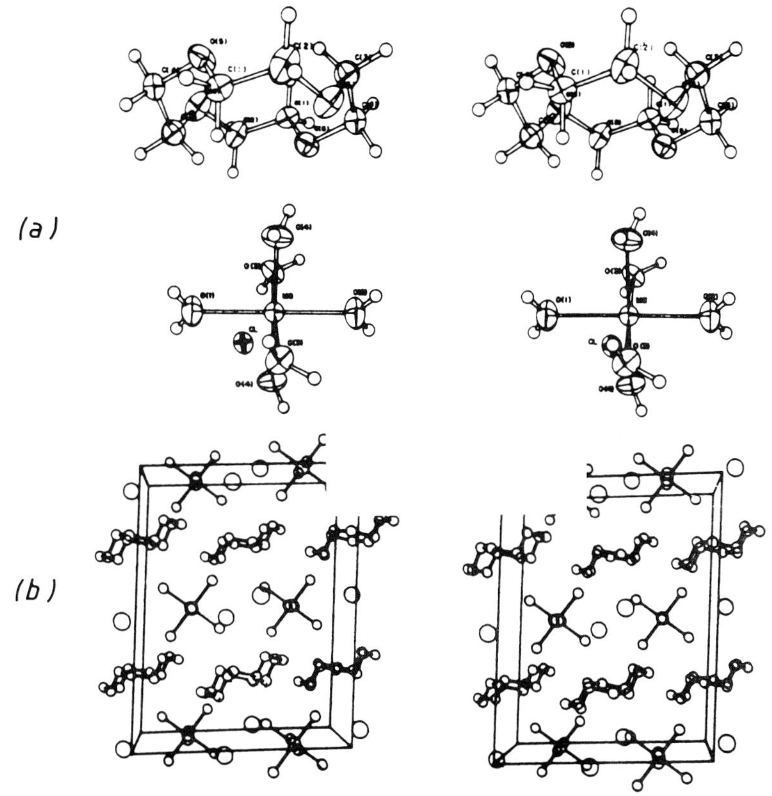

Fig. 1-24. (a) Structure of the 1/1 complex of 12-crown-4 with magnesium chloride and six water molecules, (b) assembly of six units in the crystal (three-dimensional views) (taken from F. P. van Remoortere et al., *Inorg. Chem.,* **14,** 734, 1975).

1.4 Lateral Interactions

Cation interactions with lateral appendages on the macrocycle can lead to novel possibilities for controlling complexing properties.

Two derivatives of a tetrafunctional chiral macrocycle have been synthesized, *anti*-2 and *syn*-2 (Fig. 1-25), each possessing a C_2 symmetry axis, respectively in the plane of and perpendicular to the ring [36].

The *syn* isomers were found to complex cations much more strongly than did the *anti* isomers. In the *syn* isomers, the complexed cation, Sr^{2+} or Ca^{2+}, is located inside the macrocyclic cavity and the six ring oxygens are in close contact with it. The short cation-carboxylate distance indicates that there is a strong elec-

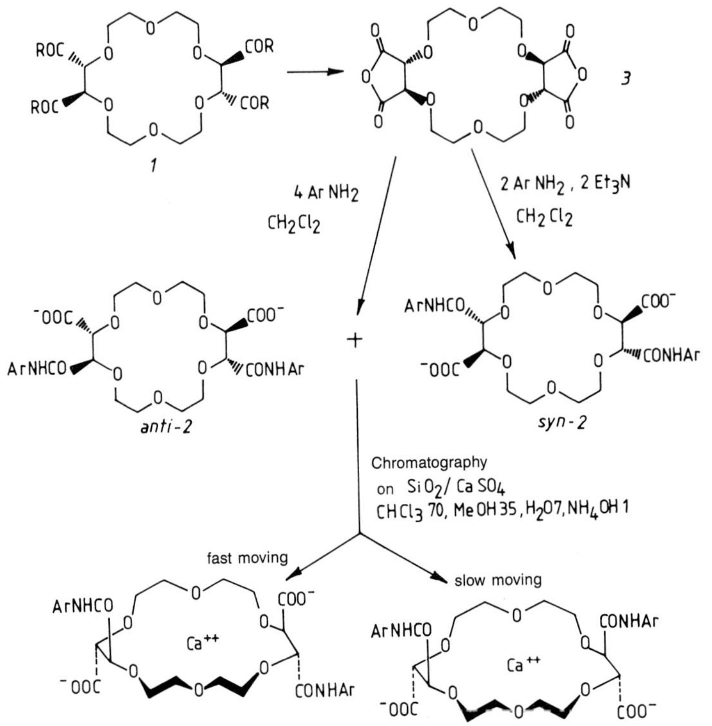

Fig. 1-25. Synthesis of the *anti* and *syn* isomers of a chiral tetrafunctional macrocycle.

trostatic interaction with both carboxylate groups. These macrocycles bind cations much more strongly than does the parent 18-crown-6.

The ring conformation in the complex of the *syn-2a* isomer (Ar = Ph) resembles that in the cation complexes of 18-crown-6 (Fig. 1-26a). The six oxygens form an almost planar hexagon surrounding the cation which is also coordinated by both carboxylates and the chloride anion.

In contrast, the complex of the *anti-2a* isomer shows a strained sigmoidal conformation for the macrocyclic ring (Fig. 1-26b). The electrostatic interaction of the complexed cation with the two carboxylate groups imposes two eclipsed C–C bond conformations and the six oxygens are displaced from their median plane. This large deformation may explain why the complex of the *anti* isomers are ca. 100-fold less stable than those of the *syn* isomers. The amide groups are directed towards the exterior and the "open" side of the macrocycle is occupied by three water molecules which are attached by hydrogen bonding.

Solid state model of a molecular channel. With relevance to the studies described above, the chiral macrocyclic potassium tetracarboxamide complex *1* has been report-

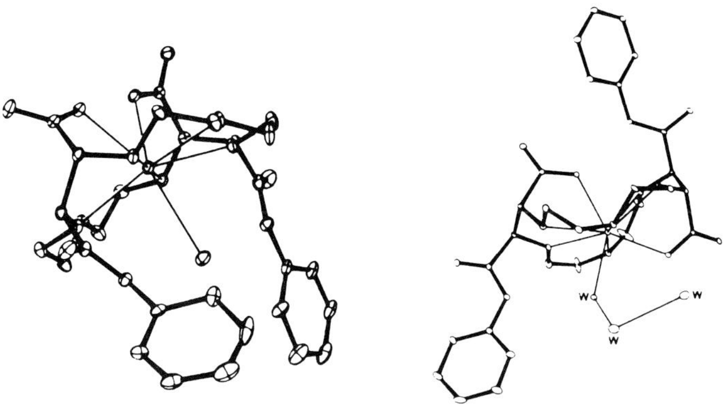

(a) (b)

Fig. 1-26. Structures of the (a) *(syn-2a)* Sr^{2+} and (b) *(anti-2a)* Ca^{2+} complexes (taken from J.M. Lehn et al., *J. Am. Chem. Soc.,* **103,** 701, 1981).

ed [37]. Its structure (Fig. 1-27) presents a solid-state model of a molecular channel in which the macrocyclic units are organized in a polymolecular stack, and the K^+ cations are located alternately inside and on top of successive macrocycles. The polymolecular organic unit results from the repetition of a stacked asymmetric dimeric unit $[(\mathit{1},K)_2(H_2O)_3]^{2+}$, and is stabilized by hydrogen bonding of the water molecules with each other ($W_2 \cdots W_3$) and with the amide groups, as well as by one ion dipole interaction $K_2^+ \cdots O(W_1)$. The $[KBr_3(H_2O)_4]^{2-}$ counterions form a polymeric chain parallel to the organic one, made up of alternating water-bridged KBr and BrBr fragments.

The system $[(Br^-,K^+,2H_2O)(2Br^-,2H_2O)]^{2-}$ is very interesting from the point of view of interactions because polymerization reinforces the hydrogen bonding.

The organic part of this structure has several remarkable features:

- Staircase-like parallel stacking of macrocyclic units suggests a channel structure.
- The conformations of both macrocycles are similar to that of the crown ether itself.
- The $-CON(CH_3)_2$ side chains are in an axial conformation and direct their polar groups into the internal space of the ligand.

As a result of the latter two features, all the nonpolar groups of the macrocyclic units are on the outside. The channel interior is therefore polar, and the exterior is lipophilic.

However, we have seen above that in related compounds with secondary amide residues (X = CONHR), the CO groups are directed outwards [36]. Consequently, variation in the nature of the amide group allows control of the orientation of the polar groups. This is of considerable interest for designing receptor molecules built on the basic unit *1*.

There are two distinct coordination sites for the K^+ cation, one inside the macrocyclic ring and the other above its mean plane. They may be considered as being two steps in the flow of a non-solvated K^+ cation through a channel.

The polar interior of the channel contains three water molecules per asymmetric unit. They solvate the K^+ and amide groups, and provide hydrogen bonding links between successive macrocyclic units.

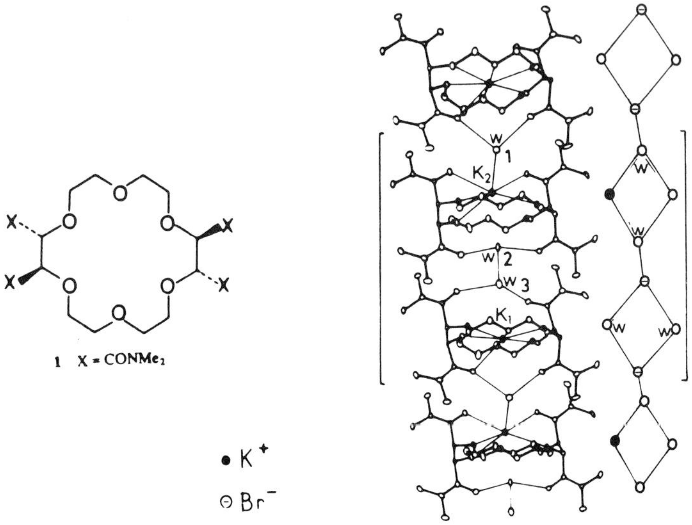

Fig. 1-27. A chiral macrocyclic tetracarboxamide and its potassium complex representing a model of a molecular channel (taken from D. Moras et al., *Nature,* **295**, 526, 1982).

Numerous macrocycles possessing lateral appendages capable of interacting with cations have been reported and we shall give a few examples. The 18-crown-6 derivative shown in Fig. 1-28 forms a complex with K^+.

It can be seen that the side arm ether oxygen is apical and interacts with K^+ ($K^+ \cdots O$ = 2.91 Å; the distances between K^+ and the ring oxygen atoms range from 2.79 to 2.91 Å). Iodide occupies the opposite apical position [38].

The participation of the side arm in cation complexation enhances binding relative to the corresponding macrocycle without side arms [38b]. A very large variety of ligands of this type has been synthesized and studied, these being 12-, 15- and 18-membered macrocyclic polyethers with side arms of various lengths [38b].

Another ligand shown in Fig. 1-29 possesses four side arms. A varying number of the latter are used depending on the nature of the complexed cation, leading to penta-, hepta- and octacoordination for Li^+, Na^+ and K^+, respectively [39].

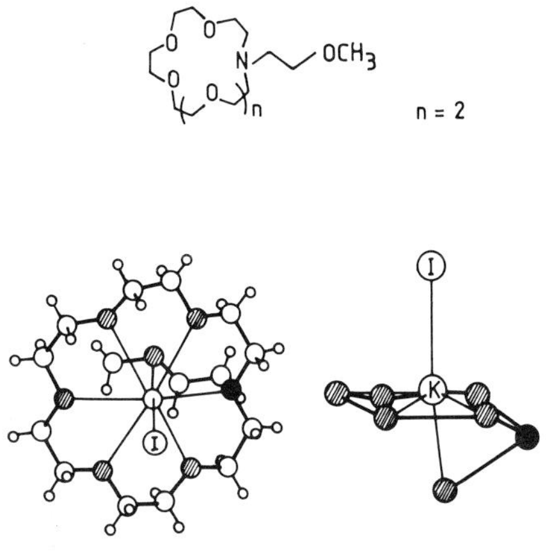

Fig. 1-28. Complexation of potassium cations by a lariat ether (taken from F. R. Fronczek et al., *J. Am. Chem. Soc.,* **105**, 6717, 1983).

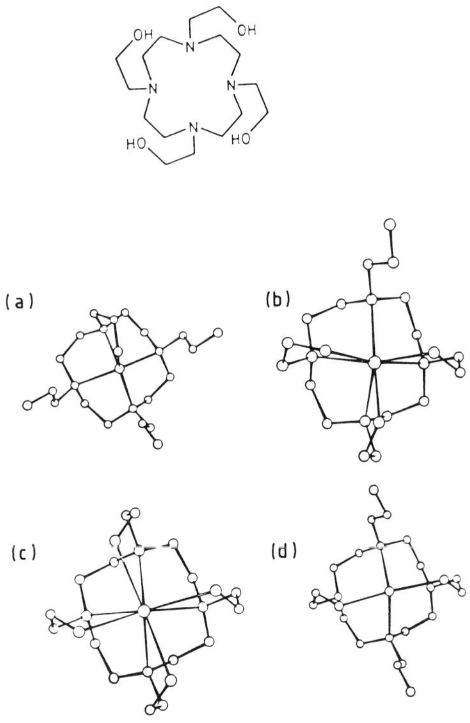

a) Li^+, b) Na^+, c) K^+, d) H_2O

Fig. 1-29. Macrocycle with four side arms (taken from J. Dale et al., *J. Chem. Soc., Chem. Comm.,* 1172, 1982).

In a further example of a macrocycle with a lateral appendage, it was found that the side arm did not directly interact with the cation, but rather with a water molecule coordinated to the cation (Fig. 1-30). The arrangement of the water molecules in the crystal is remarkable [40].

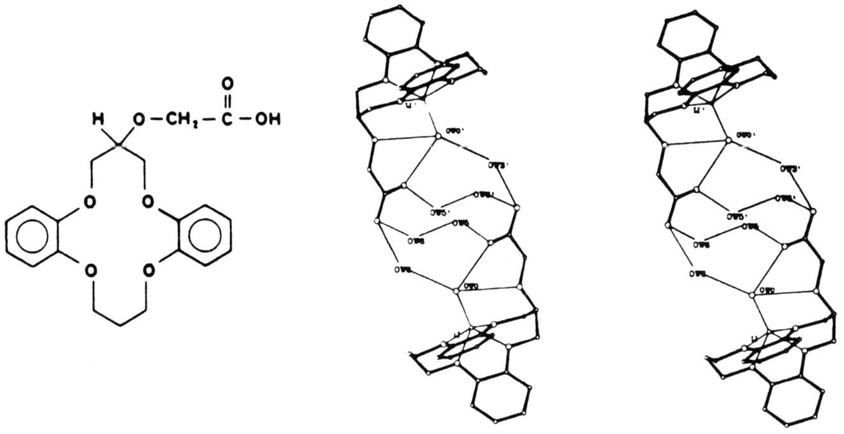

Fig. 1-30. A macrocycle possessing a side arm; structure of its complex with Li^+ (stereoview) (taken from W.H. Lipscomb et al., *J. Am. Chem. Soc.,* **106,** 1280, 1984).

Structural aspects of macrocyclic complexes are covered in detail in several books and reviews [41].

1.5 Formation of a Second Coordination Sphere

Neutral platinum complexes of the type *trans*-$PtCl_2(L)NH_3$, where L = PMe_3, PEt_3 and NH_3, form crystalline adducts with 18-crown-6 and dibenzo-18-crown-6.

The crystallographic study of the 1 : 1 adduct of *trans*-$PtCl_2(PMe_3)NH_3$ with dibenzo-18-crown-6 (Fig. 1-31 a) showed that the NH_3 ligand coordinated to the transition metal is involved in three bifurcated hydrogen bonds with the macrocycle oxygen atoms [42] (Fig. 1-31 b).

In the 2 : 1 adduct, [*trans*-$PtCl_2(PMe_3)NH_3]_2$: [18-O-6], the crown adopts its familiar complexing conformation and binds two molecules of the platinum complex, one to each face of the crown (Fig. 1-32). Thus, all six of its oxygen atoms are hydrogen bonded in an approximately trigonal fashion to the two NH_3 ligands [42].

The formation of these adducts is a type of generalization of the participation of a water molecule mentioned above. The first coordination sphere consists of relay molecules bound to the macrocycle, and the latter constitutes the second coordination sphere.

The 2 : 1 adduct in Fig. 1-32 led to the study of adducts formed between crown ethers and transition metal complexes containing several ammine ligands such as $[Cu(NH_3)_4H_2O]^{2+}$.

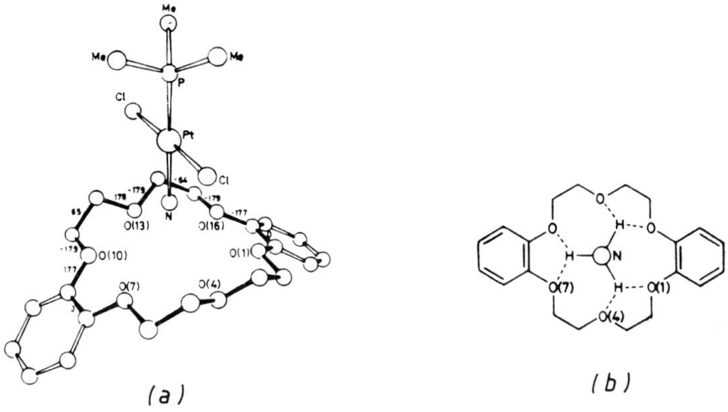

Fig. 1-31. (a) Crystal structure of the 1:1 complex of *trans*-$PtCl_2$ $(PMe_3)NH_3$ with dibenzo-18-crown-6, (b) view perpendicular to the plane of the macrocycle (taken from J. F. Stoddart et al., *J. Chem. Soc., Chem. Comm.,* 847, 1981).

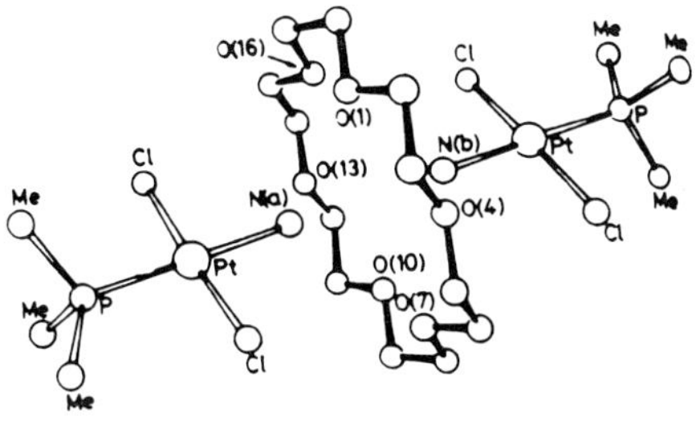

Fig. 1-32. Crystal structure of the 2:1 complex of *trans*-$PtCl_2$ $(PMe_3)NH_3$ with 18-crown-6 (taken from J. F. Stoddart et al., *J. Chem. Soc., Chem. Comm.,* 847, 1981).

The 1:1 adduct of this complex with 18-O-6 crown ether, $[Cu(NH_3)_4(H_2O)(18\text{-}O\text{-}6)]^{2+}[2PF_6]^{2-}$, was found to have the structure of a copolymer in which the macrocycle alternates with the $[Cu(NH_3)_4H_2O]^{2+}$ ions [43] (Fig. 1-33).

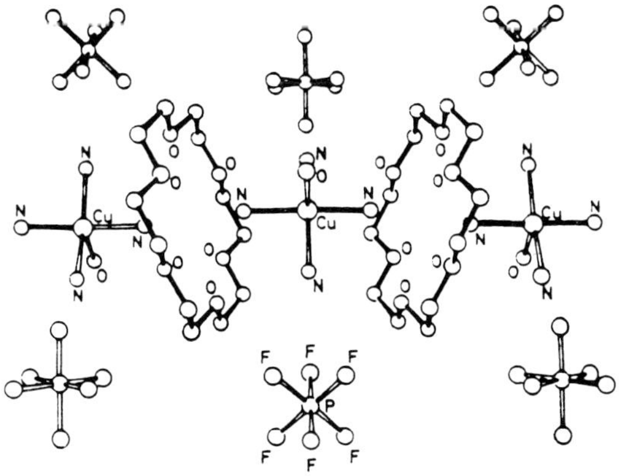

Fig. 1-33. Copolymer structure of the complex of 18-crown-6 with $[Cu(NH_3)_4H_2O]^{2+}$ $[2PF_6]^{2-}$ (taken from J. F. Stoddart et al., *J. Chem. Soc., Chem. Comm.,* 849, 1981).

Each crown molecule is involved in ten hydrogen bonds (Fig. 1-34). O(1) and O(10) each accept one hydrogen bond with trigonal geometry, and O(4), O(7), O(13), and O(16) all accept two hydrogen bonds, one each from the N(1) ligands with approximately trigonal geometry and one each from the N(2) ligands with approximately tetrahedral geometry.

The hydrogen bonds are depicted in Fig. 1-34a. The N(1a) and N(1b) ammine ligands are each involved in three hydrogen bonds; N(1a) with O(1), O(7) and O(13), and N(1b) with O(4), O(10) and O(16). In addition to these six hydrogen bonds which can be designated as being primary, four secondary hydrogen bonds are formed between the two N(2a) ammine ligands and O(7) and O(13), and the two N(2b) ammine ligands and O(4) and O(16), respectively. The network of hydrogen bonds on one face of the macrocycle is shown in Fig. 1-34b.

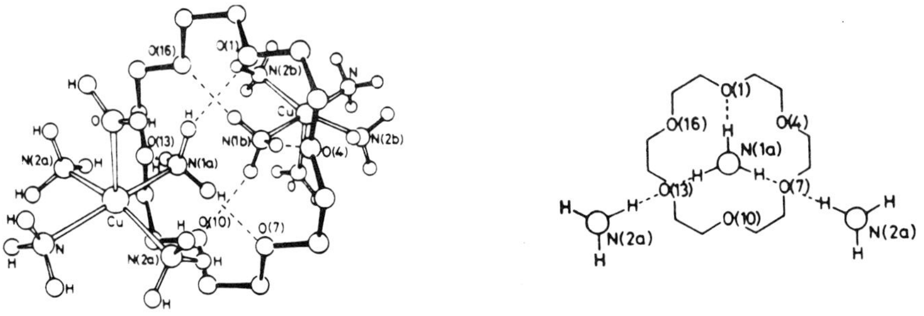

(a) (b)

Fig. 1-34. Hydrogen bonds in the complex depicted in Fig. 1-33. (Not all the hydrogen bonds are shown).

The complex $[Pt(bipy)(NH_3)_2]^{2+}[2PF_6]^{2-}$ forms a 1 : 1 crystalline adduct with dibenzo-30-crown-10. The X-ray crystal structure of this adduct has been determined [44] (Fig. 1-35). The crown ether envelops the transition metal complex in a quite remarkable manner by forming a kind of cup.

Apart from the hydrogen bonding between three of the ten oxygen atoms in the crown ether and three of the six hydrogen atoms associated with the NH_3 ligands in the complex, there are also stabilizing interactions between the two benzene rings in the crown ether and the bipyridyl ligand. These interactions have been detected by using UV-visible spectroscopy. Addition of DB-30-O-10 to a solution of the metal complex in acetonitrile gives rise to the appearance of an absorption in the charge-transfer region (λ_{max} ~ 350 nm) [44].

The field of second-sphere coordination has been substantially developed over recent years. Amongst applications are the purification of receptor molecules by formation of crystalline adducts, and the separation of a mixture of transition metal cations (e. g., copper from cobalt) [45].

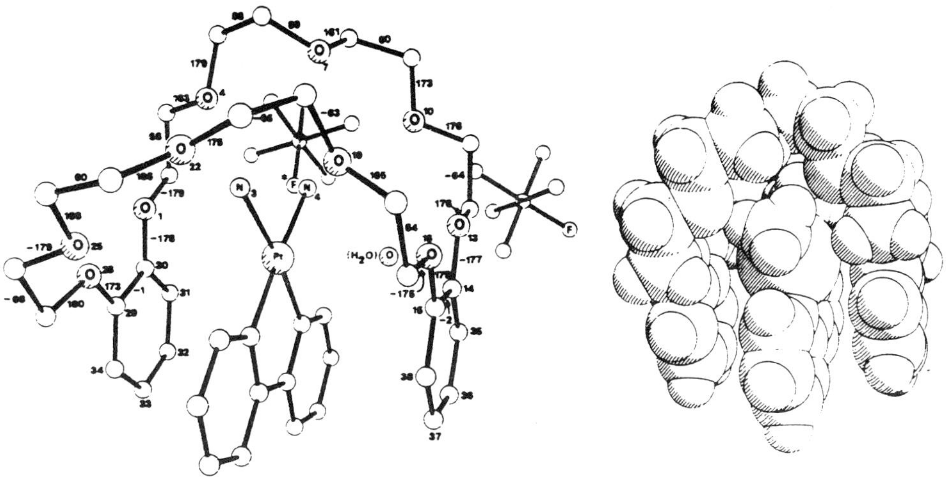

Fig. 1-35. Crystal structure of the adduct formed between dibenzo-30-crown-10 and the complex $[Pt(bipy)(NH_3)_2]^{2+}$ $[2PF_6]^{2-}$ (taken from J.F. Stoddart et al., *Angew. Chem., Int. Ed. Engl.,* **20,** 1051, 1981).

1.6 Miscellaneous Complexes

The H_3O^+ Cation. A hydroxonium – macrocyclic polyether complex. The hydroxonium ion has been isolated as a discrete entity by inclusion in a macrocyclic cavity. The X-ray crystal structure of the complex formed by tetracarboxylic 18-crown-6 with H_3O^+ was determined (Fig. 1-36). The H_3O^+ cation is anchored in the center of the cavity by three $OH^+\cdots O$ hydrogen bonds (2.67, 2.73 and 2.74 Å) formed bet-

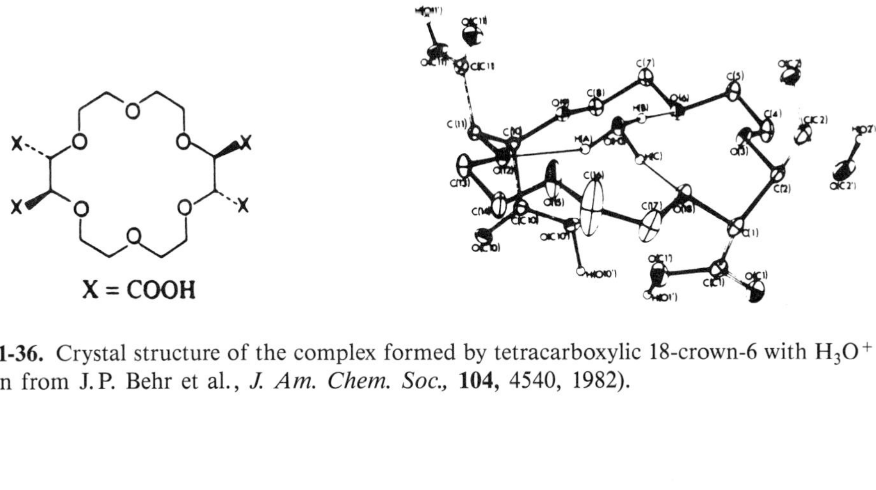

Fig. 1-36. Crystal structure of the complex formed by tetracarboxylic 18-crown-6 with H_3O^+ (taken from J.P. Behr et al., *J. Am. Chem. Soc.,* **104,** 4540, 1982).

ween the H_3O^+ cation and macrocyclic oxygens 6, 12 and 18. Further stabilization arises from ion-dipolar interactions between the H_3O^+ ion and oxygens 3, 9 and 15. The H_3O^+ ion is pyramidal [46].

A complex of 18-crown-6 containing two H_3O^+ ions in the cavity has been isolated (Fig. 1-37). The X-ray crystal structure shows that the dichloropicrate anions are hydrogen bonded to the H_3O^+ ions whereas there are no strong interactions between the macrocycle oxygen atoms and the cations ($H_3O^+ \cdots O$ = 2.8 – 2.9 Å cf. ~ 2.7 Å for the previous example) [47].

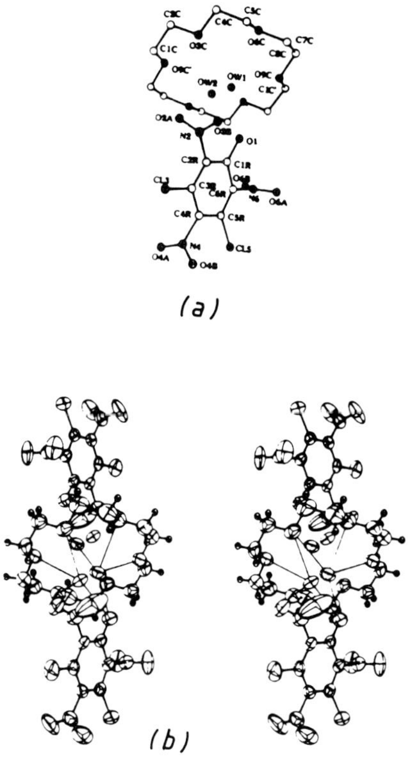

Fig. 1-37. (a) Complexation of two H_3O^+ ions by the crown [18-O-6], (b) stereodrawing of the complex $\{[H_3O^+, (PiCl_2^-)]_2, [18\text{-O-}6]\}$ (taken from I. M. Kolthoff et al., *Acta Cryst.,* **C40,** 1584, 1984. A more recent study of this complex at low temperature (181 K) revealed that its composition is ligand-water-picric acid rather than ligand-H_3O^+-picrate; I. M. Kolthoff et al., *Acta Cryst.,* **C44,** 303, 1988).

In the complexes formed between dimethylthallium picrate and two isomers of dicyclohexano-18-crown-6, the linear Me_2Tl^+ ion is equatorially surrounded by six crown ether oxygen atoms; the ion pair is "threaded" through the macrocycle (Fig. 1-38). The cyclohexane rings are in the chair conformation. There is a pseudo plane of symmetry for the *cis-syn-cis* isomer, (a), and a center of symmetry for the *cis-anti-cis* isomer, (b) [48].

It is possible to envisage synthesizing catenanes through replacement of the methyl groups by X substituents having reactive groups at the chain ends (Fig. 1-39).

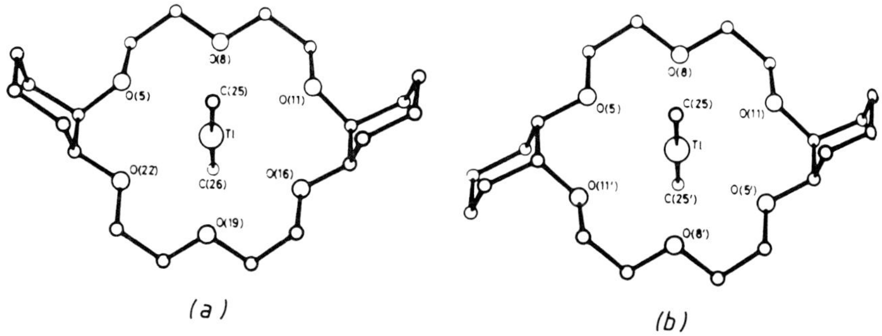

Fig. 1-38. Crystal structures of the complexes formed between two isomers of dicyclohexano-18-crown-6 (*cis-syn-cis* and *cis-anti-cis*) and dimethylthallium picrate (taken from M. R. Truter et al., *J. Chem. Soc., Chem. Comm.,* 727, 1982).

Fig. 1-39. Suggested synthesis of catenanes using dialkylthallium compounds having X functional groups.

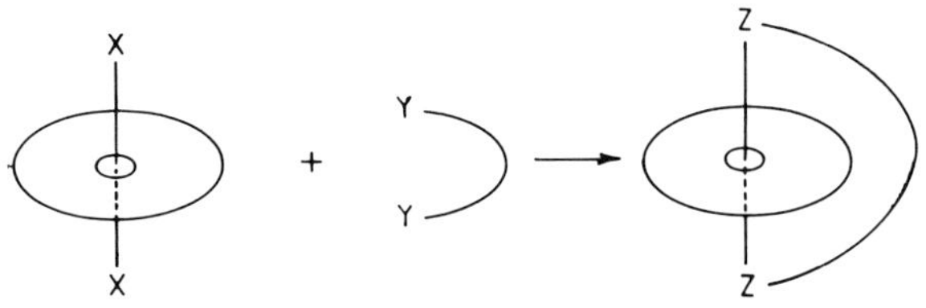

1.7 Complexation Sites other than Oxygen

1.7.1 Azaoxa Macrocycles

(a) Synthesis

The Stetter and Marx method [49]. The first examples of diazapolyoxa macrocycles (Fig. 1-40 a) were obtained by J. M. Lehn et al. who also discovered their complexing properties [50].

The method used was that of Stetter and Marx starting from a diamine and an acyl dichloride which were condensed under high dilution conditions. The monocyclic diamide obtained was then reduced (Fig. 1-40 b).

The Richman and Atkins method [51]. These workers avoided the high dilution technique by reacting bissulfonamide sodium salts with dihalides or disulfonates (Fig. 1-41).

Starting from *2* and *3 a*, the yield of cyclic product *1 b* was greater than 70% (80% for the cyclization step, and 90% for the acid hydrolysis of *1 a*). Furthermore, this method is versatile, and many amines and amine ethers have been prepared [52] of which some examples are given in Fig. 1-42.

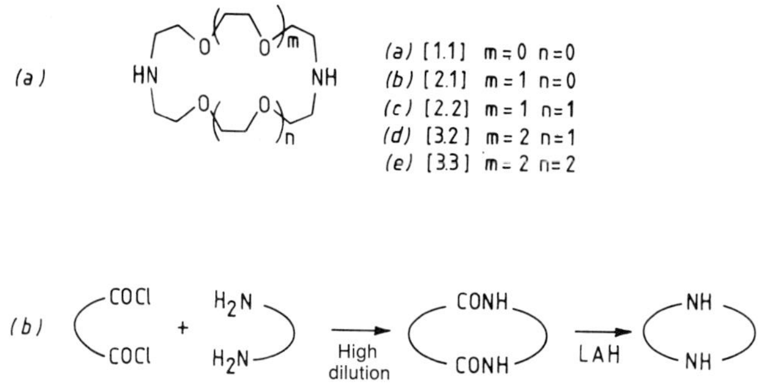

Fig. 1-40. Synthesis of diazapolyoxa macrocycles.

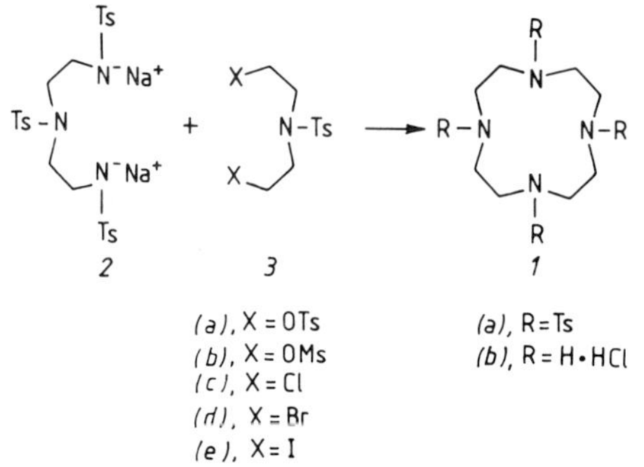

Fig. 1-41. Synthesis of tetraazamacrocycles [51]. N.B.: This method was reported in 1954 by Stetter and Roos (*Chem. Ber.*, **87**, 566, 1954), who performed the cyclization under high dilution conditions. According to Richman and Atkins, high dilution conditions are unnecessary.

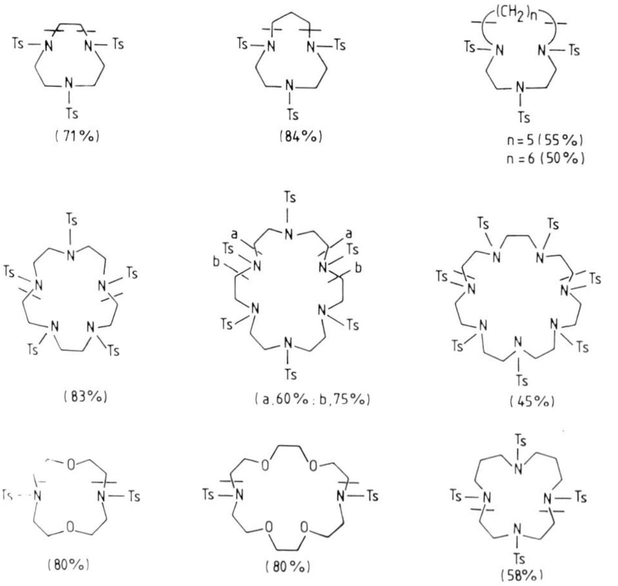

Fig. 1-42. Syntheses of polyaza and polyoxadiaza macrocycles. The bond formation sites, and hence the nature of the starting materials, are indicated by a slash.

Changing the leaving group X has a marked effect on the cyclization. Tosylates and mesylates give comparable yields these being greater than those obtained using the halides.

Preferential formation of macrocycle over that of open chain products has been attributed to the unusually small loss of internal entropy on cyclization [53]. The bulky *p*-toluenesulfonyl groups in e. g., *2* and *3a* (Fig. 1-41) restrict rotation about the bonds to the N-Ts groups. *Terminal group* and *rigid group* effects on the probability of cyclization are therefore observed.

The detosylation reaction (using hot concentrated sulfuric acid or hydrobromic acid in acetic acid in the presence of phenol) occurred in high yield leading to polyazaoxa and polyaza macrocycles. These compounds form complexes with numerous transition metal cations [54] and in their protonated form they bind anions [55a]. These two aspects fall outside the scope of this book and will not be further developed.

It should be noted that the introduction of nitrogen atoms into the macrocycle has two important consequences: (a) the acidity of the solution will have a direct influence on complexation so that pH measurements can be used to establish the stability constants, and (b) the amine group can be used as a reactant site in order to build up more complex structures, e. g., macropolycycles. A very comprehensive review concerning the synthesis of azaoxa macrocycles has recently appeared [55b].

(b) Structures

The crystal structure of 1,7,10,16-tetraoxa-4,13-diazacyclooctadecane (TDCO) has been determined [56]. The oxygen atom lone pairs are directed towards the interior of the ring, whereas the nitrogen atom lone pairs are orientated towards the exterior, the N-H bonds pointing towards the interior (Fig. 1-43a). The four N-H···O bonds ensure internal solvation of the NH bonds and stabilize the crown conformation (Fig. 1-43b) which is similar to that found in its complexes, and also that of 18-crown-6.

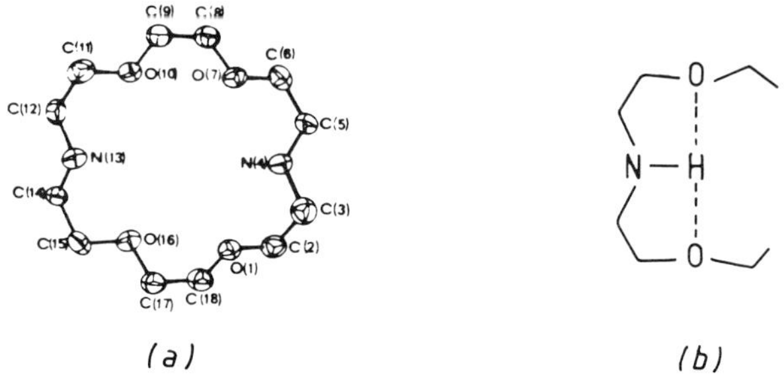

Fig. 1-43. Crystal structure of 1,7,10,16-tetraoxa-4,13-diazacyclooctadecane (TDCO) (taken from R. Weiss et al., *Bull Soc. Chim. Fr.,* 549, 1972).

The complex of the above macrocycle, TDCO, with potassium thiocyanate consists of chains formed by the complex cations $[K(TDCO)]^+$ and the SCN^- anions [57]. The K^+ cation, which is situated at the center of the cavity, is bound directly to the four oxygen atoms and the two nitrogen atoms (Fig. 1-44). The conformation of the ring in the complex can be superimposed on that of the free ring except for the orientations of the NH bonds, the nitrogen lone pair pointing towards the interior and participating in the bond with K^+. The distances between the oxygen atoms via the center of symmetry are not significantly different from those in the free ring whereas the N(4)-N(13) distance is shorter.

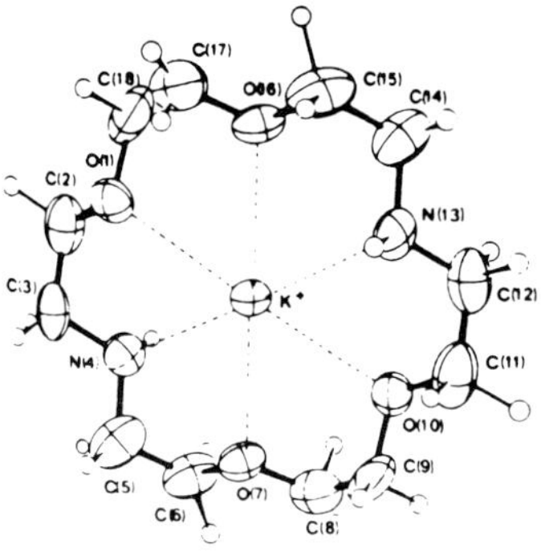

Fig. 1-44. The TDCO potassium complex (taken from R. Weiss et al., *Bull Soc. Chim. Fr.,* 551, 1972).

In the complex of copper (II) chloride with TDCO (Fig. 1-45), the copper atom is situated in the cavity and its coordination polyhedron is an octahedron with two chlorine, two nitrogen and two oxygen atoms at the corners [58]. Thus, two oxygen atoms do not participate in the coordination. This complex is an example of non-complementarity, the ring being too large and the sites too numerous and ill-placed.

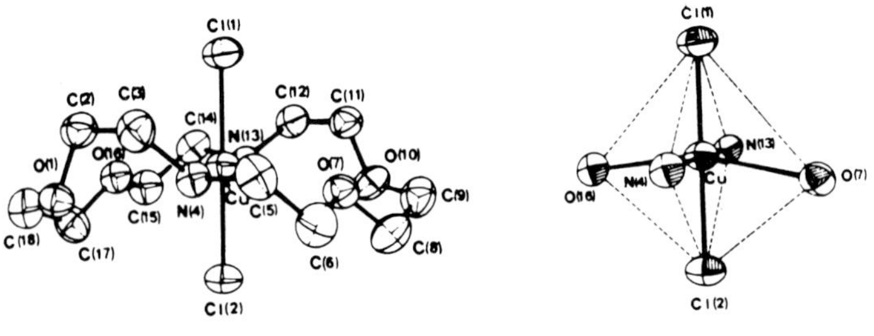

Fig. 1-45. Structure of the complex $CuCl_2$/TDCO and the coordination polyhedron of copper (taken from R. Weiss et al., *Acta Cryst.,* **B29**, 542, 1973).

1.7.2 Oxathia Macrocycles

(a) Synthesis

Three out of the many methods for the preparation of thiacrowns were considered by Bradshaw et al. for the preparation of 1-thia-[18-crown-6], as shown in Fig. 1-46 [59].

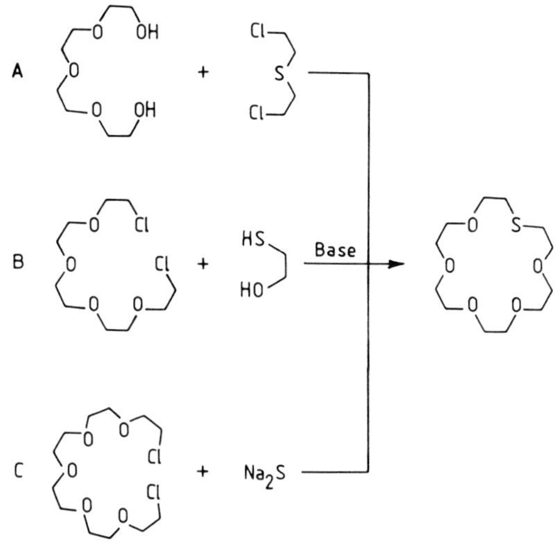

Fig. 1-46. Synthetic routes to thiacrowns.

In the first process, A, tetraethylene glycol is readily available, and the dichloride can be prepared from its corresponding glycol. The dichloride, however, is a β-halosulfide (vesicant) and is very dangerous. In addition, the glycol is not reactive unless strong bases are used.

Process B could be eliminated because the alcohol group of 2-mercaptoethanol is relatively unreactive.

Process C involves two readily available compounds of relatively high reactivity, the macrocycle being obtained in 36% yield.

Another method, more widely used, consists of reacting a dithiol with a dihalide under basic and high dilution conditions (Fig. 1-47).

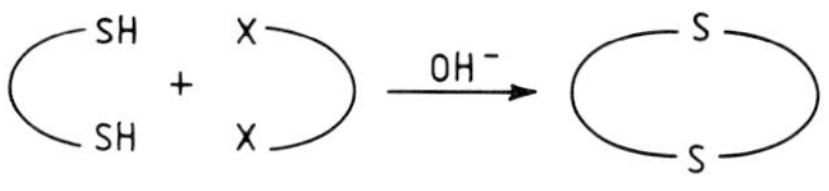

Fig. 1-47. Synthesis of a dithia macrocycle.

This method was employed for the first time by Dann et al. [60] for the preparation of 4,7,13,16-tetraoxa-1,10-dithiacyclooctadecane. It has also been used for the synthesis of a large number of other crown thioethers (Fig. 1-48).

(b) Structures

The conformation of 1,4,7,10,13-pentaoxa-16-thiacyclooctadecane was found to be centrosymmetric and isomorphous with the corresponding hexaether [61, 62]. Both the sulfur and the O(7) oxygen are directed out of the cavity and give the molecule an elliptical shape (Fig. 1-49).

Fig. 1-48. Oxathia macrocycles.

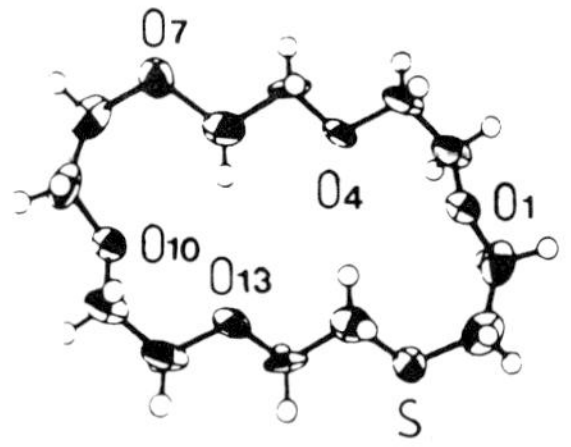

Fig. 1-49. Structure of 1,4,7,10,13-pentaoxa-16-thiacyclo-octadecane [18-O5-S], cf. Fig. 1-65 (taken from N.K. Dalley et al., *Acta Cryst.,* **B37**, 1739, 1981).

The conformations of the metal complexes differ considerably from those of the ligand [63]. Na^+, K^+ and Rb^+ cations are coordinated to the five ring oxygen atoms (Fig. 1-50).

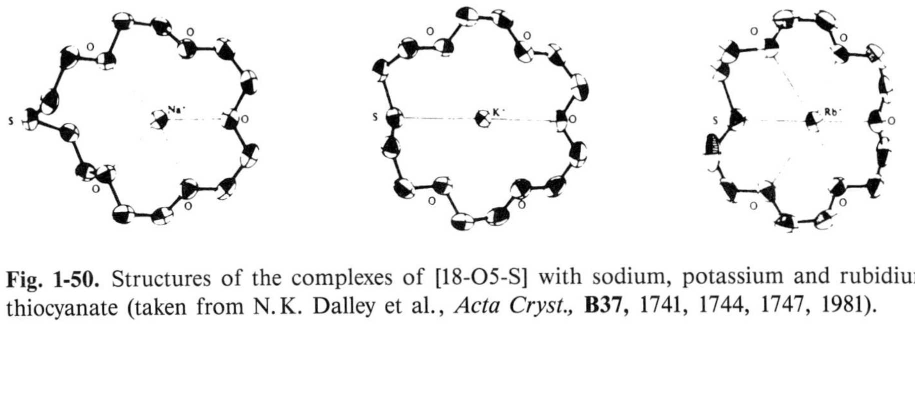

Fig. 1-50. Structures of the complexes of [18-O5-S] with sodium, potassium and rubidium thiocyanate (taken from N.K. Dalley et al., *Acta Cryst.,* **B37**, 1741, 1744, 1747, 1981).

The sulfur atom does not interact with the Na^+ ion whereas it is weakly coordinated to the K^+ and Rb^+ ions. The thiocyanate anion is coordinated through N to the cations, quite strongly in the cases of Na^+ and Rb^+, and more weakly with K^+.

In the complex with silver (Fig. 1-51), the sulfur atom is strongly coordinated to Ag^+. The Ag-S distance (~ 2.6 Å) indicates covalent character in this interaction [63].

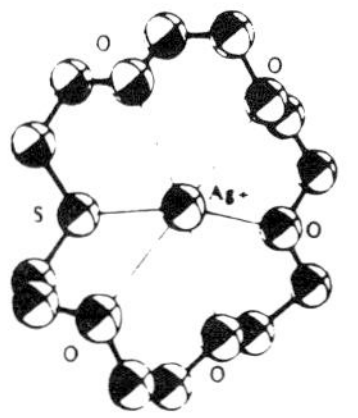

Fig. 1-51. Structure of the silver nitrate complex of [18-O5-S] (taken from N.K. Dalley et al., *Acta Cryst.,* **B37**, 1750, 1981).

The structure of 1-oxa-4,7,10-trithia[12-crown-4] (Fig. 1-52) showed that only the oxygen atom is directed towards the center of the ring, whereas the sulfur atoms are directed outwards [59, 64]. This is generally observed for thiamacrocycles.

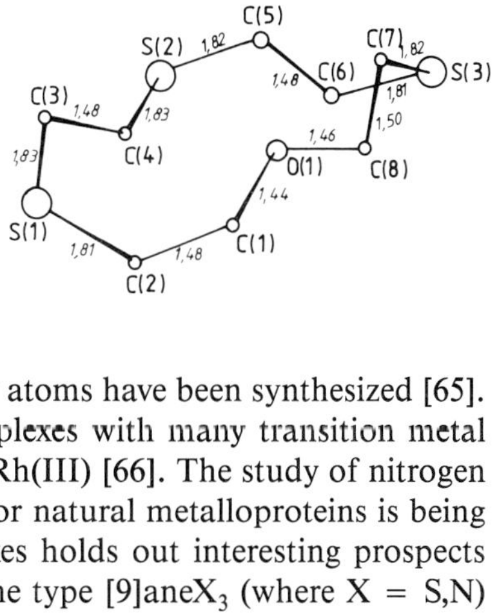

Fig. 1-52. Structure of 1-oxa-4,7,10-trithia[12-crown-4] (taken from J.S. Bradshaw et al., *J. Heterocycl. Chem.,* **11**, 45, 1974).

Numerous macrocycles containing sulfur atoms have been synthesized [65]. The macrocycle hexathia[18-crown-6] forms complexes with many transition metal cations: Pd(II), Pt(II), Co(II), Cu(II), Cu(I) and Rh(III) [66]. The study of nitrogen and sulfur containing ligands which are models for natural metalloproteins is being actively pursued; the reactivity of these complexes holds out interesting prospects [67a]. The complexing properties of ligands of the type [9]aneX_3 (where X = S,N) are of great interest [67b].

Chapter 2

Natural Macrocyclic Ionophores

2.1 Introduction

The following features of artificial receptor sites for cations have been developed in the previous chapter:

- shape of the receptor site cavity and its distortion,
- hydrogen bonds which determine the conformation, and the breaking of which can modify it,
- internal with respect to external polarity,
- the existence of a second complexation sphere,
- effects of anions and of the solvent,
- site modifications: oxygen atoms replaced by those of nitrogen, sulfur, etc.

The discovery of these phenomena in specific and simple cases has permitted determination of their effects and an understanding of the properties of more complex natural products. This has involved investigation of the occurrence of these general properties amongst the natural ionophore macrocycles.

Thus, detailed analysis of the orientation of substituents attached to the macrocycle and control of this orientation (cf. Section 1.4), leads to general concepts concerning the properties required for a channel: endopolarity, exolipophilicity, structure preservation via hydrogen bonding, and ion transport capability. Ionophores can be characterized as receptors which form stable complexes with charged species such as Na^+, K^+, and Ca^{2+}, thus being able to transport them across lipophilic phases, in particular membranes. There are several classes of natural macrocyclic ionophores [68, 69]:

- cyclodepsipeptides formed from alternating α-amino acids and α-hydroxy acids,
- cyclopeptides formed solely from α-amino acids,
- cyclodepsides which consist only of α-hydroxy acids.

2.2 Structural Elements

2.2.1 Peptides

The structural elements of peptides are the amino acids [70].

As a result of the existence of two resonance canonicals (Fig. 2-1 a), the peptide linkage has various well known features:

- It has marked double bond character leading to strongly restricted rotation (rotation barrier ~20 kcal/mol).
- It is more or less planar.
- The *trans* conformation is the more stable (by about 8 kcal/mol) (Fig. 2-1 b).

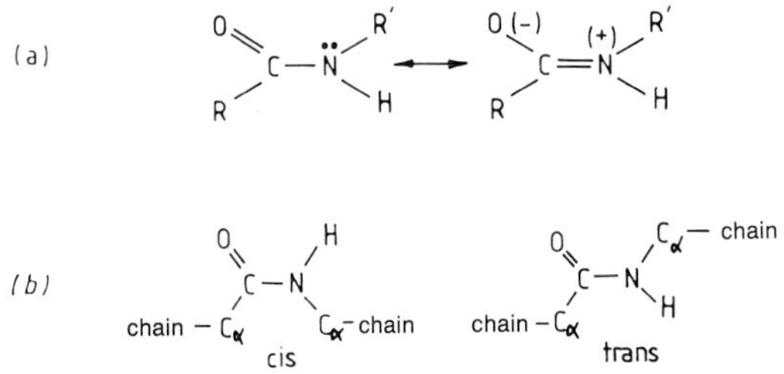

Fig. 2-1. (a) Amide bond of the peptide type, (b) *cis* and *trans* conformations of the peptide link.

Proline has the following specific and important properties:

- Rotation about the $N-C_\alpha$ bond (part of a 5-membered ring) is severely restricted. As a result, proline is relatively rigid and is of limited conformational mobility.
- The energy difference between the *trans* and *cis* forms of the peptide linkage formed with proline is small (Fig. 2-2).

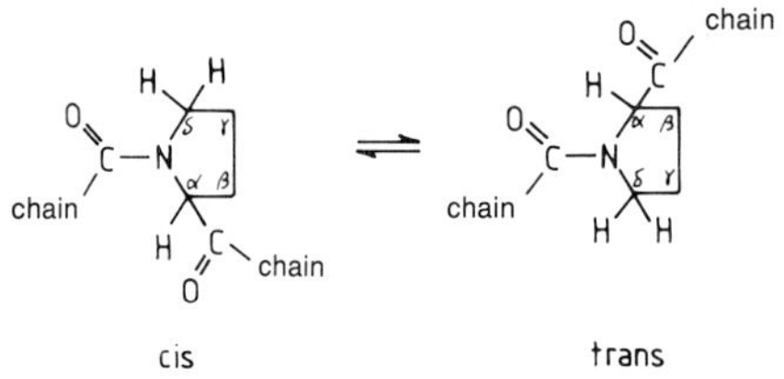

Fig. 2-2. *Cis* and *trans* conformations of the peptide link with proline.

The latter point explains the frequent presence of this residue at turn regions in proteins. Glycine, which is unsubstituted at carbon, is also common at protein turn regions.

a) Hydrogen bonds

These are of deciding influence on the general conformation of the protein since it is they which preserve the secondary structure. Depending on the number of peptide units involved, the following interactions can be distinguished:

- 5 → 1 (α-turn stabilizing the α helix),
- 4 → 1 (β-turn),
- 3 → 1 (γ-turn).

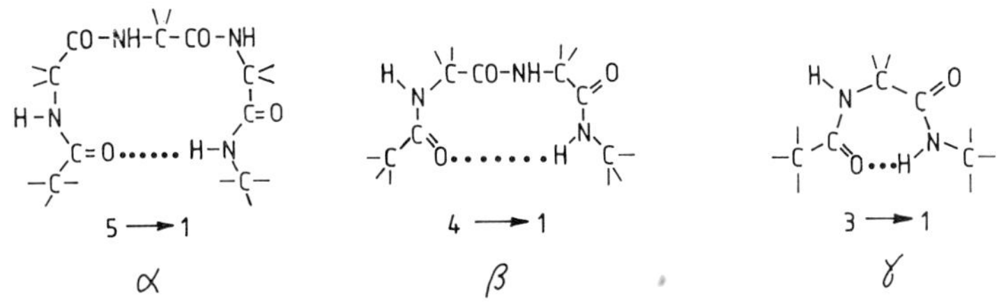

Fig. 2-3. Rings formed via intramolecular hydrogen bonding.

In the chain, β- and γ-turns are present at regions of change of direction [71]. It should be noted that the rings formed via these hydrogen bonds are 13- (α-turn), 10- (β-turn), and 7-membered (γ-turn).

b) Lateral appendages

Side chains allow control of the polarity depending on the nature of lipophilic (valine, leucine, phenylalanine), hydrophilic (serine, threonine) and ionic (lysine, aspartic acid, glutamic acid and arginine) groups.

2.2.2 Depsipeptides and Depsides

Depsipeptides contain both peptide units and ester links. Depsides have only the latter. In these compounds, the ester group ensures local planarity. Although a rotation barrier exists, it is much lower than that for amides (~5–10 kcal/mol) since the canonical form b is much less important (Fig. 2-4). The *trans* form is preferred.

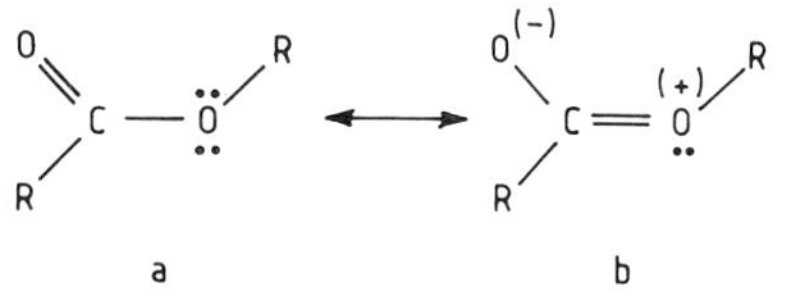

Fig. 2-4. Resonance canonicals for the ester group.

2.3 Cyclodepsipeptides and their Complexes

These macrocycles, of which several examples are shown in Fig. 2-5, are characterized by having lipid side chains and polar groups in the chain (amide, ester, ether).

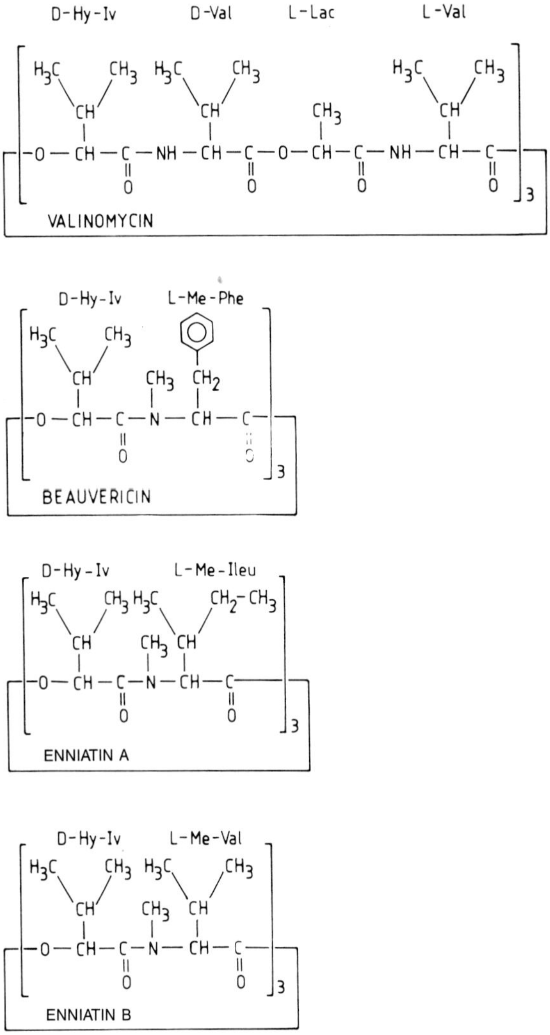

Fig. 2-5. Examples of cyclic antibiotic depsipeptides.

2.3.1 Valinomycin

This antibiotic has a molecular structure consisting of three identical tetradepsipeptide fragments, D-hydroxyisovaleric acid – D-valine – L-lactic acid – L-valine. Valinomycin is the natural macrocyclic ionophore with probably the most interesting properties. It played the most important part in the birth of the chemistry of molecular recognition.

a) Historical background

Valinomycin was isolated from *Streptomyces fulvissimus* by Brockmann and Schmidt-Kastner in 1955 [72]. It was synthesized in 1963 by Shemyakin et al. [73]. The understanding of its activity developed in several stages:

- In 1959, McMurray and Begg showed that it is a potent uncoupling agent in oxidative phosphorylation [74].
- Moore and Pressman then discovered the role of valinomycin in the transport of K^+ cations in mitochondria [1, 75]. By insertion of a K^+ ion selective electrode, they found that valinomycin gives rise to disappearance of K^+ and appearance of H^+. They concluded that the antibiotic catalyzes K^+/H^+ exchange across the mitochondrial membrane. In contrast, Na^+ was unaffected and K^+/Na^+ discrimination was thus very high.
- Since the synthetic analogs obtained by Shemyakin et al. were much less efficient than valinomycin itself, activation of a valinomycin receptor contained in a mitochondrial transport protein was proposed [76].
- However, Mueller and Rudin [77] showed that valinomycin renders artificial bilayer membranes permeable to K^+ cations. This led to rejection of the hypothesis of its acting via a receptor protein. The above workers therefore interpreted the phenomenon in terms of pore formation. However, valinomycin was also found to transport K^+ across liquid membranes which were too thick for pore formation. Consequently, the hypothesis of transport via a mobile carrier (ionophore) was finally established [78].

The ability of ionophores to complex alkali metal ions was discovered in 1951 by Harned et al. who observed the solubilization of alkali metal salts in hexane [79].

The stabilities of nonactin complexes with alkali metal cations were measured by Simon in 1967 [80a] and in the same year, the X-ray crystal structure of the K^+-nonactin complex was determined [80b].

b) Conformation

In solution. The conformation of free valinomycin in solution has been studied by various physical methods: nuclear magnetic resonance, circular dichroism, optical rotatory dispersion, and infrared and Raman spectroscopy [69, 81–90]. Calculations of conformational energy have also been made [91].

Ovchinnikov and Ivanov [82] found that the conformation depends on solvent polarity. They proposed the existence of three conformations, A, B, and C, in equilibrium, thc proportions of which depend on the solvent (Fig. 2-6).

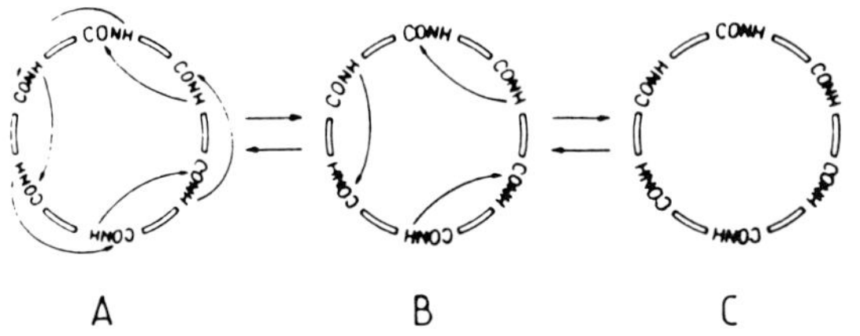

Fig. 2-6. Valinomycin conformations depending on solvent polarity. The arrows represent hydrogen bonds (taken from Y. A. Ovchinnikov et al., *Tetrahedron,* **30**, 1871, 1974).

In nonpolar solvents (CCl_4, octane), the bracelet-type A form with six intramolecular NH···OC hydrogen bonds is mainly present. The "helical" form B with three intramolecular hydrogen bonds predominates in solvents of medium polarity. In the C form found in polar solvents, there are no intramolecular hydrogen bonds since all the NH groups are hydrogen-bonded to solvent molecules.

In Fig. 2-7, the six hydrogen bonds formed under nonpolar conditions (A) and the three hydrogen bonds formed under weakly polar conditions (B) can be readily seen.

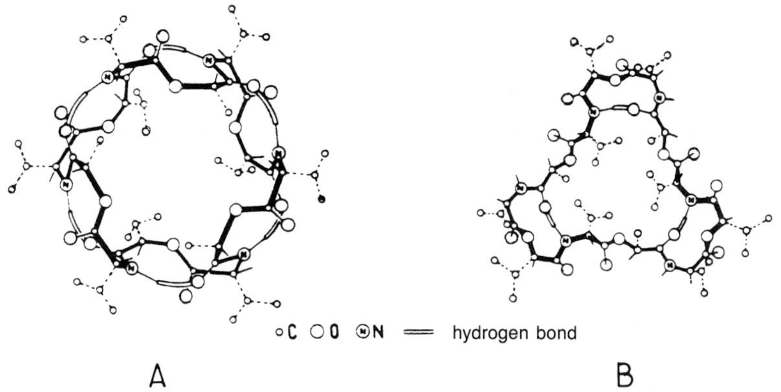

Fig. 2-7. Details of the hydrogen bonds in the A and B forms of valinomycin.

In the solid state. X-ray crystal structure studies of the valinomycin molecule have shown that it is without threefold symmetry but is of oval shape with a pseudocenter of symmetry [92–95] (Fig. 2-8a).

This form is maintained by six hydrogen bonds: the four between amide NH and CO groups are of the 4 → 1 type, and the remaining two between NH and ester CO groups are of the 5 → 1 type (Fig. 2-8b). The four 4 → 1 bonds result in a 10-membered ring and the 5 → 1 bonds give a 13-membered ring.

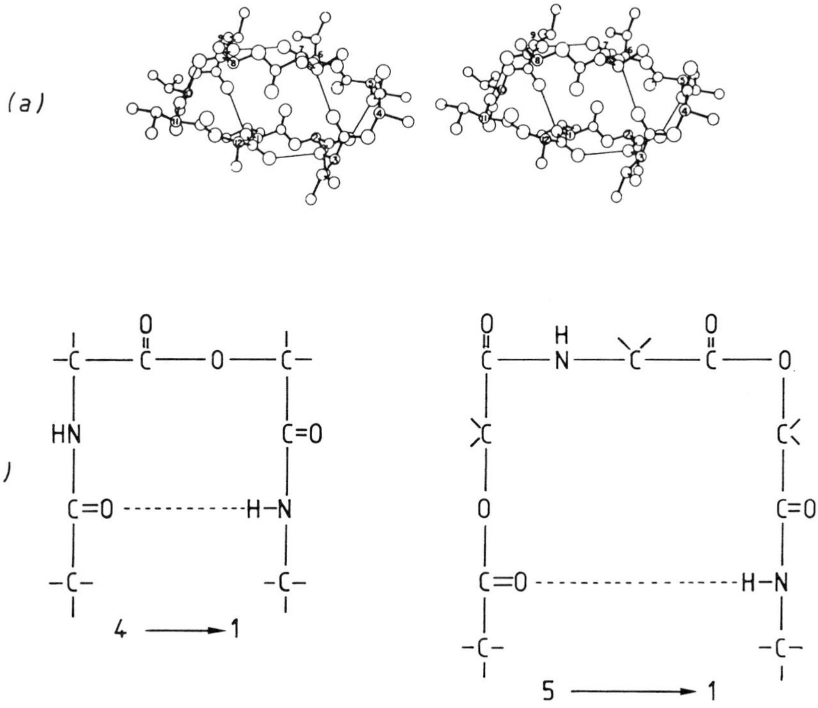

Fig. 2-8. (a) Stereodiagram of the conformation of crystalline valinomycin, (b) details of the hydrogen bonding (taken from I. L. Karle, *J. Am. Chem. Soc.*, **97**, 4379, 1975).

In solution, an equilibrium can exist between, on the one hand, this form with four 4 → 1 type bonds and two 5 → 1 type bonds and, on the other hand, the form described above with six 4 → 1 type bonds (Fig. 2-6).

c) K^+-valinomycin complex

Complexation changes the conformation and transforms the two 5 → 1 type OCO···HN bonds between ester and amide groups into 4 → 1 type NCO···HN bonds between two amide groups [95, 41].

A detailed portrayal of these transformations is given in Fig. 2-9. The 5 → 1 type hydrogen bonds 1 and 1′ (amide-ester) are broken and they are replaced by 4 → 1 type bonds (amide-amide) represented by the two arrows.

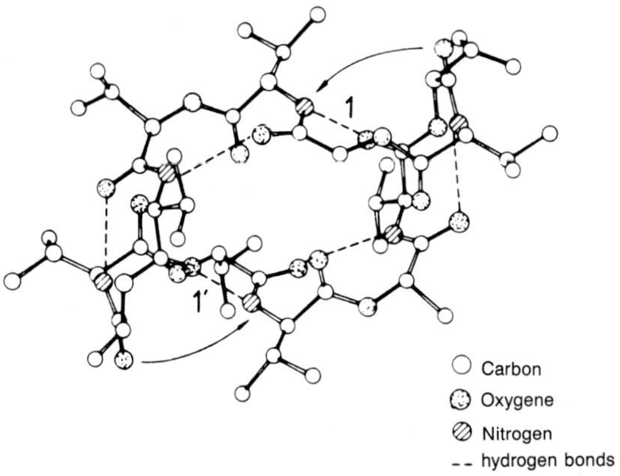

Fig. 2-9. Transformation of 1 and 1′ hydrogen bonds (5 → 1) into 4 → 1 hydrogen bonds when K^+ cation complexation occurs (taken from H. Hauptman et al., *Science,* **176,** 911, 1972).

Fig. 2-10 shows a "planar" representation of this modification of the hydrogen bond network which occurs on complexation.

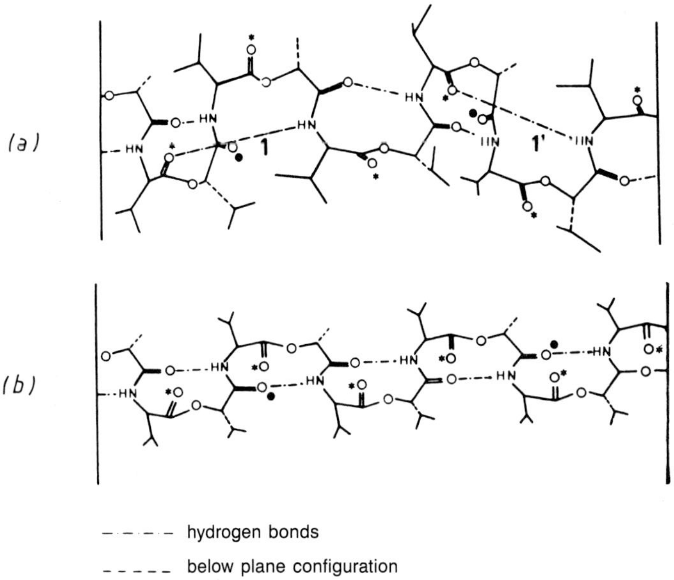

Fig. 2-10. Hydrogen bonding in (a) free valinomycin and (b) the valinomycin-K^+ complex. In the free ligand, (a), two 13-membered rings and four 10-membered rings are formed via 5 → 1 and 4 → 1 type hydrogen bonding. In the complex, (b), 10-membered rings are formed via 4 → 1 type hydrogen bonding.

The rupture of the 1 an 1′ bonds frees the two ester carbonyl groups for coordination to the cation.

In this complex having threefold symmetry, the cation is located in the center of a 36-membered macrocycle and is octahedrally coordinated to six ester carbonyl groups [94–97] (Fig. 2-11).

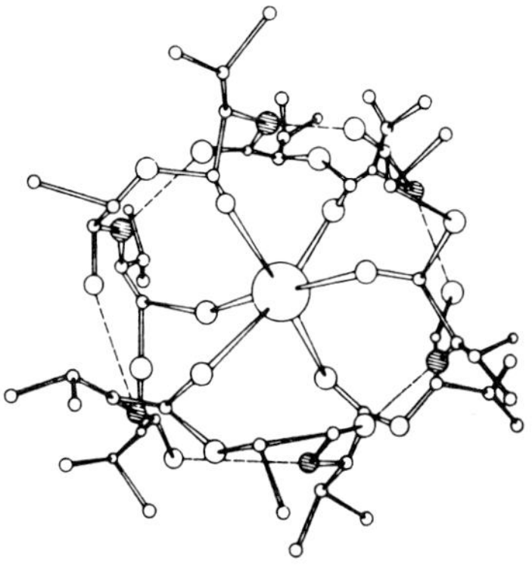

Fig. 2-11. Crystal structure of the valinomycin-K^+ complex (taken from M. Dobler et al., *Helv. Chim. Acta,* **58,** 432, 1975).

All the carbonyl oxygen atoms of the amide groups take part in hydrogen bonding (see also Fig. 2-10b) which stabilizes the ring conformation and leads to limited flexibility. As a result, the size of the cavity cannot be adjusted and complexation is therefore selective. The diameter of the cavity corresponds to the size of K^+ or Rb^+.

All the lipophilic side chains point towards the exterior of the molecule which explains the solubility of the complex in media of low polarity (membranes).

Ovchinnikov et al. have synthesized a large number of valinomycin analogs, in particular enantio-valinomycin [98]. The latter exhibits identical antibiotic activity to that of valinomycin thus excluding the possibility of interaction with an enantiospecific receptor.

Anion effect. The K^+ cation appears to be well protected inside the valinomycin cavity. However, the crystal structure of the valinomycin – potassium picrate complex indicates a weak but definite interaction between an oxygen atom of the *p*-nitro group and the K^+ cation [99] (Fig. 2-12). The fact that the picrate anion can interact with K^+ implies that the cation is not entirely protected and that access can occur from either above or below, thus suggesting torus topology.

Fig. 2-12. Stereoview of the crystal structure of a valinomycin-potassium picrate-xylene complex. The picrate anion is to the left of the figure, and the xylene molecule to the right (taken from L. K. Steinrauf et al., *J. Am. Chem. Soc.,* **103,** 5880, 1981).

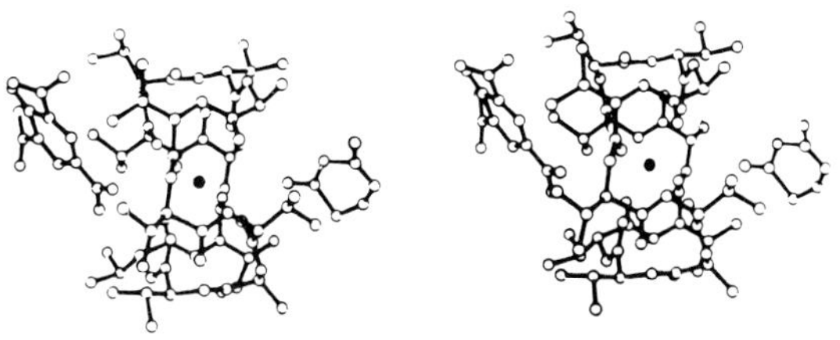

d) Na^+-valinomycin complex

The crystal structure of a complex of valinomycin with sodium picrate revealed a water molecule occupying the center of the cavity. The Na^+ ion is displaced by 2.3 Å with respect to the position of K^+ (Fig. 2-13a). It is thus situated external to the ring, being bonded to the three carbonyl groups adjacent to the lactyl residues as well as to the water molecule and the two picrate anion oxygen atoms [100].

These differences in the coordination of the Na^+ and K^+ cations (Fig. 2-13b) can explain the effects of picrate-like anions on cation transport in membranes, these effects being much greater for Na^+ than for K^+.

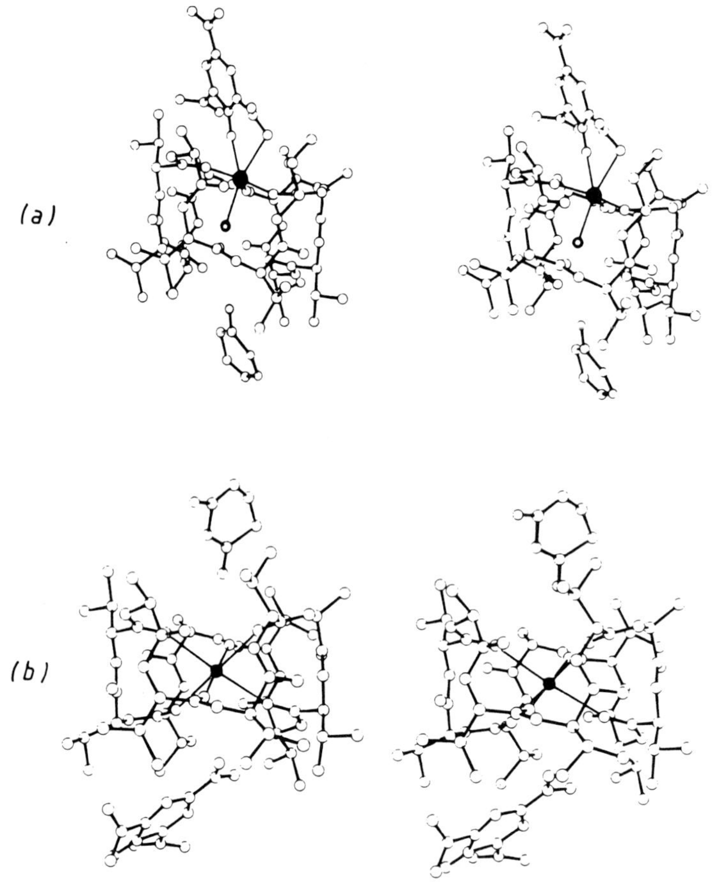

Fig. 2-13. (a) Crystal structure of a valinomycin-sodium picrate-H_2O-xylene complex. The water molecule is represented by the heavy circle, and the cation by the solid circle (taken from L. K. Steinrauf et al., *J. Am. Chem. Soc.,* **104,** 4085, 1982). (b) Crystal structure of the valinomycin-potassium picrate-xylene complex for comparison. (Stereoviews.)

They also explain the very different spectra shown by valinomycin-potassium and valinomycin-sodium complexes in solution.

In view of these results, sixfold symmetry for uncomplexed valinomycin in membranes was suggested with a water molecule in the center of the cavity [100].

2.3.2 Beauvericin. Enniatins A and B

Beauvericin and enniatins A and B are hexadepsipeptides, in other words 18-membered macrocycles with alternating *D* and *L* residues (Fig. 2-14). An important difference with valinomycin is the presence of methyl groups on the amide group nitrogen atoms. As a result, intramolecular hydrogen bonds are absent leading to greater flexibility of the macrocycle and less selective complexation. Furthermore, N-methylation affects the position of conformational equilibrium for the peptide link. Replacement of H by CH_3 makes the *cis* form less unfavorable with respect to the *trans* form resulting in greater mobility of the macrocycle.

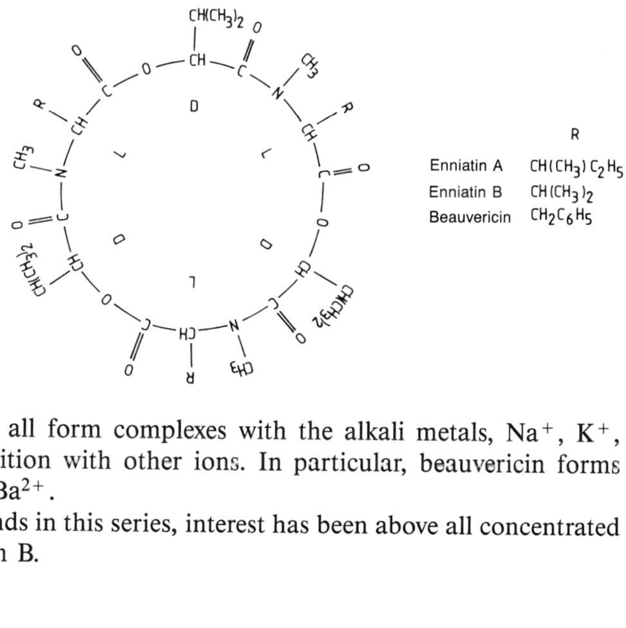

Fig. 2-14. Structures of enniatins A and B, and of beauvericin.

These compounds all form complexes with the alkali metals, Na^+, K^+, Rb^+ and Cs^+, and in addition with other ions. In particular, beauvericin forms complexes with Ca^{2+} and Ba^{2+}.

Amongst compounds in this series, interest has been above all concentrated on beauvericin and enniatin B.

a) Beauvericin complexes

Beauvericin complexes alkali metal cations [101] and, unlike valinomycin, it also complexes the alkaline earth metal cations [102]. The selectivity of transport across an artificial membrane (chloroform) is:

$$Rb^+ > K^+ > Ba^{2+} > Na^+ > Ca^{2+} > Li^+.$$

This selectivity is different with other membranes (liposomes, chromatophores):

$$Ca^{2+} > K^{+} > Cs^{+} > Li^{+} > Na^{+} \quad [103].$$

It should be noted that in the latter case, calcium is transported with an apparent positive charge of +1. Furthermore, it has been observed that the transport properties depend strongly on the anion [101]. Thus, in the presence of picrate anion, the selectivity decreases in the order $Ba^{2+} > Ca^{2+}$ whereas with chloride anion the reverse selectivity is found ($Ca^{2+} > Ba^{2+}$) [104]. The crystal structure of the complex of beauvericin with barium picrate provides an explanation for these properties [104, 105]. The adduct has a 2:2 dimer structure of the form $(Bv, Ba, Pic_3, Ba, Bv)^+$ Pic^-, being a "sandwich" complex with a central layer of three picrate anions between two Ba^{2+} cations outside which are two external layers of beauvericin (Fig. 2-15). This beauvericin-cation-anion association explains both the apparent charge of +1 and the high sensitivity of the transport properties towards the nature of the anion.

a)

b)

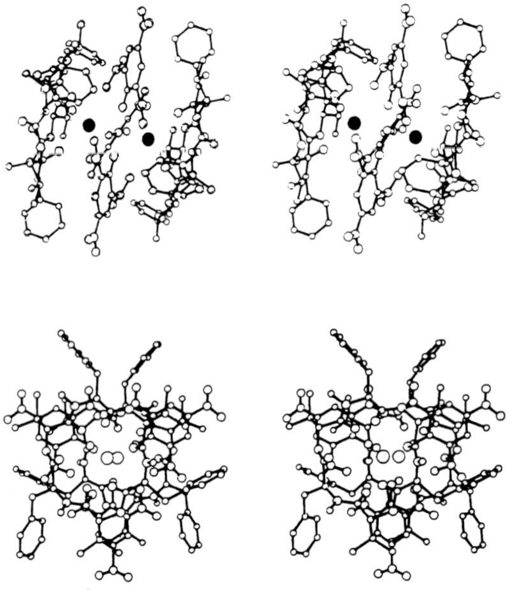

Fig. 2-15. Crystal structure of the beauvericin-barium picrate complex viewed down (a) the twofold axis and (b) the threefold axis (stereoviews) (taken from L. K. Steinrauf et al., *J. Am. Chem. Soc.*, **102**, 2704, 1980).

b) K^+-enniatin complexes

Physicochemical studies of free enniatin B in solution have shown that it exists in two conformations in equilibrium [106–108] (Fig. 2-16). The (a) form predominates under nonpolar conditions whereas the (b) form is predominant in polar media and in the solid state.

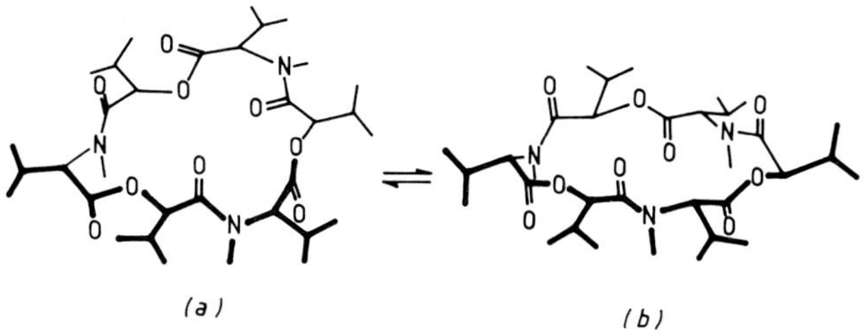

Fig. 2-16. Conformation of enniatin B in (a) nonpolar and (b) polar solvents.

The molecule is doughnut-shaped, and is polar on the inside and nonpolar on the outside.

The conformation of the potassium complex in the solid state has been determined by X-ray crystallography [109] and is identical to that of the complex in solution as inferred from NMR studies [106].

The carbonyl group oxygen atoms, arranged alternately above and below the median plane, form an approximately octahedral coordination sphere (Fig. 2-17).

In the solid, the molecules are stacked in columns. Sandwich structures have been proposed for the 2 : 1 and 3 : 2 complexes in solution [107, 110] (Fig. 2-18).

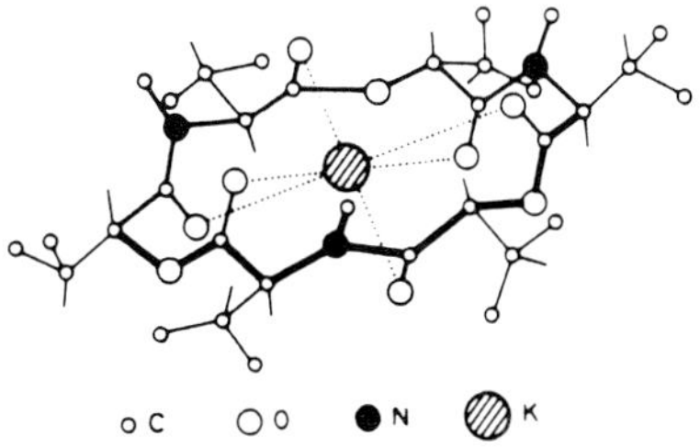

Fig. 2-17. Crystal structure of the enniatin B complex with K^+ (taken from Y. A. Ovchinnikov, *FEBS Letters,* **44**, 1, 1974).

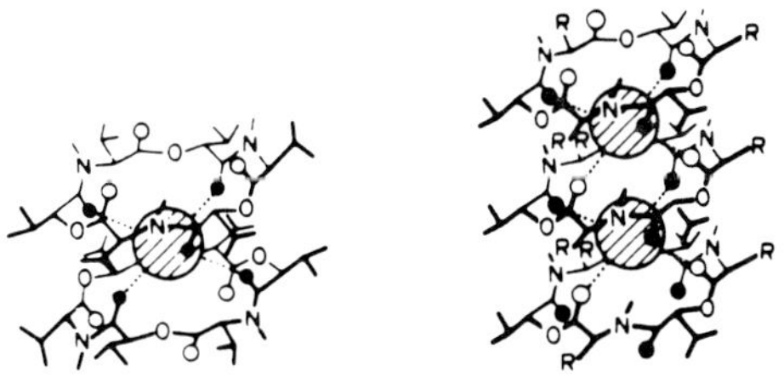

Fig. 2-18. Possible structures for enniatin B-cation complexes with stoichiometries ranging from 2/1 to 3/2.

2.4 Cyclopeptides

This other class of macrocyclic compounds is also very commonly met in biological systems. However, compared to cyclodepsipeptides, relatively few natural cyclopeptides are complexing agents for alkali and alkaline earth metal cations.

2.4.1 Antamanide

Antamanide is a cyclic decapeptide, containing exclusively *L*-amino acids (Fig. 2-19). It was isolated from *Amanita phalloides* by Wieland [111]. It counteracts the toxic effects of phalloidin, a peptide from the same toadstool, for which it is the unique antidote.

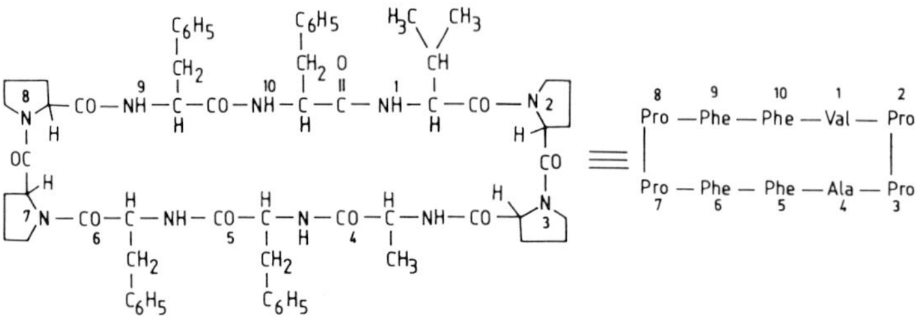

Fig. 2-19. Structure of antamanide.

The X-ray crystal structures of antamanide [112a] and of [Phe^4,Val^6]antamanide (Fig. 2-20), a symmetrical biologically active antamanide analog, have shown that the conformations of the two macrocycles are very similar [112b]. The substitution of the 4 and 6 residues therefore leads to only very minor conformational modifications.

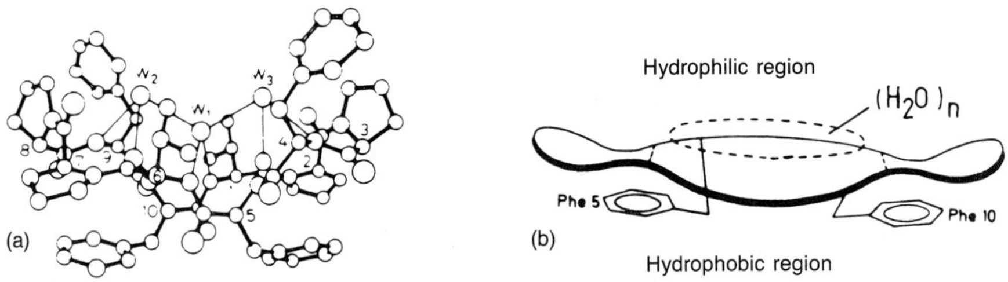

Fig. 2-20. (a) Conformation of [Phe^4,Val^6]antamanide. The water molecules are labeled W_i, and hydrogen bonds are indicated by light lines (taken from I. L. Karle et al., *Proc. Nat. Acad. Sci.,* **73**, 1782, 1976). (b) Schematic representation of the hydrophilic and hydrophobic regions (taken from H. L. Lotter, *Z. Naturforsch.,* **39 C**, 535, 1984).

In both cases the rings are elongated and relatively planar. They contain four and three water molecules, respectively, which are hydrogen-bonded to the NH groups. There are only two 5 → 1 type intramolecular hydrogen bonds. Two of the peptide bonds are *cis* (Pro^2-Pro^3, Pro^7-Pro^8), the remainder being *trans.*

An important and somewhat surprising result is to be noted, namely that the conformation of the macrocycle is independent of the solvent used for crystallization, whether polar or nonpolar. Only the number of water molecules bound is affected by the solvent [112b].

The Li^+-antamanide and Na^+–[Ph^4,Val^6]antamanide complexes were found to be isostructural. The cation is pentacoordinated, being surrounded by four carbonyl oxygen atoms and one molecule of the solvent used for crystallization (acetonitrile and ethanol, respectively) [113, 114]. The solvent molecule is located in a cavity situated above the molecule, the lower cavity remaining empty (Figs. 2-21a and b). In these complexes (Fig. 2-21c), the macrocycle conformation is very different from those for the free cyclodecapeptides (Fig. 2-21d).

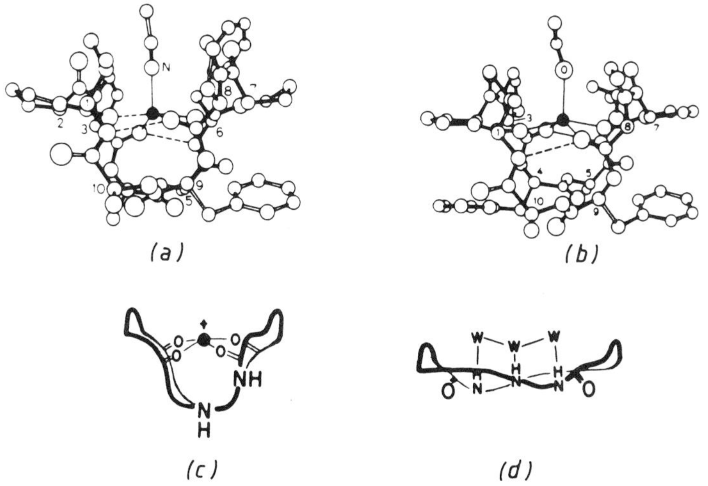

Fig. 2-21. Crystal structures of the (a) Li^+-antamanide and (b) Na^+-[Phe^4,Val^6]antamanide complexes (taken from I. L. Karle et al., *Proc. Nat. Acad. Sci.,* **70**, 1836, 1973). Schematic representation of (c) complexed and (d) free antamanide.

It should be noted that for antamanides, whether or not complexed, the conformations observed in the solid state are not always the same as those determined in solution studies [98, 115a].

A large number of antamanide analogs have been synthesized [98]. Enantioantamanide possesses only 1/20th of the antitoxic activity of antamanide which suggests that this activity is stereospecific contrary to the case for valinomycin.

The biological activity appears to depend not only on Na^+ complexation but also on other interactions. Thus, certain antamanide analogs are completely inactive although they form complexes which are more stable.

Complexation is a necessary but not sufficient condition. The presence of lipophilic chains and chirality also play a part, suggesting that the peptide becomes attached to a stereospecific receptor.

Recently, the crucial importance of lipophilicity has been demonstrated using another antamanide derivative [115 b]. This compound, cyclo[Val-Pro-Pro-Ala-Cha-Cha-Pro-Pro-Cha-Cha], was obtained by hydrogenation of the phenylalanyl residues in antamanide giving cyclohexylalanyl residues (Cha).

This derivative is without any antitoxic activity towards phalloidin in spite of its ability to form complexes with metal cations.

The X-ray crystal structure of its complex with Li^+ cation (Fig. 2-22) when compared to the structures described above provides an explanation for this absence of activity.

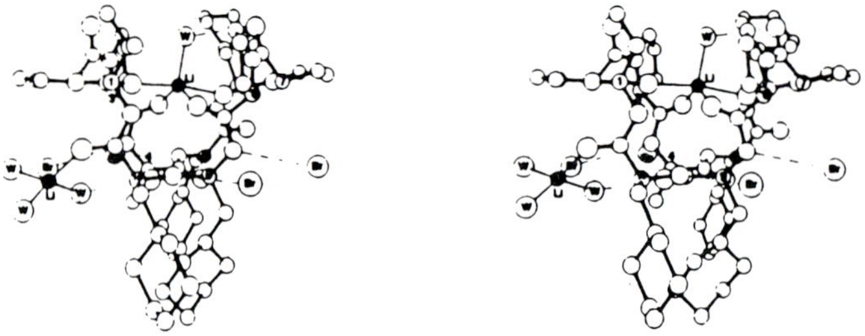

Fig. 2-22. Structure of the Li^+-[Cha^5,Cha^6,Cha^9,Cha^{10}]antamanide complex (stereopair view) (taken from I. L. Karle, *Proc. Nat. Acad. Sci.,* **82**, 7155, 1985).

As regards similarities, it can be seen that the macrocycle conformation is almost the same in all cases (Figs. 2-20 to 2-22) which explains why the derivative containing Cha residues is nonetheless a good complexing agent for cations.

The differences lie in the disposition of the Phe (Figs. 2-20 and 2-21) and Cha (Fig. 2-22) moieties with respect to the macrocycle. In the case of the Phe residues, the phenyl groups are folded against the macrocycle backbone thereby giving the complex marked lipohilic character (Fig. 2-20). With the Cha residues, the cyclohexyl groups are extended away from the macrocycle, resulting in the exposure of large portions of the polar backbone to the environment. This gives rise to:

- Binding of an additional external Li^+ ion which is coordinated to several molecules of water.
- Formation of seven hydrogen bonds (four $NH \cdots Br^-$ and three $C{=}O \cdots HOH$ bonds).

All these interactions result in the disappearance of the lipophilic character of the complex thus preventing its attachment to the lipophilic part of the membrane. The antitoxic potency is therefore lost. It should be recalled that the postulated mechanism for antitoxic activity involves attachment of the antamanide to the phalloidin receptor site.

2.4.2 Gramicidin S

Another natural cyclodecapeptide, gramicidin S (Fig. 2-23a), also behaves as an antibiotic, but in this case not because of its complexation and cation transport properties.

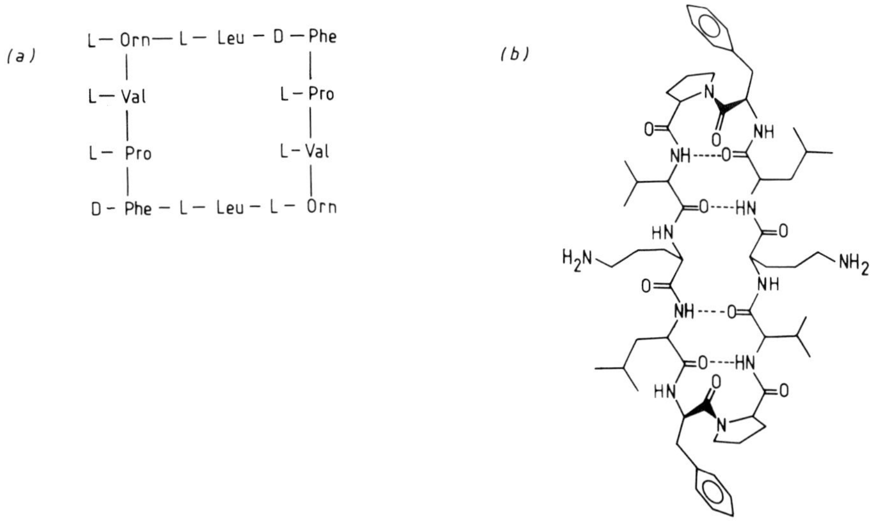

Fig. 2-23. Structure of gramicidin S; (b) shows the four hydrogen bonds which stabilize the macrocycle conformation.

Indeed, gramicidin S does not form complexes with alkali metal cations [116a]. This absence of complexing properties may appear surprising if this compound is compared to antamanide which is also a cyclodecapeptide and which strongly complexes alkali metal cations as we have just seen. This fundamental difference between the two cyclopeptides is worthy of comment.

In the first chapter devoted to synthetic macrocyclic complexes, we saw that the *size* of the macrocycle is of crucial importance. It determines the type of complex obtained. In the case of natural macrocycles, this factor also plays a role which, although not negligible, is influenced by other factors, in particular hydrogen bonding.

The structure of gramicidin S (Fig. 2-23b) reveals an extensive network of intramolecular hydrogen bonds which greatly stabilize the macrocycle conformation [116b]. It is likely that the absence of complexing properties is connected to the existence of this highly stabilized conformation since complex formation would necessitate these four hydrogen bonds being broken [115a].

2.4.3 Synthetic Cyclopeptides

We have already pointed out in Part I the importance of cyclopeptides in the understanding of the conformational states of "linear" proteins. An additional aspect of these investigations is the study of possible complexing properties of these compounds.

Another approach is to synthesize cyclopeptides which are analogous to natural cyclic compounds.

The synthesis and study of cyclopeptides are therefore carried out for many reasons, hence the interest of many laboratories in these substances. We shall only touch very briefly upon this vast field.

a) Cyclo-(L-Pro-Gly)$_3$

The Pro-Gly unit is very commonly met in proteins (e.g., collagen, elastin). The association of these two amino acids (each having, as we have already pointed out, very particular properties: rigidity of the 5-membered ring; little difference in energy between the *cis* and *trans* forms of the peptide bond formed with the proline nitrogen; low steric bulkiness of glycine) gives rise to specific effects at their location in the chain, such as reversal of chain direction.

The importance of the Pro-Gly unit is therefore the reason for its incorporation into cyclic systems, e.g., *cyclo*-(L-Pro-Gly)$_3$ [117a].

The existence of several conformers depending on the nature of the solvent has been shown by NMR spectroscopy and circular dichroism. In nonpolar solvents, the cyclopeptide has C_3 symmetry with *trans* peptide links and three $3 \rightarrow 1$ type hydrogen bonds (γ-turns) (Fig. 2-24a). The rigidity arises from these hydrogen bonds and from the proline groups.

In polar media, the conformation is no longer symmetric because of the presence of a *cis* peptide link. The structure of this conformer has been established by an X-ray study. It can be seen (Fig. 2-24b) that there are no intramolecular hydrogen bonds in this conformation but instead a dimer is formed (Fig. 2-24c) via intermolecular hydrogen bonding [117c].

Solution studies have shown that *cyclo*-(L-Pro-Gly)$_3$ forms complexes with many cations [117b]. In the case of alkali metal cations, it is selective for Li^+ and Na^+ with respect to K^+ and larger cations. The Ca^{2+} complex has a stability constant of 1.1×10^5 M^{-1} in acetonitrile. With Mg^{2+}, three types of complex are formed having Mg^{2+}: ligand stoichiometries of 2:1, 1:1 and 1:2. In all these complexes, the hexapeptide has C_3 symmetry (all the peptide links are *trans*), but the intramolecular hydrogen bonds existing in the free cyclopeptide are lost (Fig. 2-25a).

In the solid state, crystallographic studies for some isolated complexes have complemented and qualified the observed solution results.

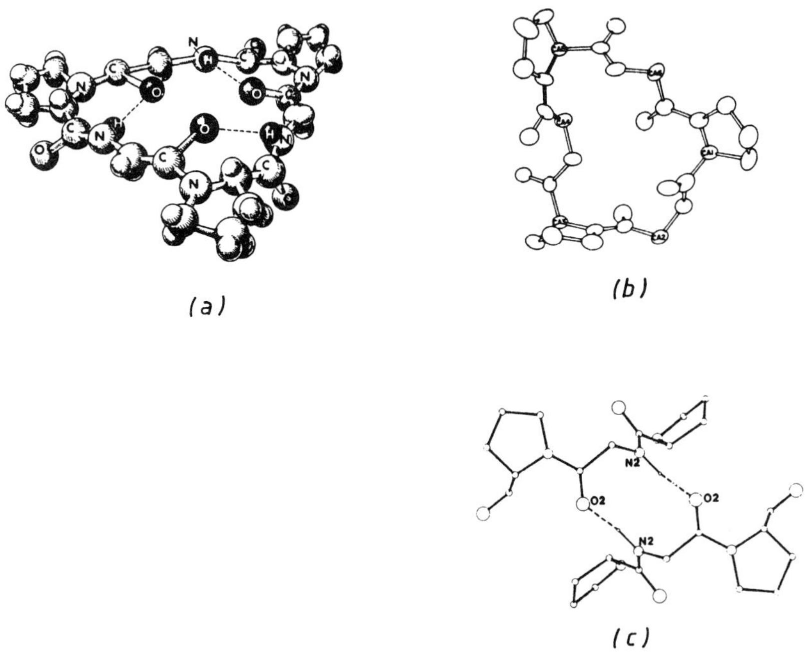

Fig. 2-24. Conformation of *cyclo*-(L-Pro-Gly)$_3$; (a) in slightly polar media, (b) in polar media. In the latter case, the X-ray crystal structure revealed the existence of dimers (c) (taken from E. R. Blout et al., "Peptides polypeptides and proteins"; Wiley, 1974, p. 266; *Acc. Chem. Res.,* **9,** 106, 1976; K. I. Varughese et al., *Proc. Natl. Acad. Sci. USA,* **79,** 4519, 1982).

Thus, with Ca^{2+}, a sandwich complex (two ligands: one Ca^{2+}) was obtained (Fig. 2-25b). The cation was found to be octahedrally coordinated by six glycyl carbonyl groups.

The two cyclopeptides have different conformations. In macrocycle A, the Gly and Pro carbonyls are orientated on either side of the macrocycle plane, and the Pro C=O groups form hydrogen bonds with several water molecules. In macrocycle B, all the carbonyls are directed towards the interior thus allowing the C=O groups (Pro) to form three hydrogen bonds with the NH groups (Gly) of macrocycle A. Furthermore, the NH groups (Gly) of ring B form hydrogen bonds with the ClO_4^- anion [117c, d]. It may be that the difference between the stoichiometry observed in solution (1 : 1) and that observed in the solid state (2 : 1) arises from the strong stabilization of the latter complex by the extensive network of hydrogen bonds.

In the case of Mg^{2+}, the X-ray study of the 1 : 1 complex has shown (Fig. 2-25c) that coordination of the cation is provided by three carbonyls (Gly) and three water molecules [117c].

Finally, it should be noted that *cyclo*-(L-Pro-Gly)$_3$ also forms complexes with ammonium salts and in particular with amino acid salts [117e].

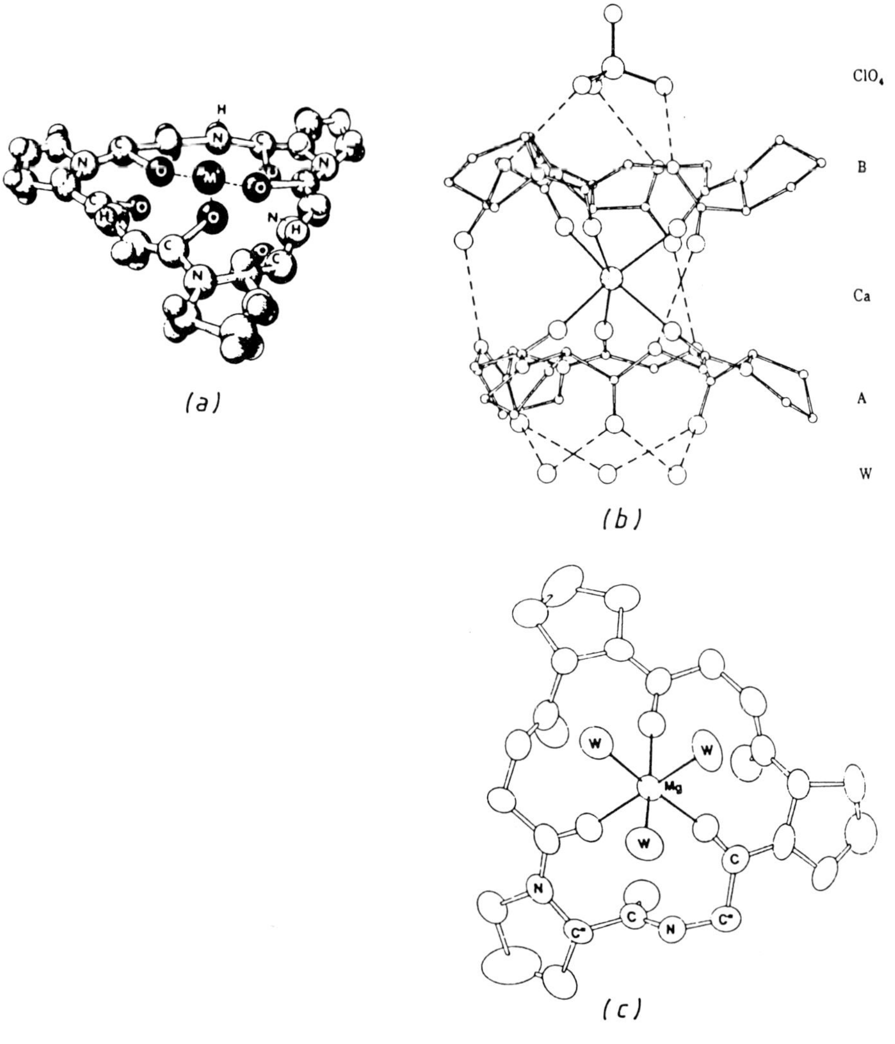

Fig. 2-25. Complexes formed with *cyclo*-(L-Pro-Gly)$_3$; (a) schematic structure in solution, (b) structure of the 2 ligands/Ca^{2+} complex, and (c) structure of the 1/1 complex with Mg^{2+} (taken from K. I. Varughese et al., *Proc. Natl. Acad. Sci. USA,* **79**, 4519, 1982).

b) Cyclo-(D-Val-L-Pro-L-Val-D-Pro)$_3$, Prolinomycin

Numerous analogs of valinomycin, the archetypal natural ionophore, have been synthesized in various laboratories.

In an article published long ago describing the synthesis of prolinomycin, the steps involved in the preparation of this compound were very clearly set out. In this cyclopeptide analog of valinomycin synthesized by Gisin and Merrifield [118], the valinomycin ester linkages are replaced by peptide linkages furnished by proline residues. Prolinomycin complexes alkali metal cations more strongly than does valinomycin but transports them more slowly through lipid membranes [119–122].

The X-ray structure of the rubidium complex (Fig. 2-26) has been determined and has shown the "structuring" effect of proline [123]. Restricted rotation for the prolyl residues decreases the conformational mobility of prolinomycin; such mobility is important for complexation and decomplexation processes. This explains the fact that although the above cyclopeptide is a better complexing agent than is valinomycin, it is a poorer carrier.

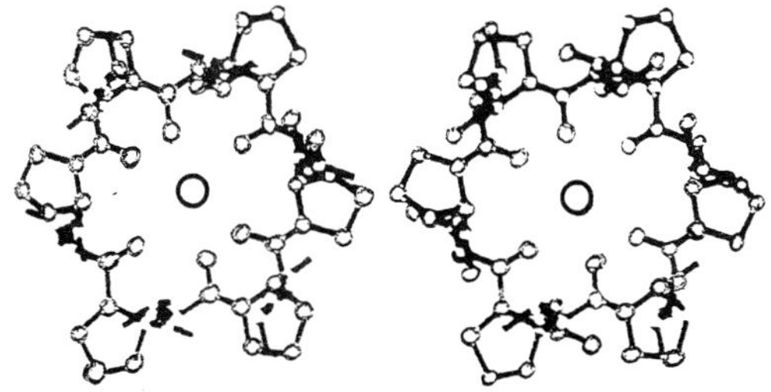

Fig. 2-26. Crystal structure of the Rb^+-prolinomycin complex (stereoview) (taken from L. K. Steinrauf et al., *Acta Cryst.*, **B36**, 1052, 1980).

Other cyclopeptide analogs of valinomycin have been synthesized and studied: cyclo-(L-Val-Gly-Gly-L-Pro)$_3$ and cyclo-(L-Ala-Gly-D-Phe-L-Pro)$_3$ [124, 125].

c) An ionic cyclopeptide

A good example of tailor-making molecules with a view to obtaining specific properties is the elaboration of the synthetic cyclic octapeptide cyclo[Glu-Sar-Gly-(N-decyl)Gly]$_2$ which is selective towards Ca^{2+} ion [126]. It has the following features (Fig. 2-27):

- Two ionizable functional groups thus allowing formation of neutral complexes with divalent cations.
- A binding site of suitable size for Ca^{2+}. It is formed from two carboxylate groups and the cyclopeptide carbonyls. The carboxylate groups geometry favors complexation of M^{2+} rather than two M^+ cations.
- Two C_{10} chains thus making the compound lipophilic.

This peptide dissolved in chloroform does in fact selectively extract Ca^{2+} ions from aqueous solution, the stoichiometry being 1 : 1. It does not extract Mg^{2+}, K^+ and Na^+ ions. Neither does it extract Cl^- which shows that both carboxylic acid groups are indeed deprotonated.

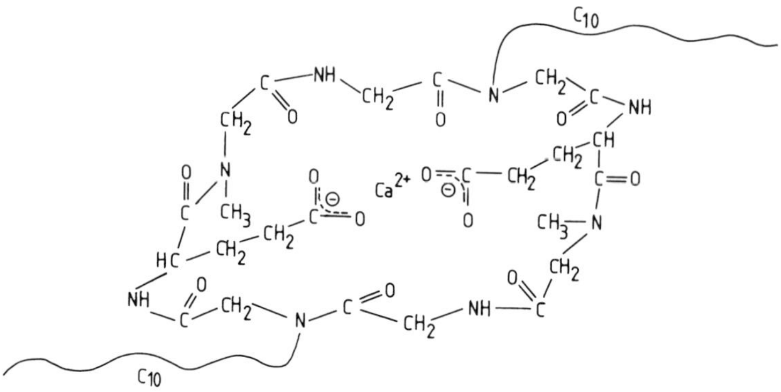

Fig. 2-27. Diagrammatic representation of a synthetic cyclic octapeptide which is selective towards Ca^{2+} cations.

d) Cyclic peptides of sarcosine

Sarcosine (N-methylglycine) is an amino acid which does not occur very commonly in natural products (one example is its presence in both 16-membered rings of actinomycin D).

The peptide link formed with sarcosine has certain distinctive features arising from the presence of a methyl substituent on the nitrogen atom:

- Both *cis* and *trans* isomers are almost equally accessible.
- No possibility of intramolecular hydrogen bonding.

A cyclopeptide of the type $(Sar)_n$ should therefore allow the intrinsic conformational properties of a polypeptide sequence to be studied without interference from the presence of C_α substituents or the formation of hydrogen bonds. In order to achieve this, a series of cyclic peptides of sarcosine with the general formula I (Fig. 2-28) has been synthesized [127, 128].

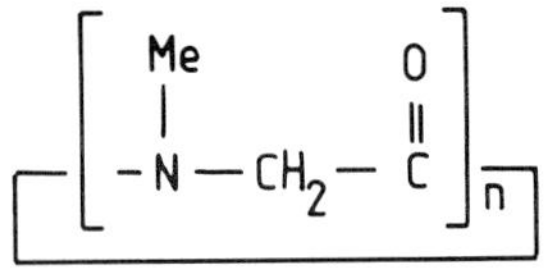

I, n = 2,3,4,5,6,7,8,10

Fig. 2-28. General formula of cyclopeptide derivatives of sarcosine.

Conformational studies in solution (NMR) and in the solid state (X-ray analysis) led to several surprising observations:

- In the case of n = 3 (*cis,cis,cis* conformation) and n = 4 (Fig. 2-29a) medium rings the most significant feature is the high barrier to ring inversion attributed to the steric influence of the methyl groups.

- For large rings (n = 5, 6, 7, 8, 10), the X-ray structures of which are given in Fig. 2-29, sequences composed solely of *cis* or solely of *trans* amide groups were observed e. g., c,c,c,t,t and c,c,t,t,c,c.
- Another surprising feature, arising from the above sequences, was the formation of helical segments which could be analyzed in terms of the torsion angles ϕ and ψ, as shown in Fig. 2-30.

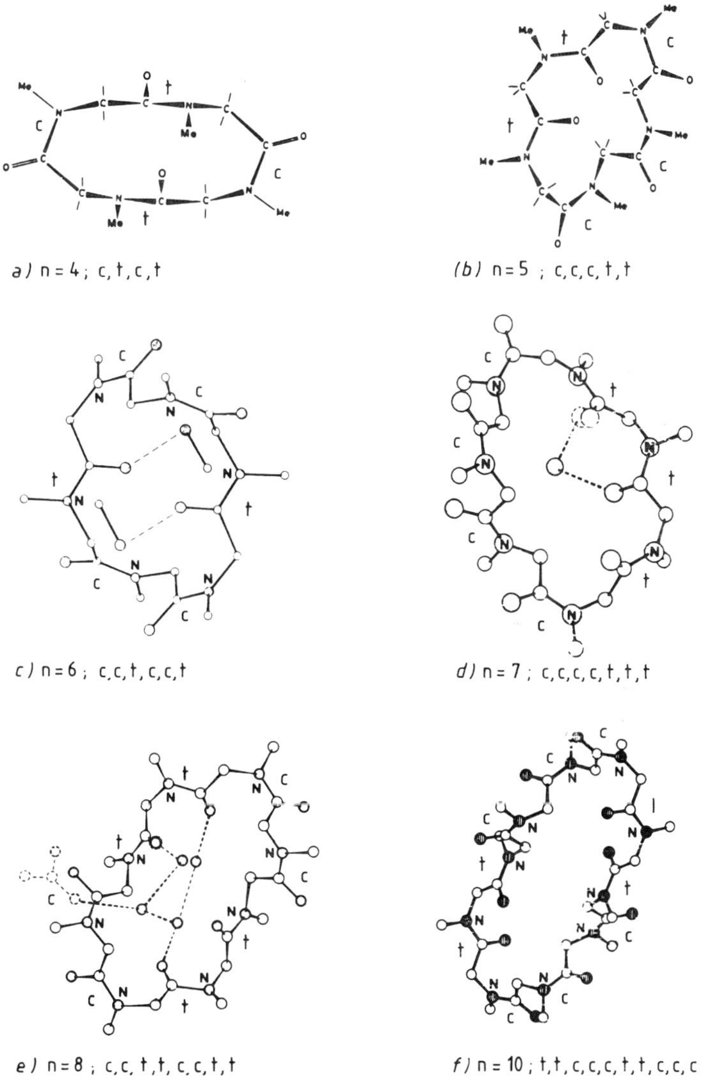

Fig. 2-29. Conformations of cyclotetra-, cyclopenta-, cyclohexa-, cyclohepta-, cycloocta-, and cyclodecasarcosyl (a–f) (taken from K. Titlestad, *Acta Chem. Scand.,* **B29,** 153, 1975; P. Groth, *Acta Chem. Scand.,* **A30,** 840, 1976; P. Groth, *Acta Chem. Scand.,* **A31,** 232, 1977).

Definition of the torsion angles ϕ and ψ. In a peptide chain, neighboring amide groups are connected to each other via a carbon atom (C_α). The only rotations possible in this unit are those about the NC_α bond (ϕ) and the $C_\alpha C$ bond (ψ). These rotations are shown in Fig. 2-30.

The torsion angles ϕ and ψ allow the local conformation to be defined, which can be represented by a point (ϕ, ψ) on a two dimensional diagram (Ramachandran diagram or conformational map). For a peptide with i residues one can thus plot all the values (ϕ_i, ψ_i) and obtain information as to the conformation of the molecule. If the (ϕ, ψ) points are very scattered, there is no regular arrangement of the chain. If, on the other hand, the points are all situated in the same area, the conformation of the polypeptide chain is of regular structure.

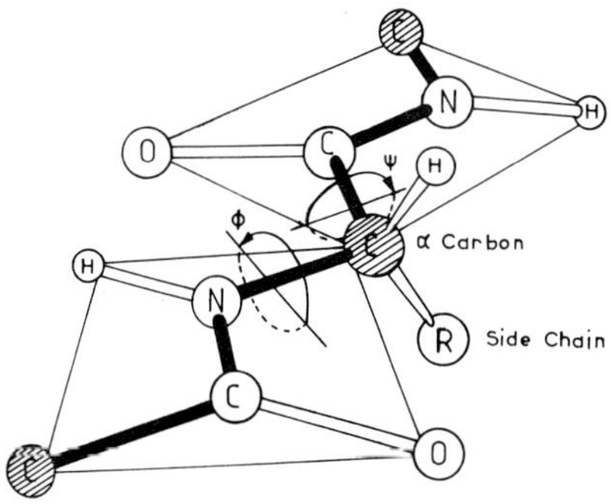

Fig. 2-30. Definition of the ϕ and ψ torsion angles in a peptide chain.

By plotting the (ϕ, ψ) values for all the compounds in Fig. 2-29 on a diagram (Fig. 2-31), Dale et al. observed that the points all lay in the same region, ϕ being between $-70°$ and $-95°$, and ψ around $180°$ (an explanation was given for the three points far from this region).

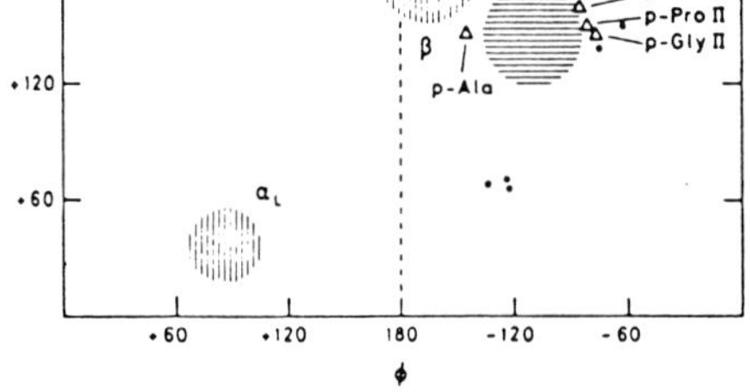

Fig. 2-31. Conformational map (Ramachandran diagram) for cyclic sarcosine peptides (see Figs. 2-28 and 2-29). The (ϕ, ψ) values for these compounds are represented by black dots. The hatched areas correspond to the regions for certain chain types or conformations (taken from J. Dale et al., *Acta Chem. Scand.,* **B31,** 523, 1977).

The group of points is close to the region for polyproline I (helix with *cis* amide bonds) and polyproline II (helix with *trans* amide bonds) as well as the region for polyglycine II (helix without hydrogen bonds) [129].

Thus, in spite of the great difference in ring size, the rings are conformationally very homogeneous. The formation of helical segments explains the high barrier to ring inversion observed for $(Sar)_n$ when n = 5, 6, 7, 8, 10.

To summarize, conformational studies have shown an astonishing conformational uniformity for the rings in spite of their containing both *cis* and *trans* amide links and the fact that they are not stabilized hy hydrogen bonds. The barriers to ring inversion are high. In the case of small rings, this has been attributed to steric hindrance caused by the N-methyl groups, and for large rings to folding of the chains in helical segments.

Certain of the rings (n = 6, 8) show complexing properties, e.g., the octamer was found to dissolve in acetone on addition of KNCS [127].

2.5 The Depsides and their Metal Complexes

The macrocycles in this class of natural compounds, unlike all those mentioned above, contain no peptide bonds but only ester type links (lactone) together with ether groups.

2.5.1 Nactins

The nactins (Fig. 2-32) are 32-membered cyclic tetralactones isolated from various species of *Actinomyces.* Five homologs are known: nonactin, monactin, dinactin, trinactin and tetranactin.

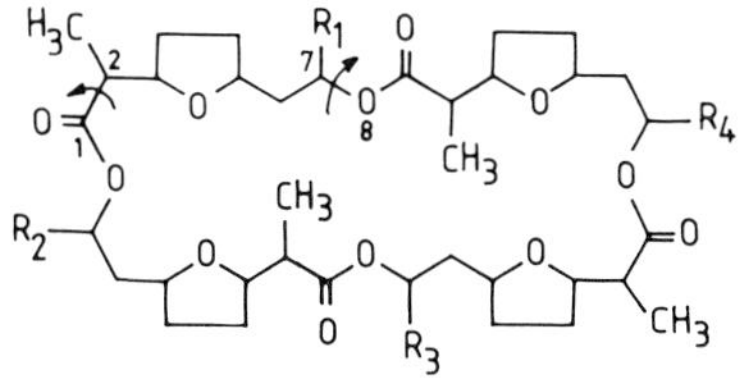

$R_1 = R_2 = R_3 = R_4 = CH_3$		Nonactin
$R_1 = R_2 = R_3 = CH_3$	$R_4 = C_2H_5$	Monactin
$R_1 = R_3 = CH_3$	$R_2 = R_4 = C_2H_5$	Dinactin
$R_1 = CH_3$	$R_2 = R_3 = R_4 = C_2H_5$	Trinactin
	$R_1 = R_2 = R_3 = R_4 = C_2H_5$	Tétranactin

Fig. 2-32. Depsides.

The most detailed structural studies have been performed on nonactin and tetranactin [130]. X-ray analysis has revealed considerable conformational differences which are somewhat surprising given that the difference between these two compounds concerns only the R substituents (CH_3 or C_2H_5). Whereas nonactin (Fig. 2-33a) has a conformation with S_4 symmetry, tetranactin is elongated (Fig. 2-33b). As regards dinactin, this has an intermediate asymmetric structure and its conformational energy is the lowest of those for the three nactins [131].

(a)

(b)

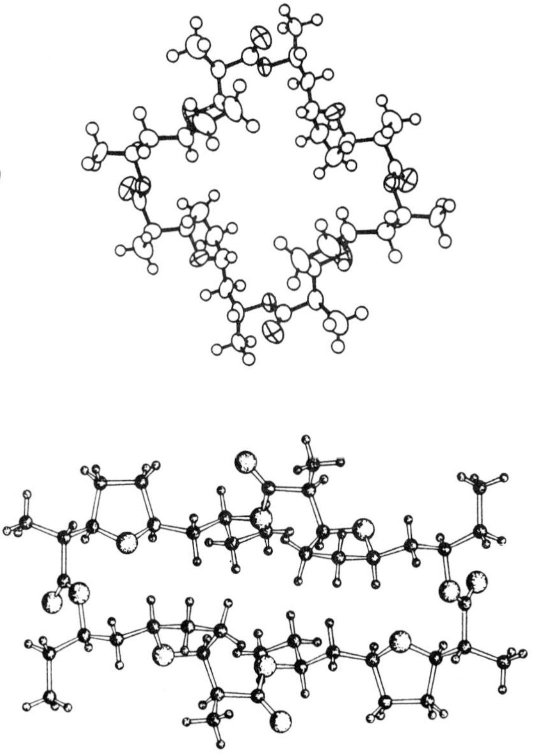

Fig. 2-33. Structure of (a) nonactin and (b) tetranactin (taken from M. Dobler, *Helv. Chim. Acta,* **55,** 1371, 1972; and Y. Nawata et al., *Acta Cryst.,* **B30,** 1047, 1974).

Spectroscopic studies of the conformations in solution have indicated appreciable rotational freedom around most of the macrocycle bonds. As a result, numerous conformers are present in solution [132].

Nactin ionophores complex alkali metal cations with decreasing selectivity in the following order:

$$K^+ > Rb^+ > Cs^+ \sim Na^+ \gg Li^+.$$

NH_4^+ is also strongly complexed [80, 133].

Complexation brings about notable conformational changes resulting from rotation of the carbonyl oxygen atoms towards the interior of the molecule.

Although the conformations of nonactin and tetranactin are different in the uncomplexed state, those for their alkali metal complexes are almost identical [80b, 134].

The alkali metal ion is surrounded by four oxygen atoms from tetrahydrofuran rings and by four ester carbonyl oxygen atoms in approximately cubic 8-coordination, as illustrated in Fig. 2-34 for the K^+-nonactin complex [80b].

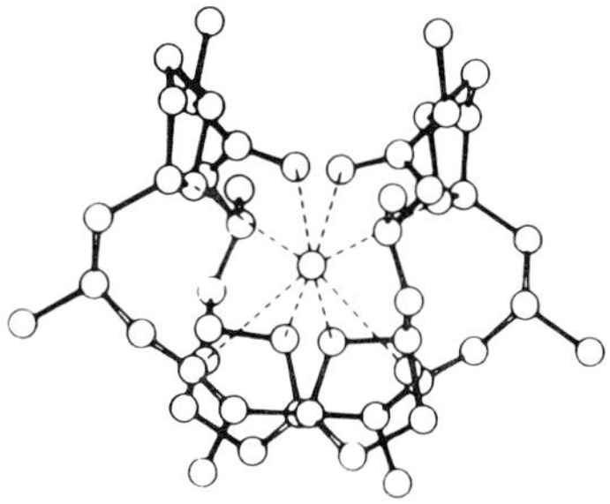

Fig. 2-34. Conformation of the K^+-nonactin complex (taken from J. D. Dunitz et al., *J. Mol. Biol.,* **30,** 559, 1967).

This structure reveals the features necessary for an efficient ionophore:

- The cation should be well protected by the ligand.
- The molecular surface should be lipophilic in nature (particularly marked for tetranactin).

Although the alkali metal complexes are all of similar shape, the selectivities differ greatly. An explanation is provided in terms of the degree of fit between the substrate and the receptor. This is determined on the one hand by the distances between the alkali metal ion and the ligand donor atoms, and on the other by the constraints imposed on the ligand.

In the complex of potassium with nonactin, the distances between the cation and the ether oxygen atoms ($K^+\cdots O$ = 2.81–2.88 Å) are slightly greater than those between the cation and the carbonyl oxygen atoms ($K^+\cdots O$ = 2.73–2.81 Å), the latter being approximately equal to the sum of the K^+ ionic radius (1.33 Å) and the oxygen van der Waals radius (1.40 Å) [135].

In the case of the complex with sodium, the carbonyl oxygen atoms are again nearer the cation ($Na^+\cdots O$ = 2.40–2.44 Å), but for the ether oxygens the $Na^+\cdots O$ distances (2.74–2.79 Å) are much greater than the sum of the ionic and van der Waals radii (2.35 Å). The ether electrostatic interaction is therefore much weaker leading to deformed cubic coordination [136]. The cavity is only partly able to adapt itself to the smaller cation which explains the lower stability of the Na^+ complexes. The Li^+ ion is much too small to interact favorably with the ligand.

For the complexes of sodium and potassium with tetranactin, identical findings to those for nonactin have been reported [134]. This similarity between the two nactins is illustrated graphically in Fig. 2-35. The plots show the variation in the distances between the alkali metal cation and the carbonyl or ether oxygen atoms as a function of cation ionic radius [69]. Incorporation of ions larger than K^+, such as Rb^+ and Cs^+, brings about cavity expansion [134, 137] with a slight displacement of the carbonyl oxygens towards the exterior resulting in change of the angles of torsion about the ester bonds. These deformations, which are largest in the case of Cs^+, explain the lower stability of the Cs^+ complex.

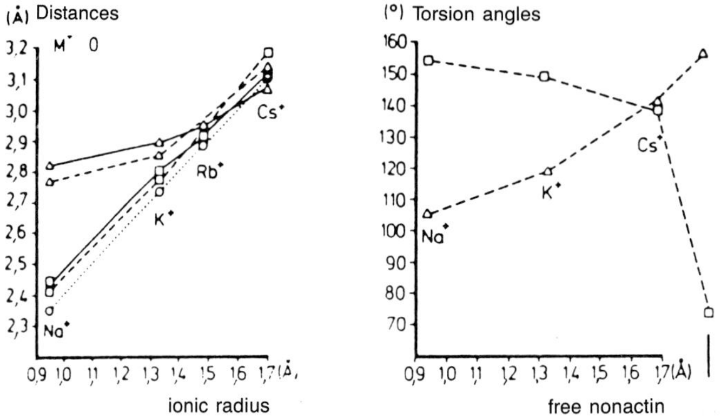

Fig. 2-35. (a) Distances between the alkali metal cation and the ligand donor atoms in complexes of nonactin (broken lines) and tetranactin (solid lines) vs. cation ionic radius. △: Mean distance between the cation and the ether oxygen atoms. □: Mean distance between the cation and the carbonyl oxygen atoms. ○: sum of the ionic radius of the central cation and the van der Waals radius of the oxygen atom (theoretical $M^+ \cdots O$ distance). (b) Torsion angles about ester linkages in nonactin complexes as a function of cation ionic radius. Between C^7 and O^8 (△), and between C^1 and C^2 (□). (The relevant torsion angles are shown in Fig. 2-32). (Taken from Y. Nawata et al., *Acta Cryst.,* **B33,** 52, 1977).

Whereas in alkali metal complexes the distances between the cation and the ether oxygen atoms are greater then the distances between the cation and the carbonyl oxygen atoms, the reverse is so for ammonium cation complexes [138, 140]. Fig. 2-36 shows the structure of the NH_4^+-nonactin complex.

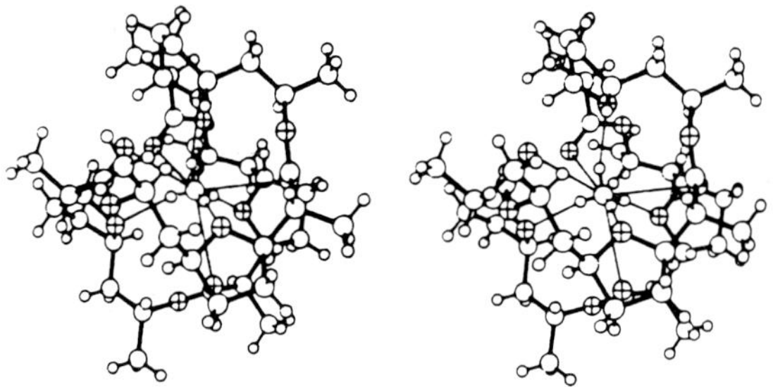

Fig. 2-36. Stereoscopic view of the NH_4^+ complex of nonactin (taken from M. Dobler et al., *Helv. Chim. Acta.,* **59,** 614, 1976).

The ammonium group forms four strong hydrogen bonds with the ether oxygen atoms (NH···O = 2.86 Å) but it interacts more weakly with the carbonyl oxygen atoms (N···O = 3.08 Å) [138]. The same features have been found for the NH_4^+-tetranactin complex (ether O···HN = 2.89 Å; carbonyl O···N = 3.00 Å) [139–140]. The hydrogen bonding interactions explain the high stabilities of these complexes.

2.5.2 Synthetic Nactin Analogs

A series of macrotetrolides of structure related to that of the nactins has been synthesized [141] (see also ref. 165, Part I).

Two different models have been developed (Fig. 2-37). Because model B contains alternating ether and ester groups, it is more similar to the nactins than is model A.

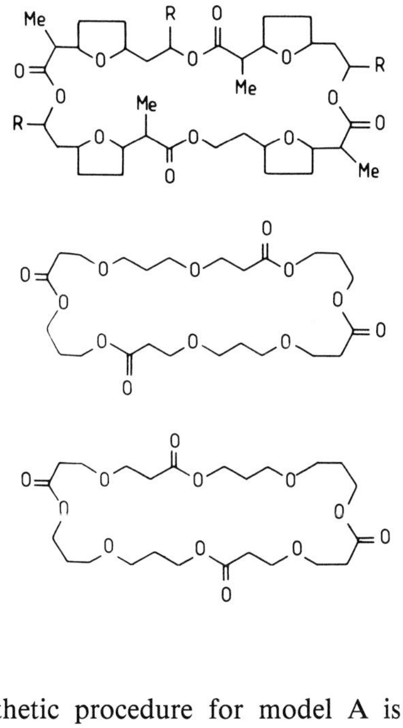

Fig. 2-37. Synthesis of synthetic nactin analogs (A and B).

The scheme showing the synthetic procedure for model A is given in Fig. 2-38.

Depending on the nature of Z, products A_1, A_2, A_3 and A_4 were obtained. The yields for the cyclization step (under high dilution conditions) were between 50 and 80%, but because of purification difficulties, the pure products could be obtained in only 25 to 50% yield. This was also so for synthesis of type B compounds (Fig. 2-39).

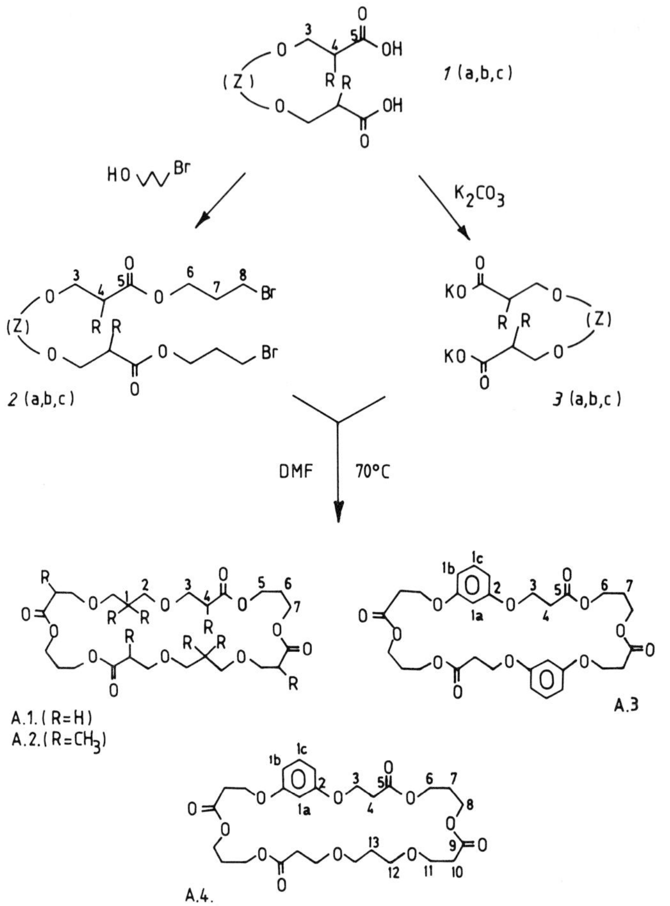

Fig. 2-38. Scheme depicting synthesis of type A nactin analogs.

Their ionophoric properties, though less important than those for crown ethers, showed that some of them possess ion transport abilities. The K^+/Na^+ selectivity is, however, inferior to those of nactins and classical crown ethers. Conversely, quite good Rb^+/K^+ selectivity has been observed for several of these compounds, e.g., macrocycle B.2(d) in Fig. 2-39.

Numerous other nactin analogs in the B series have been recently described. In order to introduce some rigidity into the macrocycles, heterocyclic moieties have been incorporated [142].

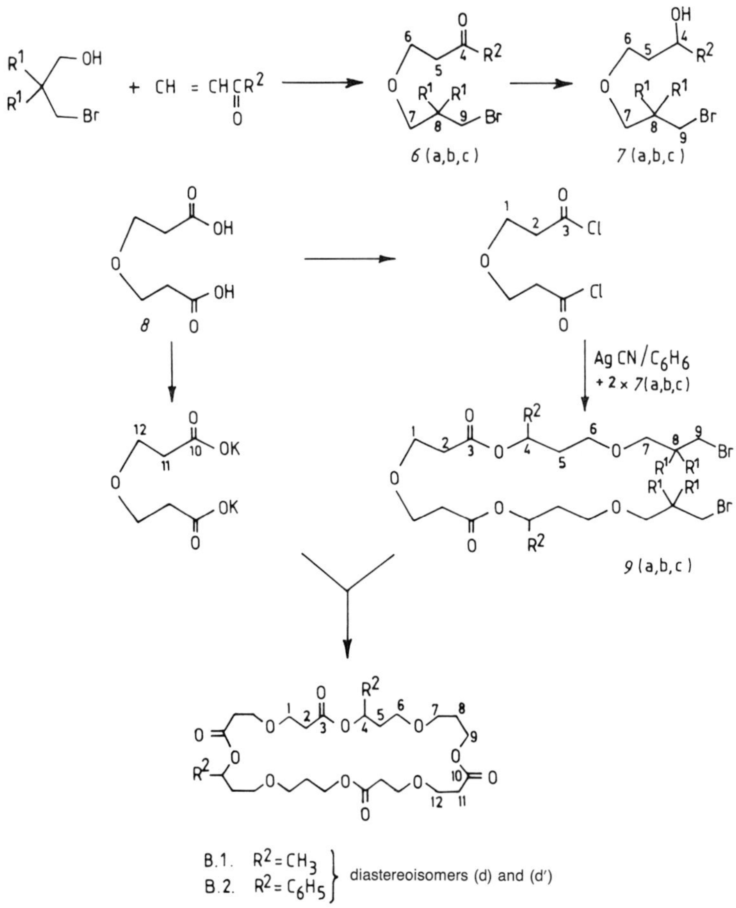

Fig. 2-39. Scheme depicting synthesis of type B nactin analogs.

2.6 Transmembrane Molecular Channels

Continuous movement of ions takes place between the cytoplasm and the extracellular medium. Cell membranes, consisting of lipid bilayers of low dielectric constant, present a high barrier to ion penetration. As a result, simple diffusion can not be considered as being the mechanism operating in biological systems. Two processes lead to lowering of the barrier to ion penetration. The first involves a *carrier* which diffuses across the membrane. The second occurs via formation of a transmembrane channel [143].

2.6.1 The Gramicidins

Gramicidins A, B and C (Fig. 2-40) have been isolated from the antibiotic mixture produced by *Bacillus brevis.* They are linear pentadecapeptides with the end groups blocked by formylation of the amino terminus and by amide-bond linkage to ethanolamine at the carboxy terminus [144].

Valine-gramicidin A:

HCO-L-Val-Gly-L-Ala-D-Leu-L-Ala-D-Val-L-Val-D-Val-L-Try-D-Leu-L-Try-D-Leu-L-Try-D-Leu-L-Try-$NHCH_2CH_2OH$

Valine-gramicidin B:

HCO-L-Val-Gly-L-Ala-D-Leu-L-Ala-D-Val-L-Val-D-Val-L-Try-D-Leu-L-Phe-D-Leu-L-Try-D-Leu-L-Try-$NHCH_2CH_2OH$

Valine-gramicidin C:

HCO-L-Val-Gly-L-Ala-D-Leu-L-Ala-D-Val-L-Val-D-Val-L-Try-D-Leu-L-Tyr-D-Leu-L-Try-D-Leu-L-Try-$NHCH_2CH_2OH$

Fig. 2-40. The gramicidins.

a) Formation of transmembrane channels

The presence of gramicidins in both biological and artificial membranes gives rise to substantial changes in ionic conductivity which can be explained by the formation of channels along which the ions flow [145]. This mode of ion transport is much more efficient than that involving mobile carriers. At a potential difference of 100 mV, 3×10^7 K^+ ions per second could be transported through a channel but only 2.5×10^4 using valinomycin [146].

On the other hand, the K^+/Na^+ selectivity of the channel was found to be much lower [147a]. Divalent cations were transported with difficulty there being a high barrier to their penetration midway through the channel [147b].

b) Mechanism of the effect of gramicidins

The active species appears to be a dimer formed in the membrane. The Urry model [145] involves a helix formed from two gramicidin molecules attached head-to-head by hydrogen bonding (Fig. 2-41). Spectroscopic studies have indicated the existence of another type of gramicidin dimer consisting of an antiparallel-β double helix [148] (Fig. 2-42).

The involvement of a dimer in ion transport is confirmed by the variation of membrane conductance which increases with the square of gramicidin concentration [149]. A head-to-head "covalent" dimer, [*N,N′*-(dideformyl gramicidinA)-malonamide], has been synthesized by Urry et al. [150]. Ion transport by this compound is analogous to that by gramicidin A, except that in this case, the conductance increases linearly with concentration.

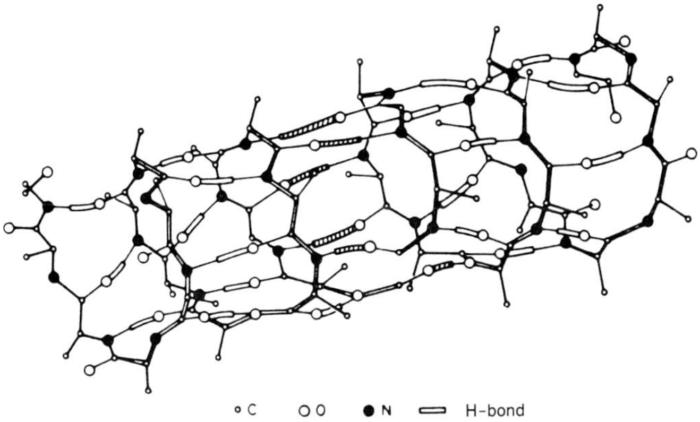

Fig. 2-41. Helical head-to-head dimer of gramicidin A (taken from D. W. Urry et al., *Proc. Nat. Acad. Sci. USA,* **68**, 1907, 1971; and Y. A. Ovchinnikov et al., "Membrane-Active Complexones" BBA Library Vol. XII, Elsevier, Amsterdam, 1974).

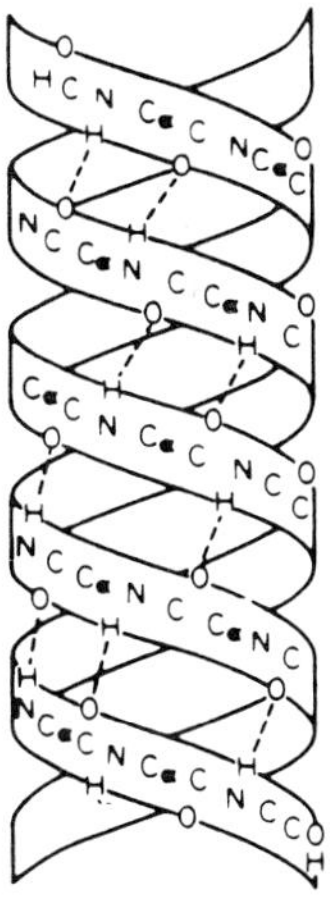

Fig. 2-42. Schematic representation of the antiparallel-β double helix of gramicidin A dimer with seven residues per turn. Hydrogen bonds (- - -) link the two chains (taken from E. R. Blout et al., *Biochem.,* **13**, 5249, 1974).

Recent work has shown that the Urry head-to-head model is the biologically active form [151].

Crystallographic studies (X-ray, neutron diffraction) have been performed on both free gramicidin A and its complexes with Cs^+ and K^+. In the native state the channel is 32 Å long with a diameter of 5 Å. Complexation shortens the channel (26 Å) and widens its diameter (6.8 Å) [152].

Synthetic polypeptides, e. g. poly(Leu-Leu-D-Phe-Pro), also form channels [153].

2.6.2 Alamethicin and Suzukacillin

Alamethicin (Fig. 2-43 a) is a linear nonadecapeptide with the N-terminus blocked by an acetyl group and the eighteenth residue, glutamic acid, linked by an amide bond on its side chain to phenylalaninol [154].

```
(a) ALAMETHICIN:

Ac-Mea-Pro-Mea-Ala-Mea-Ala-Gln-Mea-Val-Mea-Gly-Leu-Mea-Pro-Val-Mea-Mea-Glu-Gln
                                                                            |
                                                                         NH-CH-CH2OH
                                                                            |
                                                                            CH2-C6H5

(b) SUZUKACILLIN:

Ac-Mea-Pro-Val-Mea-Val-Ala-Mea-Ala-Mea-Mea-Gln-Mea
                                                |
   HOGln-Glu-Mea-Mea-Val-Pro-Mea-Leu-Gly-Mea-Leu
         |
      NH-CH-CH2OH
         |
         CH2-C6H5
```

Fig. 2-43. Formulae for alamethicin and suzukacillin.

Suzukacillin (Fig. 2-43b) is a linear peptide with a sequence of 23 amino acids including one phenylalaninol residue as in alamethicin [155]. Both peptides exhibit analogous properties and have very similar compositions.

The striking feature of these two compounds is the profusion of 2-methyl alanine residues (Mea, still referred to as Aib by many authors: α-aminoisobutyric acid), there being eight in alamethicin and ten in suzukacillin.

The disubstitution at the Mea C_α results in the peptide link formed with this amino acid being very rigid. This is manifested on the conformational map (ϕ, ψ) by an "allowed" region of very limited size corresponding to the 3_{10} and α (or $3,6_{13}$) helical conformations. These two helices are in fact encountered in alamethicin as we shall see below (Fig. 2-46).

Initial studies on cation transport across lipid membranes using alamethicin soon showed that the mechanism involves channel formation. Furthermore, it was found that ion transport in membranes is strongly dependent on the applied potential, and that the conductance is proportional to concentration raised to the power of 6 or 9. Consequently, association of 6 to 9 molecules is necessary to form a transmembrane channel [156].

Mueller suggested a model (Fig. 2-44) to explain the process of molecular aggregation to form channels on application of a potential [157]. The molecules, which initially lie at the surface of the membrane (a), are tilted by the applied poten-

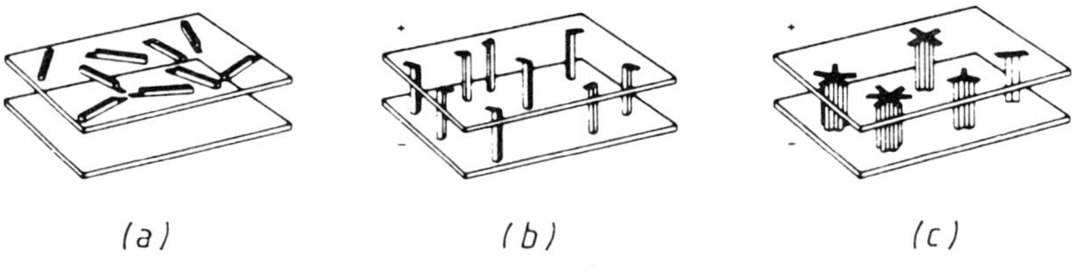

Fig. 2-44. Model for the process of channel formation in a membrane (taken from P. Mueller et al., *J. Supramol. Struct.,* **2,** 538, 1974; and P. Mueller et al., *Ann. N. Y. Acad. Sci.,* **264,** 247, 1975).

tial (b) and then aggregate into trimers or oligomers with the central opening forming a channel (c).

Various methods have been used to elucidate the mode of formation and the structure of the channels [158]. A hexameric aggregate model has been proposed (Fig. 2-45).

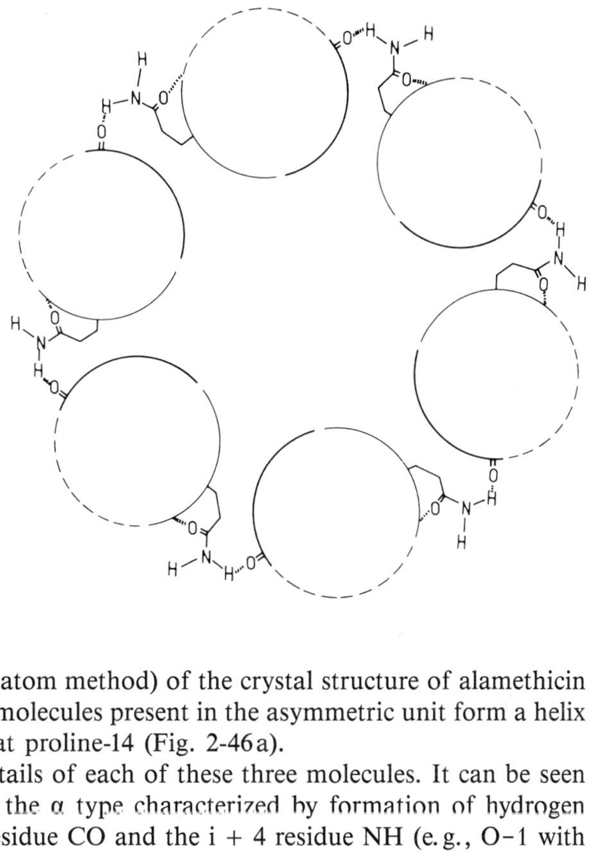

Fig. 2-45. Model for a hexameric aggregate of polypeptides such as alamethicin or suzukacillin (taken from P. Balaram et al., *Acc. Chem. Res.,* **14,** 356, 1981).

X-ray analysis (heavy-atom method) of the crystal structure of alamethicin [159] has shown that the three molecules present in the asymmetric unit form a helix with a bend in the helix axis at proline-14 (Fig. 2-46a).

Figure 2-46b gives details of each of these three molecules. It can be seen that the helices are mostly of the α type characterized by formation of hydrogen bonds (5 → 1) between the i residue CO and the i + 4 residue NH (e.g., O-1 with N-5). However, at the C-terminal domain certain fragments of the helix are of the 3_{10} type with hydrogen bonds (4 → 1) between the i residue CO and the i + 3 residue NH (e.g., O-16 with N-19). The discontinuity at residue 14 is equally visible in all cases. The structure also reveals a crucial point for channel formation this being the distribution of lipophilic and hydrophilic regions. In Fig. 2-46c, it can be seen that the solvent-accessible polar groups lie on a narrow strip of surface parallel to the helix axis. These groups are the side chains at Gln-7 and Glu-18 as well as the carbonyl groups of the Mea-10 and Gly-11 residues (these latter two residues are not involved in intramolecular hydrogen bonding except for Mea-10 in the III form). Finally, is should be noted that the polar groups are almost entirely localized in the C-terminal region.

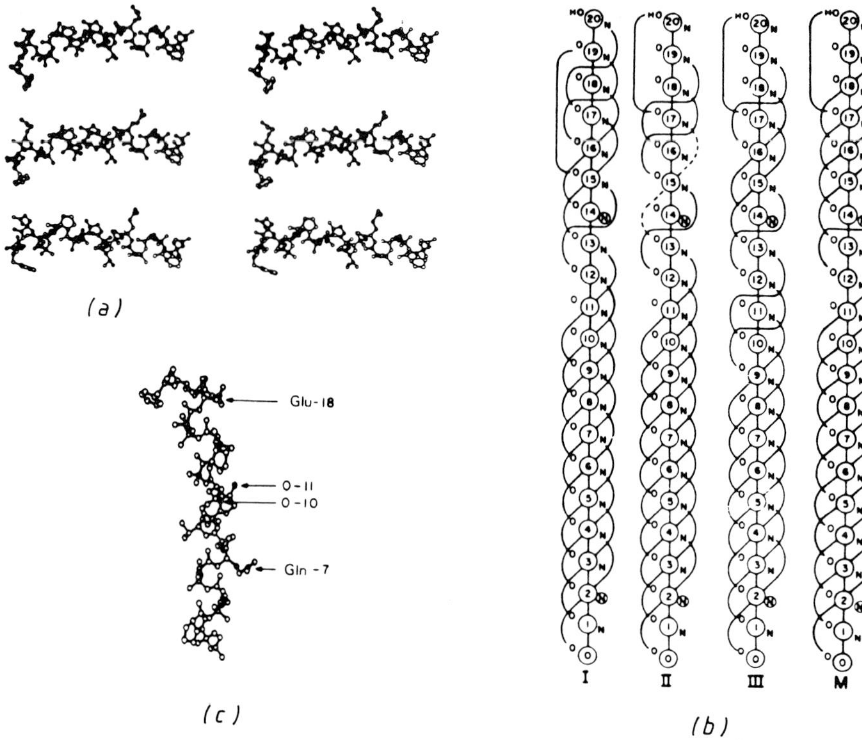

Fig. 2-46. (a) Conformation of the three independent alamethicin molecules forming a helix and (b) the hydrogen-bonding pattern (M is a representation of the average structure used in the channel model); (c) location of the lipophilic (left) and hydrophilic (right) regions (taken from R. O. Fox et al., *Nature,* **300,** 325, 1982) (a: stereo diagram).

The X-ray crystal structure, solved by direct methods, of an alamethicin fragment (segment 12-20) has confirmed the conformation of this alamethicin C-terminal segment [160].

Channel model. The channel is formed by the association of several molecules (ten in Fig. 2-47) [159]. The polar faces orientated towards the interior give rise to a hydrophilic pore, whereas the lipophilic surfaces pointing towards the exterior favor solubility in the membrane. The channel structures are stabilized by an internal annulus of hydrogen bonded glutamine residues (Gln-7). The same type of interaction takes place at the top of the channel between the glutamine groups (Gln-19).

Model for the effect produced by an applied voltage. In the absence of an electric field, the hydrophobic N-terminal segment of alamethicin inserts into the alkyl chain

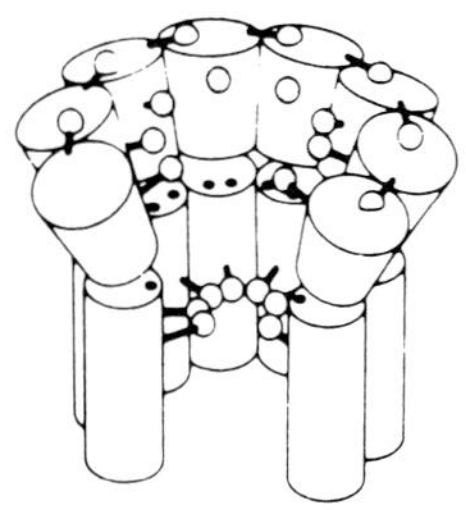

Fig. 2-47. Schematic representation of an alamethicin pore (to facilitate the view of the pore interior, two molecules are not represented) (taken from R. O. Fox et al., *Nature,* **300,** 325, 1982).

region of the lipid bilayer (bottom part of Fig. 2-48 a). The C-terminal portion is modelled either as a non-helical random coil or as an α-helical segment bent away from the channel axis, and partly inserted into the lipid layer (top part of Fig. 2-48 a).

The application of an electric field favors either a coil/helix transition or a reorientation of the C-terminal segment, both resulting in the intermediate shown in Fig. 2-48 b, followed by formation of a channel with a hydrophobic upper part and a hydrophilic interior (Fig. 2-48 c). Furthermore, the dipole moment of the peptide groups (Gln-7), initially perpendicular to the field, may be oriented in such a way as to align the dipole with the field thus enlarging and opening the channel.

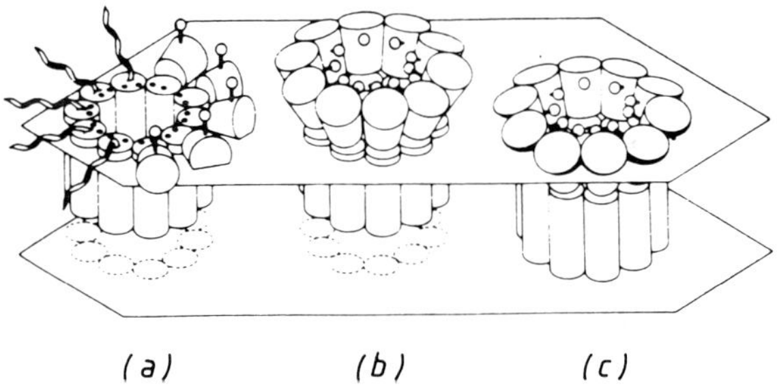

Fig. 2-48. Model of the channel formed by the alamethicin molecules under the effect of an applied voltage (taken from R. O. Fox et al., *Nature,* **300,** 325, 1982).

2.6.3 A Synthetic Channel

An isocyanide polymer $[R-N=C<]_n$ in which R is benzo-18-crown-6 has been synthesized. Its molecular weight was about 15000. The polymer forms a helix, and this structure leads to stacking of the crown ethers thus forming four channels (Fig. 2-49). This polymer complexes alkali metal cations but its ability to transport ions through the channels has not been fully established [161].

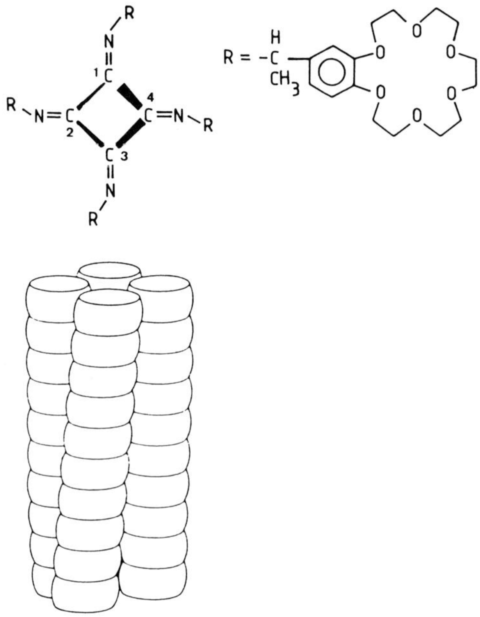

Fig. 2-49. A helical isocyanide polymer: formation of four channels (taken from R. J. M. Nolte et al., *Rec. Trav. Chim. Pays-Bas,* **103,** 260, 1984).

Chapter 3

Macrobicyclic Ligands. Cryptands

3.1 Introduction

3.1.1 General Points

Compared to the usual macrocycles of the crown type which have a two dimensional framework, the cryptands are three dimensional in structure (Fig. 3-1) containing an intramolecular cavity (or "crypt") [162, 163].

Fig. 3-1. Two-dimensional structure of a crown-type macrocycle, and three-dimensional structure of a macrobicycle or cryptand, together with representations of their complexes.

Cryptands form inclusion complexes, the cryptates, in which the substrate is located inside the cavity.

Comparison between macrocycles and macrobicycles illustrates the quantitative just as much as qualitative differences in properties between these two types of ligands. These are reflected in synthetic strategies, conformational properties and structure, as well as in the physical, structural, complexing and chemical properties.

Several synthetic strategies can be envisaged (Fig. 3-2): (a) stepwise, (b) tripode coupling, (c) single capping, (d) double capping, (e) tripode capping, (f) internal template, (g) external template.

Although critical evaluation of these synthetic strategies is difficult without further specifying the chemical nature of the system under study, it is possible to formulate some simple rules:

- paths with the minimum number of steps are generally preferable;
- steps involving creation of more than two connecting bonds are to be avoided since polymer formation increases rapidly with the number of connecting bonds;

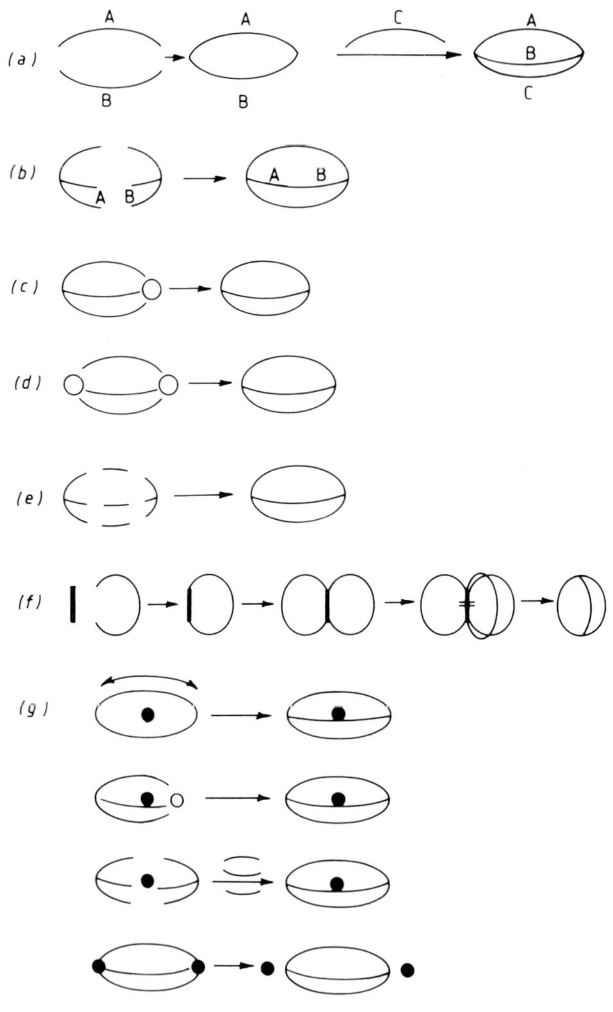

Fig. 3-2. Synthetic strategies for cryptands: (a) stepwise, (b) tripode coupling, (c) single capping, (d) double capping, (e) tripode capping, (f) internal template, (g) external template.

- convergent paths, in which separately elaborated systems are condensed, are preferable to linear paths;
- steps making use of systems which contain sites more reactive than those to be condensed, give rise to additional protection-deprotection operations;
- paths where the last step occurs in the highest possible yield, are to be preferred;
- paths with the minimum number of cyclization steps are preferable;
- certain intermediate structures can have interesting complexing properties;
- use of templates gives rise to better control of the condensation step when the three components a, b, and c are different or when the lefthand part is different from the righthand part (Fig. 3-3).

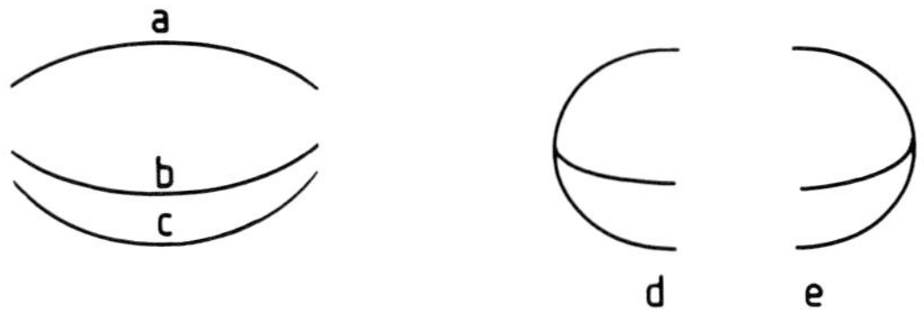

Fig. 3-3. Condensation reactions in unsymmetric systems: a ≠ b ≠ c; d ≠ e.

In summary, the degree of control decreases with increasing multiplicity for a given strategy, that is the number of components involved in a single step.

Certain specific synthetic problems arise from the nature of the macrocycle:

- Efficient synthesis generally requires the use of high dilution conditions to minimize the formation of oligomers and polymers;
- The cyclization steps should be very fast in order to maintain high dilution conditions (low stationary concentration) throughout the reaction;
- Rigid components considerably increase the yield. They reduce the number of conformational degrees of freedom thus increasing the probability of a favorable conformation and hence also the cyclization rate;
- The chains undergoing condensation should have similar lengths;
- Use of templates increases the cyclization yields by keeping the chains together during the reaction;
- Both the reactants and the products should be stable under the reaction conditions since high dilution requires slow addition of the reactants and a long reaction time.

3.1.2 Nature of the Bridgeheads

a) Carbon bridgeheads

Carbon bridgeheads are unable to undergo inversion, hence configurational isomers can be isolated. However, there may be problems of separation and purification. The compounds are rigid and difficult to deform. They are unaffected by changes in the pH of the medium and their liposolubility is high. C–C bonds are not readily formed, and often only in reactions which are relatively slow compared to those involving formation of C–N bonds.

b) Nitrogen bridgeheads

Nitrogen bridgeheads can participate in complexation and increase the stability of the complexes relative to those with carbon bridgeheads. They invert rapidly in the case of unstrained alkylamines (inversion rate constant ~10^3 sec^{-1} at 25 °C) and there is no problem with the separation of isomers since thermodynamic equilibrium is rapidly established. Nitrogen bridgeheads are easily deformed thus diminishing the rigidity of the ligand. They can be protonated and so the complexation properties are pH dependent. In general, the complexes are decomposed in strong acids. Bridgehead nitrogens confer water solubility. C–N bond forming reactions may be very fast and relatively easy to perform.

3.2 Synthetic Strategies

3.2.1 Stepwise Synthesis

a) Macrobicyclic diamines. Cryptands

The first macrobicyclic diamines were synthesized by Simmons and Park [164] using the procedure of Stetter and Marx (Fig. 3-4) [165].

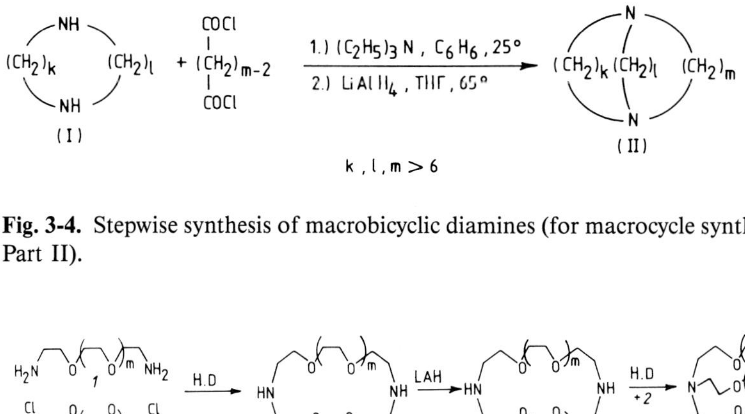

Fig. 3-4. Stepwise synthesis of macrobicyclic diamines (for macrocycle synthesis, see Fig. 1-40, Part II).

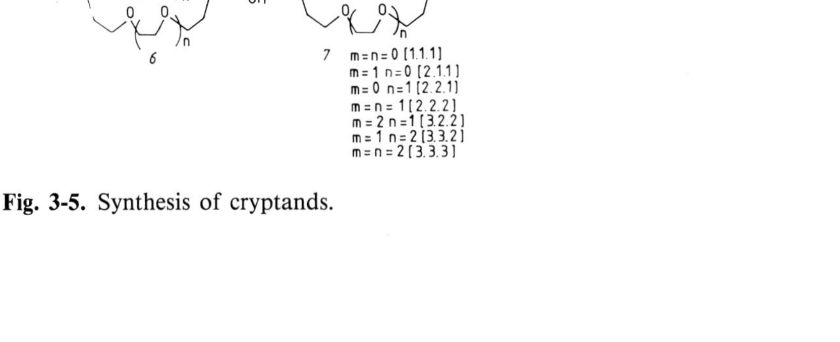

Fig. 3-5. Synthesis of cryptands.

Bridgehead nitrogen *macrobicyclic ligands* with oxygens in the chains have been synthesized according to the scheme in Fig. 3-5. The formation of the macrocycles *3* (in 65-75% yield) and macrobicycles *5* (in 40-60% yield) takes place under high dilution conditions. The only compound to be formed in low yield is *5* (in 10% yield) with m = n = 0 ([1.1.1]diamide) as a result of a dimerization reaction leading to the macrotricycle (see Fig. 3-94). Diborane is employed to reduce the diamides *5* (in 95% yields) [166, 167].

By using a flow cell of the type used for fast kinetic studies (stopped-flow), Dye et al. succeeded in considerably reducing cyclization reaction times whilst maintaining the high yields [168].

b) Specific conformational properties of nitrogen-containing bicyclic systems

These ligands display a particular form of isomerism since the lone pair at the bridgehead nitrogen may be directed either inwards or outwards with respect to the cavity thus giving rise to three forms [164, 169]: exo-exo or out,out = o,o; exo-endo or out,in = o,i; endo-endo or in,in = i,i (Fig. 3-6).

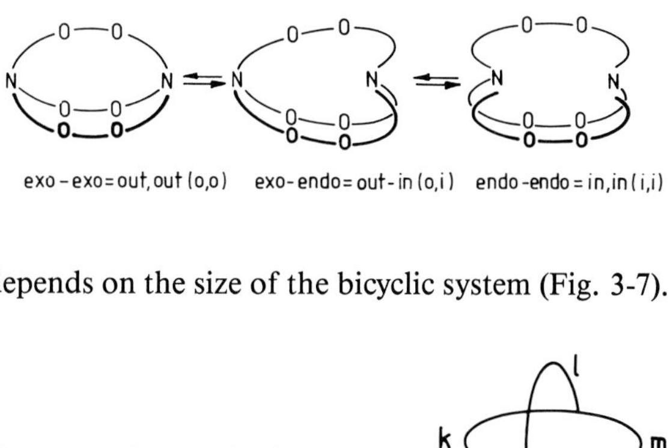

Fig. 3-6. Exo-exo, exo-endo and endo-endo isomers of macrobicyclic diamines.

The predominant form depends on the size of the bicyclic system (Fig. 3-7).

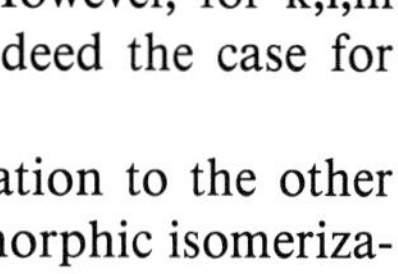

Fig. 3-7. Schematic representation of a macrobicyclic diamine.

When the number of atoms, k, l and m, in the three bridges is two or three, only the out,out form exists, e.g. 1,4-diazabicyclo[2.2.2]octane. However, for k,l,m > 3, the other forms become energetically accessible. This is indeed the case for k,l,m = 5 [170], and even more so with larger systems [171].

Two mechanisms of transformation from one conformation to the other can be envisaged [167], namely bridgehead inversion and homeomorphic isomerization.

For bridgehead inversion, three forms can interconvert at one or two nitrogen sites by reversal of the direction of the lone pair (see Fig. 3-6):

$$o,o \rightleftharpoons o,i \rightleftharpoons i,i$$

Homeomorphic isomerization (Fig. 3-8) takes place *via* movement of one bridge of the bicyclic system through the macrocycle formed by the other two bridges, thus leading to interconversion of either the o,o and i,i or the i,o and o,i forms, depending on the case.

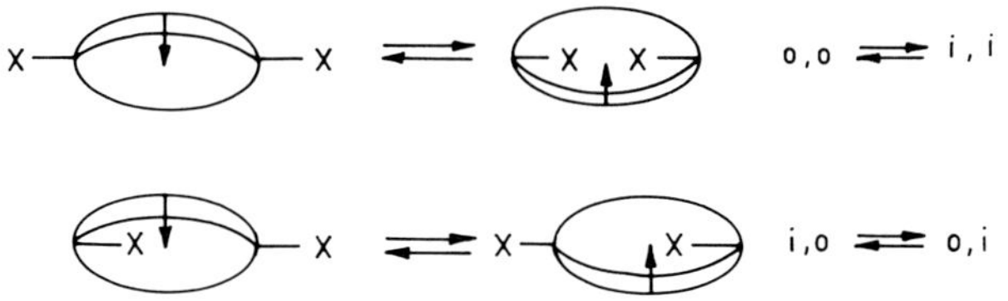

Fig. 3-8. Homeomorphic isomerization in macrobicyclic systems (X = nitrogen lone pair).

The X-ray crystal structure determinations of one representative of each type of isomer have shown that the [2.2.2] cryptand has the i,i conformation, its derivative H_3B[2.2.2]BH_3 is o,o and the derivative H_3B[1.1.1] is o,i [172].

The low temperature NMR spectra of the [2.2.2] cryptand indicate that a conformational type *kinetic process* takes place [167]. The 251 MHz ^{1}H NMR spectrum at ca. −125 °C shows formation of a doublet for N−CH_2 whereas the ^{13}C spectrum at 63 MHz remains unchanged. These results can be attributed to movement in the chains but not at the bridgehead.

c) Other complexation sites in the chains

Sulfur atoms. The synthetic method for oxathia and thia macrocyclic and macrobicyclic diamines (Fig. 3-9) is analogous to that previously described for oxygen-containing cryptands and presents no particular problems [173] except for the low solubility of the compounds containing a large number of sulfur atoms.

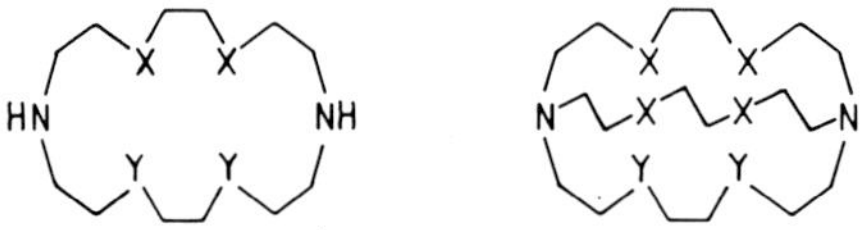

1 X = O, Y = S
2 X = Y = S

3 X = O, Y = S
4 X = S, Y = O
5 X = Y = S

Fig. 3-9. Oxathia and thia macrocyclic and macrobicyclic diamines.

Nitrogen atoms. Various polyaza-polyoxa macrobicyclic ligands (Fig. 3-10) have been synthesized using the same general scheme [174a]. The particular problem in this case is that of protection of the nitrogen sites in the chains by using a group allowing formation and reaction of acyl chlorides, and which can be subsequently removed in the final step. The tosyl group has been found to be entirely satisfactory. The polyaza analogue of cryptand [2.2.2] has also been synthesized. The hexaprotonated form of this octaazamacrobicycle complexes fluoride anion [174 b].

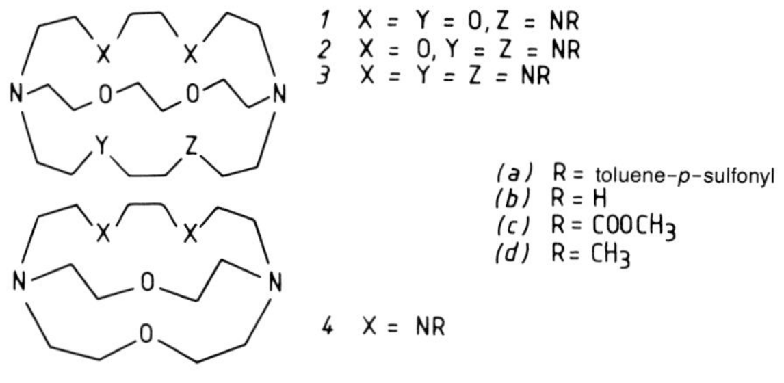

Fig. 3-10. Polyazapolyoxy-macrobicycles.

d) Substituted cryptands.

[2.2.2] Cryptands with long chain substituents or with cyclohexyl rings incorporated in the bridges (Fig. 3-11) have been synthesized for use as phase transfer catalysts [175].

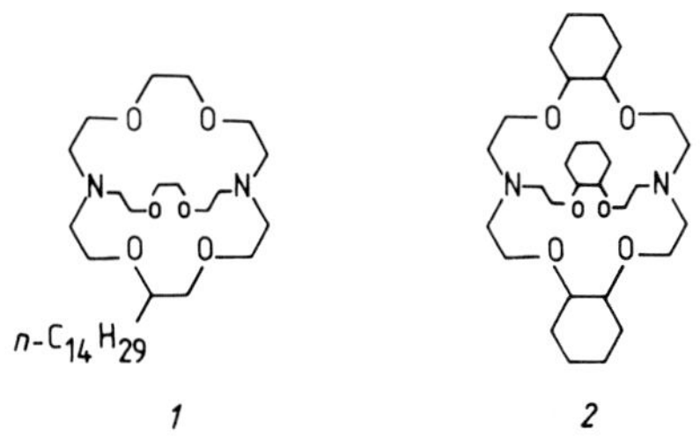

Fig. 3-11. Chain-substituted cryptands (lipophilic cryptands).

These cryptands are quite lipophilic, but in the case of *2* the presence of cyclohexane rings makes the system more rigid thus causing a drastic decrease in the complexation constants.

Macrobicyclic compound containing a ferrocene unit. The ferrocene bis-acyl chloride derivative *1* has been condensed with diaza-18-crown-6, *2*, giving the bicyclic compound *3* (Fig. 3-12) in 30% yield [176 a].

This compound opens the way to the formation of cryptates with specific properties. Thus, the cryptand *3* complexes many metal cations in acetonitrile, viz. the alkaline earth metal cations, Eu^{3+}, Th^{3+}, and Dy^{3+} ($K_s \sim 10^3-10^4$). Complexation affects the redox potential of the ferrocene unit. The oxidized form of *3* only very weakly complexes the cations listed above (electrostatic repulsion) [176 b].

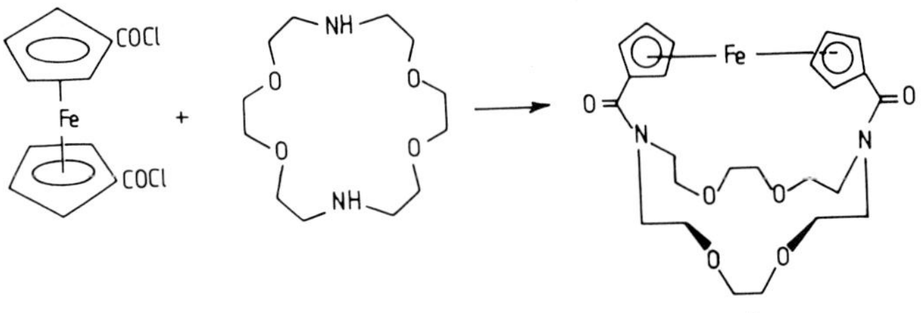

1 2 3

Fig. 3-12. Macrobicyclic compound containing the ferrocene unit.

The field of macrocyclic receptor molecules containing redox sites within their structures is currently undergoing rapid expansion [176 c].

Attachment of azobenzenes to a diazacrown. Two mechanisms (Fig. 3-13) have been proposed for the thermal *cis*-to-*trans* isomerization of azobenzene derivatives, namely a rotational mechanism involving rotation about the N=N bond and an inversion mechanism involving inversion of one of the nitrogen atoms. The essential difference between the two processes is that the rotational mechanism proceeds via a dipolar transition state accompanied by a large volume change ($\Delta V^{\ddagger}$) with respect to the reactants. The rotational mechanism is considered to be unlikely since change in pressure has little effect on the kinetics of isomerization.

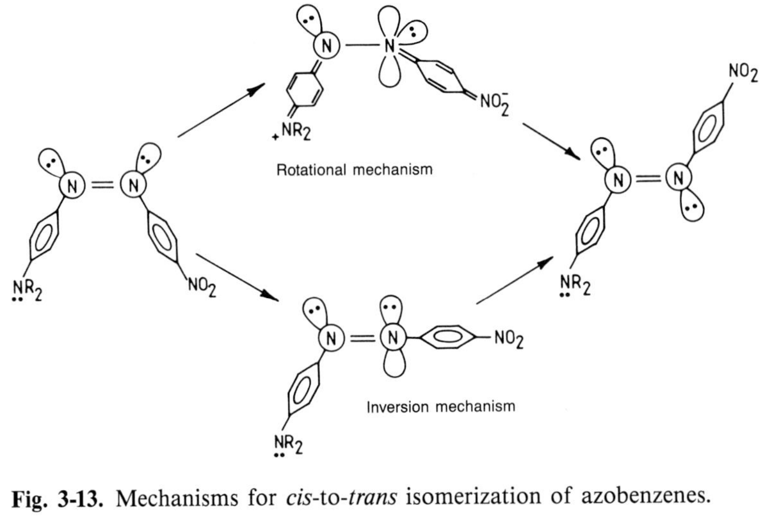

Fig. 3-13. Mechanisms for *cis*-to-*trans* isomerization of azobenzenes.

In order to elucidate the mechanism of *cis*-to-*trans* isomerization of azobenzene derivatives, an azobenzene-bridged [2.2] crown ether has been synthesized (Fig. 3-14). In this compound, the two benzene rings are covalently bonded to the macrocycle, and *cis*-to-*trans* isomerization by the rotational mechanism is therefore highly unlikely [177].

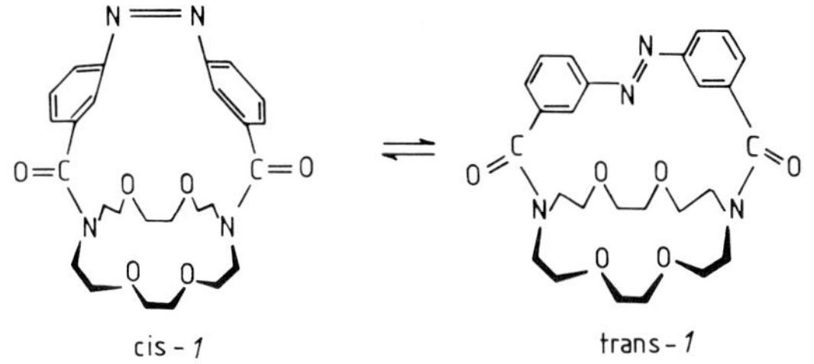

Fig. 3-14. *Cis*-to-*trans* isomerization of azobenzene bonded to a diaza-crown ether.

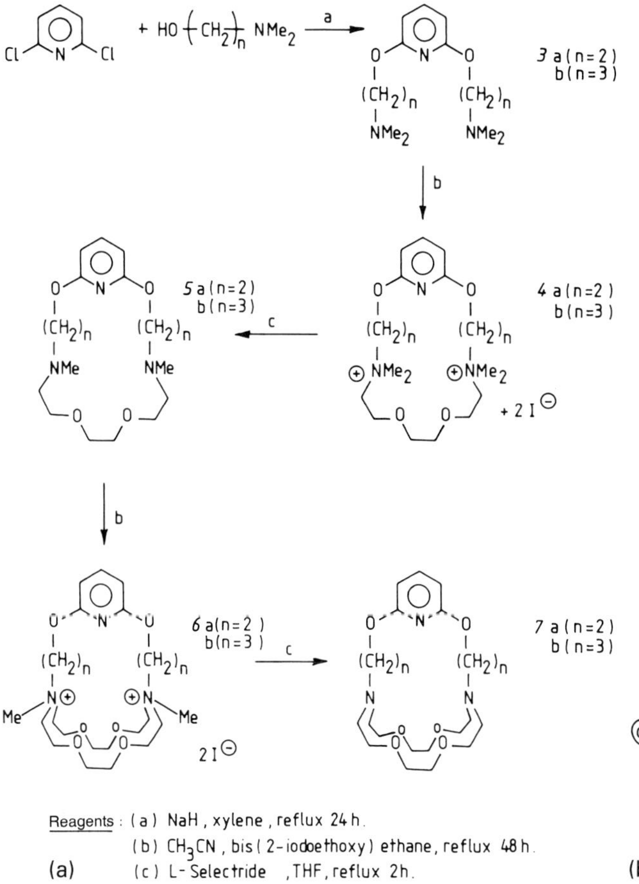

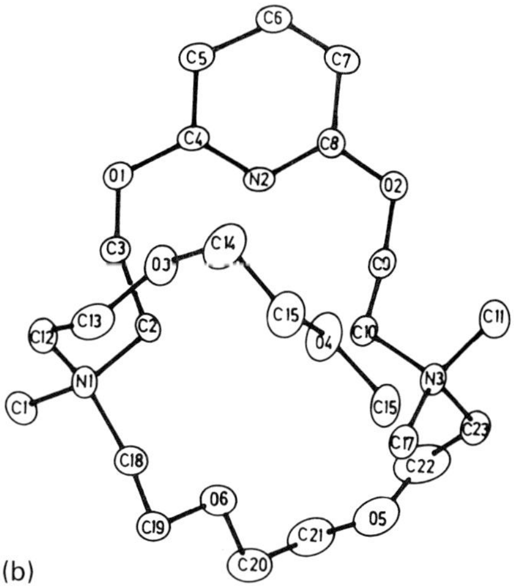

Fig. 3-15. (a) Synthesis of crown ethers and cryptands incorporating a pyridine group in one chain; (b) Structure of cryptand *6a* (taken from G. R. Newkome et al., *Tetrahedron Letters,* **22,** 3039, 1981).

Indeed, the volume of activation obtained from the kinetic effect of pressure on isomerization of *cis*−*1* is small ($\Delta V^{\ddagger}$ = 2.0 ml/mol) thus excluding the rotational mechanism.

The product *1* has therefore been considered as being the standard azobenzene for studying the mechanism of isomerization. The results obtained with most of the *cis*-azobenzenes studied (values of $\Delta H^{\ddagger}$, $\Delta S^{\ddagger}$ and $\Delta V^{\ddagger}$) show that isomerization takes place via the inversion mechanism. However, in the case of azobenzenes with push-pull substituents, e.g., 4-dimethylamino-4′-nitroazobenzene, the rotational mechanism may be operative as a result of stabilization of the dipolar transition state.

Cis-to-*trans* isomerization of the azo group thus allows photocontrol of the properties of compounds such as *1* (see Section 3.3.4.k below).

Cryptands with a pyridine group in one of the chains. These cryptands have been obtained by a quaternization-demethylation sequence (Fig. 3-15a) [178 a]. This type of sequence allows the synthesis of different and/or functionalized bridges.

Demethylation is achieved using L-selectride (lithium tri-*sec*-butylborohydride).

The NMR data favor an *exo-exo* configuration for the diammonium salt *6a,* and this has been confirmed by the X-ray structure determination (Fig. 3-15b).

Tris catechol cryptands. Enterobactin, which is a naturally-occurring siderophore, contains three catechol units within its structure, and it is these groups which are responsible for the complexation of ferric ion. Such units have been incorporated into macrobicyclic ligands; some examples are shown in Fig. 3-16. The syntheses were

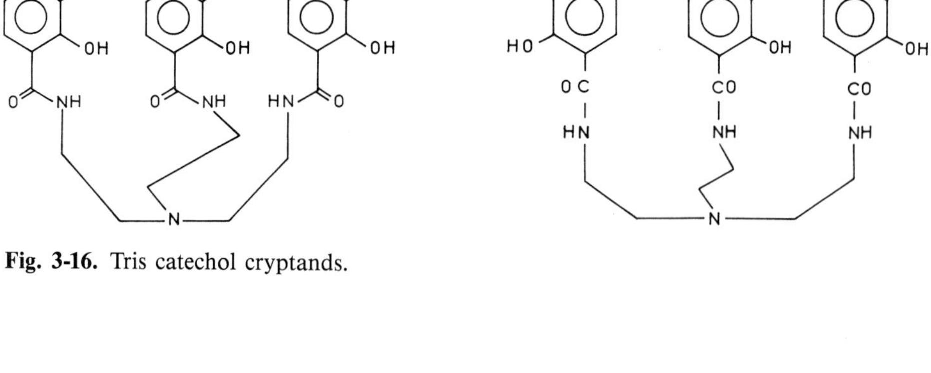

Fig. 3-16. Tris catechol cryptands.

performed by using either high dilution techniques or an external template [178b]. These macrobicycles are good complexing agents for Fe(III).

Anthraceno-cryptands. Cryptands containing the photoactive anthracene group have been synthesized (Fig. 3-17) [179]. The two compounds shown have different properties both in the free state and when complexed.

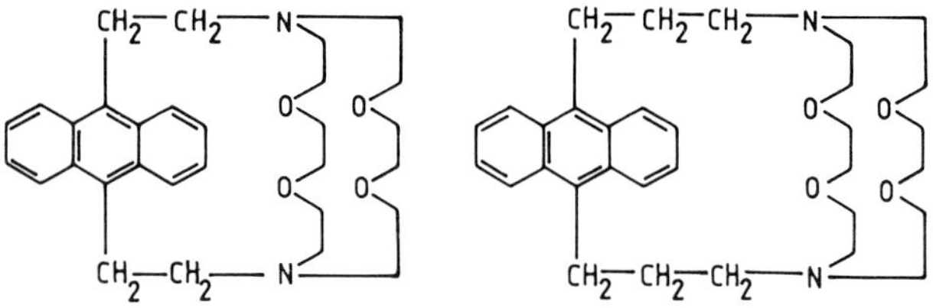

1 (A_{22}) *2* (A_{33})

Fig. 3-17. Anthraceno-cryptands.

Compound *1* displays an intense fluorescence (ϕ = 0.68) in methanol solution without any detectable exciplex emission. Compound *2*, on the other hand, fluoresces weakly (ϕ = 0.04) but forms an internal exciplex (tertiary amine – aromatic system). Conversely, in acetonitrile, both *1* and *2* exhibit almost exclusively intramolecular exciplex formation (ϕ = 0.20 and 0.12, respectively). To explain this, it has been suggested that whereas in methanol the lone pairs in *1* are *out-out* and hence solvation by methanol occurs and formation of an exciplex is impossible, in the case of *2* the lone pairs are oriented inwards. This is to be compared to the situation in acetonitrile where the nitrogen lone pairs are *in-in* for both *1* and *2*. However, these hypotheses are contradicted by the X-ray crystal structures [180] which reveal that in the solid state the lone pairs in *1* are oriented inwards and those in *2* outwards (Fig. 3-18). Nevertheless, the explanation put forward above remains valid if there is substantial conformational mobility in solution.

Addition of complexable cations (Na^+ or K^+) or protonation also reveals differences between *1* and *2*. In the case of ligand *1*, the fluorescence spectra are little affected by cation addition or protonation. Conversely, in the case of ligand *2*, both cation addition and protonation give rise to increased fluorescence. This is because amine-anthracene exciplex formation is inhibited by involvement of the nitrogen lone pairs in both complexation and protonation, and fluorescence therefore occurs due to the anthracene chromophore [179].

e) High pressure synthesis of cryptands

High pressure chemistry (piezochemistry) finds applications in many fields including organic synthesis and kinetics [181]. It has proved particularly effective in cryptand synthesis. We recall here very briefly the principles of the method. All reactions are

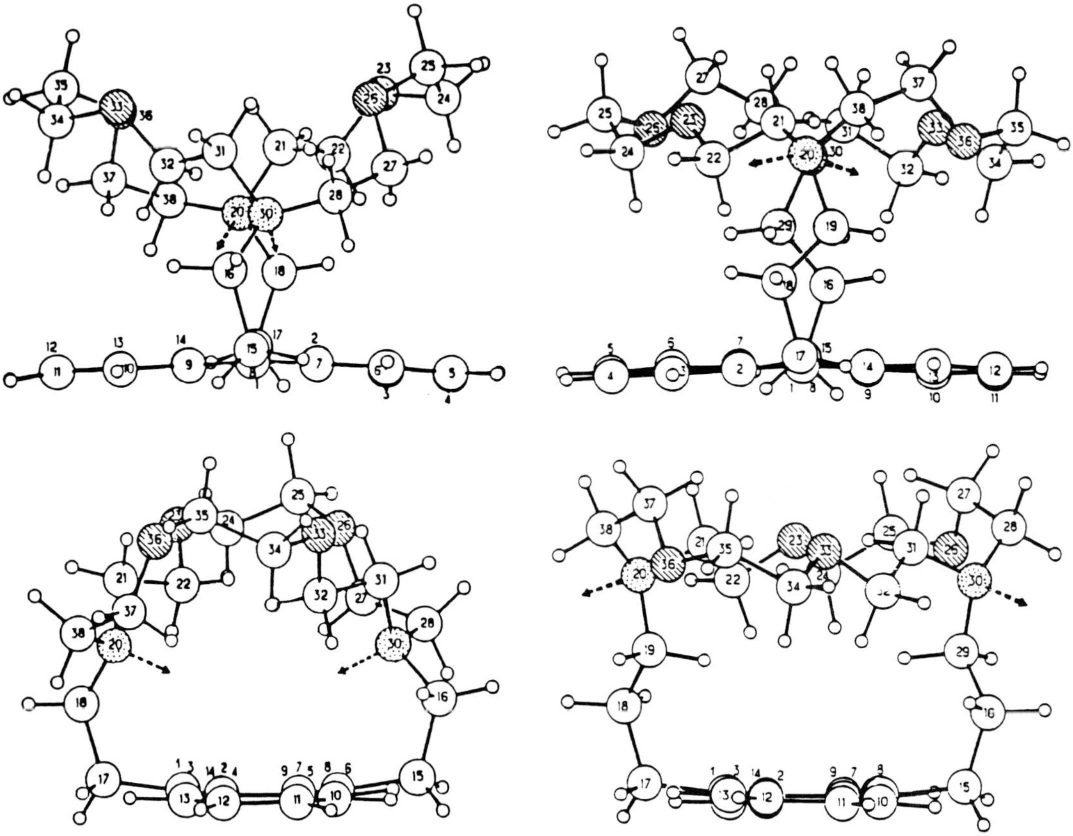

1 *2*

Fig. 3-18. Structures of anthraceno-cryptands *1* and *2*. (We thank J. P. Desvergne for supplying these figures).

characterized by an activation volume $\Delta V^{\neq}$ (expressed in units of cm^3/mol). Depending on the type of reaction, $\Delta V^{\neq}$ can be positive, negative or close to zero. Reactions occurring with $\Delta V^{\neq} < 0$ are favored by increase in pressure. Formation of a quaternary ammonium salt from a tertiary amine and an alkylating agent is a reaction with a very negative activation volume. Jurczak et al. have made use of this property in the synthesis of cryptands (Fig. 3-19) [182].

Experimentally, an equimolar solution of *1* and *2* (0.08 mol/1 in acetone) is placed in a teflon-lined container which is sealed and placed under pressure (10 kilobars, 25 °C, 20h). This method has been used for the synthesis of numerous cryptands. The yields of diquaternary salts are quite remarkable (Table 3-1).

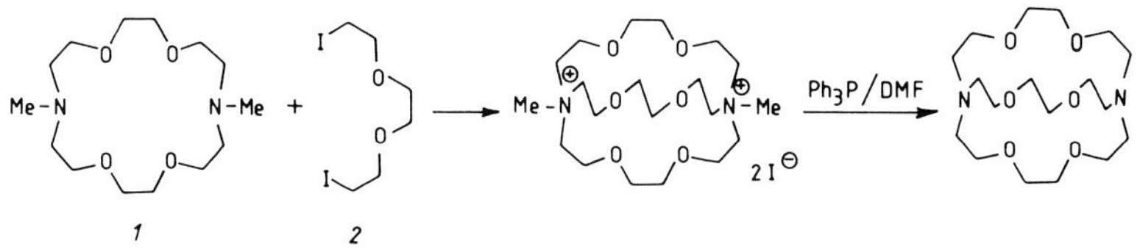

Fig. 3-19. Reactions used for the high-pressure synthesis of cryptands.

Table 3-1. High-pressure synthesis of cryptands.

Cryptand	Quaternization yield %	Demethylation yield %
[1.1.1]	95	59
[2.1.1]	81*	51
[2.1.1]	91*	72
[2.2.1]	82**	62
[2.2.1]	91**	80
[2.2.2]	94	65
[2_B.2.1]	100	87
[2_B.2.2]	100	71

* [2.1.1] formed from different starting macrocycles and diiodo compounds
** [2.2.1] formed from different starting macrocycles and diiodo compounds
(Taken from J. Jurczack et al., *Topics in Current Chem.* **130,** 183, 1985.)

Demethylation is carried out by treatment of the quaternary salt with triphenylphosphine in DMF, this step occurring in somewhat lower yield.

The method's main drawback, which is inherent to present high pressure technology, is the small volume of the reactor (2–100 ml). However, the technique is still useful because of the high yields.

Chiral cryptands have also been synthesized using this procedure [183].

f) Macrobicyclic systems with carbon bridgeheads

Macrobicyclic polyethers derived from glycerol. o,i-i,o Isomerism. Macrobicycles containing carbon bridgeheads have been synthesized from achiral and chiral glycerol derivatives [184]. Compounds *1* and *2* (Fig. 3-20) were obtained from the achiral derivative, and macrobicycles *3, 4* and *5* (Fig. 3-21) were synthesized stereospecifically from a chiral glycerol derivative.

The synthetic approach is represented in Fig. 3-22. The intermediate *syn* macrocycle affords the *exo,exo* (or *endo,endo*) macrobicycle, whereas the *anti* macrocycle leads to the *exo,endo* derivative.

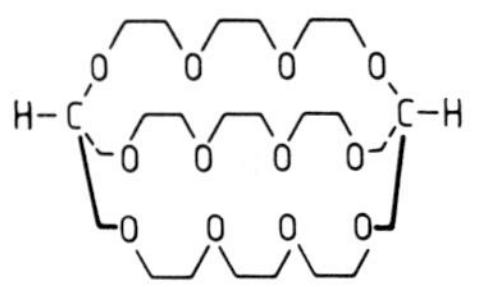

1

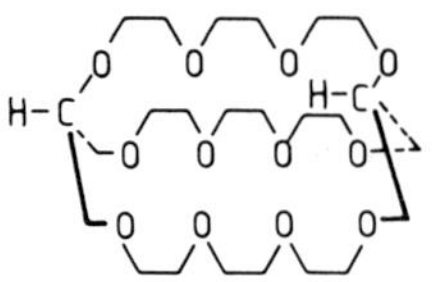

2

Fig. 3-20. Exo-exo *1* and exo-endo *2* stereoisomers of carbon bridgehead macrobicyclic compounds.

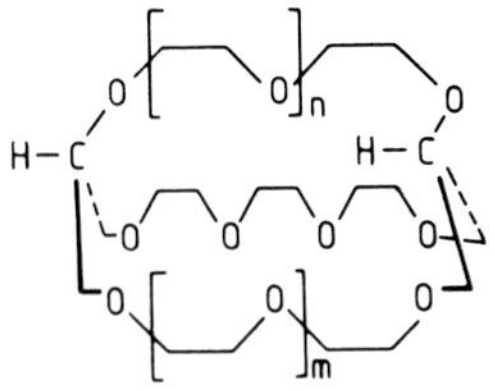

3 n = 2 , m = 1
4 n = 1 , m = 1
5 n = 1 , m = 2

Fig. 3-21. Carbon bridgehead macrobicycles; exo-endo compounds.

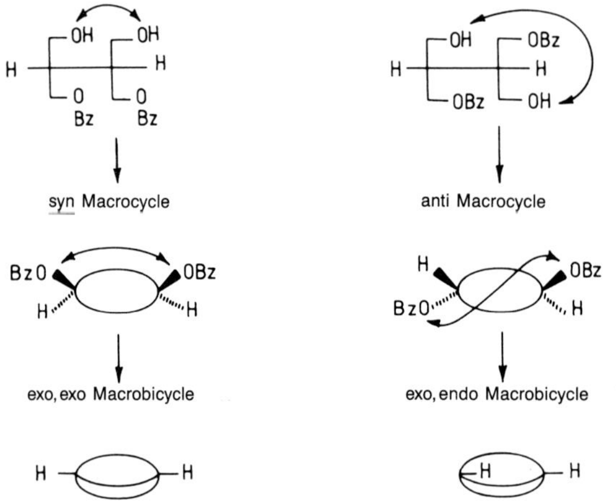

Fig. 3-22. Stereospecific synthesis of macrobicyclic polyethers.

The room temperature ^{13}C NMR spectra of compounds *1, 2, 4* and *5* reveals the presence of a single narrow peak for the bridgehead carbon atoms. For compound *1* (o,o), this indicates either the presence of only the o,o (or i,i) form, or rapid o,o $\rightleftharpoons$ i,i isomerization on the NMR time scale. In the cases of *2, 4* and *5* (all i,o), the presence of a single resonance signal implies rapid i,o $\rightleftharpoons$ o,i isomerization since in the compound o,i-bicyclo[8.8.8]hexacosane the bridgehead carbon atoms give two separate signals (the presence of oxygen atoms in the chain thus facilitates isomerization).

For compound *3* (n=2, m=1), interconversion is sufficiently slow at low temperatures for separate resonances to be observed for the bridgehead carbon atoms whereas at room temperature a single broad signal is seen. The coalescence temperature is around −8 °C which corresponds to a value of 54 kJ/mol, or ca. 13 kcal/mol, for $\Delta G^{\neq}$.

With bridgehead carbon atoms, only homeomorphic isomerization, that is to say one chain being passed through the ring formed by the other two chains, can take place. Thus, for the o,i ⇌ i,o isomerization, the mechanism of interconversion is given in Fig. 3-23.

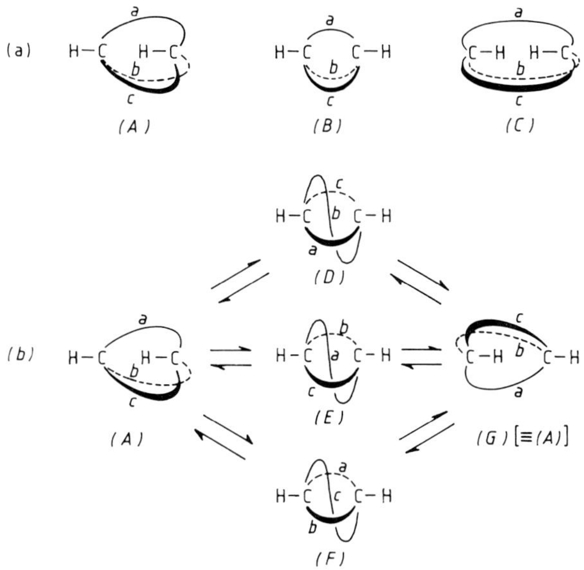

Fig. 3-23. (a) Out-in, out-out, and in-in stereoisomers, (b) representation of preferred conformations (A and G) of the macrobicycle [(A) ≡ (G)], and of possible transition states for the interconversion A ⇌ G.

It should be noted that in all the systems studied (compounds *1-5*) the bridgehead carbon atom is adjacent to a carbon atom in two of the chains, and an oxygen atom in the third chain. Thus, around the bridgehead, two successive oxygen atoms are separated in one case by three carbon atoms and in the two other cases by two carbon atoms, the latter separation being the same as that in the chains (we shall see below that this separation is very important for obtaining efficient complexing agents).

Macrobicycles *1–5* have cavities which are too large to complex potassium cations efficiently. Complexation of K^+ by these ligands has nevertheless been demonstrated [184].

Macrobicyclic hydrocarbons. The first structures of this type containing carbon bridgeheads and hydrocarbon chains were obtained and studied by Park and Simmons (Fig. 3-24) [185].

Two of the three isomers of bicyclo[8.8.8]hexacosane, namely *2* and *3* have been synthesized from 1,10-cyclooctadecadione in ca. 20% overall yield, and charac-

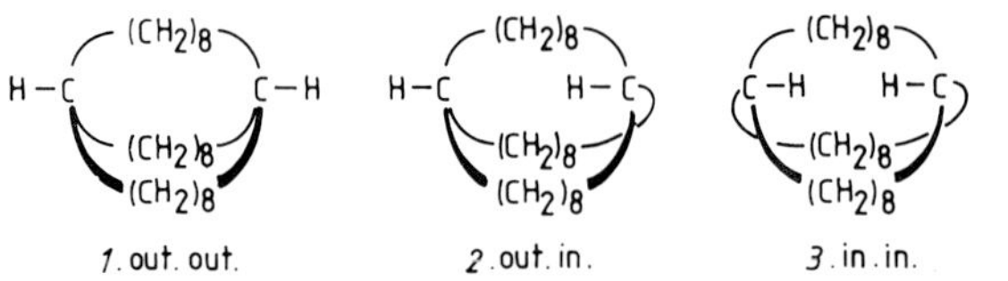

Fig. 3-24. Isomers of bicyclo[8.8.8]hexacosane.

terized (Fig. 3-25). The intermediates *5* and *6* were obtained as *cis-trans* mixtures. The dibromides *cis-* and *trans-7* were separated by fractional recrystallization and then individually converted into the bicyclic hydrocarbons *2* and *3*.

The diester *trans-8* can give only compound *2* (o,i) but the diester *cis-8* can lead to both macrocycles *3* (i,i) and *1* (o,o). The stereochemistry of compound *2* obtained from *trans-8* has been assigned on the basis of the ^{13}C NMR spectrum. The stereochemistry of the hydrocarbon derived from the diester *cis-8* has been assigned by considering two factors: the order of thermodynamic stablility as calculated using molecular mechanics (i,i > o,o > o,i), and the inertness to bromination exhibited by the macrobicycle formed from *cis-8*. Tertiary hydrogen atoms in normal environments are rapidly attacked by bromine. Thus, atropisomer *2* (o,i) gives the bromo derivative. On the other hand, because of its lack of reactivity, the isomer derived from *cis-8* must therefore be i,i *3*, with the bridgehead protons abnormally shielded from attack by bromine.

Macrobicyclic analogues to Pedersen crown ethers. Two macrobicyclic polyethers with carbon bridgeheads have been synthesized with the aim of producing rigid ligands of suitable cavity sizes capable of strongly binding Group 1 metal cations [186]. The synthesis of these two compounds is given in Fig. 3-26.

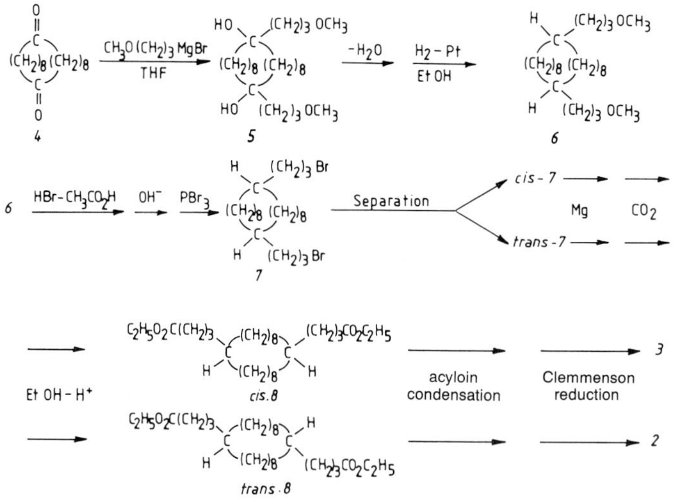

Fig. 3-25. Synthesis of endo-endo *(3)* and exo-endo *(2)* isomers of bicyclo[8.8.8]hexacosane.

Fig. 3-26. Synthesis of macrobicyclic analogues of Pedersen crown ethers.

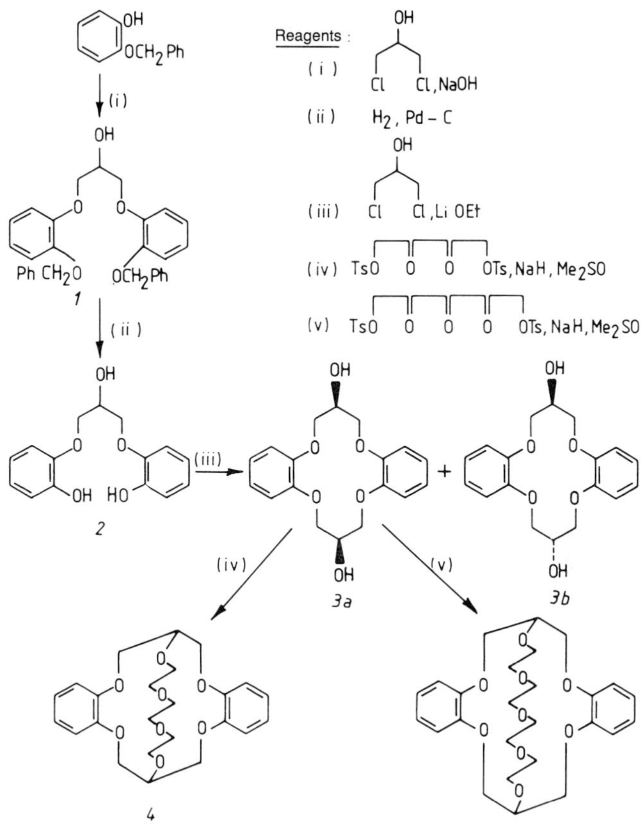

The *cis* and *trans* diols *3a* and *3b* were obtained as a mixture and were separated by column chromatography. Bicyclic compounds *4* and *5* were afforded by the isomer *3a*.

Molecules *4* and *5* strongly complex alkali metal cations. The stability constants (Table 3-2) are comparable to those for [2.2.2].

Table 3-2. Stability constants (log K_s) in water as solvent.

	4	*5*	Dibenzo-18-crown-6	18-crown-6	Cryptand [2.2.2]
Na^+	5.4	3.5	1.1	0.3	3.9
K^+	5.7	4.3	1.6	2.06	5.4
Rb^+	3.8	4.4	1.1		4.35

(Taken from D. G. Parsons, *J. Chem. Soc. Perkin I,* 451, 1978).

The crystal structures of macrobicycle *4* and its smaller homolog *6* (Fig. 3-27) have been determined [187].

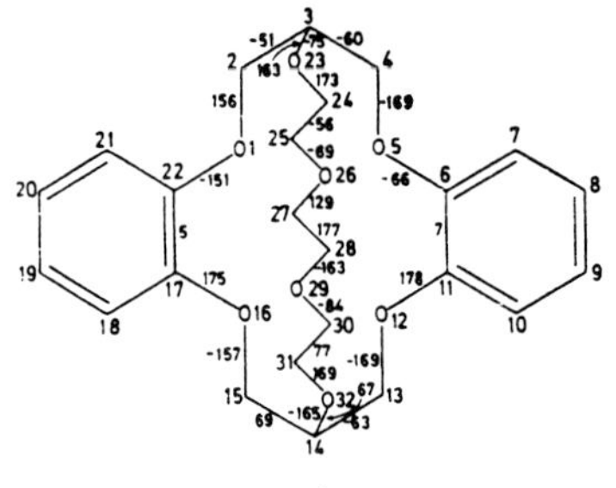

4

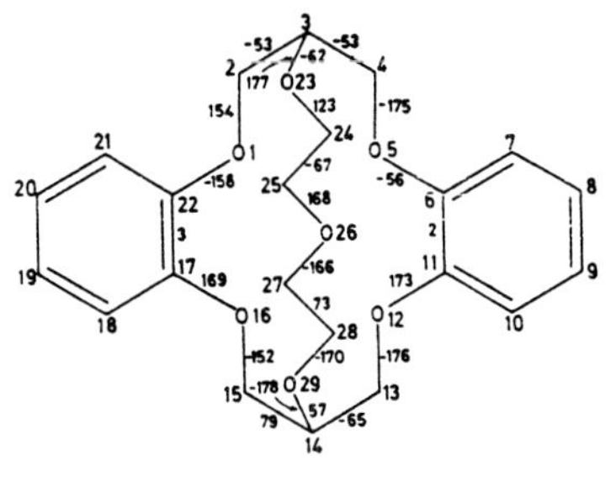

6

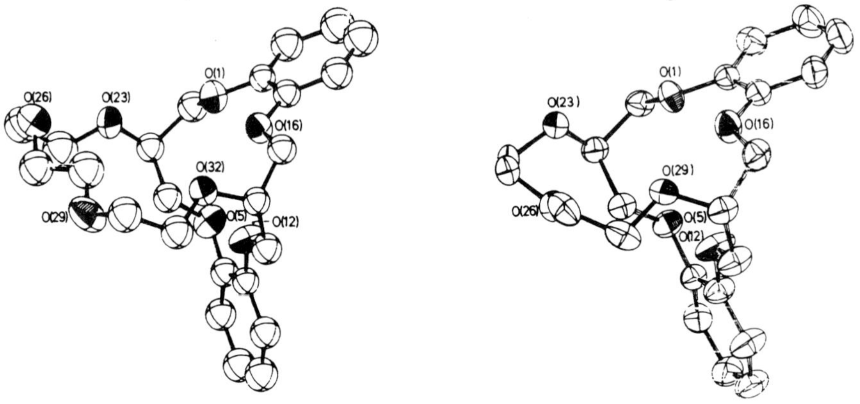

Fig. 3-27. Structures of macrobicyclic analogues of benzo-18-crown-6 *(4)* and benzo-15-crown-5 *(6)*. (Taken from, J. D. Owen, *J. Chem. Soc., Perkin II*, 12, 1981.)

The similarity between the two structures should be noted and also the fact that each of the macrobicyclic compounds shows an open face containing several oxygen atoms available to complex with cations.

Other macrobicyclic polyethers with bridgehead carbon atoms have been synthesized (Fig. 3-28) [188].

These compounds form extremely weak complexes with alkali metal cations (in methanol, log K_s lies between 1.1 and 2.4). This is because of the presence at the bridgehead of three carbon atoms between two successive oxygen atoms, which is not so for the compounds described above (see discussion page 255). The large stability constants listed in Table 3-2 illustrate well the importance of this point.

A macrobicycle with the same number of oxygen atoms in all three chains (Fig. 3-29) has been synthesized stereospecifically [189].

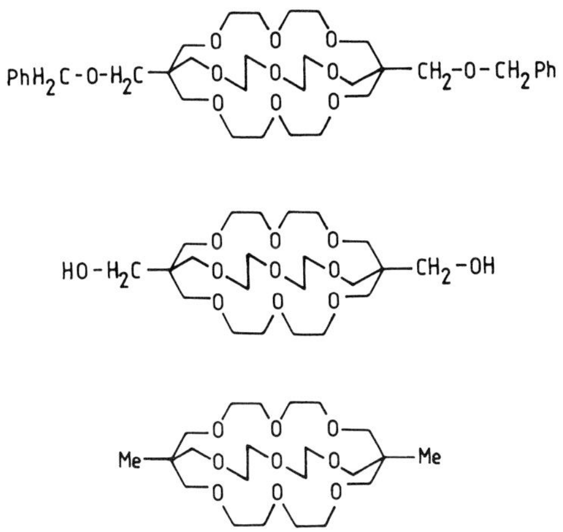

Fig. 3-28. Macrobicyclic carbon bridgehead polyethers.

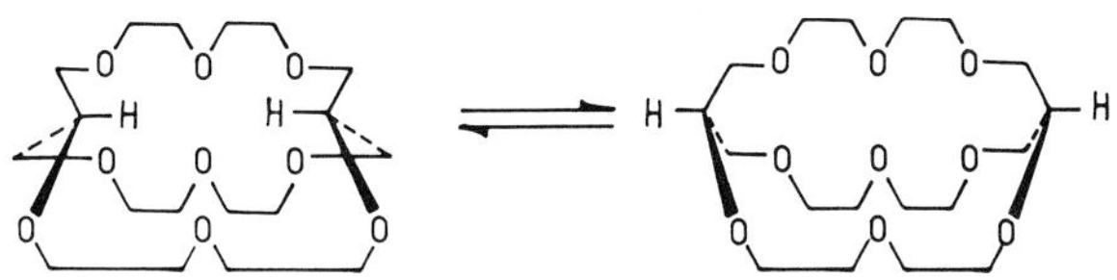

i,i *o,o*

Fig. 3-29. i,i ⇌ o,o Interconversion of the conformational diastereoisomers of [3.3.3] cryptand synthesized stereospecifically.

The structure of the Ba^{2+} complex with this ligand (in the o,o form) shows that the cation is coordinated to eleven atoms (nine oxygens, and two nitrogens belonging to the thiocyanate groups).

Several other macrobicyclic polyethers (Fig. 3-30), having the required separation between the oxygen atoms, form stable complexes with Na^+ and K^+ cations [190]. High K^+/Na^+ selectivity in the case of compound *1* and high Na^+/K^+ selectivity for ligand *3* are to be noted (Table 3-3).

Two cryptands each with one carbon and one nitrogen bridgehead have been synthesized (Fig. 3-31). They are well adapted in size for the Li^+ cation [191].

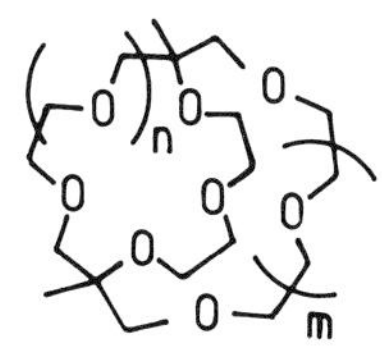

1 n = 2, m = 1
2 n = 1, m = 1
3 n = 1, m = 0

Fig. 3-30. Macrobicyclic polyethers.

Table 3-3. Stability constants (log K_s) in methanol as solvent.

Compound	Ring size	log K_s Na^+	K^+	Selectivity $K_s(K^+)/K_s(Na^+)$
1	20.18.18	4.33	7.06	540
2	18.17.15	5.38	5.94	3.6
3	15.15.14	4.26	2.66	0.025
18-crown-6		4.30	6.02	52
15-crown-5		3.31	3.34	1.1

(Taken from M. Okahara et al., *J. Chem. Soc. Chem. Comm.*, 1045, 1984).

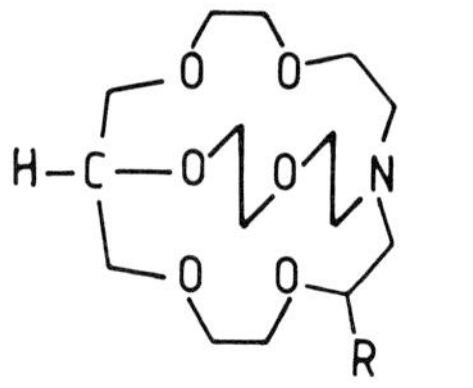

1 R=H

2 R=n-$C_{12}H_{25}$

Fig. 3-31. Mixed bridgehead macrobicyclic polyethers: carbon-nitrogen.

g) Functionalized macrobicyclic compounds

Functionalization of a macrocycle (or macrobicycle) permits the following:

- adjustment of its lipophilicity and hence its ability to insert into membranes;
- the possibility of attaching it to a support thus affording polymers containing units capable of complexation.

The macrobicyclic compound *6* possessing in one of its chains a 1,3-diol group available for further reactions, has been synthesized from the oxetane *3* (Fig. 3-32), the latter being obtained by reaction of 3,3-bis(chloromethyl)oxetane *1* with the amino-alcohol *2* [192a].

The availability of macrocycles containing bis(spirooxetane) moieties present the possibility of preparing bicyclic derivatives with bridgehead carbon atoms. These macrocycles are obtained using the reaction shown in Fig. 3-33. They form weak complexes with alkali metal cations [192b].

The above macrocycles offer numerous possibilities for synthesizing bicyclic systems. Thus, compound *1* leads via hydrolysis to formation of the bicyclic product *2*, as well as the tetraol *3* (Fig. 3-34). Compound *2* can be obtained in yields of ca. 30%. Alcoholysis of *1* in ethyleneglycol or diethyleneglycol gives diols *4a* or *4b*, respectively (in 15-25% yield).

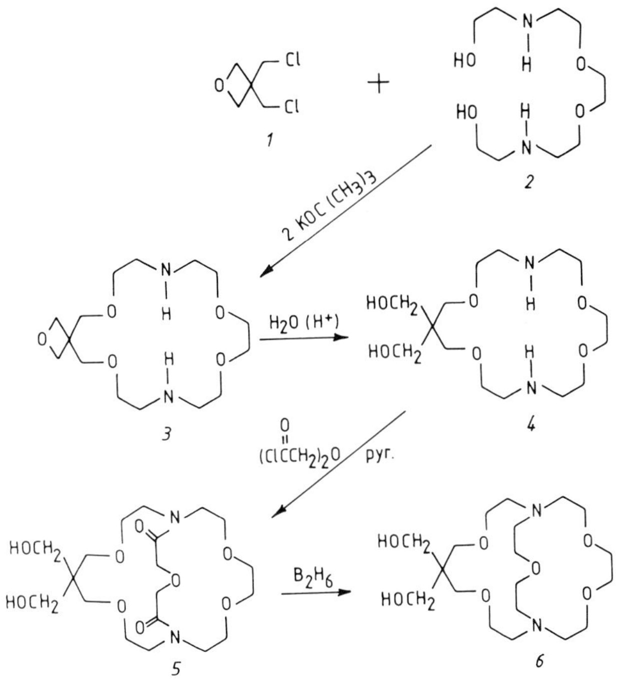

Fig. 3-32. Synthesis of functionalized macro-bicyclic compounds with nitrogen bridgeheads.

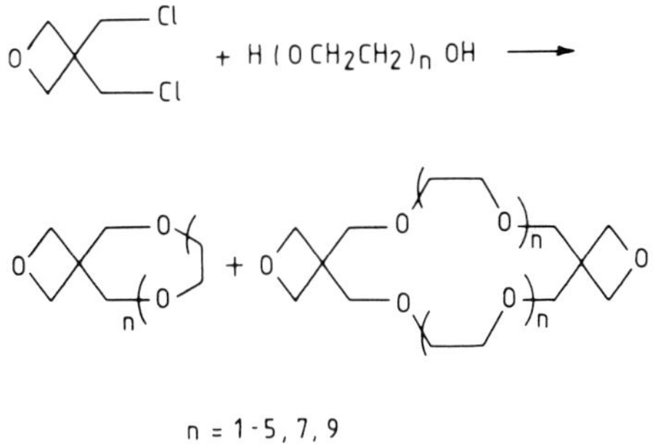

Fig. 3-33. Synthesis of bis(spirooxetane) macrocycles.

Addition of ammonia to compound *1* at 200 °C affords bicyclic amine 5 as the main product (Fig. 3-35). Cleavage of one oxetane ring gives rise to an amino-methyl group which attacks and opens the second oxetane ring. Transannular cyclization is sterically favored, hence the appreciable yield (50%) of product *5*.

Macrocycle *6* is obtained by treatment of *1* with hydrazine at high temperatures (66% yield). Condensation of this diamine *6* with diacid chlorides (Fig. 3-36) can give new macroheterobicyclic compounds, as was demonstrated by condensation of *6* with diglycolyl dichloride to form *7* as a mixture of isomers. Isomers *7a* and *7b* were separated by fractional crystallization [192a].

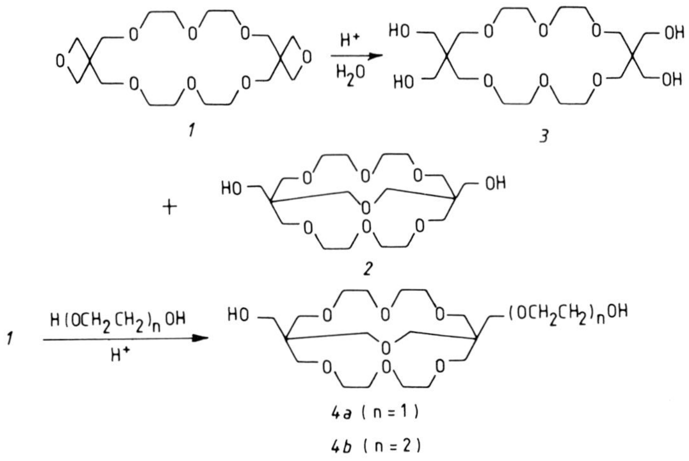

Fig. 3-34. Synthesis of functionalized macrobicyclic compounds with carbon bridgeheads.

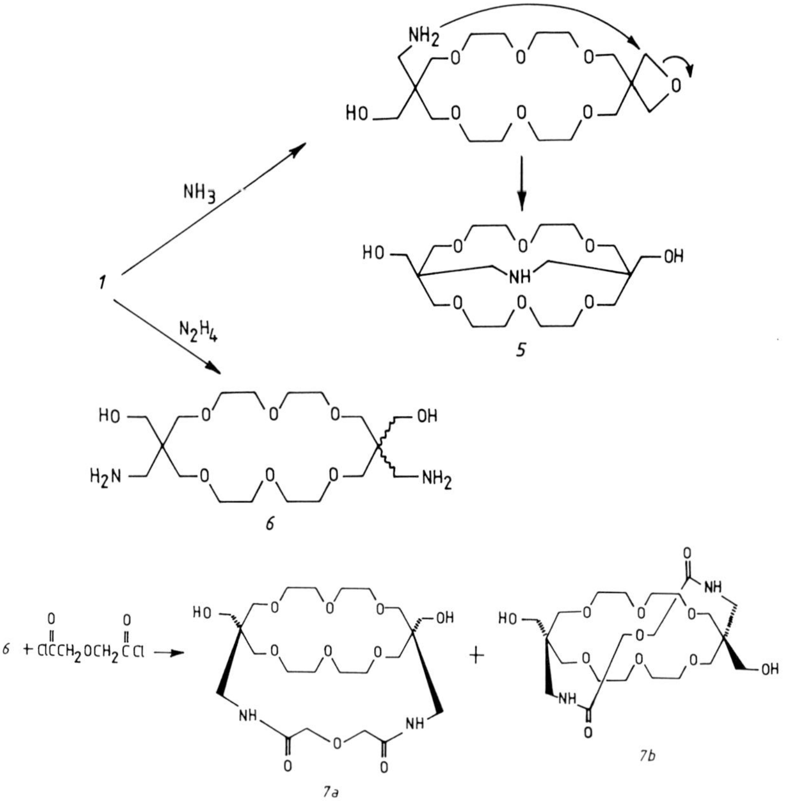

Figs. 3-35 and 3-36. Synthesis of functionalized macrobicyclic compounds with carbon bridgeheads and one *(5)* or two *(7)* nitrogen atoms in one of the chains.

h) Macrobicycles with redox sites

The introduction of a ferrocene unit, which can be oxidized to ferricinium, has been mentioned above (Fig. 3-12, compound *3*).

Quaternization of the N-4 positions of the pyrazine rings in heteromacrocycle *1* using, 1,4-diiodobutane (Fig. 3-37) afforded the new series of cyclophanes, *2* [193].

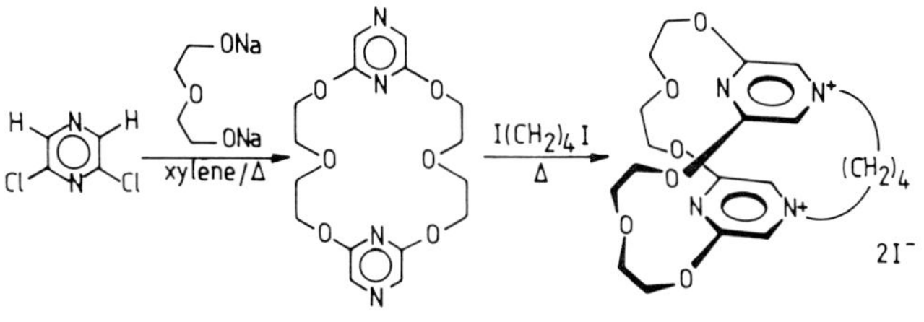

1 2

Fig. 3-37. Synthesis of macroheterobicyclic compounds possessing pyrazino subunits.

This type of compound is particularly interesting because of the possibility of combining its complexing and redox properties, the latter resulting from the pyrazino sites. Internal electron transfer as a result of the interaction distance between the sites can be envisaged.

3.2.2 Condensation of Two Tripodes

Formation of a bicycle by this process (Fig. 3-38) is statistically disfavoured.

Fig. 3-38. Condensation of two tripodes.

Intermolecular condensation competes with the intramolecular condensation. Fig. 3-39 gives a schematic representation of all the possible intra- and intermolecular reactions which can take place after formation of the first bond between the two tripodes. Low yields for this type of cyclization carried out in a single step are therefore to be expected.

Compound *1a* in Fig. 3-40 has been obtained in 31% yield by condensation of 1,3,5-tris-[(*p*-bromomethyl)phenyl]benzene with 1,3,5-tris-[(*p*-mercaptomethyl) phenyl] benzene under high dilution conditions [194].

The same coupling method has been applied to the preparations of compounds *2a* and *3a* which occur in yields of 9% and 7%, respectively. Compounds

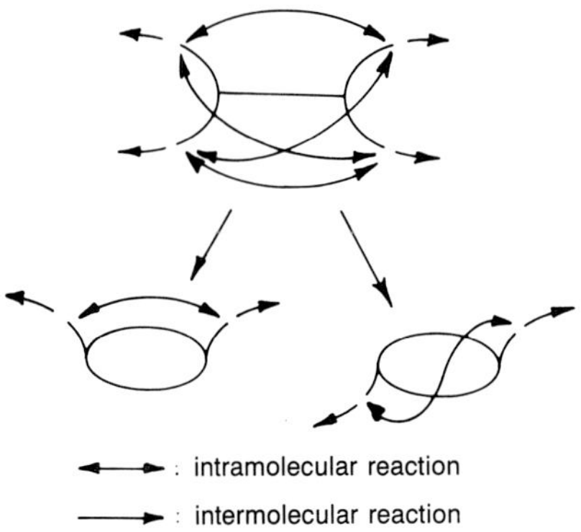

Fig. 3-39. Intra- and intermolecular condensation of two tripodes.

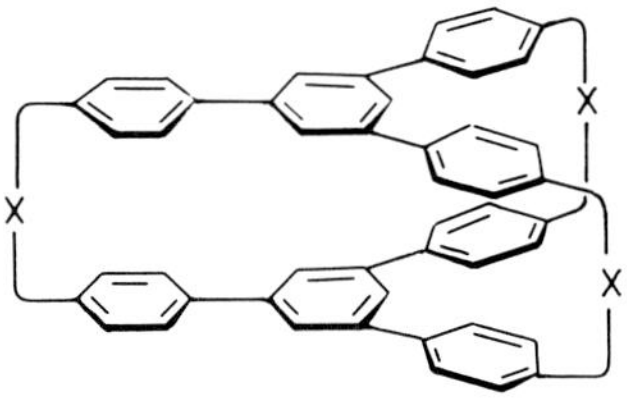

1a - 1f

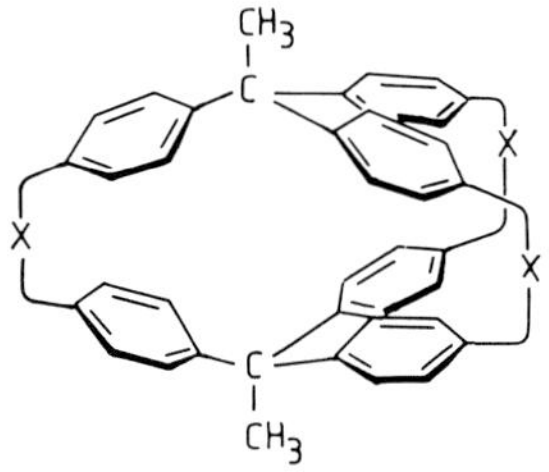

2a - 2f

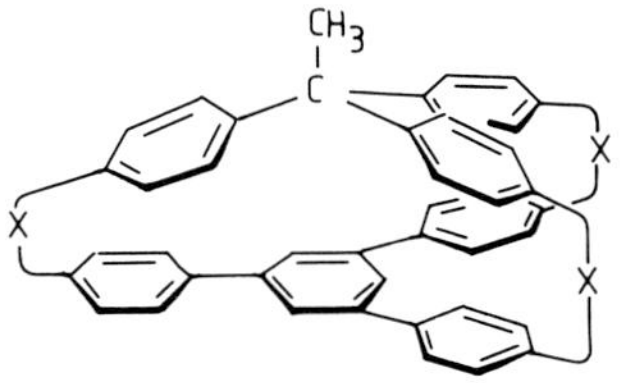

3a

(a), X = $-CH_2-S-CH_2-$

(b), X = $-CH_2-\overset{\oplus}{S}(CH_3)-CH_2-$

(c), X = $-CH_2-CH(SCH_3)-$

(d), X = $-CH_2-CH[\overset{\oplus}{S}(CH_3)_2]-$

(e), X = $-CH=CH-$

(f), X = $-CH_2-CH_2-$

Fig. 3-40. Macrobicyclic compounds possessing benzene rings in the chains.

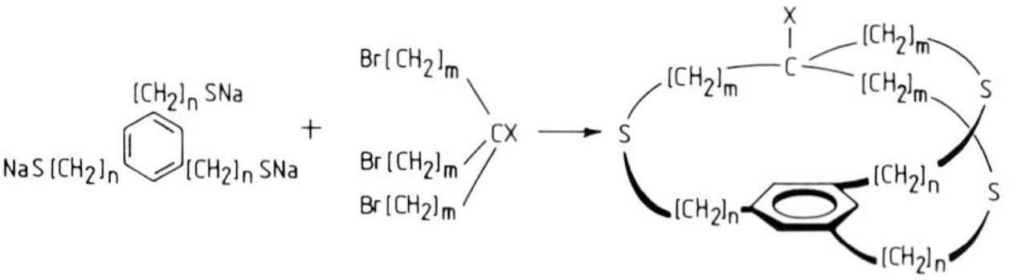

		Yield
1	n = 1, m = 3, X = H	44%
2	n = 1, m = 2, X = H	17%
3	n = 2, m = 3, X = H	23%
4	n = 2, m = 2, X = H	40%
5	n = 2, m = 1, X = Me	3%

Fig. 3-41. Macrobicyclic compounds with a benzene bridgehead.

1a and *2a* were converted into the trienes *1e* and *2e*, and then into the hydrocarbons *1f* and *2f* via the reaction sequence (a) → (f) [195] (these reactions are detailed in Fig. 3-104, Part. I).

The macrobicycles *1–5* (Fig. 3-41) have been prepared by applying the high dilution principle in the condensation of aromatic sodium trithiolates and appropriate trihalogeno derivatives [196].

For compound *5*, the yield was found to be higher (15%) when starting from 1, 1, 1-trismercaptomethylethane and 1, 3, 5-trisbromoethylbenzene. The yields for these direct bicyclization reactions are acceptable.

Various [k.k.k](1, 3, 5)cyclophane macrobicyclic polylactones have also been synthesized by tripode-tripode coupling (Fig. 3-42). Tri(acyl choride)-triol condensation is performed under high dilution conditions and in the presence of Ag^+ ion which favors esterification reactions. The bicyclization yields are low.

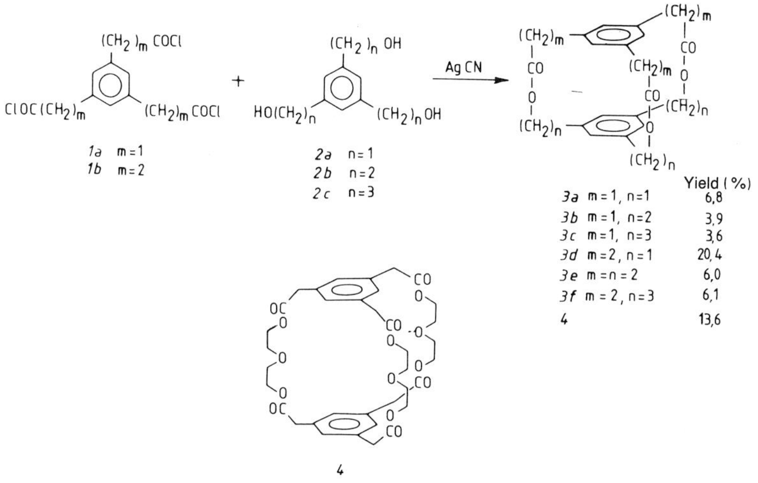

Fig. 3-42. Synthesis of macrobicyclic polylactones.

Although compound *4* is rich in donor atoms, it does not form complexes with metal cations because its cavity is too large [197a].

The (1, 3, 5) cyclophane *1* is obtained by condensation of the triamine *3* with the tri(acyl chloride) *5* (Fig. 3-43). The low yield (11%) in this reaction is not too much of a drawback since the starting materials are very readily obtained.

The X-ray crystal structure of compound *1* shows that the three amide N – H groups point towards the interior of the cavity and thus constitute interaction sites for anionic species such as F^- [197b].

In all the above examples of tripode-tripode coupling, at least one of the reagents contains a 1, 3, 5-benzene ring (rigid group). This type of coupling can also be applied in the case of reagents with a bridgehead atom.

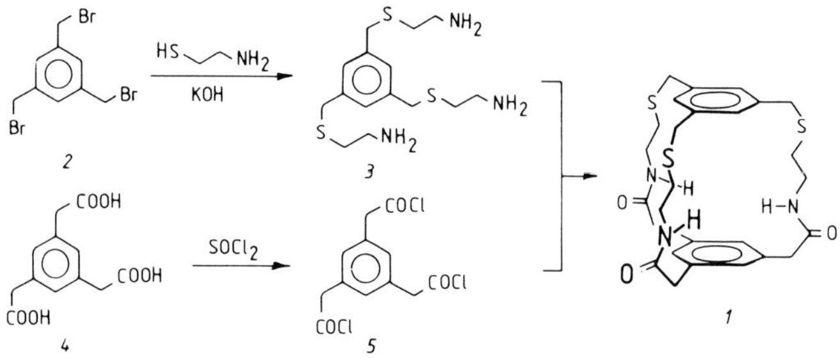

Fig. 3-43. Synthesis of a (1,3,5)cyclophane via amide bond formation.

Condensation of a tosylamide salt with a mesylate is very suitable for the formation of macrocycles, since the bulkiness of the tosyl groups makes the intermolecular reaction very unfavorable. This reaction, previously mentioned concerning the preparation of monocycles, has been applied to macrobicyclization. In this way, a whole series of compounds has been obtained [198].

The syntheses of compounds *1* and *2* as shown in Fig. 3-44 illustrate the method. The tripode compound *3* is the starting material for all the syntheses. It gives access to the trimesylate *6* and to the hexatosyl compound *8*. Condensation between *3* and *6* carried out in hot DMF and in the presence of a large excess of Cs_2CO_3 afforded the macrobicycle *9* (in 50% yield). Likewise, condensation between *8* and *6* gave *10* (in 27% yield).

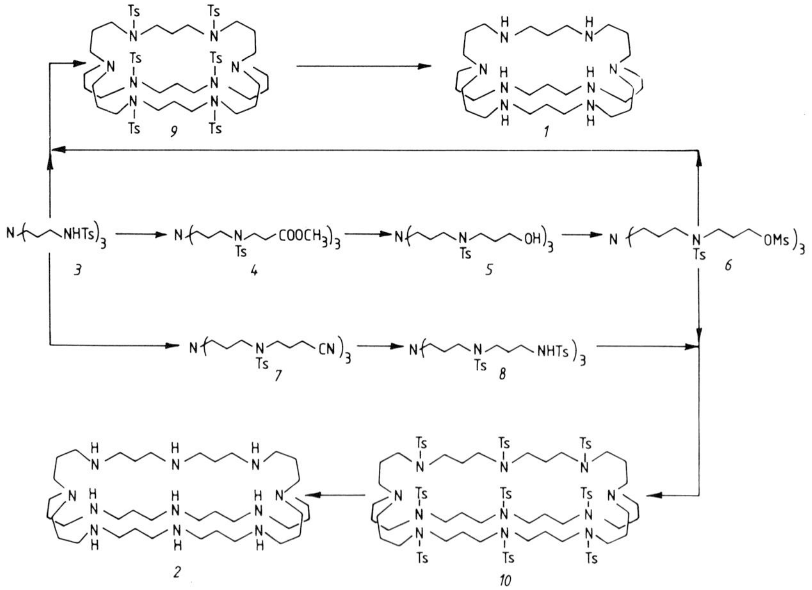

Fig. 3-44. Synthesis of polyaza macrobicycles by coupling of two tripodes.

The macrobicycles *9* and *10* are very readily detosylated [198]. When protonated, the polyaza compounds obtained are extremely good complexing agents for anions [199].

Cyclophane type polyaza-macrobicycles have also been synthesized using this condensation method [200].

Tripode-tripode coupling has been used to prepare complexones for neutral molecules in high yield. The synthesis of these compounds is given in Fig. 3-45, the bicyclization yield being exceptionally high (79%). The out-out *5* and out-in *5* products were separated by chromatography and then detosylated using an HBr/phenol mixture.

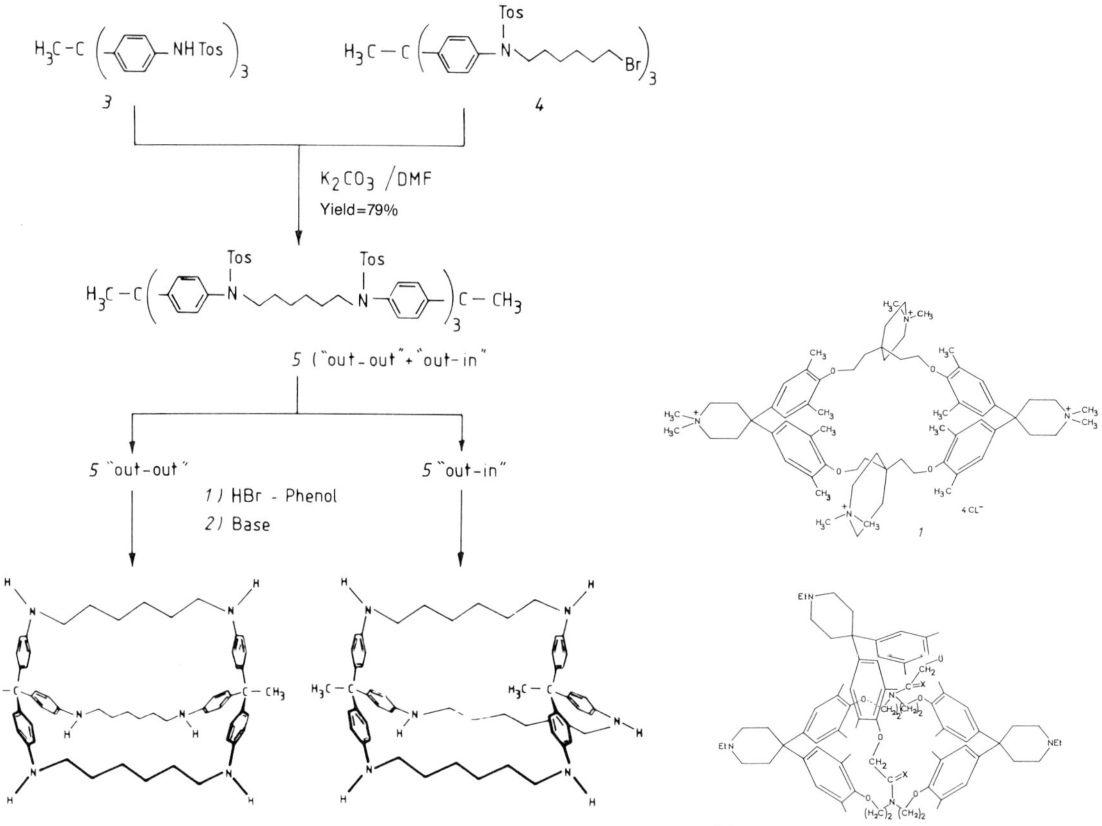

(a) *1* "out-out" *2* "out-in" (b)

Fig. 3-45. (a) Synthesis of bistriphenylethane via tripode-tripode coupling; (b) Cyclophane hosts with enforced hydrophobic cavity binding sites.

In aqueous acid, compounds *1* and *2* complex naphthalene. The dissolution of adamantane in water in the presence of ligand *1* is remarkable [201 a]. In these complexes, the organic substrate held in the cavity is totally isolated from the exterior aqueous medium, and the entire complex is rendered soluble by means of the protonated amine groups.

The complexation of neutral molecules in aqueous or organic solution has been greatly developed since the early 1980s [201 b]. The complexing agents are chiefly of the cyclophane type [201 c]. The compounds in one particularly efficient class possess a cavity of enhanced hydrophobicity with respect to that for the cyclophanes shown in Fig. 3-45 a. This property results from the greater distance from the cavity center of the charged sites responsible for water solubility (Fig. 3-45 b). These mono- and bicyclic systems complex a wide variety of aromatic substrates (naphthalene, pyrene etc.) including their functionalized derivatives [201 d].

Another type of reaction, the oxidative coupling of alkynes, has been used for the efficient synthesis of cage compounds [202]. In this case, the coupling reaction consists of dimerization of the starting tripode *1* (Fig. 3-46).

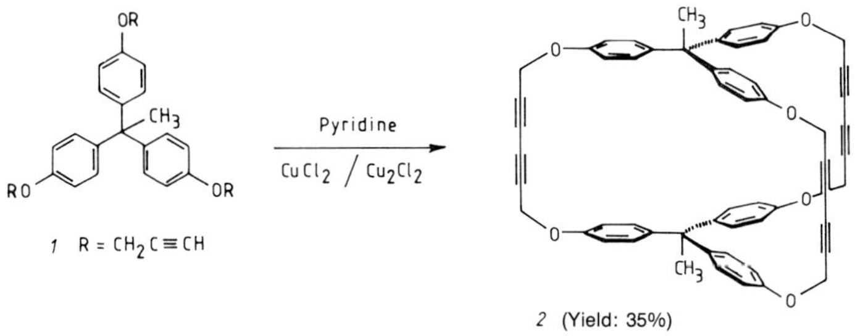

Fig. 3-46. Macrobicyclization via oxidative coupling of trialkynes.

3.2.3. Variant of the Condensation of Two Tripodes (Tripode Capping)

Compared to tripode-tripode coupling which in itself presents a low probability of macrobicycle formation, tripode capping (Fig. 3-47) appears to defy reason.

It involes the assembly, in a single operation, of five bi-or trifunctional fragments (the number of reactive centers is twelve). Consequently, the yields obtained are almost invariably very low with, however, some remarkable exceptions.

This method has been used to synthesize the triply bridged phanes *1–4* shown in Fig. 3-48. As expected, the yields are low [194].

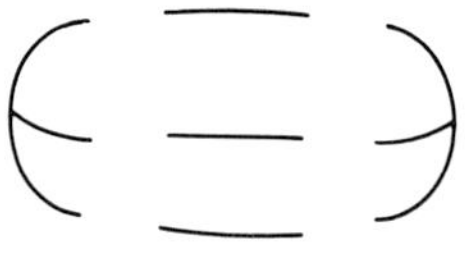

Fig. 3-47. Scheme showing tripode capping.

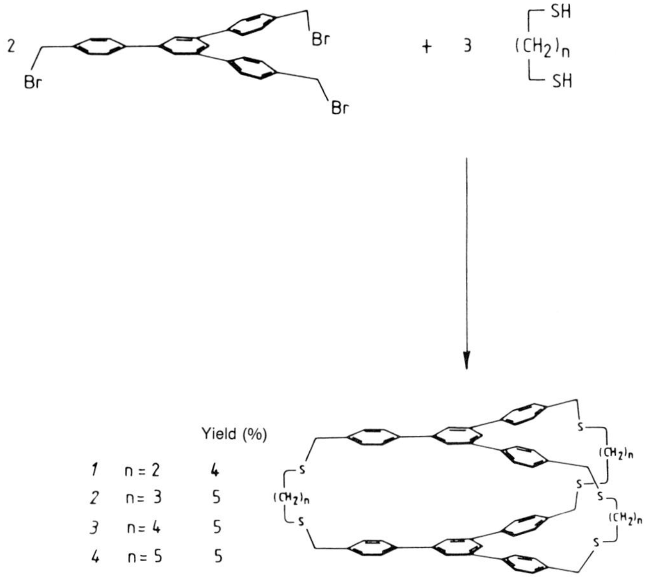

Fig. 3-48. Macrobicyclization using the tripode capping method.

Condensation of triethanolamine with 2,6-bis(chloromethyl)pyridine in dimethylformamide in the presence of sodium hydride gave, in 2% yield, a bicyclic compound containing a pyridine ring in each chain (Fig. 3-49a) [203].

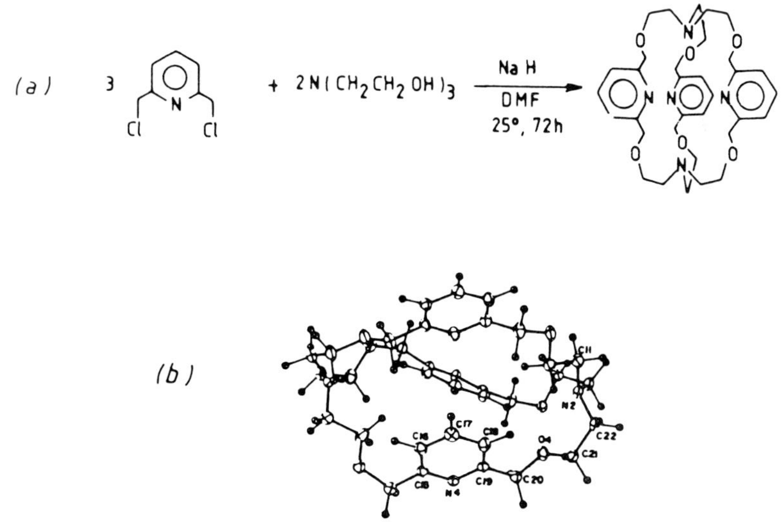

Fig. 3-49. Synthesis and structure of a cryptand containing a pyridine ring in each chain (taken from G. R. Newkome et al., *Tetrahedron Letters,* **22,** 3035, 1981).

The X-ray crystal structure (Fig. 3-49b) showed that in this bicyclic molecule, the bridgehead nitrogen lone pairs are in the i,i configuration and that one of the pyridine rings is curiously located within the cavity.

The same reaction, carried out starting from 2,6-dichloropyridine and triethanolamine in xylene (Fig. 3-50) affords a cryptand in which the bridgehead nitrogens possess a planar configuration with crystallographically equivalent 120° C–N–C bond angles [204]. The planarity of the bridgehead nitrogen atoms can be clearly seen in Fig. 3-51b. These flattened planar configurations are extremely surprising.

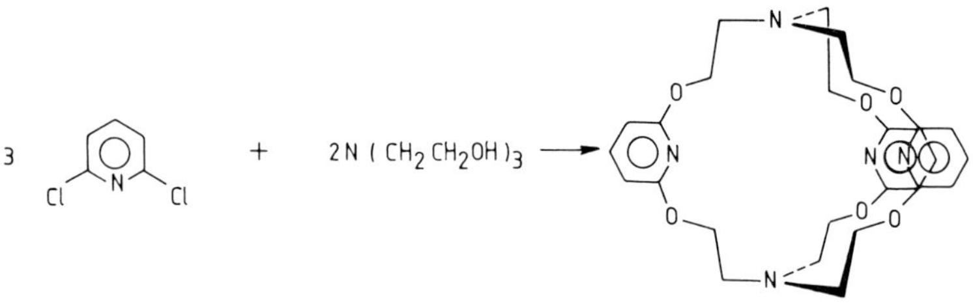

Fig. 3-50. Synthesis of a cryptand using the tripode capping method (5% yield).

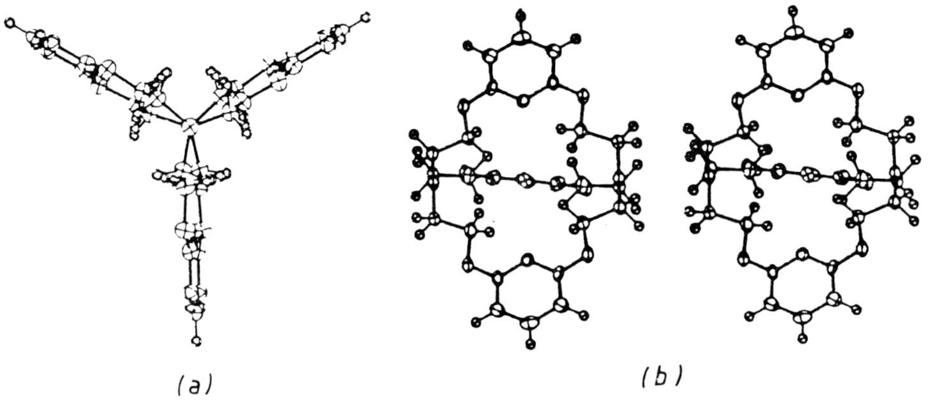

Fig. 3-51. Structure of a cryptand in which the bridgehead nitrogens possess a planar configuration (taken from G. R. Newkome et al., *J. Am. Chem. Soc.,* **101,** 1047, 1979) (b: stereoview).

Tripode capping can be a very efficient method as illustrated by the remarkable synthesis of compound *1* (Fig. 3-52). The starting materials are easily accessible and this twelve center reaction takes place in high yield (60%).

Fig. 3-52. One-step synthesis of a hexa-imine cryptand.

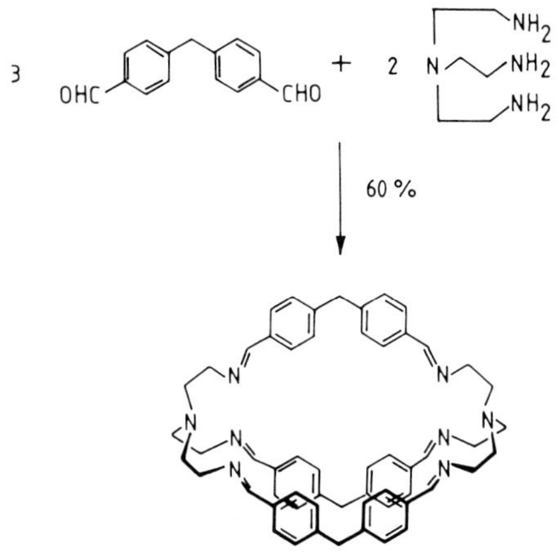

1

Binuclear complexes of this compound, $[Cu_2 \subset 1]^{2+}$ and $[Ag_2 \subset 1]^{2+}$, have been characterized [205]. Reduction of the imine groups gives the polyamine which in the protonated form complexes anions.

3.2.4 Cyclizations Using an Internal Template

The stepwise synthesis (cf. Fig. 3-2) of bicyclo[k.l.m] or diazabicyclo[k.l.m] systems* becomes difficult with k, l, m in the range 3-5 (this corresponds to formation of 8- to 12-membered medium rings). In this connection, it should be recalled that the synthesis of [1.1.1] cryptand is very laborious (cf. Sections 3.2.1 a and 3.3.2 d).

Specific synthetic strategies have been elaborated for this type of compound using internal templates.

a) N – N Templates

The synthesis of medium ring diazabicyclo[k.l.m]alkanes has been carried out (Fig. 3-53 a) using as internal template an N – N bond which is subsequently eliminated from an intermediate dication. This dication has a propellane structure and is in fact a hydrazinium dication in which the N – N bond readily undergoes reductive cleavage [206].

* The nomenclature of bicyclo systems and cryptands can lead to confusion. For bicyclo systems, the integers k, l, m designate the total number of carbon and heteroatoms in each of the three chains. In the case of cryptands, the integers [n, o, p] designate the number of oxygen atoms in each bridge, the chain with one oxygen atom being $-CH_2CH_2OCH_2CH_2-$ and that with two oxygen atoms $CH_2CH_2(OCH_2CH_2)_2-$, etc. [1.1.1] Cryptand thus corresponds to diazabicyclo[5.5.5]heptadecane.

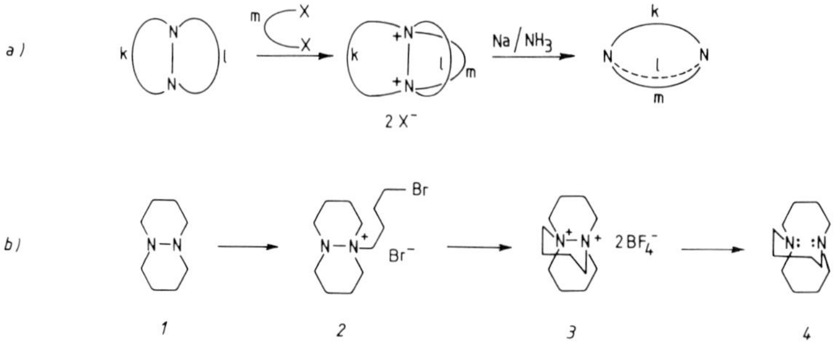

Fig. 3-53. Synthesis of diazabicyclo[k.l.m]alkanes using an internal template.

The preparation by Alder et al. of 1,6-diazabicyclo[4.4.4]tetradecane is shown in Fig. 3-53b. Treatment of 1,6-diazabicyclo[4.4.0]decane, *1* with 1,4-dibromobutane gives *2*, which is cyclized to *3* on treatment with $AgBF_4$ in aqueous HBF_4. The dication *3* is reduced using Na/NH_3 to give *4* in 84% yield [206b]. Using this method, various diamines, notably [3.3.2], [3.3.3], [4.3.3] and [4.4.3], have been obtained. The route is therefore quite general and very useful for the formation of medium diazabicycles [206c].

The chemistry of these compounds is extremely rich. The proximity imposed on the bridgehead nitrogens gives rise to specific properties due to their mutual interactions. Because of the small size of the cavities in these macrobicycles, their protonation is also very remarkable (see below).

We shall comment briefly on the outstanding properties of macrobicycles *1–4* (Fig. 3-54) concentrating particularly on the most widely studied compound *4*.

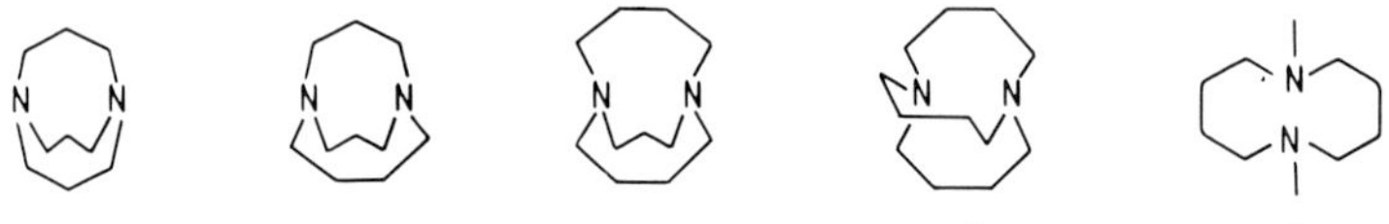

1 *2* *3* *4* *5*

Fig. 3-54. Examples of diazabicyclo[k.l.m]alkanes, *1–4*.

On oxidation, these compounds give successively the radical cation *B* followed by the dication *C* (Fig. 3-55). The number of electrons involved in the interactions between the two bridgehead nitrogen atoms therefore proceeds from 4 *(A)* to 3 *(B)* to 2 *(C)*.

The first ionization potentials (I_1) of these compounds as determined by photoelectron spectroscopy are exceptionally low, being between 6.64 and 6.85 eV (cf. values of ca. 8 eV for tertiary amines such as NEt_3, NPr_3 and NBu_3). This facile oxidation to the radical cation $(N{-}{-}{-}N)^{\dot{+}}$ is explained by interactions between the lone

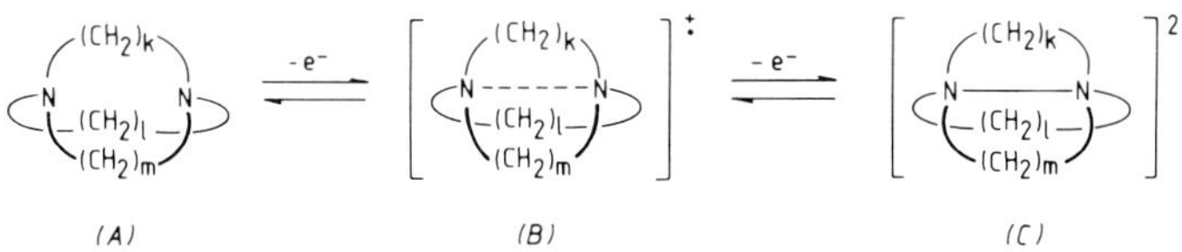

Fig. 3-55. Formation of radical cation B and dication C by oxidation of a diazabicyclo-[k.l.m]alkane.

pairs, the loss of an electron greatly decreasing these repulsive interactions [207a]. The second ionization potentials (I_2) are between 7.77 and 7.90 eV. The large value of ca. 1.1 eV for the difference I_2-I_1 lends support to there being a strong interaction between the lone pairs. Compound *5* (see Fig. 3-54) in which the nitrogen lone pairs are free to avoid each other shows a single ionization peak at 7.81 eV [207b].

The radical cations formed are extremely stable. In acetonitrile solution at room temperature, the radical cation $1^{+\cdot}$ has a half-life of 1 day. Under the same conditions, $t_{1/2}$ for the radical cation $4^{+\cdot}$ is 1 year, whereas that for $5^{+\cdot}$ is only 5 milli-seconds.

The unusual stability of these radical cations can be attributed to formation of a three electron bond consisting of two σ electron and one σ^* electron [207 c]. Other examples of this type of "exotic" bond exist, and numerous theoretical studies have been devoted to it [208].

The bond order is low (1/2), but in the case of bicyclic systems, the conformational pressure imposed by the carbon bridges favors overlap of the nitrogen orbitals and ensures both the formation and stability of the three electron bond. In the case of $5^{+\cdot}$, since this conformational pressure is absent, rupture of the $(N{-}{-}{-}N)^{+\cdot}$ bond readily occurs giving $(N^{+\cdot})$ and (:N). A proton α to the radical cation is subsequently transferred to the amine group. This explains the short lifetime of the radical cation $5^{+\cdot}$ [207b, d].

Crystallographic studies on compound *4* and its derivatives $4^{+\cdot}$ and 4^{2+} have elucidated several points [209a, b]. In the diamine *4* (Fig. 3-56) the repulsion of the nitrogen lone pairs results in a very long N---N distance (2.806 Å).

This large separation and the consequent high lone pair – lone pair repulsion can be better appreciated when one takes into account that the conformational pressure of the $(CH_2)_4$ bridges tends to impose an intrabridgehead distance of ca. 1.5 Å. At this distance observed for the 4^{2+} dication, the bridges adopt a relaxed conformation. The three rings about the N---N axis in 4^{2+} and about the C---C axis in the analogous case of [4.4.4]propellane [209c], are in the chair conformation.

The structure of radical cation $4^{+\cdot}$ is of paramount importance. Firstly, the N---N distance is greatly shortened (by ca. 0.5 Å) resulting in change from an overall repulsive system *(4)* to a weakly attractive system ($4^{+\cdot}$). Secondly, the most important point is the increased inward pyramidalization of the nitrogen atoms (decrease in CNC bond angle from 115.5 to 114°). In fact, a radical cation located on only one of the two nitrogen atoms would have a CNC angle of 120° whereas the

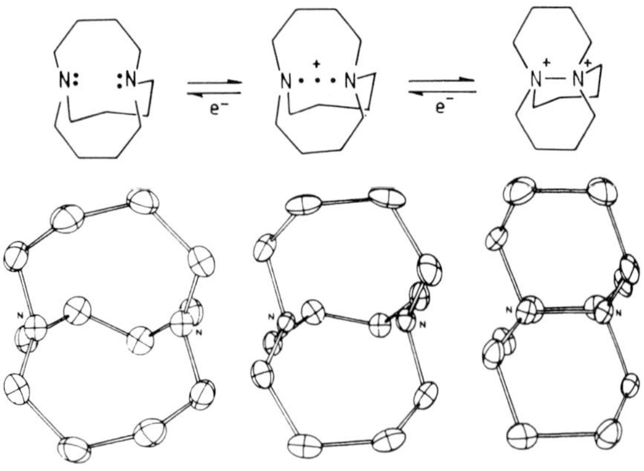

N····N Distance (Å)	2,806	2,295	1,532
C-N-C Band angle	115,5	114,0	109,0

Fig. 3-56. Structure of 1,6-diazabicyclo[4.4.4]tetradecane and of the radical cation and dication formed on oxidation (taken from R. W. Alder et al., *J. Chem. Soc., Chem. Comm.,* 949, 1985).

other nitrogen would be pyramidal. The symmetry group of *4*$^{+\cdot}$ is very close to D_3, reflecting well the symmetrical distribution of the three electrons on the two nitrogen atoms. Finally, changes in the bond and torsional angles (see [209b]) tend to relieve strain in the carbon chains on proceeding from *4* to *4*$^{+\cdot}$ to *4*$^{2+}$.

Various solution spectroscopic studies provide further evidence for a symmetrical N---N bond in the radical cation [209d].

b) N – H Templates

We have already pointed out (Fig. 3-58, Part I) that intramolecular hydrogen bonds can assist macrocycle synthesis. Using the NH hydrogen as internal template, Montanari et al. have developed an efficient synthesis of [1.1.1] cryptates *4a* and *4b* (Fig. 3-57). These compounds have been obtained in ca. 40% yield by condensation of the monometallated derivative of diazacoronand *1* with diethylene glycol bis-(methanesulfonate) or its alkyl substituted derivatives [210]. The macrobicyclic systems *4a* and *4b* were isolated as their proton cryptates. As we shall see below, the proton cannot be removed from the cavity in these macrobicycles.

The mechanism for assistance by the intramolecular hydrogen bond is shown in intermediate *3*. This mechanism is based on the following experimental data: (a) the yield is 40% with 1 equivalent of base, whereas with 2 equivalents, unre-

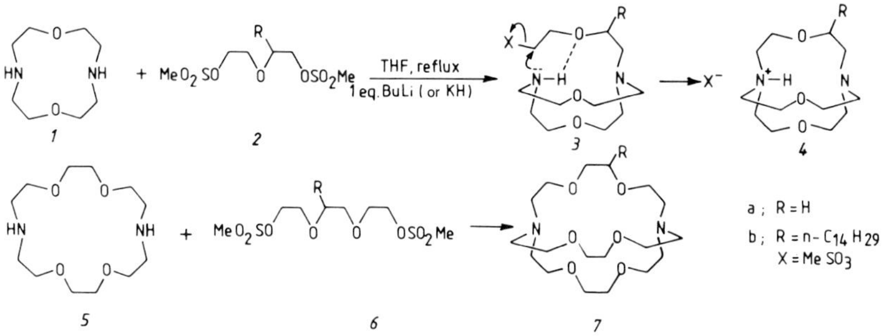

Fig. 3-57. Cryptand synthesis assisted by intramolecular hydrogen bonding.

solvable polymeric mixtures are obtained; (b) the yield is independent of the metal cation (Li^+ or K^+) and therefore it is unlikely that the cation acts as an external template.

The method may be extended to larger macrobicycles: *5* + *6* → *7a* or *7b*. However, it seems that in these cases assistance due to templating by the metal cation is more important than that due to templating by the hydrogen bond.

3.2.5 Cyclization Using an External Template

The use of metal cations as external templates has been mentioned previously (cf. Section 3.3.4 in Part I). An important class of macrobicycles has been elaborated using the same method. The role of the metal cation is firstly to hold a certain number of reactive fragments in its coordination sphere, thus implying slow ligand-metal exchange. Ideally, none of the ligand-metal bonds should be broken during the synthesis. Secondly, the metal cation affects the reactivity of the functional groups involved. Obviously, an activating effect is desirable. Finally, as we shall see, the metal cation can also play a part by protecting reactive groups.

This method is most useful when the starting materials are simple and therefore available in large quantities, and when the bicyclization reaction occurs in a single step in high yield.

a) Sepulchrates

The above ideal conditions pertain for the synthesis by Sargeson et al. of the octaazacryptate shown in Fig. 3-58, which was obtained in greater than 95% yield by condensation of tris- (ethylenediamine)cobalt(III) ion with formaldehyde and ammonia [211]. Several tens of grams can thus be easily prepared in a single operation [211 c].

Behind the simplicity of the overall reaction, lies of course a series of intermediate steps all taking place in the coordination sphere of the metal cation. A sche-

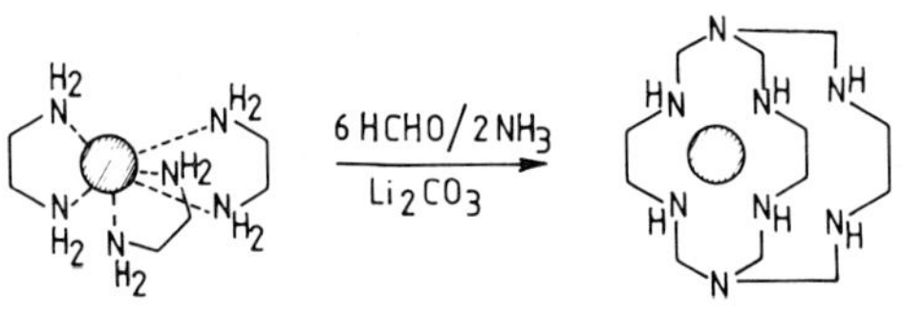

Fig. 3-58. Synthesis of an octaazacryptate (sepulchrate), the cobalt ion acting as an external template.

matic representation of the postulated mechanism [211b] is given in Fig. 3-59 (the charges and the parts of the molecule not directly involved have been omitted).

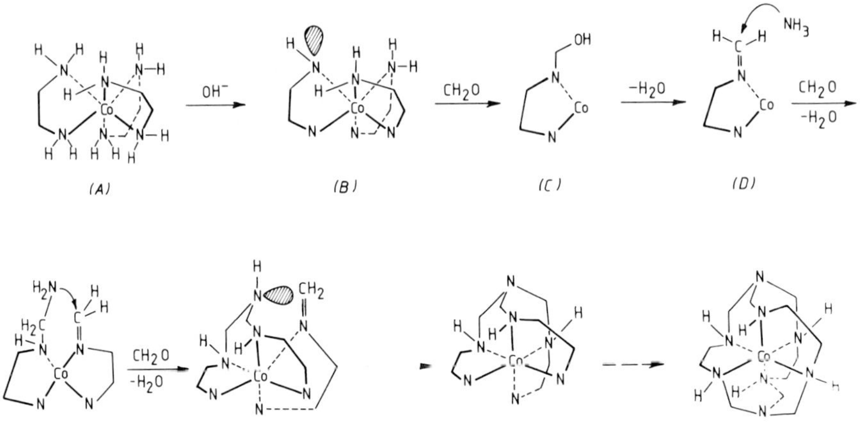

Fig. 3-59. Mechanism for the formation of the sepulchrate.

The complex $Co(en)_3^{3+}$ *(A)* is deprotonated to give *B* which reacts with formaldehyde to afford the carbinolamine *C*. The latter is dehydrated to give the imine *D* which then undergoes attack by ammonia. It is in this step that the role of the metal cation is crucial. Firstly, its coordination to the nitrogen atom makes the C=N bond similar in character to an iminium ion thus facilitating addition of a nucleophile to carbon. Secondly, the metal cation protects the imine group from protonation and hence from being destroyed.

After addition of NH_3, the same sequence of operations (formaldehyde addition followed by dehydration) is repeated for the second and then the third strand to give successively *E* and *F*. In each case, the uncoordinated amine group adds to the imine formed, ending with formation of *G*. The same sequence of operations is then carried out on the opposite face to give finally the macrobicyclic complex *H*.

Sample withdrawal during the reaction led to the isolation of certain intermediates hence suggesting the proposed reaction mechanism.

Finally, it should be noted that by starting with one or other of the chiral forms of $Co(en)_3^{3+}$, a single optically active isomer is obtained with retention of chirality of the $Co(en)_3^{3+}$ ion. This stereospecificity implies that for each intermediate step there is a preferred orientation of the reacting groups [212].

Sargeson has given the name "sepulchrate" (sep) to this type of bicyclic system. The structure of $Co(sep)^{3+}$ (Fig. 3-60) shows that the complex is of the cryptate type, that it has approximately D_3 symmetry, and that six nitrogens interact with the metal cation.

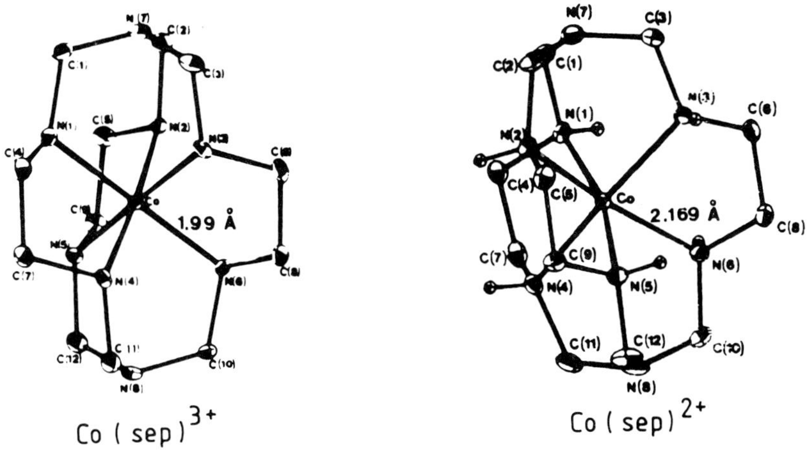

Fig. 3-60. Structures of $Co(sep)^{3+}$ and $Co(sep)^{2+}$ sepulchrates (taken from A. M. Sargeson, *Chem. in Britain,* **15**, 23, 1979).

The poetical description applied to this complex suggests it should exhibit unusual complexing properties. The "sep" ligand cannot be removed from the $Co(sep)^{3+}$ complex. The complex survives prolonged boiling in 12 mol/1 aqueous HCl. Even the bridgehead nitrogen atoms, although uncoordinated, are not protonated. The Co(II) complex, formed by reduction of $Co(sep)^{3+}$ with zinc under acid conditions, is also very stable. Thus, racemization of chiral $Co(sep)^{2+}$ is extremely slow suggesting that there is very little, if any, decomplexation. This has been confirmed by exchange experiments with radioactive cobalt, as follows:

$$Co(sep)^{2+} + {}^{60}Co^{2+} \rightleftharpoons {}^{60}Co(sep)^{2+} + Co^{2+}.$$

After 24h, less than 1% of ^{60}Co had been incorporated into the cavity [211 d].

The stability of $Co(sep)^{2+}$ has allowed its X-ray crystal structure to be determined (Fig. 3-60; [211 d]).

It should be noted that treatment of $Co(sep)^{2+}$ with O_2 under aqueous conditions leads to reoxidation of the complex to $Co(sep)^{3+}$ with greater than 99%

retention of absolute configuration. This is accompanied by formation of hydrogen peroxide:

$$2\ Co(sep)^{2+} + O_2 + 2H^+ \rightarrow 2\ Co(sep)^{3+} + H_2O_2.$$

The reaction is quantitative and very fast [211 d].

The electrochemical properties of $Co(sep)^{3+}$ have been studied in detail. Cyclic voltammetry of cobalt sepulchrate shows a reversible reduction wave at -0.27 V vs. NHE. This reduction potential is more negative than those observed for $Co(en)_3^{3+/2+}$ (-0.13V) and $Co(NH_3)_6^{3+/2+}$ ($+0.01$ V).

Comparison of these values with reduction potentials observed for other amine ligands giving hexacoordinated complexes (Fig. 3-61) has formed the basis for an explanation of the variation in reduction potential of $Co^{3+/2+}$ systems with the nature of the ligand [213 a].

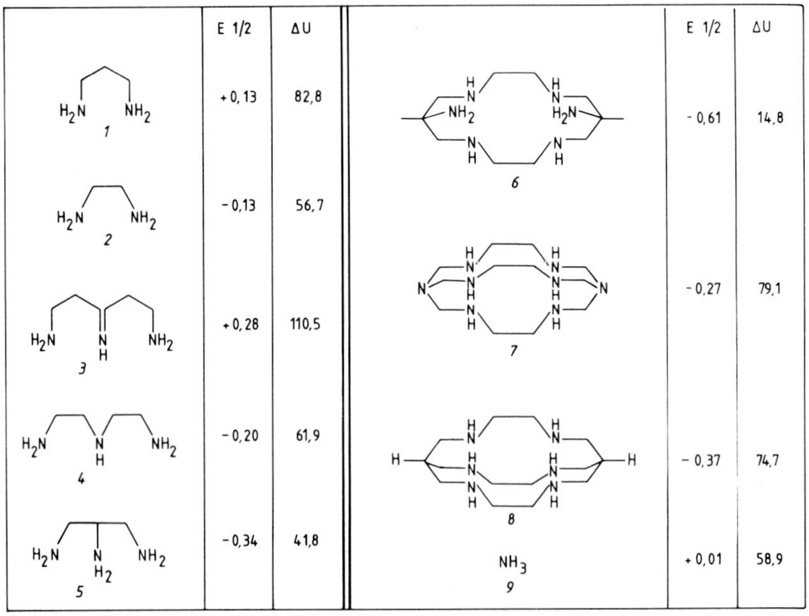

Fig. 3-61. Correlation between Co(III)/Co(II) reduction potentials and the strain relaxation energy (ΔU) accompanying this process (taken from T. W. Hambley, *Inorg. Chem.*, **27**, 2496, 1988).

Molecular mechanics calculations have been used to determine the *U* Co(III) and *U* Co(II) strain energies for all the complexes formed with Co(III) and Co(II). The resulting Co(III)–N and Co(II)–N bond lengths are in excellent agreement with the crystallographic values, the latter having been determined for all the Co(III) complexes and some of those of Co(II). In all cases, the strain energy of the Co(II) state of any given complex is lower than of the Co(III) state. This is explained by the fact that the Co(II)–N bond is longer (by about 0.2 Å) than the Co(III)–N bond. The intraligand interactions are therefore weaker than for the Co(III) complexes. The reduction of Co(III) to Co(II) thus results in a decrease in strain energy (the strain or steric relaxation energy) expressed as $\Delta U = U$ Co(III) $-$ U Co(II) [213b]. The magnitude of this relaxation energy depends on the system (Fig. 3-61). With diamino ligands *1* and *2*, ΔU is found to be larger for *1* than for *2*. This means that the steric relaxation energy for the reduction Co(III) → Co(II) is greater with *1* than with *2*, and hence Co(III) is less stable with *1* ($E_{1/2}$ = +0.13 V) than it is with *2* ($E_{1/2}$ = −0.13 V). These results are confirmed by the crystallographic data for the complexes $Co(1)^{3+}$ and $Co(2)^{3+}$. Because of steric strain, the Co–N distance is slightly greater in the complex with *1* (1.98 Å) [213c] than in the complex with *2* (1.96 Å) [213b]. In other words, it is more difficult for ligand *1* to coordinate to the small Co(III) cation than it is for ligand *2* to do so. Conversely, *1* coordinates more readily than does *2* with the larger Co(II) cation.

The same trends are observed for the triamino compounds *3-5*. The strain relaxation energy for the reduction Co(III) → Co(II) is very great for *3* which gives rise to a high value of $E_{1/2}$ (+0.28 V vs. NHE). Compounds *4* and *5* have smaller ΔU values and therefore more negative values of $E_{1/2}$. In *5*, the interaction sites are very close together and so the ligand is perfectly adapted for coordination to Co(III) ($E_{1/2}$ = −0.34 V vs. NHE). The crystallographic results are in complete agreement with these trends, the Co(III)–N bond lengths being 2.03 Å for *3*, 1.97 Å for *4*, and 1.96 Å for *5* [213d].

Compound *6* (which, as it were, contains two units of *5* in the ring) has an exceptionally low reduction potential ($E_{1/2}$ = −0.61 V vs. NHE), and one of the shortest Co(III)–N distances known (1.94 Å) [213e]. The very small value of ΔU is therefore not surprising.

Because of their structures, the macrobicyclic compounds *7* and *8* have reduced degrees of freedom giving rise to high values of U Co(III) and U Co(II). As a result, the reduction Co(III) → Co(II) occurs in both cases with substantial strain relaxation. However, the $E_{1/2}$ values remain negative.

All the above examples show that strain relaxation is correlated with reduction potential. Other factors are also involved, e.g., the degree of substitution of the amine groups [213b]. It should be noted that Busch et al. have come to identical conclusions for the relation between ligand strain and reduction potential in the case of polyaza-macrocyclic compounds [213f].

The $Co(sep)^{2+/3+}$ system also exhibits very interesting electron transfer properties [211a,d]. The electron transfer rate has been measured using the rate of racemization of an equimolar mixture of Δ-$[Co(sep)]^{3+}$ and Λ-$[Co(sep)]^{2+}$ in the following way*:

$$\Lambda\text{-}[Co(sep)]^{2+} + \Delta\text{-}[Co(sep)]^{3+} \rightleftharpoons \Lambda\text{-}[Co(sep)]^{3+} + \Delta\text{-}[Co(sep)]^{2+}.$$

* The configurations Λ and Δ are defined according to IUPAC rules, *Pure and Appl. Chem.* **28,** 1 (1971).

It should be noted that the racemization process is due to electron transfer and not to ligand-metal exchange which would give the same end result (see the $Co(sep)^{2+}$ + $^{60}Co^{2+}$ exchange experiments described above). The rate constant for this self-exchange reaction is $\sim 10^5$-fold faster than that for $[Co(en)_3]^{2+/3+}$. Several hypotheses have been put forward to explain the fast electron transfer. Molecular mechanics calculations have shown that there is relief of internal ligand strain in the transition state thus reducing the activation energy for the transfer reaction. In other words, this result suggests that the cavity is a little too large for Co(III) and slightly too small for Co(II). The strain existing for both oxidation states could be relieved in the transition state hence favoring reaction [214a]. In systems such as $Co(en)_3{}^{2+/3+}$ having many more degrees of freedom, the strain described above cannot exist.

Other calculations emphasize the conformational changes which occur on Co(II) → Co(III) oxidation. In the case of cobalt sepulchrate, these calculations show that the *same* conformational isomer is the most stable for both Co(III) and Co(II). Electron transfer therefore occurs without major structural modifications. This situation is exceptional since, in general, for less organized systems electron transfer occurs via several intermediate states, the thermodynamic and kinetic properties of each of these influencing the overall process observed [214b].

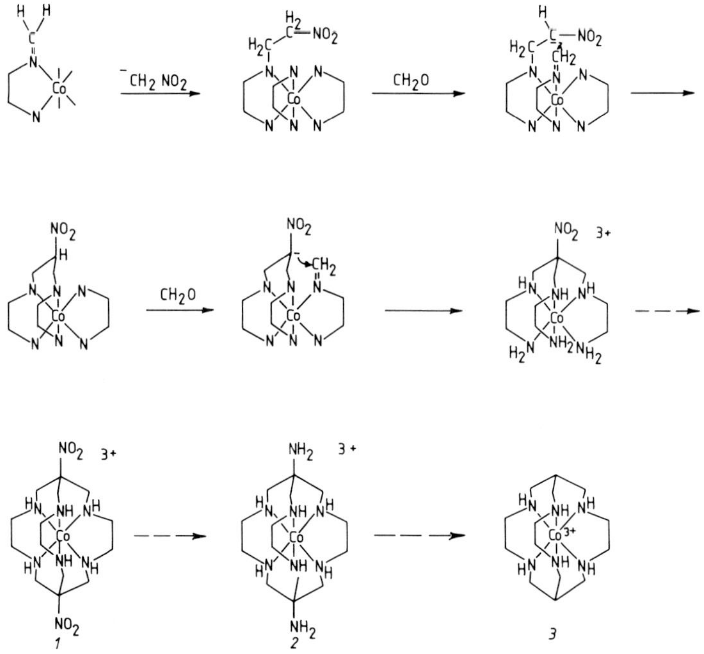

Fig. 3-62. Scheme showing sarcophagine synthesis.

Compound *8*, shown in Fig. 3-61, has been named "sarcophagine". Its synthesis makes use of the reactivity of the carbon atom in nitromethane. The mechanism of formation (Fig. 3-62) is similar to that for $Co(sep)^{3+}$. The compound $Co[di(NO_2)sar]^{3+}$ (*1*) is reduced to $Co[di(NH_2)sar]^{3+}$ (*2*) which in turn is converted to $Co(sar)^{3+}$ (*3*) [214a, c].

Other ligands containing both sulfur and nitrogen atoms have been obtained from the Co(III) complex of the ligand (ten) prepared by condensation of ethylenimine and 1, 1, 1-tris(mercaptomethyl)ethane (Fig. 3-63) [215a].

Encapsulation of Co(III) was effected by condensing $[Co(ten)]^{3+}$ with formaldehyde and either ammonia or nitromethane giving, respectively, $[Co(azacapten)]^{3+}$ or $[Co(NO_2capten)]^{3+}$. Reduction of the latter affords the corresponding amino derivative, $[Co(NH_3capten)]^{4+}$.

The introduction of sulfur atoms gives rise to a somewhat larger macrobicyclic cavity than in the hexaamine analogues. Furthermore, thioether donors are softer than amine donors thus increasing the stability of the lower oxidation states ($E_{1/2}$ = −0.01 V vs. NHE) [215b]. The rate of electron self-exchange Co(II) ⇌ Co(III) for $Co(azacapten)^{2+/3+}$ is very rapid, being ca. 10^3-fold as fast as that for $Co(sep)^{2+/3+}$ [215c].

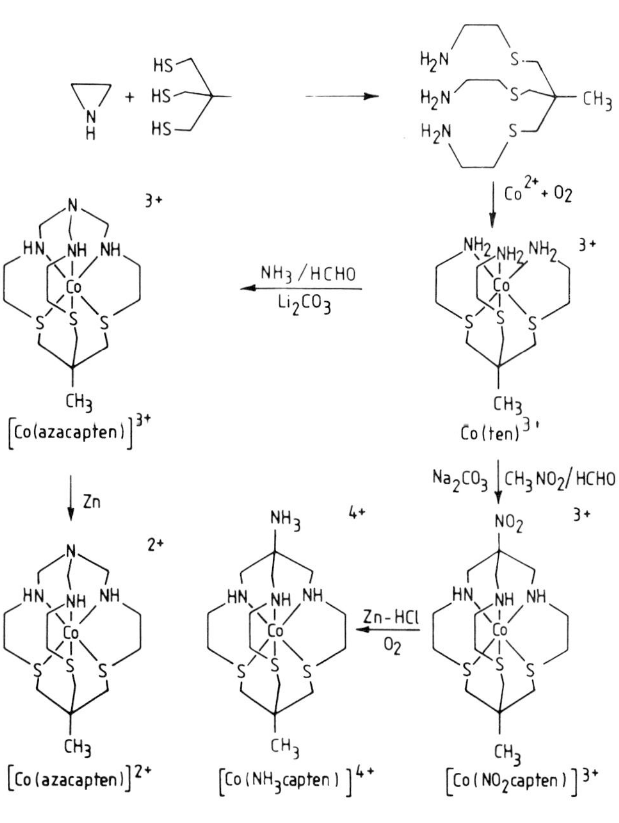

Fig. 3-63. Synthesis of polyaza polythia macrobicycles.

Macrotricyclic polyazacryptates have been synthesized in ca. 50% yield by using metal-templated tripode capping involving condensation of bis[1,1,1-tris-(aminomethyl)ethane]Co(III) with formaldehyde and nitromethane (Fig. 3-64) [216a].

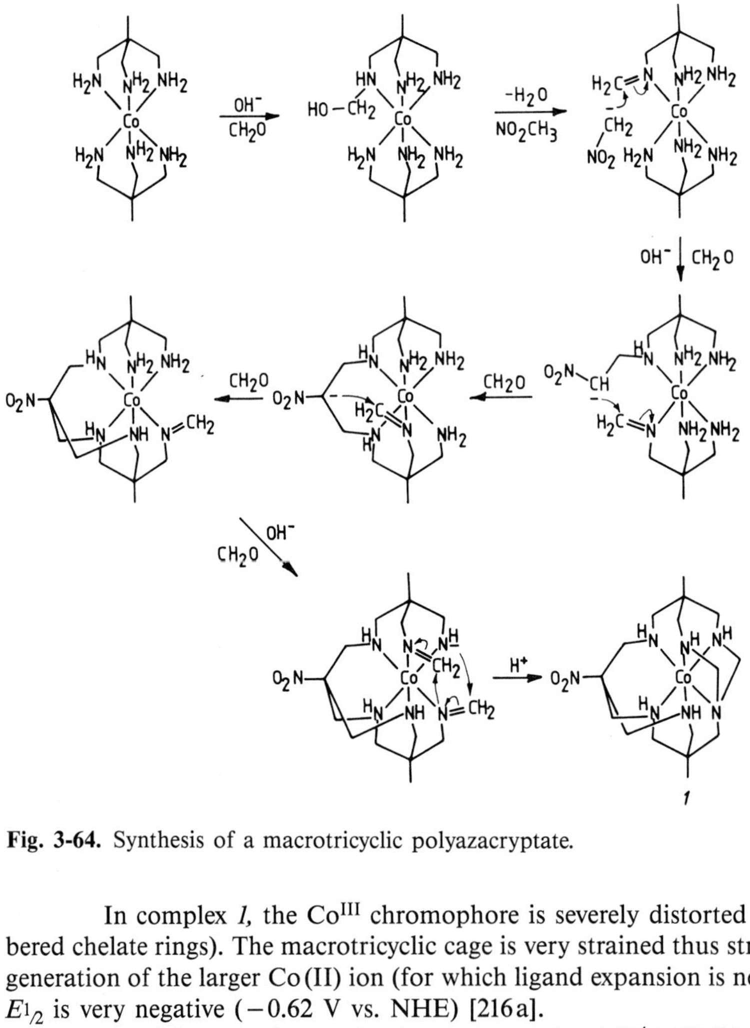

Fig. 3-64. Synthesis of a macrotricyclic polyazacryptate.

In complex *1*, the Co^{III} chromophore is severely distorted (4- and 6-membered chelate rings). The macrotricyclic cage is very strained thus strongly inhibiting generation of the larger Co(II) ion (for which ligand expansion is necessary). Hence $E_{1/2}$ is very negative (−0.62 V vs. NHE) [216a].

Sepulchrate and sarcophagine compounds of Pt^{4+} [216b] and Ni^{2+} [216c] have been synthesized starting from the corresponding $M(en)_3^{n+}$ complexes. In the case of Ni^{2+}, the yield was very low.

The ligands in the $Co(sar)^{3+}$ family of complexes can be obtained in the free state by decomplexation (the experimental method for this important reaction has yet to be published). Complexes of other cations, V(IV), Cr(III) etc., can thus be formed [216d, e].

The cobalt complexes have found use in the photoreduction of water. They possess several of the properties necessary for efficient electron-transfer agents. In particular, the stability of the Co(III) and Co(II) oxidation states, and the wide variety of ligands, should be noted since these allow control of both the redox potential and the rate of electron transfer [217].

b) Photoactive cryptands

Macrobicyclic ligands containing the photoactive bipyridine and phenanthroline groups (Fig. 3-65) have been synthesized [218].

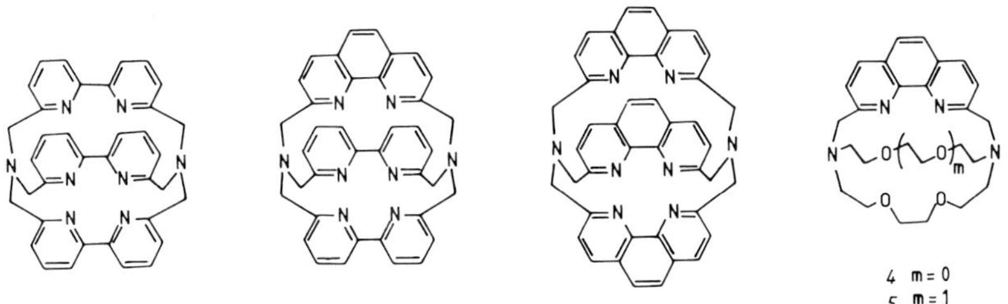

1 *2* *3* *4* m = 0 *5* m = 1

Fig. 3-65. Cryptands containing bipyridine and phenanthroline groups.

Symmetrical cryptands *1* and *3* can be obtained in one step from *6* and *9*, respectively (Fig. 3-66).

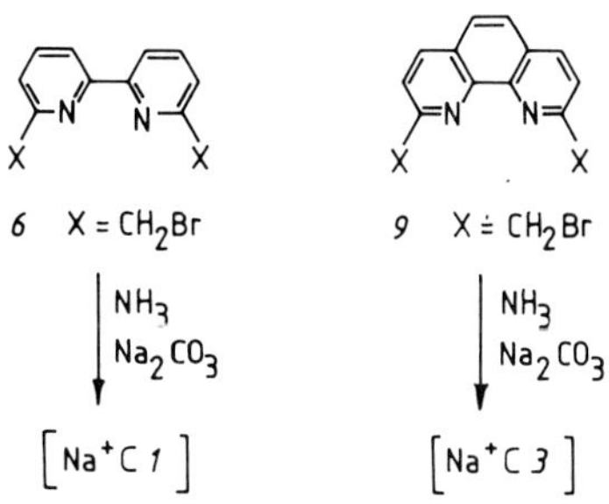

Fig. 3-66. Synthesis of symmetrical cryptands *1* and *3* using an external template.

Thus, treatment of *6* (or *9*) with NH_3 in the presence of Na_2CO_3 (external template) gives directly macrobicycle *1* (or *3*) as its complex with sodium cation (in 25–30% yield).

The complexes obtained are very stable. The X-ray crystal structure of the $[Na^+ \subset 3]$ complex shows that the sodium ion is octacoordinated (Fig. 3-67) [219]. The rigidity of the phenanthroline units provides an explanation for the difficulties encountered in removing the ligand from the complex.

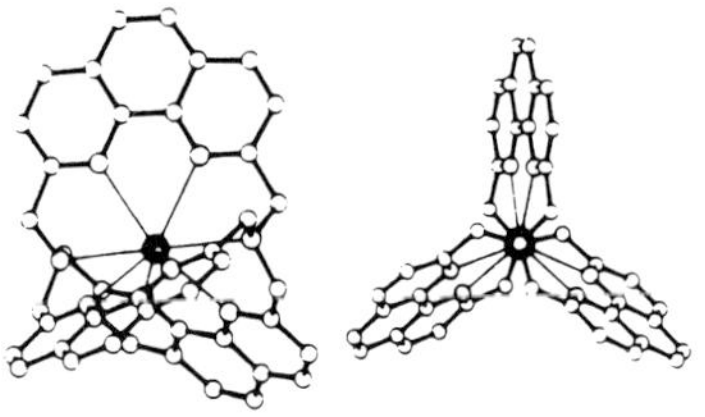

Fig. 3-67. Crystal structure of the sodium cryptate of the tris(phenanthroline) macrobicyclic ligand (*3* in Fig. 3-65) (taken from C. Pascard et al., *Helv. Chim. Acta,* **68,** 1577, 1985).

The complexes formed by these ligands with Eu(III) and Tb(III) have very interesting photochemical properties. In water, the complexes are luminescent with long emission lifetimes, whereas the free aqua ions, because of deactivation by water, are not luminescent. These properties may allow medical applications of the cryptands as luminescence immunoassays [220].

c) Macrobicycles and macrotricycles as porphyrin models

A family of ligands (Fig. 3-68) possessing certain porphyrin properties has been synthesized (Fig. 3-69) [221 a].

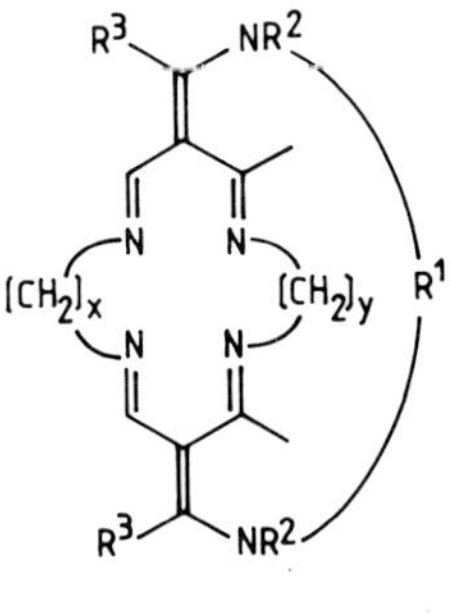

$R^1 = (CH_2)_n$, n = 4–8; m-, p-xylyl, etc.

R^2 = H, CH_3, alkyl, $-CH_2C_6H_5$

R^3 = H, CH_3, alkyl, C_6H_5, etc.

x, y = 2 or 3

Fig. 3-68. Superstructure macrocyclic ligands (or cyclidenes) (taken from D. M. Busch et al., *Pure Appl. Chem.,* **52,** 2477, 1980; "Progress in Macrocyclic Chemistry," vol. 3, Wiley, 1981).

It should be recalled (see Section 3.3.4d, Part I) that the presence of an R^1 chain allows formation of a cavity endowing the framework thus obtained with various interesting properties. For instance, the complex of the ligand *1* (x = y = 3; $R^1 = (CH_2)_6$; $R^2 = R^3 = CH_3$, Fig. 3-68) with Co(II) (Part I, Fig. 3-59 and 3-60) binds O_2 very efficiently [221 b, c]. The presence of the R^1 chain is very important. Its main role is to protect the bound molecule from interactions with the surrounding medium (other metal centers and the solvent). Furthermore, its affinity for O_2 can be adjusted by varying the R^1 chain length [221 c]. An Fe(II) complex with a similar ligand also binds dioxygen [221 d].

Formation of bicyclic systems *1* is frequently accompanied by a dimerization reaction giving *2* (Fig. 3-69). Dimerization becomes important when the R^1 chains are (a) short, and (b) of the type containing a rigid group (aromatic).

The dimers can also be obtained with better control using the route shown in Fig. 3-70.

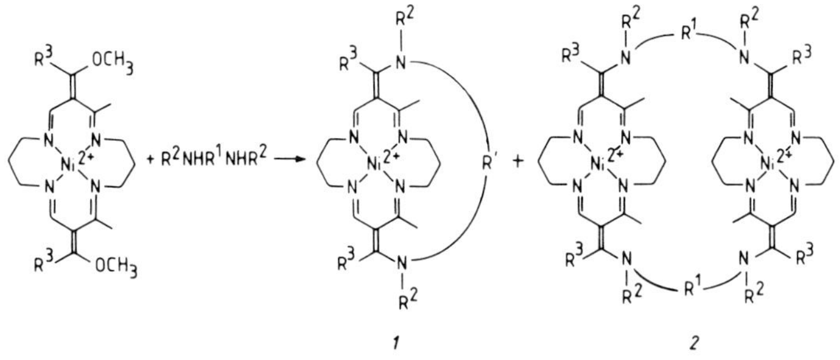

Fig. 3-69. Synthesis of cyclidene *1* and its dimer *2*.

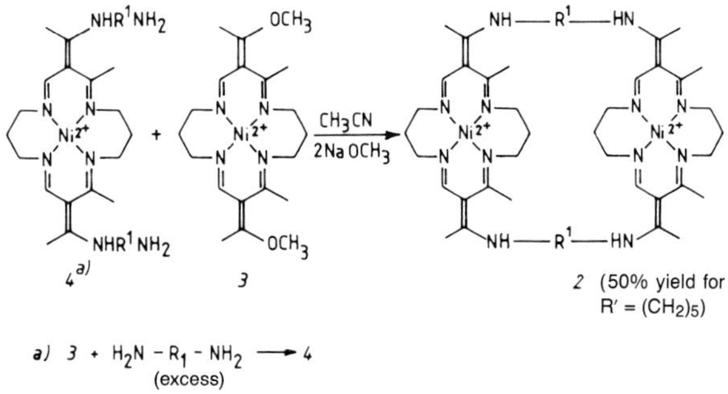

Fig. 3-70. Specific synthesis of dimer *2*.

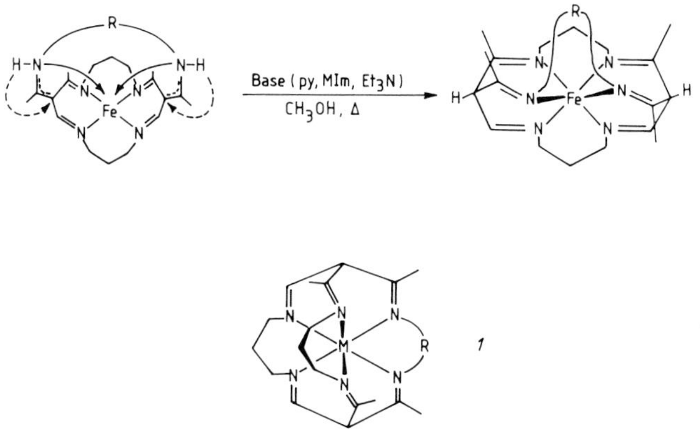

Fig. 3-71. Conversion, via tautomerization, of a cyclidene into a clathrochelate (e.g., *1*).

Such dimerization affords ligands enclosing two identical metal ions separated by a persistent void [221 a, e, f]. When $R^1 = -CH_2-C_6H_4-CH_2-$ (m-xylyl), the Ni---Ni distance is 13.6 Å (X-ray crystal structure determination). Electrochemical studies show that two metal cations are independent (a single two electron wave is observed at +0.9 V) as expected in view of the large separation of the ions.

Intramolecular tautomerization of certain ligands gives clathrochelate type structures (*I*) in which the ion is totally encapsulated by the ligand (Fig. 3-71) [221 a].

More recently, macrotricyclic systems have been synthesized (Fig. 3-72).

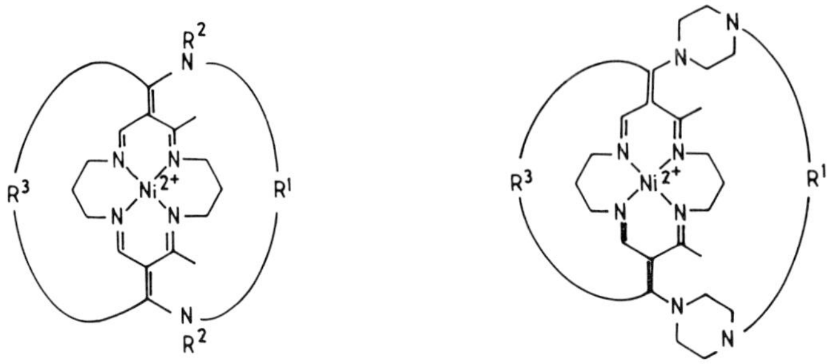

Fig. 3-72. Complexes of doubly bridged cyclidenes.

Busch has proposed that one of the chains determines the O_2 affinity, and the other chain increases the lipophilicity of the cavity thus allowing the attachment of small molecules capable of reacting with the bound dioxygen. These ternary molecules may be oxygenase models constituting a new class of catalyst [222].

3.2.6 Miscellaneous

a) Porphyrin-quinones

Condensation of dialdehydes with diamines has been very widely used to form macrocyclic di- and tetraimines (cf. Part I) (Fig. 3-73).

One of the interesting aspects of this reaction is its reversibility, thus providing the possibility for the mixture to reach thermodynamic equilibrium. In this case, the thermodynamically most stable product under the conditions used is obtained, and dilution should favor formation of the shortest polycondensation species. Using

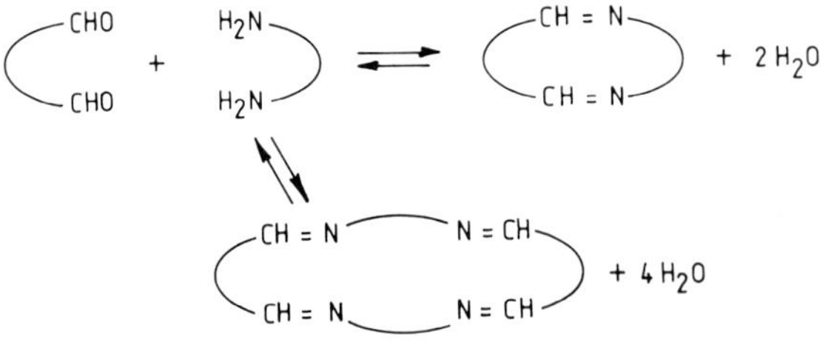

Fig. 3-73. Synthesis of macrocyclic di- and tetraimines.

these principles, a molecule containing both a porphyrin and a quinone unit held at a distance of 10 Å from each other in parallel planes has been synthesized [223a].

2,3,5,6-Tetrakis[2-(4-formylphenoxy)ethoxy]benzoquinone, *2*, was condensed with the α,α,α,α-atropisomer of *meso*-tetrakis(o-aminophenyl)porphyrin, *3* (Fig. 3-74). Equilibration for 20h followed by reduction gave the porphyrin-quinone, *4*, in 80–95% yield, which is remarkable for such tetrapode-tetrapode coupling.

Electron transfer between porphyrin and quinone is an important process in photosynthetic systems. Model compounds such as *4* allow more precise study of the mechanism, in particular the influence of structural factors. Photochemical studies on the complex of *4* with Zn(II) have shown that electron transfer occurs at 10 Å separation [223b]. Other compounds in which porphyrin and quinone units are asso-

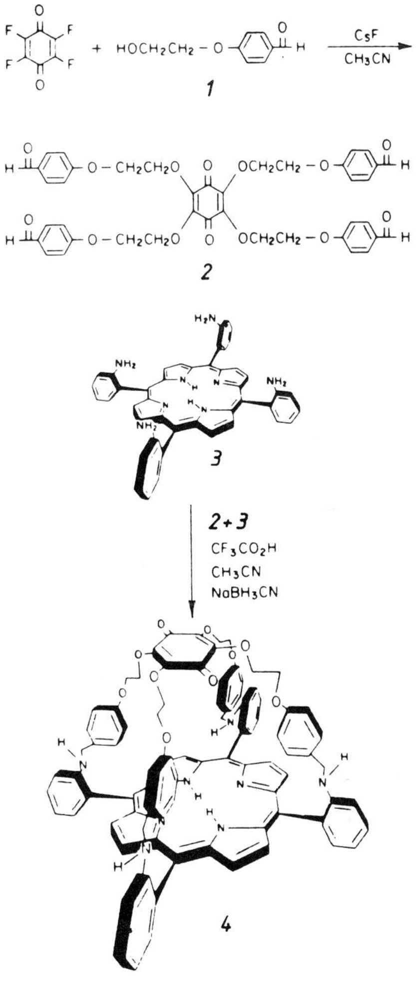

Fig. 3-74. Synthesis of a porphyrin-quinone via tetrapode-tetrapode coupling.

ciated have been described. The synthetic approaches are different and longer [224a]. The field of superstructures based on porphyrin is extremely large and rich in potential [224b].

b) Bicyclic polypeptides

Various natural bicyclic polypeptides are known (amanitin, phalloidin, etc.).

The first synthetic bicyclic peptide was obtained starting from the nonapeptide *1*, whose preparation we do not describe. This active ester (*1*) was used in the initial cyclization step (Fig. 3-75) involving deprotection with trifluoroacetic acid (TFA) of the amine α to the lysine residue, followed by cyclization under basic conditions to give the monocycle *2*.

The azide method was employed for the second cyclization step. This involved deprotection of the amine ε to the lysine residue, conversion of the hydrazide protecting group into an azide protecting group, and finally cyclization under basic conditions to give *3* whose structural formula is given in Fig. 3-75. This bicyclic peptide binds various cations including Zn^{2+} [225].

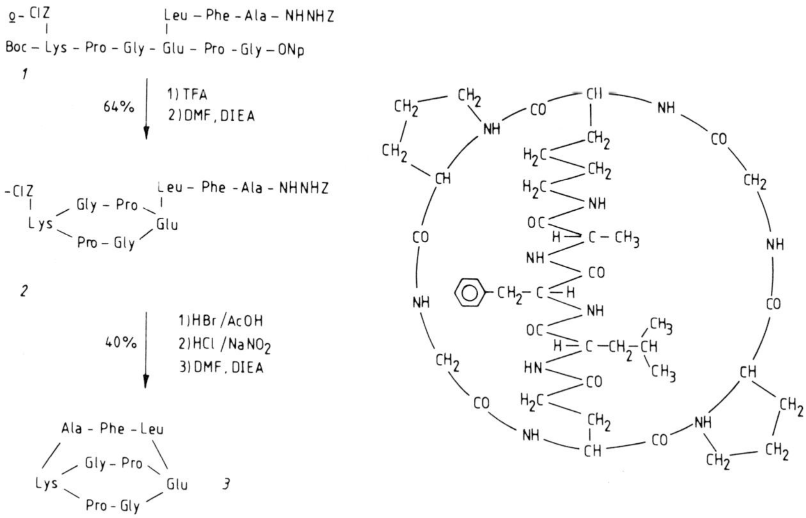

Fig. 3-75. Synthesis of a bicyclic nonapeptide (Z: benzyloxycarbonyl; DIEA: Diisopropylethylamine).

3.3 Macrocyclic and Macrobicyclic Complexes. Cryptates

In this final section, we discuss the complexing properties of macrocyclic and macrobicyclic systems. Since these compounds are tailor-made with a view to specific objectives, the expected properties are usually obtained, but with varying degrees of success. Failure sometimes occurs, of course, revealing design imperfections and lack of understanding of certain interactions, as well as neglect of some specific effects. It is obvious that the synthesis of homologous compounds starting from an efficient complexing agent is relatively risk-free. Nevertheless, it allows refinement of certain properties of the ligand. The elaboration of a new family of compounds is plainly more uncertain. The nature of the interaction sites and their relative arrangement in the molecular framework are generally chosen using very simple models whose basic properties (dipole moment, polarizability, affinity for a given substrate, interaction orientation, size, etc.) are relatively well known (from experiment) and understood (from theory). Incorporation of these interaction sites in the molecular system modifies their behavior. Unexpected effects may occur (e.g., hydrogen bonding) hence compromising the desired properties of the ligand.

3.3.1 Proton Transfer

Internal protonation of a small diaza bicyclic system is a very specific phenomenon since it involves the formation of an isolated hydrogen bond inside the cavity (Fig. 3-76). Such a system is the site for proton transfer occurring under conditions of protection and isolation from the surrounding medium. It is of interest to study the general features of proton transfer and to illustrate them with specific examples.

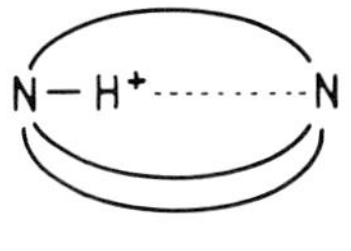

Fig. 3-76. Internal protonation in a diaza bicyclic system.

a) General points

Proton transfer is a reaction occurring over a short distance, and is of great importance in chemistry, biology and physics. It forms the basis for all acid-base equilibria [226]. It is involved in transport phenomena, photosynthesis, enzyme reactions, vision, etc. Transfer occurs via a hydrogen bond, e.g. with a Y^- anion:

$$\underset{}{X}-\underset{a}{H}---\underset{b}{Y^-} \rightleftharpoons X^- --- H-Y$$

(a and b are the X–H and H---Y^- distances, respectively).

If X=Y, the process is symmetrical, but in general it is unsymmetrical. The length of bond a is less than that of b except in cases where the hydrogen bond is very strong.

The rate of transfer depends on thermodynamic and steric factors, and on whether or not there is a tunnel effect.

b) Proton transfer between nitrogen sites: $[NH{-}{-}{-}N]^+ \rightarrow [N{-}{-}{-}HN]^+$

In general, the bonds are not symmetrical. They are linear, and proton transfer is reversible.

The potential energy curve for the transfer process depends on the N---N distance. The barrier to transfer decreases with the distance between the two nitrogen atoms. The rate of transfer is thus greater in the case of short bonds [226b]. For sufficiently short N---N distances, the system may become symmetrical.

c) Transfer mechanism for $[H_3NH^+{-}{-}{-}NH_3] \rightarrow [H_3N{-}{-}{-}HNH_3^+]$

Proton transfer occurs in the following three steps:

$$NH_4^+ + NH_3 \xrightarrow{a} H_3NH^+{-}{-}{-}NH_3 \xrightarrow{b} NH_3{-}{-}{-}HNH_3^+ \xrightarrow{c} NH_3 + NH_4^+.$$

In this mechanism, a and c are, respectively, the association and dissociation steps, proton transfer occurring in the extremely fast step b ($k_2 \sim 1.2 \times 10^9$ mol^{-1} s^{-1} at 25 °C).

An *ab initio* SCF–LCAO–MO study of the reaction path for this sequence [227a] has provided the potential energy curves for the system $[NH_3{-}{-}{-}H{-}{-}{-}NH_3]^+$ as a function of the position z of the exchanging proton, for various N---N distances (*d*) (Figs. 3-77 and 3-78). The N---N distance was optimized to a value $d_m = 2.587$ Å by keeping the proton midway between the two nitrogens ($z = 0$). With this N---N distance, the minimum energy is obtained with the proton at $z = 0.1835$ Å (point P_0 on curve 2). The transfer barrier is very low at this point (1.13 kcal/mol). Curves 1-7 represent the energy of the system as a function of the N---N distance and of the position of the proton. The minimum energy of the system is achieved with $d_e = 2.778$ Å and $z = 0.316$ Å (points Q_0 and Q_1 in Fig. 3-78).

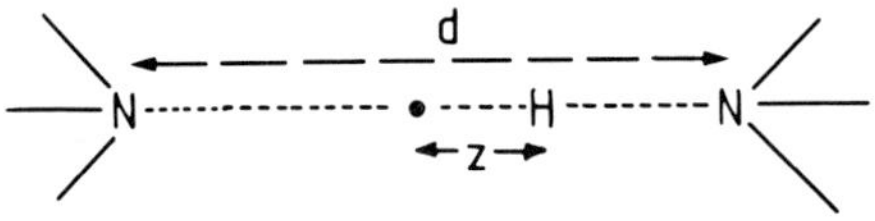

Fig. 3-77. Distances in the $(H_3N{-}{-}{-}H{-}{-}{-}NH_3)^+$ system.

The reaction path may be visualized by the curves C_0 and C_1 joining successive minima. The reaction proceeds in three steps; association, transfer and dissociation. Proceeding from the right side of Fig. 3-78, the overall process can be described in the following way: (a) the two entities approach each other up to point Q_0 representing the energy minimum for the associated species; (b) decrease in the N---N distance up to P_0; (c) proton transfer to P_1; (d) increase in the N---N distance as far as Q_1 and then dissociation. It should be noted that the transition from Q_0 to P_0 is favored energetically since the barrier to direct transition from Q_0 to Q_1 is 6.1 kcal/mol (curve 6) whereas by proceeding via P_0 the energy barrier is only 2.5 kcal mol^{-1} (the difference between the potential energies of Q_0 and S). This low barrier in the latter case explains the ease of the proton transfer reaction in this system.

More recent calculations [227b] give very comparable results: a minimum in energy for the system when d_e = 2.731 Å and z = 0.279 Å, corresponding to N---H bond lengths of 1.087 Å and 1.644 Å. These calculations show also that for a very short N---N distance (2.45 Å) the proton is midway between the two nitrogen

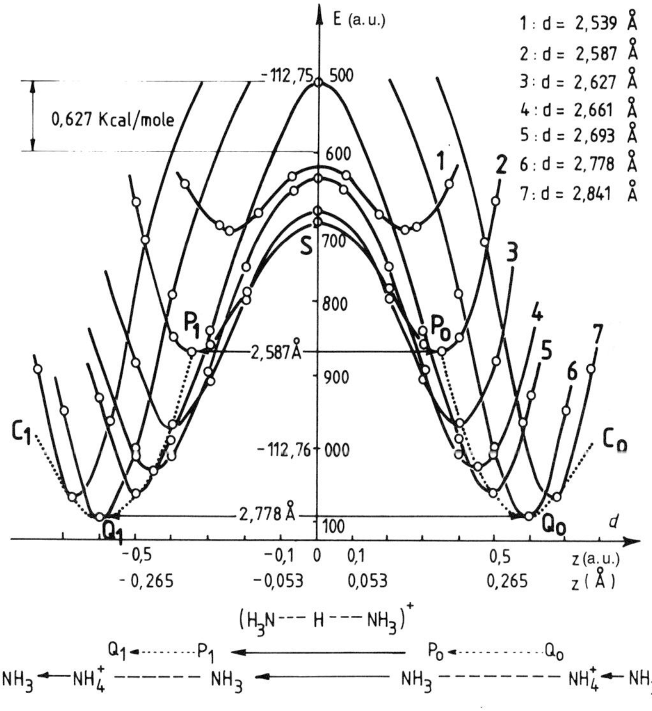

Fig. 3-78. Potential energy curves for the system $(H_3N---H---NH_3)^+$ as a function of the position z of the exchanging proton for various N–N distances d (taken from A. Veillard et al., *Molecular Physics,* **29**, 849, 1975).

atoms ($z = 0$). The N---H bond is thus very greatly stretched (1.23 Å). In this situation, the potential energy curve no longer shows a double minimum but tends towards a symmetrical curve with a single minimum [226, 227b]. Practical examples illustrate these theoretical calculations.

d) Linear symmetrical hydrogen bonds of the type $[N---H---N]^+$

Low temperature IR spectroscopy (90 K) has shown that the $[N---H---N]^+$ hydrogen bond in quinuclidin-3-one hemiperchlorate (Fig. 3-79a) is very strong.

A 120K neutron diffraction study has enabled the location of the proton to be determined [228a]. The crystal structure determination has provided the N---N and N---H distances, respectively 2.635 Å and 1.317 Å, with the hydrogen atom located at the crystallographic symmetric center (Fig. 3-79b). This system thus presents the features described above, viz. a short N---N distance and long N---H bonds. It is therefore legitimate to use a symmetrical representation $[N---H---N]^+$ for this system. In this particular case, the concept of proton transfer loses its meaning to some extent, since the mechanism is virtually reduced to the two steps of association and dissociation. At room temperature (298K) the location of H is, however, no longer symmetric [228b].

a)

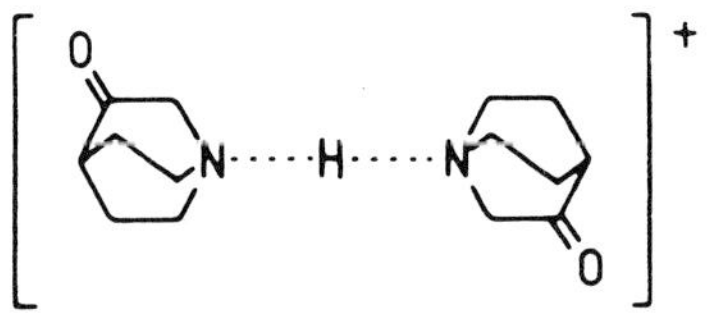

b)

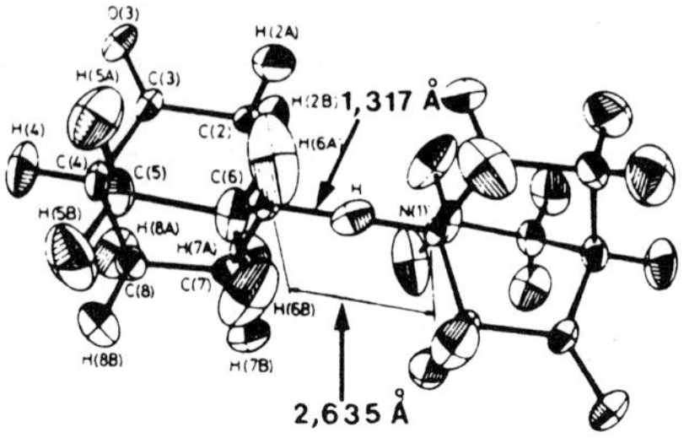

Fig. 3-79. Structure of the bis(quinuclidin-3-one)hydrogen(1+) cation (taken from J. Rozière et al., *J. Chem. Soc., Chem. Comm.*, 388, 1982).

e) Rate of proton transfer

The rate depends on numerous factors: length and geometry of the H bond, symmetry of the energy diagram, the tunnel effect, thermodynamic factors (stability, bond strength), steric factors (hindered approach of the reagent as in the case of proton cryptates) [229].

Tunneling very greatly increases the rate of transfer, but the tunnel effect is only appreciable in the case of a symmetrical energy barrier with a double minimum.

f) Role of symmetry in tunneling in proton transfer reactions [230]

Symmetric potential energy curves. The inversion of ammonia was the first system for which the tunnel effect was put forward to explain the appearance of doublets in its vibration spectra [231]. The potential energy curve for this inversion reaction is symmetrical with a double minimum and an energy barrier V_0 (Fig. 3-80).

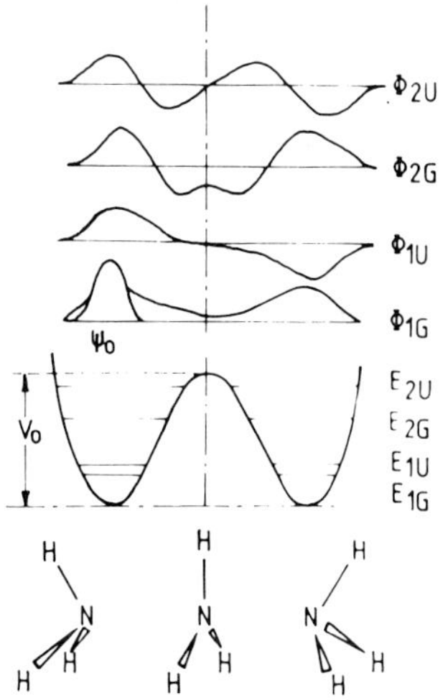

Fig. 3-80. Potential energy curve for the inversion of ammonia (V_o: inversion barrier), together with symmetric (Φ_{1G}, Φ_{2G}) and antisymmetric (Φ_{1U}, Φ_{2U}) linear combinations of the starting wave function, ψ_0, the latter being identical on each side of the barrier (ψ_o is only shown to the left of the barrier). The vibrational energy levels E_{1G} to E_{2U} corresponding to these combinations are also shown (taken from J. R. de la Vega, *Acc. Chem. Res.,* **15**, 185, 1982).

The splitting of the vibrational levels is due to overlap of the vibrational wave functions on each side of the energy barrier. This effect becomes more marked as the interaction becomes stronger, that is as the height and thickness of the barrier are reduced.

The frequency of the tunnel effect, ν_t, is related to the overlap of the wave functions, and it is proportional to the energy difference, ΔE, between the symmetric and antisymmetric levels of a given vibrational state:

$$\nu_t = 2\,\Delta E/h$$

The rate of inversion by tunneling decreases exponentially with the reduced mass, μ, of the system, and also the barrier thickness, s, and height, V_0.

Unsymmetrical potential energy curves. In this case, the vibrational levels are of different energies on either side of the barrier. There is therefore poor overlap and a reduced tunnel effect. Intramolecular proton transfer represented in Fig. 3-81 illustrates this type of reaction.

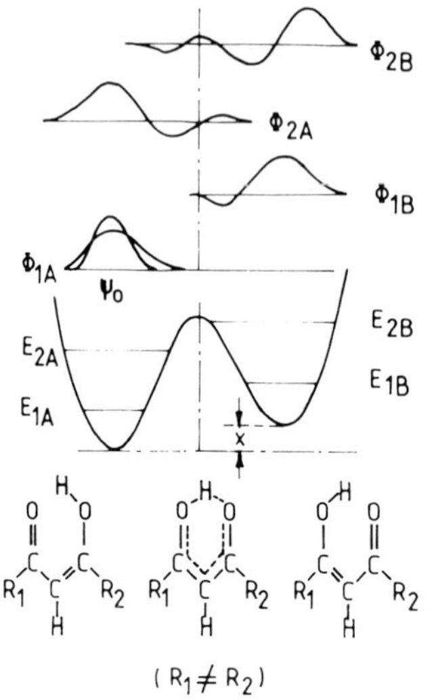

Fig. 3-81. Asymmetric potential energy curve. Vibrational wave function combinations and vibrational energy levels.

In the case of methanol-methoxide proton transfer, $[CH_3OH\cdots{}^{-}OCH_3]$, the profile for the proton exchange is symmetric for the relative orientation of the methyl groups shown in Fig. 3-82 A. However, rotation of one of the methyl groups destroys this symmetry and inhibits the tunnel effect (Fig. 3-82 B). From these results it can be deduced that the rate of proton exchange in this system is determined by the rate of rotation of the methyl group. The frequency of the latter rotation is ca. 10^{10} sec^{-1}, which means that the symmetric profile in Fig. 3-82 A is produced 10^{10} times per second. Each time the profile is obtained, a proton crosses the barrier at a rate of 10^{12}–10^{13} $sec.^{-1}$ [230].

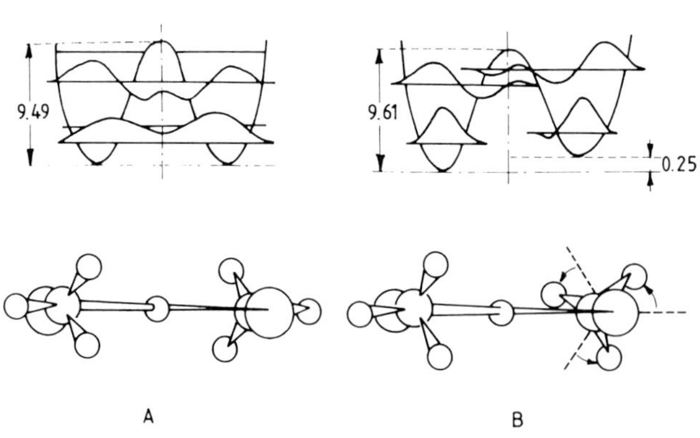

Fig. 3-82. Methanol-methoxide proton transfer. Depending on the relative orientation of the methyl groups, the profile for the potential energy curve and that for the wave functions are symmetric (A) or asymmetric (B). Proton transfer via the tunnel effect only occurs in the case of profile (A) (taken from J. R. de la Vega, *Acc. Chem. Res.,* **15,** 185, 1982).

g) Mechanism for intramolecular transfer of two protons

Naphthazarin and its monomethyl and dimethyl substituted derivatives constitute excellent models for the study of the part played by the symmetry of the potential energy curve profile [230]. These systems have two intramolecular hydrogen-bonded protons. Transfer of the two protons can be concerted (sequence A → C → B) or stepwise (sequence A → D → B or A → G → B) (Fig. 3-83a). As a result, interconversion of the A and B tautomers is no longer represented by a curve but rather by a potential energy surface (Fig. 3-83b) which has been studied by *ab initio* SCF-MO calculations for naphthazarin, monomethylnaphthazarin, and 2,7- and 2,6-dimethylnaphthazarin. The potential energy surface in Fig. 3-83b is representative of

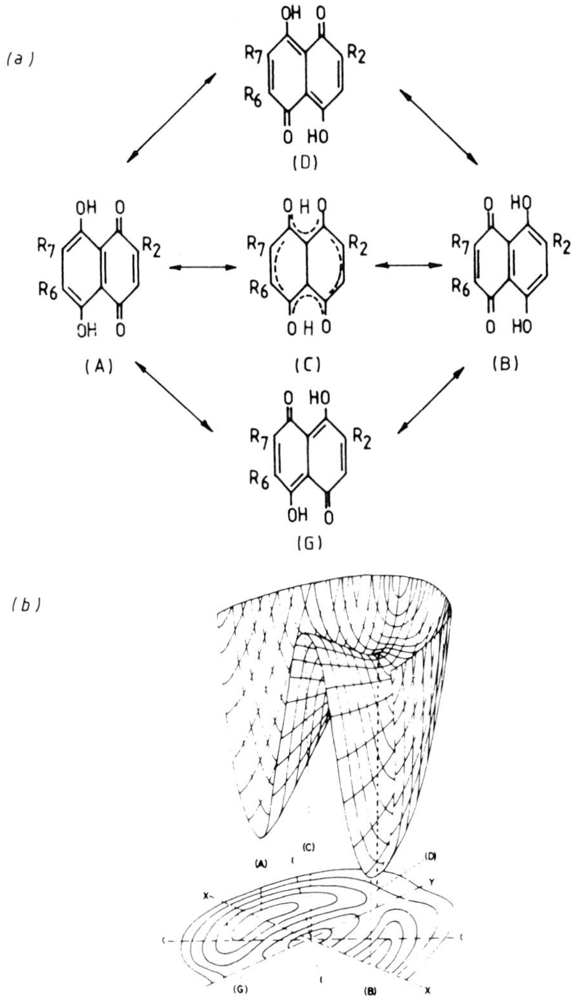

Fig. 3-83. (a) Schematic representation of the various proton transfer pathways for naphthazarin and its methyl derivatives, (b) potential energy surface for interconversion of the tautomers (A) and (B).

the four compounds studied (only a few slight modifications connected to the potential energies at points A, B, C, D and G differentiate the four surfaces). The surface shows two minima, A and B, corresponding to the two forms with both hydroxyl groups in the same ring (see Fig. 3-83a). Point C corresponds to the intermediate in which each of the two hydrogen atoms is midway between the two oxygen atoms, thus giving the height of the transfer barrier (28 kcal/mol for naphthazarin, Fig. 3-84a). Points D and G correspond to the intermediates for the process leading stepwise from A to B. These two points give the transfer barrier for the stepwise process (25 kcal/mol for naphthazarin). The potential energy surface gives the variation in energy of the system on proceeding from A to B.

For naphthazarin and its two dimethyl derivatives, both A and B have the same energy (Fig. 3-84a), whereas 2-methylnaphthazarin is more stable than 7-methylnaphthazarin by 0.9 kcal/mol, resulting in the potential energy curve being asymmetric (Fig. 3-84b). Proton transfer takes place with a frequency of 20–40 MHz in the case where the potential curve is symmetric. However, no exchange is observed when the curve is asymmetric. The absence of proton exchange for monomethylnaphthazarin is due to both suppression of the tunnel effect because of asymmetry in the potential energy profile, and also the high energy barrier which prevents transfer by the classical process.

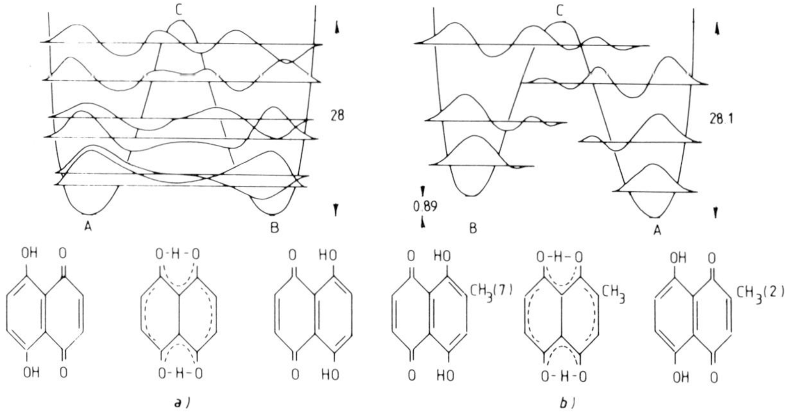

Fig. 3-84. Potential energy profile for concerted exchange of both protons for (a) naphthazarin and (b) monomethylnaphthazarin. Symmetric and antisymmetric linear combinations of the vibrational wave functions.

To summarize, the influence of the symmetry of the profile on the contribution of the tunnel effect is particularly marked when the barriers exceed 10 kcal/mol. The symmetry can be removed by a very small perturbation such as that brought about by rotation of a methyl group in the system $[CH_3OH{-}{-}{-}^{-}OCH_3]$.

h) Very slow proton transfer

1,8-Bis(dialkylamino)naphthalene compounds (products *4-7* in Fig. 3-85) exhibit particularly remarkable basicities and proton transfer rates [232] resulting in their description as "proton sponges".

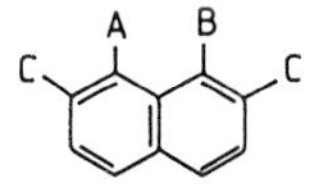

	A	B	C
1	NH_2	NH_2	H
2	NH Me	NH Me	H
3	N Me_2	NH Me	H
4	NMe_2	NMe_2	H
5	NEt_2	NEt_2	H
6	NMe_2	NMe_2	OMe
7	NEt_2	NEt_2	OMe

Fig. 3-85. 1,8-Diaminonaphthalenes with various substituents.

These compounds are extremely strong bases, their pK_a being 12–16 whereas the pK_a for the "normal" first protonation of compound *1* (1,8-diaminonaphthalene) is ca. 5. It is instructive to examine the dependence of the pK_a on the degree of N-methylation (Table 3-4). Di- and trimethylation results in only slight in-

Table 3-4. Equilibrium constants for protonation, and rate constants for deprotonation.

Compound	Solvent V/V	pK_a [a)]	k_{OH^-} [b)] $dm^3\ mol^{-1}\ s^{-1}$
CH_3NH_2	H_2O	10.6	3.7×10^{10}
1	H_2O	4.6	–
2	H_2O	5.6	–
3	H_2O	6.4	–
4	H_2O	12.1	1.9×10^5
4	30% Me_2SO – H_2O	12.1	6.1×10^5
5	30% Me_2SO – H_2O	12.7	1.6×10^4
6	60% Me_2SO – H_2O	16.1	4.4×10^2
7	60% Me_2SO – H_2O	16.3	3.3

a) pK_a (= log K_a) for monoprotonation: $B + H^+ \rightleftharpoons BH^+$ $K_a = [BH^+]/[H^+]\ [B]$;
For compounds *4–7*, diprotonation does not occur even in the presence of a large excess of strong acid
b) Rate constants for the deprotonation of monoprotonated forms of various compounds (see Fig. 3-87)
(Taken from F. Hibbert, *Acc. Chem. Res.* **17**, 115, 1984).

creases in basicity. On the other hand, tetramethylation gives rise to a sudden jump in pK_a of ca. six units. This result suggests that steric factors play an important role, and this is confirmed by the increase in pK_a observed when the methyl groups are replaced by ethyl substituents. The second important factor is the very strong intramolecular hydrogen bond formed on monoprotonation.

X-ray crystal structure studies on the diamine *4* and its monoprotonated derivative corroborate this interpretation. The structure of the free diamine [232b] shows that because of very strong 1,8 repulsion (methyl-methyl, lone pair-lone pair), the two dimethylamino groups are pushed well out of the mean plane of the naphthalene unit which itself is distorted (Fig. 3-86a).

Resonance is therefore sterically inhibited resulting in loss of aromatic character for the amine groups and substantial increase in their basicity.

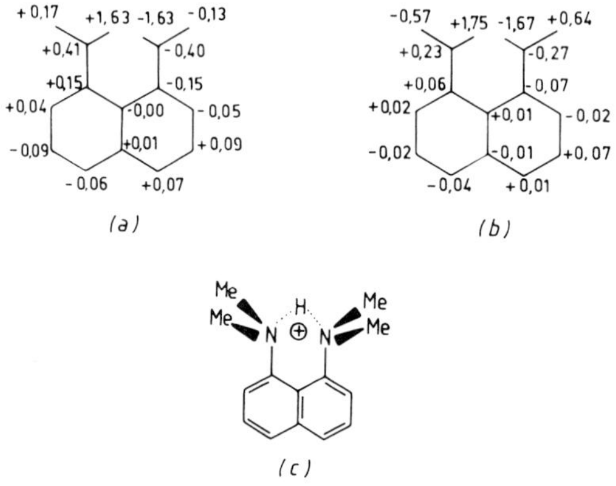

Fig. 3-86. Structures of (a) 1,8-bis(dimethylamino)naphthalene and (b and c) its monoprotonated derivative. The values in (a) and (b) represent the deviations (in Å) from the mean plane passing through the naphthalene carbon atoms. In the protonated compound (b), the proton is located slightly outside the plane (−0.28 Å) (taken from J. D. Roberts et al., *Acta. Cryst.,* **B29**, 1611, 1973; and M. R. Truter et al., *J. Chem. Soc., Dalton,* 395, 1972).

To this are added other factors connected with protonation. The latter gives rise to substantial steric release, the deviation from planarity of the naphthalene ring system and the distance between the dimethylamino groups being significantly less than in the free diamine (Fig. 3-86b) [232c]. The structure of this monoprotonated compound reveals that the N---N distance is shorter (2.60 Å) than in the free diamine (2.79 Å). This shortening is explained by the loss of lone pair repulsion, and the formation of the intramolecular hydrogen bond. However, the latter is not linear, the angle N---H---N being 134° in the structure described (Fig. 3-86c). Other more recently developed systems ensure improved N---H---N linearity [233].

Proton transfer from protonated "proton sponges" to hydroxide ion is particularly slow (Fig. 3-87 and Table 3-4).

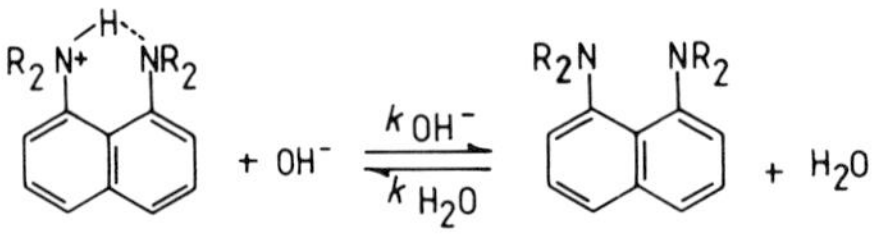

Fig. 3-87. Reaction of proton transfer to hydroxide ion.

These results are even more striking in view of the fact that this type of reaction usually occurs at a rate which is diffusion controlled (e.g., see methylammonium in Table 3-4). The low rate in the case of protonated "proton sponges" is due to both the presence of a strong intramolecular N---H---N hydrogen bond and also steric hindrance to the attack of hydroxide ion because of the presence of methyl or ethyl groups [234].

Mechanistic studies [235] have shown that proton transfer to hydroxide ion occurs in two steps: (a) rupture of the hydrogen bond, $N---H^+---N \rightarrow N-H^+ N$, followed by (b) proton abstraction. The rates are low for two reasons. Firstly, because of the strength of the hydrogen bond, the more reactive intermediate open form produced in step (a) is only present in extremely low concentration. Secondly, owing to steric factors, the second step involving proton removal from the open form occurs at a rate that is lower by a factor of ca. 10^3 than the diffusion limit observed for normal $N-H^+$ bonds. The combination of these two effects explains the observed rates which are lower by a factor of between 10^5 and 10^{10} than those for normal amines.

The crystallographic studies referred to above do not allow one to distinguish between symmetrical and unsymmetrical $[N-H^+---N]$ bonds. Other investigations have provided more precise information concerning this important point. The ESCA spectrum for the monoprotonated diamine (Fig. 3-88) shows two maxima indicating that the two nitrogen atoms are non-equivalent [236]. This is consistent with an unsymmetrical hydrogen bond. However, the energy difference between the two peaks is only 1.4 eV whereas that observed between piperidine and its hydrochloride is 2.6 eV. This might indicate that the bond is not far from being symmetric.

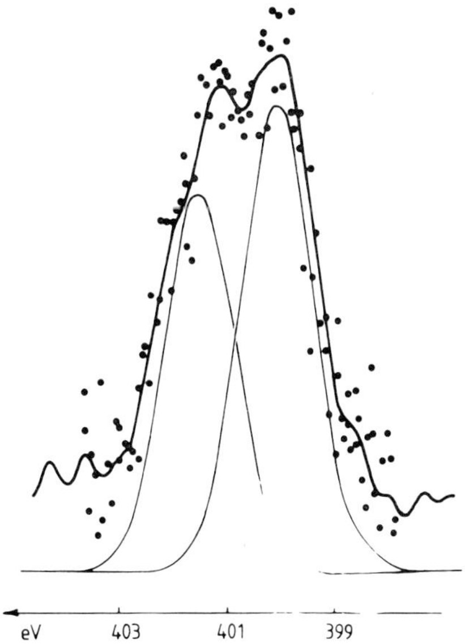

Fig. 3-88. ESCA spectrum of monoprotonated 1,8-bis(dimethylamino)naphthalene in the region of N_{1s} binding energies (taken from E. Haselbach et al., *Helv. Chim. Acta,* **55,** 1757, 1972).

3.3.2 Proton Transfer in Proton Cryptates

External and internal proton exchange were first studied by Park and Simmons [169] using diazabicycloalkaneammonium salts. Numerous studies involving a variety of cryptates followed. It proved interesting to be able to determine the acid-base properties of the various species and to study transfer both inside and outside the cavity. This involved identifying the forms, measuring the equilibria and studying the mechanisms.

a) Acid-base properties of cryptands

Before examining proton transfer in cryptands in their protonated form, it is useful to review the acid-base properties of these compounds.

Cryptands possess two sites capable of being protonated (Fig. 3-89):

$$L + H^+ \rightleftharpoons LH^+; \qquad K_1 = c(LH^+)/c(L) \cdot c(H^+)$$
$$LH^+ + H^+ \rightleftharpoons LH_2^{2+}; \quad K_2 = c(LH_2^{2+})/c(LH^+) \cdot c(H^+).$$

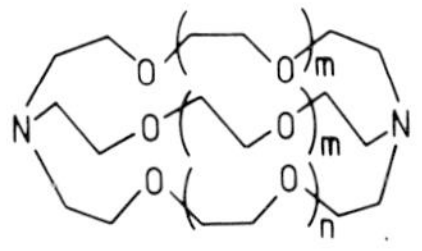

m = n = 0 [1.1.1]
m = 0; n = 1 [2.1.1]
m = 1; n = 0 [2.2.1]
m = n = 1 [2.2.2]
m = 1; n = 2 [3.2.2]
m = 2; n = 1 [3.3.2]
m = n = 2 [3.3.3]

Fig. 3-89. Structures of diaza polyoxa macrobicycles and abbreviated nomenclature.

The equilibrium constants ($pK_n = \log K_n$) are collected in Table 3-5. These values are worth discussing since, in spite of the fact that emphasis is usually placed on the remarkable complexing properties of cryptands, protonation of these compounds in itself reveals very specific properties resulting from their structures ([1.1.1] cryptand, because of its untypical behavior towards the proton, will be discussed later).

First, it is appropriate to recall some data on the variation of *protonation constants for diamines* as a function of the number of carbon atoms, n, separating the two amine groups (Table 3-6).

The first protonation constant steadily increases with increasing n. The low pK_1 value for ethylenediamine is explained by the inductive effect of the second amine group, the two groups being separated by only two carbon atoms. This effect

Table 3-5. Protonation constants[a)] for cryptands in water at 25 °C.

Ligand	pK_1	pK_2	$pK_1 + pK_2$	$pK_1 - pK_2$	I
[2.1.1]	10.64	7.85	18.49	2.79	0.05 NMe_4Br
[2.2.1]	10.54	7.50	18.04	3.04	"
[2.2.2]	9.60	7.28	16.88	2.32	"
[3.2.2]	8.50	7.33	15.83	1.17	0.05 LiCl
[3.3.2]	8.16	7.31	15.47	0.85	"
[3.3.3]	7.70	6.96	14.66	0.74	"

a) Values reported by J. M. Lehn and J. P. Sauvage [265]. Some of the constants have also been measured by other groups [237h, 284, 285].

Table 3-6. Protonation constants for some α,ω-diamines in water at 25 °C, $I = 0.1$ M (NaCl).

$H_2N-(CH_2)_n-NH_2$	pK_1	pK_2	$pK_1 + pK_2$	$pK_1 - pK_2 = \Delta pK$
n = 2	9.87	7.15	17.02	2.72
3	10.31	8.71	19.02	1.60
4	10.67	9.49	20.16	1.18
5	10.77	9.92	20.69	0.85
6	10.80	10.12	20.92	0.68
7	10.88	10.16	21.04	0.72
8	10.85	10.27	21.12	0.58
10	10.94	10.35	21.29	0.59

(Values published by H. Ogino, *Inorg. Chem.*, **19**, 1619, 1980).

falls off very rapidly. When n is very large, for statistical reasons the value of pK_1 becomes even larger than that for a typical primary amine, 10.77 [237a]. An identical evolution is observed for pK_2. However, even when the two groups are widely separated (and therefore do not influence each other) the pK_2 value is always less than that for pK_1 (for statistical reasons*) by ca. 0.6 pK units.

It can be seen in Table 3-6 that the difference $\Delta pK = pK_1 - pK_2$ is very great when the two nitrogen atoms are near to each other (n = 2 and 3). This large value of ΔpK is an electrostatic effect, the approach of H_3O^+ to $-NH_2$ being rendered difficult by the proximity of $-NH_3^+$. The repulsion decreases rapidly as n increases, and it is almost imperceptible above n = 6 (ΔpK tends to 0.6).

In the case of *cryptands,* considering firstly the ΔpK values (Table 3-5), it can be noted that ΔpK tends to 0.6 only in the [3.3.3] system in which the two nitro-

* Statistical considerations were used very long ago to account for the protonation constants of polybasic (or polyacid) systems [237b–d]. For dibasic systems, it was shown that in the case where the two sites are very far from each other, the K_1/K_2 ratio tends to 4, and hence $\log K_1/\log K_2 = \log 4 = 0.6$.

gen atoms are separated by eleven atoms (rather than six as in the case above). On proceeding to [2.2.2] (separation by eight atoms), the ΔpK value becomes very large arising from the extreme proximity of the two sites being protonated. This proximity is imposed by the bicyclic structure. For smaller cryptands, the ΔpK values increase even more.

Proximity effects explain the values of and changes in the pK_1 and pK_2 protonation constants. A relatively simple and precise method of calculation provides an estimate for amine basicity [237a]. Using this method, the protonation constant of the compound TDA1 (Fig. 3-90) in water is estimated to be $pK \sim 7$ (see note in Fig. 3-90).

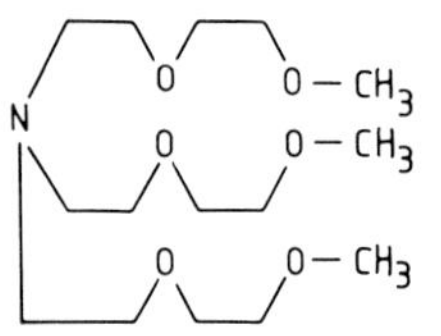

Fig. 3-90. Tris(3,6-dioxaheptyl)amine (TDA1) with a p*K* of 7.8 in methanol/water (45:6) (taken from F. Vögtle et al., *Chem. Ber.,* **112,** 1392, 1979).

This amine is a good model for all the cryptands since the main inductive effects are those of the three oxygen atoms located in position 4 with respect to the nitrogen atom (the contribution from the oxygen atoms in position 7 is very weak). Based on the p*K* of TDA1, that for the first protonation of cryptands should be ca. 7. The value of 7.7 found for the [3.3.3] compound is close to that for TDA1 thus confirming that the two amine groups in the cryptand are more or less independent of each other. The pK_1 value increases with decreasing cryptand size. For instance, the pK_1 of [2.1.1] is greater than that of [3.3.3] by ca. 3 p*K* units (Table 3-5). This regular increase in pK_1 is explained by the increasing stabilization of the monoprotonated form by the free amine group. In addition, the oxygens in position 4 contribute to this stabilization by forming hydrogen bonds to the proton held in the cavity, as evidenced by the X-ray crystal structure determinations for 18-N_2-O_4 (Fig. 1-43, Part II) [56] and [2.1.1]H_2^{2+} *(vide infra),* and also by molecular mechanics calculations [237g].

Finally, it should be noted that the overall basicity ($\log K_1 + \log K_2$) of [2.1.1] is greater than that of [3.3.3], this being the reverse trend to that observed for linear diamines (see Table 3-6). This increase in overall basicity shows that for small cryptands the large value of ΔpK is more than compensated by the stabilization of the monoprotonated form (it can also be noted that the pK_2 of [2.1.1] is slightly greater than that of [3.3.3]).

From the thermodynamic point of view, the protonation of cryptands presents many singular features. The proposed interpretations are still largely speculative since few quantitative studies have been made as yet. In particular, there is a lack of good models for comparison (linear and macrocyclic).

b) Methods for studying proton transfer

Empirical method for determining the extent of shielding of a basic site. The orientation of nitrogen sites in 1, (k + 2)-diazabicyclo[k. k. k]alkanes has been studied in the cases where k = 7, 8, 9, 10, 12 and 14, using the IR spectrum of deuteriochloroform as a probe [238 a]. This solvent has the advantage of binding only weakly to amines without affecting their structure:

$$\geqslant N{-}{-}{-}D-CCl_3$$

Hydrogen bond formation is detected in the IR spectrum by the presence of a ν_{C-D} satellite peak at a lower frequency than that for free $CDCl_3$ (ν_{C-D} = 2254 cm^{-1}) (Fig. 3-91).

These satellite peaks are clearly visible for quinuclidine and diazabicyclooctane (ν_{C-D} = 2170 and 2190 cm^{-1}, respectively). In the case of macrobicycles, the frequency shift is to be expected only for the exo form where the nitrogen atoms are directed outwards ($CDCl_3$ is too large to diffuse into the molecular cavity). The absence of hydrogen bonding in all cases when k is even is shown by these results. For [7.7.7] and [9.9.9], the presence of a hydrogen bond is indicated by a satellite peak at 2170 cm^{-1}. However, the latter is very weak suggesting either an essentially

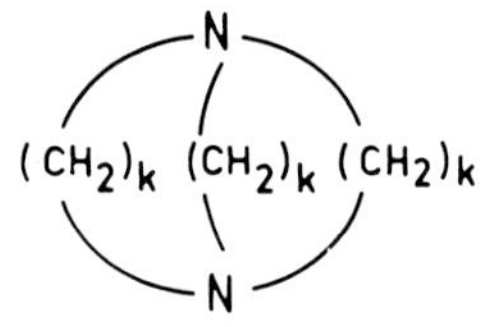

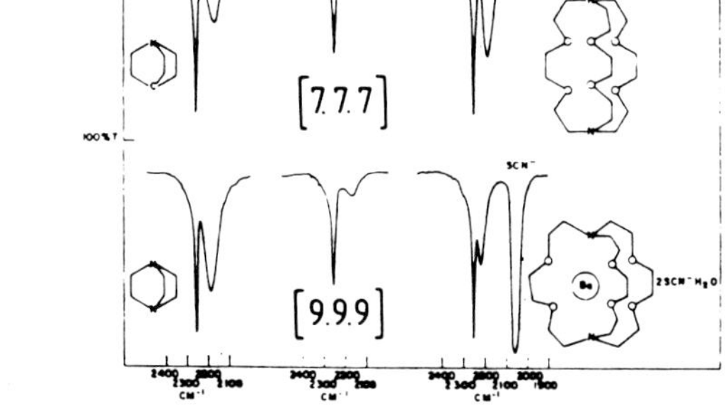

Fig. 3-91. Determination of the conformation of macrobicyclic diamines by means of an IR study of the hydrogen bonds formed between $CDCl_3$ and the amine groups. In the spectrum of the barium cryptate, the band at 2215 cm^{-1} is due to the hydrogen bond between the SCN^- anion and $CDCl_3$.

"endo, endo" conformation or steric strain in the association with $CDCl_3$ even in the "exo" conformation. In the case of the [9.9.9] compound, the satellite peak at 2170 cm^{-1} is slightly more intense. However, for [2.2.2] cryptand (which in fact is a bicyclo[8.8.8] system, see footnote in the introduction to Section 3.2.4, page 271) the satellite peak appears at 2190 cm^{-1} and is much more intense. The conformation of the cryptand in $CDCl_3$ solution is therefore either "exo, exo" or "exo, endo". The satellite peak at 2190 cm^{-1} completely disappears in the spectrum of the barium cryptate, the nitrogen atoms being bound to the cation and consequently in the "endo, endo" conformation.

Kinetic Studies. The kinetic properties of *monoprotonated forms* have been determined using the protolysis reaction:

$$L + H_2O \underset{k_{-1}}{\overset{k_1}{\rightleftharpoons}} LH^+ + HO^-$$

$$\text{where } K = c(LH^+) \cdot c(HO^-)/c(L) = k_1/k_{-1} = K_b.$$

The equilibrium constants, K, which are necessary for the interpretation of kinetic studies, have been determined using a conductiometric technique [239a]. It can be noted that there is good agreement between the pK_a values thus obtained (Table 3-7) and those measured by the potentiometric method (see Table 3-5), although the former are slightly greater.

For each of these four compounds, the kinetics of the proton transfer reaction between the monoprotonated cryptand and hydroxide ion have been studied in aqueous solution using the pressure-jump [239a] or temperature-jump [239b] techniques.

The most striking result, obtained using the pressure-jump technique, is the low rate of proton transfer from the protonated cryptand to HO^- observed in every case, and which is especially marked for [2.1.1] (Table 3-7).

Table 3-7. Protolysis of cryptands at 25 °C in water.

Cryptand	pK_a	pK_b	$10^4\ K_b$, mol dm^{-3}	k_1, s^{-1}	k_{-1}, $mol^{-1}\ dm^3\ s^{-1}$
[2.1.1]	11.17	2.83	14.7	1.59	1.08×10^3
[2.2.1]	10.91	3.09	–	–	–
[2.2.2]	9.86	4.14	0.732	7.31×10^2 a,b)	9.98×10^6 a,b)
[2.2.2B] c)	9.69	4.31	0.486	3.29×10^2 a)	6.76×10^6 a)

a) Obtained by extrapolation from lower temperatures (0–20 °C).
b) As measured using the temperature-jump method; the rate constants are $k_1 \leqslant 10^3\ sec^{-1}$ and $k_{-1} \leqslant 10^8\ mol^{-1}\ dm^3\ s^{-1}$ [239b].
c) The formula of this cryptand is given in Fig. 3-131.
(Taken from B. G. Cox et al., *J. Am. Chem. Soc.* **100**, 6002, 1978).

The rate constants, k_{-1}, are 3-7 orders of magnitude lower than those observed for the reactions between HO^- and protonated aliphatic amines (diffusion controlled reaction; $k_{-1} \sim 10^{10}\ mol^{-1}\ dm^3\ s^{-1}$). Any significant participation by exo forms of the free ligand, or exo^+ forms of the protonated ligand, can therefore be excluded. Except for [2.2.1], the simplest scheme consistent with the results involves direct transfer of an $(endo, endo)^+$ proton to HO^-, the equilibrium for protolysis being:

$$(endo, endo) + H_2O \underset{k_{-1}}{\overset{k_1}{\rightleftharpoons}} (endo, endo)^+ + HO^-$$

In the case of the [2.2.1] ligand, it is necessary to invoke a more complex reaction scheme involving inversion of the nitrogen atoms.

c) Kinetics of formation and crystal structure of an "endo, endo" diprotonated cryptand

In acidic solution, the monoprotonation of [2.2.1] is fast, but the second protonation step is sufficiently slow to be studied conveniently using stopped-flow methods in a number of solvents. The kinetics of this second protonation have been studied in methanol using various carboxylic acids [240a, b]. The equilibrium involved is:

$$LH^+ + HA \underset{k_{A^-}}{\overset{k_{HA}}{\rightleftharpoons}} LH_2^{2+} + A^-$$

hence $-dc(LH^+)/dt = k_{HA}c(LH^+) \cdot c(HA)$ (in an excess of acid HA).

The rates of protonation and deprotonation (Table 3-8) depend on the strength of the acid and on that of its conjugate base [235]. They are, however, much lower than those which might be expected for this type of reaction, the same being the case, as we have seen, for the monoprotonated compound.

Table 3-8. Rate constants for protonation of $[2.1.1]H^+$ in methanol at 25 °C.

Acid (HA)	$(pK_a)^{a)}$	$k_{HA}/mol^{-1}\ dm^3\ s^{-1}$	$k_{A^-}/mol^{-1}\ dm^3\ s^{-1}$ b)
$CHCl_2CO_2H$	(6.36)	2.3×10^5	7.8×10^3
CH_2ClCO_2H	(7.76)	8.0×10^4	7.0×10^4
CH_3CO_2H	(9.56)	4.8×10^3	2.8×10^5

a) solvent: methanol
b) deprotonation of $[2.1.1]H_2^{2+}$
(Taken from B. G. Cox et al., *J. Chem. Soc., Chem. Comm.*, 377, 1982).

Numerous equilibria for proton transfer can be envisaged (Fig. 3-92). They are all based on the assumption that proton transfer and conformational change are independent, and that inversion can occur only at unprotonated nitrogen.

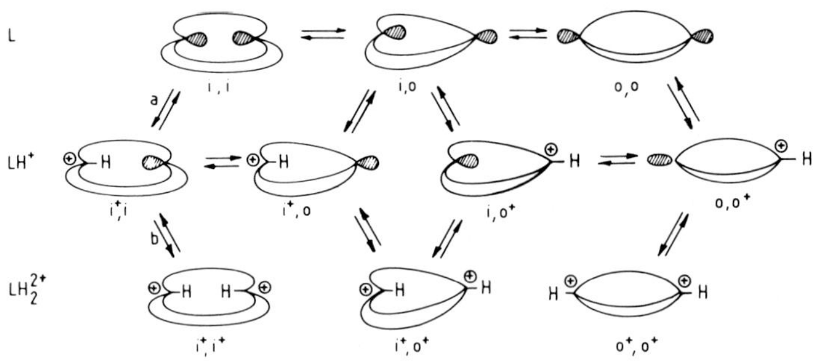

Fig. 3-92. Possible interconversion equilibria for proton transfer between [2.1.1] cryptand and its mono- and diprotonated forms.

By analogy with the structure of the ligand [2.2.2] in the solid state, the conformation of the free ligand [2.2.1] (a liquid) is expected to be "endo, endo" or in, in (i, i) (slow rate of monoprotonolysis giving rise to i^+, i, see Table 3-7). Since the second protonation step is also very slow, it presumably likewise involves endo-protonation giving i^+, i^+. Furthermore, the step involving deprotonation by A^- is also extremely slow. It therefore seems that, in Fig. 3-92, only routes (a) and (b) are consistent with results obtained.

The crystal structure of diprotonated cryptand [2.1.1], LH_2^{2+} (Fig. 3-93 a) [240a], is very similar to that of the lithium cryptate [240c].

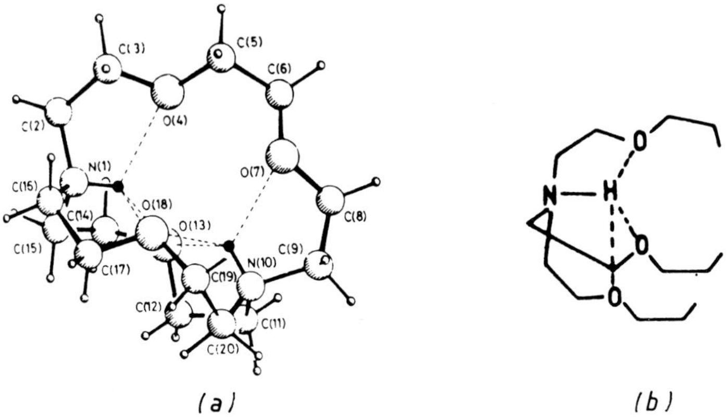

Fig. 3-93. Crystal structure of diprotonated [2.1.1] (taken from B. G. Cox et al., *J. Chem. Soc., Chem. Comm.,* 377, 1982).

The presence of two formal charges in the cavity, which might be thought to show severe electrostatic repulsion, has little effect on the geometry. This may, in part, be due to the structure being stabilized in an "endo, endo" conformation by the two groups of three hydrogen bonds formed between the protons on both nitrogens and the oxygen atoms (Fig. 3-93b).

d) Protonation of [1.1.1] cryptand

The [1.1.1] cryptand possesses exceptional acid-base and proton transfer properties which we shall discuss below after having recalled the route used for its synthesis.

Condensation of the macrocyclic diamine, *3*, with diglycolic acid dichloride under high dilution conditions [170, 241] gives a mixture of *4* (10–15%) and *6* (30%) (Fig. 3-94).

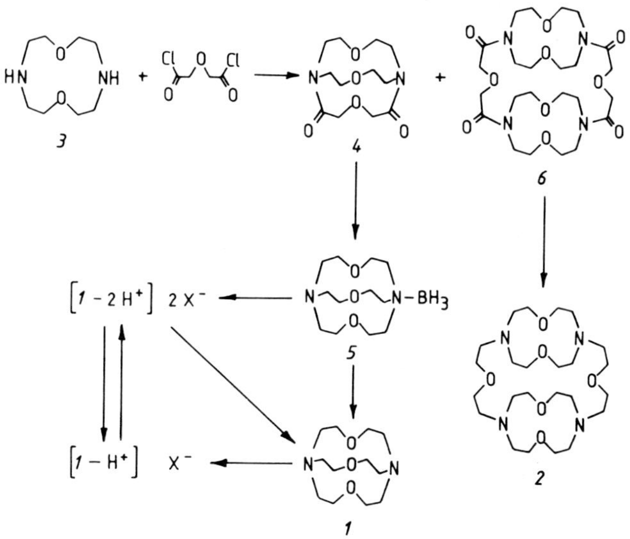

Fig. 3-94. Synthesis of [1.1.1] cryptand, and formation of macrotricycle *2*.

Reduction of *4* with diborane affords the monoborane derivative *5*. Conversion of *5* to *1* is not straightforward. The best method (KOH/methanol) takes place in only 30% yield.

Treatment of *5* with boiling 6N hydrochloric acid for two hours gives the dihydrochloride [*1*-2H^+]2Cl^-. This compound is very resistant to deprotonation and does not afford the diamine even after prolonged treatment (18 days) with 5N KOH at RT. Heating this mixture to 60 °C for 80 h results in only partial conversion to the monoprotonated species [*1*-H^+]. A very small reactant such as the solvated electron should be capable of abstracting the protons located inside the cavity. Indeed, reaction with sodium in liquid ammonia was found to give rise to a small amount of the diamine *1* (10%). The residue consisted of decomposition products with the complete absence of any [*1*-H^+]. Consequently, the deprotonation appears to be a reduction rather than just a simple reaction with a base.

Treatment of *1* or *5* with acid at RT gave only [*1*-H^+]. Refluxing in acid solution was necessary in order to obtain the diprotonated species.

When [*1*-H^+] was refluxed with a concentrated D_2O-DCl solution for 90 minutes, the [*1*-H^+D^+] mixed species was obtained, thus showing that no appreciable exchange occurs for the internal proton in [*1*-H^+].

Structure of proton cryptates. Spectral observations. The [1.1.1,2H⁺] diprotonated species. The qualitative observations described above suggests protons which are very efficiently shielded from the environment and thus "cryptated". The location of two protons in the cavity has been confirmed using NMR. The ^{13}C and ^1H spectra are uniquely consistent with a symmetric structure. Since both protons are very efficiently shielded from the environment, representation B in Fig. 3-95 is the most probable [241].

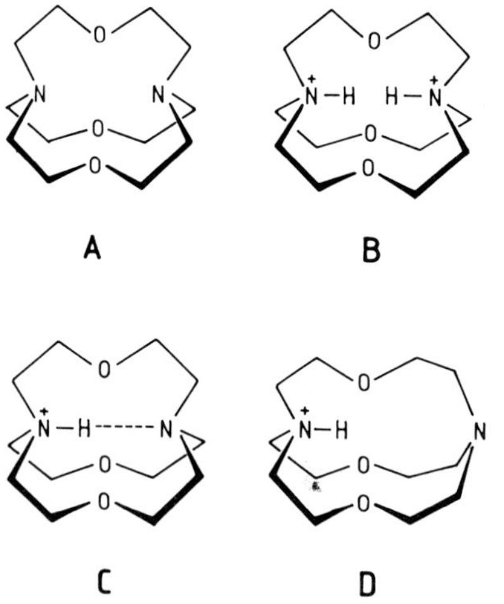

Fig. 3-95. [1.1.1] Cryptand and its protonated forms.

[1.1.1,H⁺] Monoprotonated species. In the case of this species, NMR shows that the proton is located inside the cavity and is exchanging rapidly back and forth between the two nitrogen atoms. Two structures are possible: i^+,i (C) and i^+,o (D) (Fig. 3-95). In structure C it is unlikely that the proton is located symmetrically since the separation of the two nitrogens is too great. The asymmetrical location is confirmed by the IR spectrum of [*1*-H^+] which shows a ν_{NH} band at 3400 cm^{-1} as for the compound [*1*-$2H^+$]. A strong $(N-H---N)^+$ bond gives rise to a ν_{NH} band in the 2000–2500 cm^{-1} region.

In conformation D, the following rapid exchange can occur:

$$i^+,o \rightleftharpoons i^+,i \rightleftharpoons i,i^+ \rightleftharpoons o,i^+.$$

Kinetic and thermodynamic study of protonation of [1.1.1] cryptand [242]. Proton transfer into the cavity is very slow. As a result, it is possible to observe separately both external (pH titrations, shift of CH_2N NMR resonances on protonation) and internal protonation (direct observation of the species formed, using NMR).

External protonation: mono- and diprotonated species, [1.1.1,H^+] and [1.1.1,2H^+]. The addition of trifluoroacetic acid to a solution of [1.1.1] diamine in d^6-acetone at 270 K was found to give rise to a compound with NMR spectrum very similar to that of [1.1.1]BH_3. Since the crystal structure of the latter aminoborane compound showed that its conformation is i, o-BH_3 (Fig. 3-96a) [241], it is likely that the conformation of the externally monoprotonated form is i, o^+ (Fig. 3-96b).

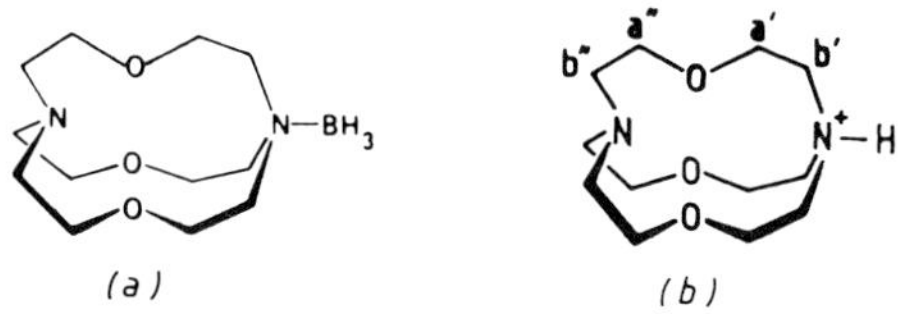

Fig. 3-96. i,o conformation of (a) the amino-borane and (b) the monoprotonated form obtained by protonation of [1.1.1] (external protonation).

A pH titration of [1.1.1] with HCl gave a value of 7 · 1 ± 0.2 for the pK_{a1} of i, o^+. However, the titration curve showed no second break, thus indicating that the value of pK_{a2} is less than 1.5 for the o^+, o^+ species. The relative amounts of the i,i, i,o^+ and o^+,o^+ species present at various of pH have been calculated from pH titrations and the NMR spectra. The equilibrium constant for $o^+,i \rightleftharpoons o^+,o$ is less than 10^{-6}, and for $i,i^+ \rightleftharpoons i^+,o$ it is less than 10^{-7}.

Internal protonation: first internal protonation of [1.1.1]. The rate of the first internal protonation of the diamine [1.1.1] has been studied in D_2O as a function of pH and temperature. The rates were measured by following the integrated NMR intensities of the CH_2O protons for the two forms, i, i and i^+, i. The rate of internal protonation was found to be pH dependent and appeared to correlate with the relative amounts of the i,i, i,o^+ and o^+,o^+ species present in the solution, the rate increasing in the order $i,i < i,o^+ < o^+,o^+$.

Second internal protonation of i^+, i. The species i^+,i can be internally protonated a second time to give i^+,i^+, but only under much more rigorous conditions than those used for the first internal protonation. For example, at pH 1 the half-time for conversion of the i, i cryptand into the i^+, i species at RT is 7 minutes, but no detectable amount of the diprotonated species is formed. In 1M HCl, the second internal protonation requires several weeks at RT and is accompanied by considerable decomposition of the ligand.

At elevated temperatures, the rates are much faster and the effect of decomposition is minimized. The rate has been measured using NMR. At 343 K in 1M hydrochloric acid, the pseudo-first-order rate constant is equal to 0.8 x 10^{-4} sec^{-1}. The activation energies for the first and second protonations have the identical value of 110 ± 8 kJ/mol, thus suggesting that the barrier to entry of the proton into the cavity involves the same type of process in both cases (deformation of the macrobicycle).

Deprotonation of i^+,i^+. The rate of deprotonation has been studied in 5M KOH in D_2O. Proton abstraction ($i^+,i^+ + HO^- \rightleftharpoons i^+,i + H_2O$) is very slow ($2.55 \times 10^{-5} sec^{-1}$ at 349K) and the activation energy is 105 ± 8 kJ/mol, a value which is very close to that for the two internal protonation steps.

Conclusion. The kinetic results of this study lead to an estimate of 17.8 for the internal pK_{a1} of cryptand [1.1.1], and ~8.3 for the second internal protonation, pK_{a2}. Thus, thermodynamically [1.1.1] is a very strong base whereas kinetically it is extremely sluggish because of the very slow proton transfers for the species i^+,i^+ and i^+,i [242]. The thermodynamic and kinetic properties of [1.1.1] are collected in Fig. 3-97.

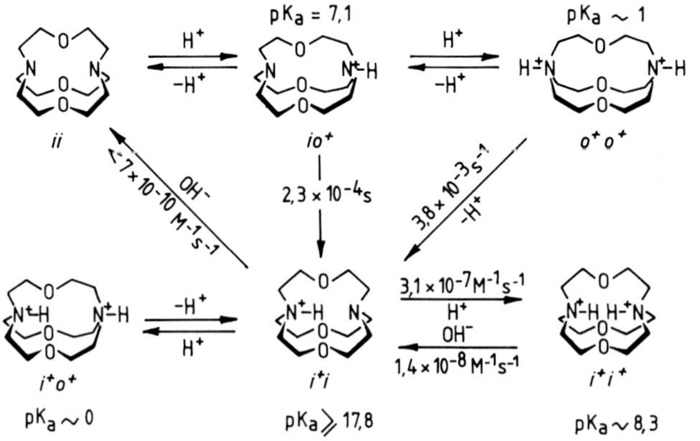

Fig. 3-97. Equilibrium and rate constants for the interconversion of various protonated forms of [1.1.1] cryptand.

e) X-ray crystal structures of proton cryptates

Crystallographic studies of the compounds [1.1.1], [1.1.1,H^+] and [1.1.1,2H^+] have confirmed and clarified the results described above (Fig. 3-98) [243].

The [1.1.1] macrobicycle exists in the i, i conformation and the N---N distance is 3.88 Å. The oxygens are directed away from the cavity because of repulsion between the oxygen and nitrogen lone pairs. A theoretical study (molecular mechanics calculations) of the conformation of this ligand has shown that the optimized i, i conformation is more stable by ca. 22 kcal/mol than the o, o conformation [244].

In the [1.1.1,H^+] monoprotonated compound, the i, i conformation is maintained with a decreased N---N distance (3.60 Å). This long distance excludes

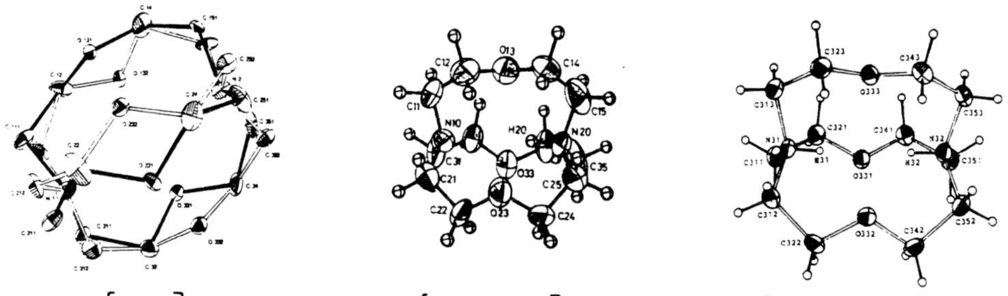

[1.1.1] [H⁺C 1.1.1] [2 H⁺C 1.1.1]

Fig. 3-98. X-ray crystal structures of [1.1.1] cryptand and of its mono- and diprotonated forms (taken from A. Knöchel et al., *J. Am. Chem. Soc.,* **108,** 107, 1986).

the proton being symmetrically located in the cavity. In spite of protonation, the electron density in the cavity remains high and the oxygen atoms are still directed away from the cavity (however, in a less marked fashion than in [1.1.1]).

The crystal structure of the diprotonated form, [1.1.1,2H$^+$], confirms the location of the protons in the cavity. The two ammonium cations are equivalent and the conformation is i, i. The presence of two protons increases the N---N distance (3.91 Å). In this case, the oxygen atoms point into the cavity.

The X-ray crystal structures of 1,6-diazabicyclo[4.4.4]tetradecane and its monoprotonated derivative have also been determined [245]. In both the free and monoprotonated forms, the conformation is i, i (Fig. 3-99). As in the case of [1.1.1],

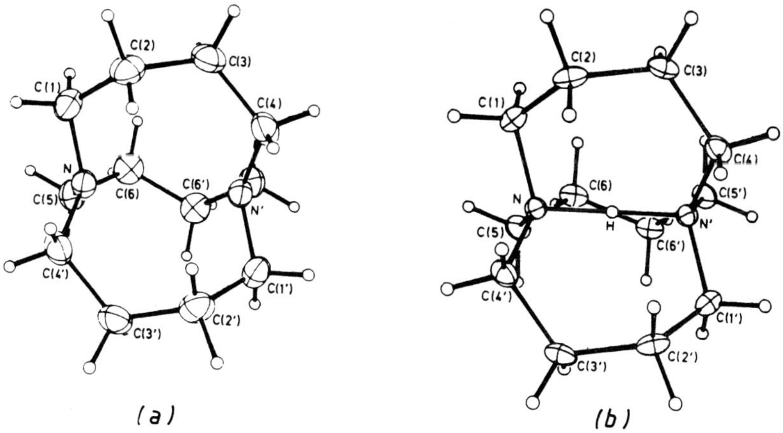

(a) (b)

Fig. 3-99. X-ray crystal structure of (a) 1,6-diazabicyclo[4.4.4]tetradecane and (b) its monoprotonated derivative (taken from R. W. Alder et al., *J. Chem. Soc., Chem. Comm.,* 999, 1983).

protonation leads to a decrease in the N---N distance from 2.81 to 2.53 Å. This constitutes the shortest known distance for an N---H---N bond (that in the 3-quinuclidone dimer is 2.63 Å). An NMR study suggests a symmetrical hydrogen bond [246a]. Diazabicyclo[k.l.m] macrocycles with k, l, m $\leqslant$ 6 exhibit exceptional basic properties similar to those of 1,8-bis(dimethylamino)naphthalenes. A comparative study of these two classes of compounds has shown that certain of the diazabicyclic compounds possess basicities which are equal to or greater than those of 1,8-bis-(dimethylamino)naphthalene derivatives [246b].

3.3.3 Metal Cryptates: Macrobicyclic Inclusion Complexes

a) General points

The first macrobicyclic inclusion complexes to be described were the chloride anion katapinates reported by Park and Simmons in 1968 [169], and the potassium cryptate of the [2.2.2] ligand obtained in 1968 and published in 1969 [247].

Since then, a very great number of complexes has been prepared and their X-ray crystal structures determined. Amongst the macrobicyclic ligands synthesized up until now, the polyoxa, azaoxa and azaoxathia macrobicycles have been the most studied as regards their complexing properties towards metal cations.

b) Complex formation

Salts such as KCl or $Ba(SCN)_2$ dissolve in a deuteriochloroform solution of [2.2.2] cryptand. Formation of the complex between the [2.2.2] cryptand and the metal cation gives rise to substantial changes in the NMR spectrum (Fig. 3-100) [163, 247].

Such differences between the NMR spectra for the cryptand and cryptate are observed with all types of cryptands. The rate of change of the spectrum depends on both the ligand and the nature of the salt (physical state, hydration, type of associated anion, temperature, stirring, etc.).

With thallium salts, the NMR spectrum sometimes shows doublets for all the proton signals as a result of proton-thallium coupling.

Formation of a cryptate in aqueous solution (D_2O) can also be observed by use of NMR. The increase in solubility of $BaSO_4$ arising from complexation is ca. 10^4-fold [162]. Unlike crown ether complexes, cryptates are very stable in water.

c) Cryptate conformations

Inclusion of an M^{n+} metal cation in the intramolecular cavity of the three possible forms of cryptands should give rise, in principle, to the following three topographies: exo-exo, exo-endo and endo-endo (Fig. 3-101).

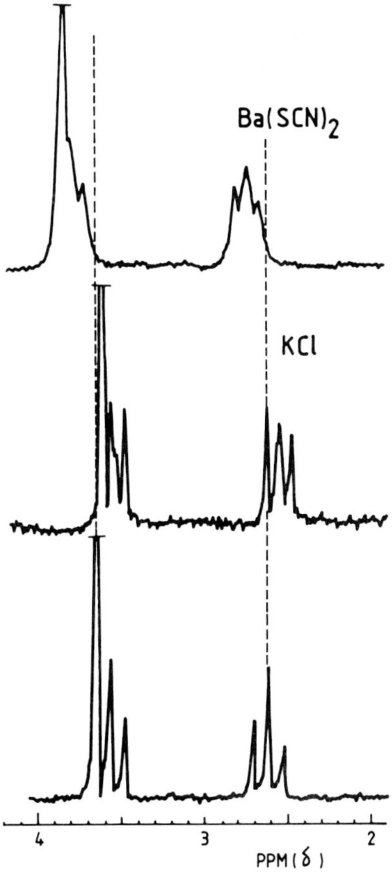

Fig. 3-100. Observation, using NMR, of the formation of [$K^+ \subset 2.2.2$] and [$Ba^{2+} \subset 2.2.2$] cryptates.

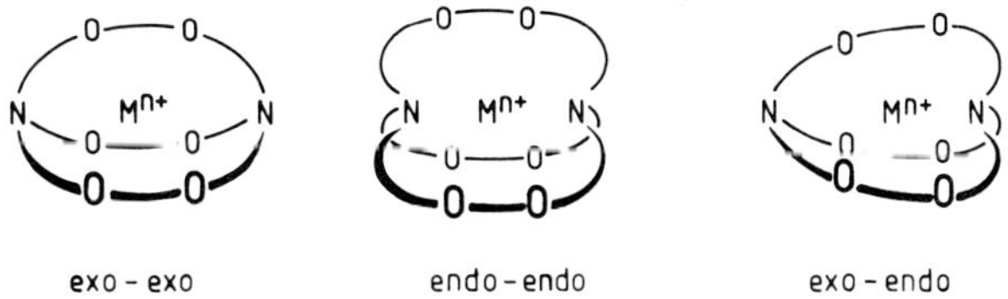

Fig. 3-101. Possible topographies of cryptates.

The endo-endo conformation is greatly favored because in this case both amine groups participate in stabilizing the complex.

d) Comparative study of cryptate structures

The X-ray crystal structure determinations for numerous cryptates of [2.2.2] [248, 249a] have confirmed both inclusion of the cation in the cavity and also, in particular, the endo-endo conformation. The symbol for inclusion, ⊂, denotes this structure, e.g. the complex of [2.2.2] with K^+ ion is [K^+ ⊂ 2.2.2]. Cryptate cavities are of well defined shape and slightly variable size arising from rotation about the N---N axis. The conformational changes, which allow the ligand to match the size of the cavity to that of the cation undergoing complexation, have been analyzed for a series of cryptates of alkali metal cations of increasing size: Na^+, K^+, Rb^+, Cs^+.

The coordination polyhedron is defined by the six oxygen atoms located at the vertices of the two triangles O(4), O(16), O(21), and O(7), O(13), O(24) (Fig. 3-102a), together with the two nitrogen atoms (on the D_3 axis) separately located above each of the two oxygen triangles. Change in the cavity size occurs via ligand torsion about the N---N axis, this being equivalent to rotation of one oxygen triangle with respect to the other through an angle α (Fig. 3-102b). Change in α has a very great effect on the *d* and N---N distances (Fig. 3-102a and Table 3-9) and hence on cavity size.

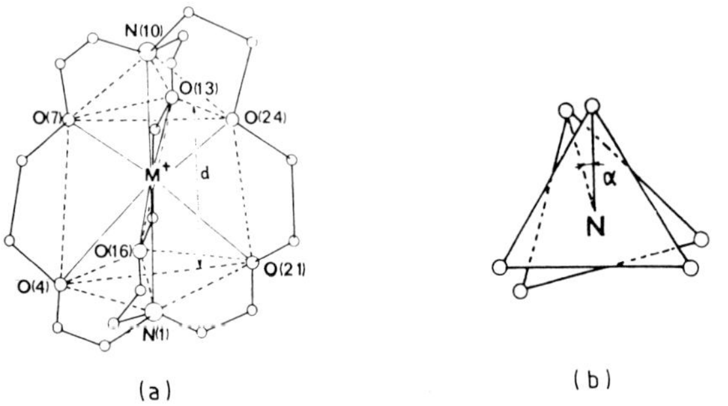

Fig. 3-102. Structure of the alkali metal cryptates with Na^+, K^+, Rb^+, and Cs^+, where *d* is the distance between the two triangular faces O(4), O(16), O(21) and O(7), O(13), O(24); *α* is the angle of twist between the two O_3 triangular faces of the bicapped trigonal prism (taken from R. Weiss et al., *J. Chem. Soc., Chem. Comm.*, 444, 1971).

Table 3-9. Fitting of the [2.2.2] cavity size to that of the cation*.

Complex	α (degrees)	d[a)] (Å)	$M^+\cdots O$ $(rM^+ + rO)^b = S_1$	$M^+\cdots N$ $(rM^+ + rN)^{b)} = S_2$	$N\cdots N$
NaI	45	2.10	2.57 (2.41)	2.77 (2.51)	5.50
KI	21	2.65	2.79 (2.78)	2.87 (2.88)	5.74
RbNCS	15	2.81	2.90 (2.93)	3.00 (3.03)	6.00
CsNCS	15	2.87	2.97 (3.14)	3.03 (3.24)	6.07

* Unit of length: Å
a) defined in the text and in Fig. 3-102.
b) rM^+: Pauling ionic radius corrected for 8-coordination [249b]; rO, rN: van der Waals radii for oxygen (1.4 Å) and nitrogen (1.5 Å).
(Taken from R. Weiss et al., *J. Chem. Soc. Chem. Comm.*, 444, 1971).

The cation-heteroatom distances indicate the degree of cation-ligand complementarity. For K^+, the observed distances are in each case almost identical to the sum (S_1 and S_2) of the ionic and van der Waals radii. In this complex, the conformation of the ligand is least strained. For Na^+, the M^+---O and M^+---N distances are significantly longer than S_1 and S_2, indicating that the ligand cannot compress itself sufficiently to ensure effective coordination to the cation. Conversely, in the case of Cs^+ the M^+---O and M^+---N distances are shorter than S_1 and S_2, revealing the existence of substantial repulsive interactions in the complex (the Cs^+ ion is too large for the cavity in the [2.2.2] ligand). Complexation of Na^+ and

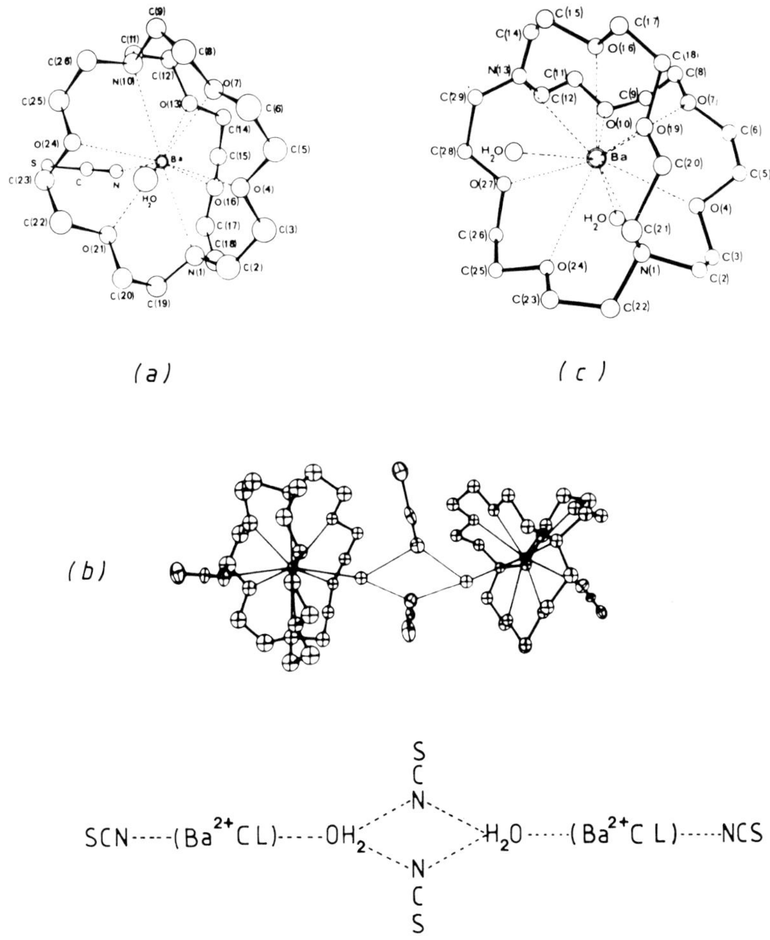

Fig. 3-103. Crystal structures of the barium cryptates (a) $[2.2.2 Ba(SCN)(H_2O)]^+$ and (c) $[3.2.2 Ba(H_2O)_2]^{2+}$; (b) represents the arrangement of the $[Ba^{2+} \subset 2.2.2]$ complex in the crystal lattice (taken from R. Weiss et al., *J. Am. Chem. Soc.,* **93**, 1806, 1971).

Cs^+ therefore necessitates ligand compression and expansion, respectively. In both cases, strain is created resulting in reduced stability of the complex formed. Finally, it should be noted that in the cryptates that we have just described, the anion is very far from the complexed cation (M^+---$X^- \geqslant 3.5$ Å).

The crystal structures of the cryptates of Ba^{2+} with [2.2.2] and [3.2.2] (Fig. 3-103) [249] have shown the involvement of water molecules or anions. The latter species complete the cation coordination spheres in the [$Ba^{2+} \subset$ 2.2.2] and [$Ba^{2+} \subset$ 3.2.2] complexes. In these complexes, the interactions of the cation with the solvent molecules and with the counterion are strong, unlike the case with alkali metal cations.

In the complex with [2.2.2], the Ba^{2+} coordination sphere is completed by a water molecule and the nitrogen atom of a thiocyanate anion. Cohesion of the crystal lattice is ensured by an intricate network of hydrogen bonds formed between the bound water molecules and two thiocyanate anions (Fig. 3-103 b).

In the complex with [3.2.2], two water molecules are bound to Ba^{2+} (Fig. 3-103 c) giving a coordination number of 11 for the cation. These water molecules are themselves hydrogen bonded to thiocyanate anions. However, none of the anions is directly coordinated to the cation.

3.3.4 Stability and Selectivity of Macrocyclic Complexes

Complexation of a metal cation, M^+, by a ligand, L, in a solvent, S, can be represented by the following equilibrium.

$$(L)_{solv} + (M^{n+}, mS) \underset{k_d}{\overset{k_f}{\rightleftharpoons}} (L, M^{n+})_{solv} + m'S$$

where k_f and k_d are the rate constants for complex formation and decomposition, respectively [250].

The stability constant, K_s, expressed in terms of concentration, and the thermodynamic stability constant, K_{th}, are given, respectively, by the following expressions:

$$K_s = c(L,M^{n+})/c(L) \cdot c(M^{n+})$$

and $$K_{th} = f_C \cdot c(L,M^{n+})/f_L \cdot c(L) \cdot f_M \cdot c(M^{n+})$$

where f_C, f_L and f_M are the three activitiy coefficients for the species present. Since the ligand L is neutral and both the free and complexed cations carry the same charge, it may be assumed that $f_L = 1$, $f_C = f_M$ and $K_s \simeq K_{th}$.

A variety of different methods has been used to determine K_s experimentally: potentiometry using cation selective electrodes, pH measurements, conductiometry, calorimetry, temperature-jump techniques, NMR, solvent extraction, osmometry, UV spectroscopy, etc.

a) Cation/cavity complementarity

Examination of the respective diameters of the alkali metal cations and of 18-crown-6 shows that the K^+ cation just fills the cavity (Table 3-10), and, indeed, it is found to be the cation which is the most strongly complexed. The relationship between the stability constant of the complex and the degree of cavity-cation complementarity is demonstrated by the plot of K_s vs. the relative sizes of the cation and the cavity [69, 251]. The highest stability constant is observed for the complementary pair K^+, [18-O-6] (Fig. 3-104). The plot exhibits a selectivity peak and it is found that:

$$K_s(K^+)/K_s(Na^+) \gg K_s(K^+)/K_s(Rb^+).$$

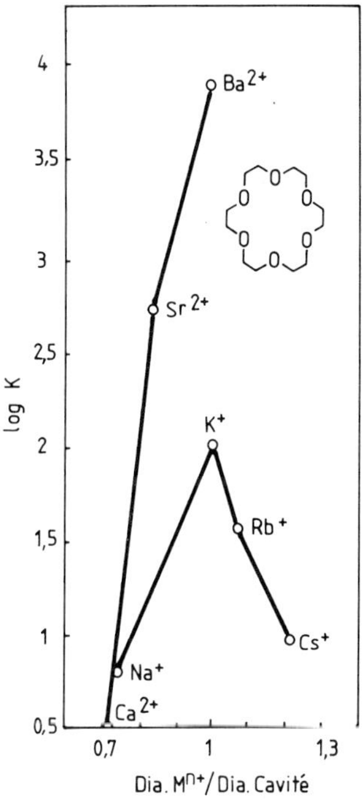

Fig. 3-104. Plot of stability constants in water for complexes of 18-crown-6 with alkali and alkaline earth metal cations, as a function of the ratio of cation to cavity size.

Based only on the cation/cavity size ratio, this selectivity peak can be readily understood since the difference in ionic diameters is ca. 0.7 Å for Na^+/K^+ and only ca. 0.3 Å for K^+/Rb^+ (Table 3-10). It can be noted in this Table that the sizes of Rb^+ and the [18-O-6] crown are compatible whereas Na^+ is too small. Furthermore, a charge effect can be seen. The stability constant for the divalent Ba^{2+} cation is much greater than that for K^+.

A calorimetric titration study in aqueous solution of the stabilities of complexes of uni- and divalent cations with crown ethers [15-O-5], [18-O-6] and the *cis*-

Table 3-10. Univalent cation diameters and hole sizes in crown ethers (Å).

Cation	Ionic diameter	Crown	Cavity diameter
Li^+	1.36	14-O-4	1.2–1.5
Na^+	1.94	15-O-5	1.7–2.2
K^+	2.66	18-O-6	2.6–3.2
Rb^+	2.94	21-O-7	3.4–4.3
Cs^+	3.34		
NH_4^+	2.86		
Ag^+	2.52		

(Taken from H. K. Frensdorff, *J. Am. Chem. Soc.*, **93**, 600, 1971.)

syn-cis and *cis-anti-cis* isomers of dicyclohexyl-[18-O-6] has been made. This has resulted in an analysis of the effect of various parameters on the stabilities of the complexes formed [252 a]. The important factors are: size, charge and nature of the cation; cavity size, and nature and number of the interaction sites; ligand conformation. The results are collected graphically in Fig. 3-105. The stability constants for the alkali metal cations in water are low ($K_s \sim 100$), whereas those for the divalent cations are significantly higher.

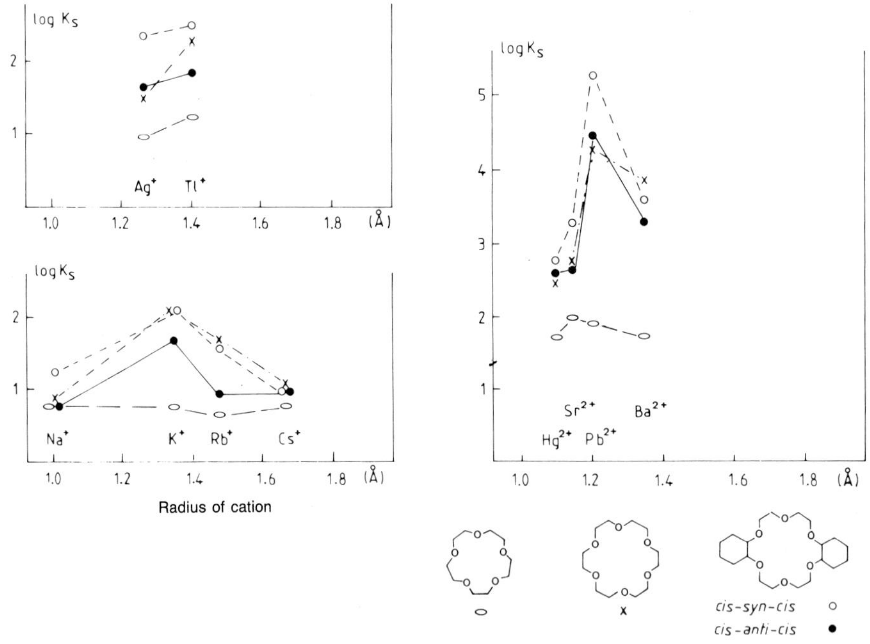

Fig. 3-105. Plot of log K_s vs. cation radius, for the reaction $M^{n+} + L \rightleftharpoons ML^{n+}$ in aqueous solution.

The stability of the complexes is different for the two isomers of the dicyclohexyl crown [DC-18-O-6]. For both uni- and divalent cations, the most stable complexes are formed with the *cis-syn-cis* isomer. The X-ray crystal structure of the complex of this isomer with Ba^{2+} shows that the two cyclohexyl rings are almost perpendicular to the plane formed by the macrocycle oxygen atoms [252b].

Both the stability and selectivity are very low in all cases for complexes of [15-O-5]. The weak complexing ability of this crown is explained by the unfavorable combination of two factors: (a) the number of interaction sites is lower, and (b) although the cavity is well adapted to small cations, the latter are more strongly solvated than are larger cations.

The stability constants of cyclic polyether complexes of various sizes (Fig. 3-106) with three cations (Na^+, K^+, Cs^+) have been determined by potentiometry with cation-selective electrodes in both aqueous and methanol solution [253], and also by titration calorimetry in methanol [254].

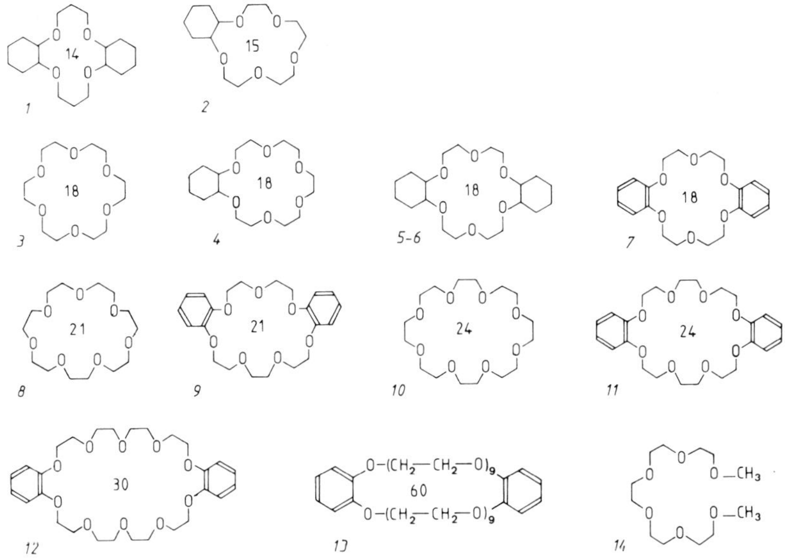

Fig. 3-106. 14- to 60-membered macrocyclic polyethers.

The stability constants listed in Table 3-11 show the influence of cation and ring size as well as that of ring substituents.

The selectivity order depends on the nature of the crown:

$Na^+ > K^+$ for DC-[14-O-4];
$Na^+ \sim K^+ > Cs^+$ for C-[15-O-5];
$K^+ > Cs^+ > Na^+$ for C-[18-O-6];
$K^+ \sim Cs^+ \gg Na^+$ for DB-[21-O-7];
$Cs^+ > K^+$ for DB-[24-O-8].

Table 3-11. Stability constants (log K_s) in methanol.

	Na^+	K^+	Cs^+
Dicyclohexyl-14-O-4 *(1)*	2.18	1.30	
Cyclohexyl-15-O-5 *(2)*	3.71	3.58[a)]	2.78[a)]
18-O-6 *(3)*	4.32	6.10	4.62
Cyclohexyl-18-O-6 *(4)*	4.09	5.89	4.30[a)]
Dicyclohexyl-18-O-6 (cis-syn-cis) *(5)*	4.08	6.01	4.61
Dicyclohexyl-18-O-6 (cis-anti-cis) *(6)*	3.68	5.38	3.49
Dibenzo-18-O-6 *(7)*	4.36	5.00	3.55[a]
21-O-7-*(8)*		4.41	5.02
Dibenzo-21-O-7 *(9)*	2.40	4.30	4.20
24-O-8 *(10)*		3.48	4.15
Dibenzo-24-O-8 *(11)*		3.49	3.78
Dibenzo-30-O-10 *(12)*	2.0	4.60	
Dibenzo-60-O-20 *(13)*		3.90	
Pentaglyme *(14)*	1.52	2.20	

a) 2:1 Ligand:metal complexes formed. (Same source as Table 3.10.)

The optimum ring size is 15- to 18-membered for Na^+, 18-membered for K^+ and 18- to 24-membered for Cs^+, in accordance with the relative sizes of the cation and crown.

The K^+ complex with [30-O-10] crown is more stable than that with [24-O-8]. This is explained by the ligand being wrapped round the cation in the former complex (see structure shown in Fig. 1-4b). The incorporation of cyclohexyl or aryl groups into macrocycle [18-O-6] modifies to some extent the complexing power of the parent crown ether [253].

The calorimetric titration study performed in methanol [254] with [15-O-5], [18-O-6] and [21-O-7] crown ethers has provided additional evidence for the low stability and selectivity of [15-O-5], the cavity being too small for Na^+ and the larger cations (Fig. 3-107). The selectivities of [18-O-6] for K^+, and of [21-O-7] for Cs^+, are also confirmed.

b) Effect of substituents

The effect of crown substitution depends on the sizes of the cavity and cation. The presence of two benzo groups decreases the stability constant by a factor of 10 for the potassium complex of [18-O-6], and by a factor of ca. 6 and 3 for the cesium complexes of [21-O-7] and [24-O-8], respectively (Table 3-11) [253].

Furthermore, substitution in the aromatic ring also affects the stability of the complex. This effect has been determined conductiometrically for a series of monobenzo-[15-O-5] and monobenzo-[18-O-6] complexes in acetone at 25 °C (Fig. 3-108) [255].

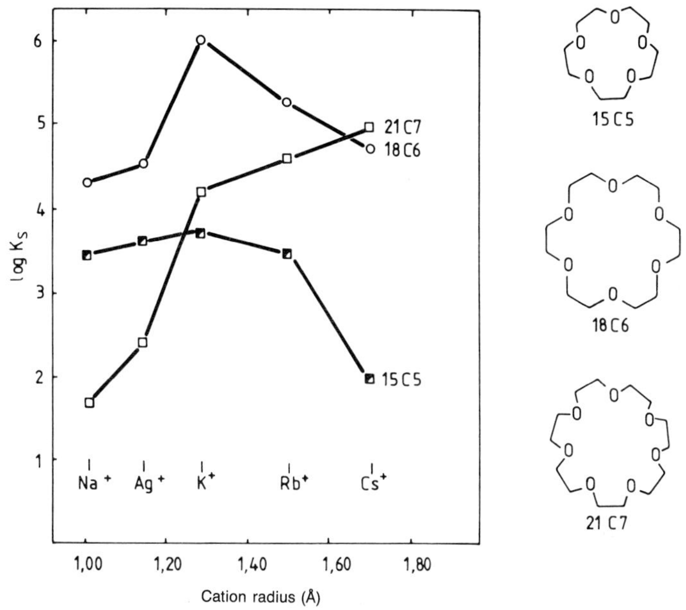

Fig. 3-107. Plot of log K_s vs. cation radius for complexes of [15-O-5], [18-O-6] and [21-O-7] macrocycles with univalent cations, in methanol at 25 °C.

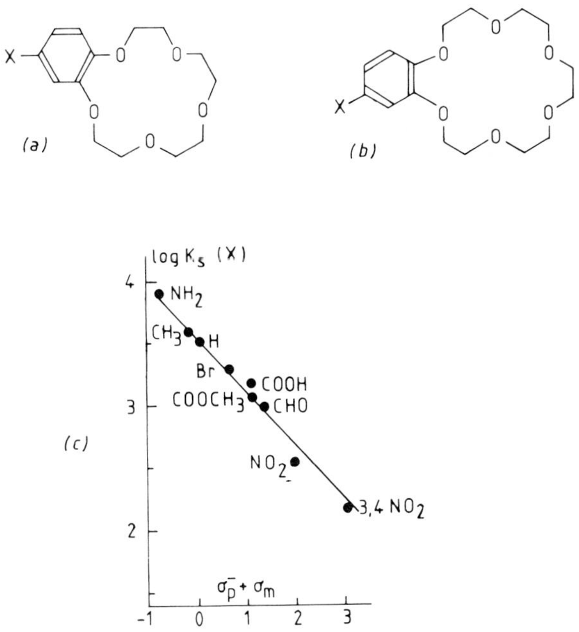

Fig. 3-108. Substituent effects in benzo-[15-O-5] and benzo-[18-O-6] macrocycles, and Hammett plot for complexes of benzo-[15-O-5] crown ether.

For the complexes of Na^+ with benzo-[15-O-5] crowns, a pronounced decrease in stability constant is observed with increasing electron withdrawing power of the substituent (Table 3-12). It can be seen in Fig. 3-108c that a reasonable Hammett correlation is obtained as a function of $\sigma_p^- + \sigma_m$.

Table 3-12. Stability constants of Na^+ complexes of 4′-X-benzo-[15-O-5] in acetone at 25 °C.

X	10^{-3} K_s (M^{-1})
NH_2	8.21
CH_3	3.96
H	3.44
Br	2.05
CO_2H	1.61
CO_2CH_3	1.23
CHO	1.11
NO_2	0.365
3′,4′-Dinitro	0.15

(Taken from J. Smid et al., *J. Am. Chem. Soc.,* **98**, 5198, 1976).

On the other hand, the effect of substituents is much smaller for benzo-[18-O-6]. The stability constant in the case of a CH_3 substituent is greater than that for an NO_2 substituent by a factor of only three in the case of Na^+, and six in the case of K^+ (Table 3-13). However, with the Na^+ complex of benzo-[15-O-5] crown ether there is an elevenfold increase in stability constant for the same change in substituent (Table 3-12). No Hammett correlation is found in the case of benzo-[18-O-6] crown ether.

Table 3-13. Stability constants of Na^+ and K^+ complexes of 4′-X-benzo-[18-crown-6] in acetone at 25 °C.

	10^{-5} K_s(M^{-1})	
X	Na^+	K^+
CH_3	1.23	3.83
H	0.52	1.27
CHO	0.39	0.77
$CONHC_4H_9$	0.32	0.56
NO_2	0.47	0.63

(Same source as Table 3.12).

Substituent effects, which are marked in the case of benzo-[15-O-5], have been attributed to change in aromatic oxygen basicity with the donor and attractor power of the substituent.

The different behavior of benzo-[18-O-6] has been explained as being the result of the presence of four instead of three aliphatic oxygens thus decreasing the relative extent of Na^+ (or K^+) interactions with the phenolic oxygens, and consequently the effect of substituents. Furthermore, the size of the cavity and its flexibility can also play a part by changing the distance between the cation and the phenolic oxygens [255].

c) Macrocyclic effect

The specific effect of macrocycles has been shown by comparing the stability constants for pentaglyme and [18-O-6] complexes (compounds *14* and *3* in Fig. 3-106; see Table 3-11). The potassium complex of the crown is 10^4-fold more stable than that of pentaglyme. Moreover, the crown is more selective by a factor of about ten [253].

d) Formation of complexes between two ligand molecules and a cation (2L: M^+): large cation + small cavity

Addition of a second ligand molecule to the 1:1 complex gives rise to the following complexation equilibria:

$$L + M^+ \rightleftharpoons LM^+ \; (K_1')$$
$$LM^+ + L \rightleftharpoons L_2M^+ \; (K_2').$$

The K_1' and K_2' stability constants for various macrocycles are collected in Table 3-14 [253].

The formation of a 2:1 complex in methanol has been confirmed for only the first three examples, although in other cases partial 2:1 complexation cannot be totally excluded. It should be noted that the [18-O-6] crown and its derivatives with one or two cyclohexyl lateral groups have little ability to form 2:1 complexes with Cs^+. A ring which is too small to accommodate the cation is possibly a necessary condition, but one which is not sufficient for formation of sandwich-type 2L:M^+ complexes in which both crowns are involved together. This observation has given rise to the idea of joining the two ligands by two bridges thus affording macrotricyclic compounds.

Several reviews treat in detail the various aspects of complexation of cations which we have just briefly mentioned [256]. Theoretical studies have helped in defining the many parameters involved in molecular recognition. The role and the nature of the solvent are particularly important as regards selectivity [257].

Table 3-14. Stability constants of 1 : 1 and 2 : 1 ligand : metal complexes in methanol at 25 °C.

Polyether[a)]	Cation	Log K_1' (M^{-1})	Log K_2' (M^{-1})	K_2'/K_1'
Cyclohexyl-15-O-5 *(2)*	K^+	3.58	1.88	0.02
Cyclohexyl-15-O-5 *(2)*	Cs^+	2.78	1.91	0.13
Dibenzo-18-O-6 *(7)*	Cs^+	3.55	2.92	0.23
18-O-6 *(3)*	Cs^+	4.62	1.30	0.0005
Cyclohexyl-18-O-6 *(4)*	Cs^+	4.30	1.52	0.002
Dicyclohexyl-18-O-6 (cis-syn-cis) *(5)*	Cs^+	4.61	0.59	0.0001
Dicyclohexyl-18-O-6 (cis-trans-cis) *(6)*	Cs^+	3.49	–	–
21-O-7 *(8)*	Cs^+	5.02	–	–
Dibenzo-21-O-7 *(9)*	Cs^+	4.20	1.9	0.005

a) See Fig. 3-106.
(Same source as Table 3.10).

e) Elaboration of a specific receptor for the Li^+ cation

None of the macrocycles described above efficiently complexes the lithium cation [253]. The strong hydration of this very small cation (ionic radius = 0.78 Å) constitutes a serious hurdle to its complexation.

The [12-O-4] crown forms a 1:1 complex with LiSCN. The X-ray crystal structure (Fig. 3-109) shows that the Li^+ cation is pentacoordinated and located well outside the mean plane of the macrocycle [258 a].

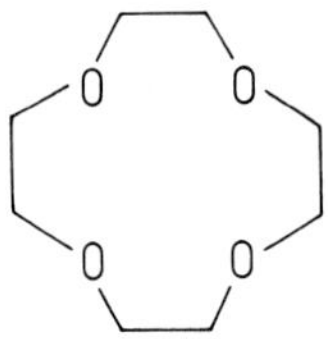

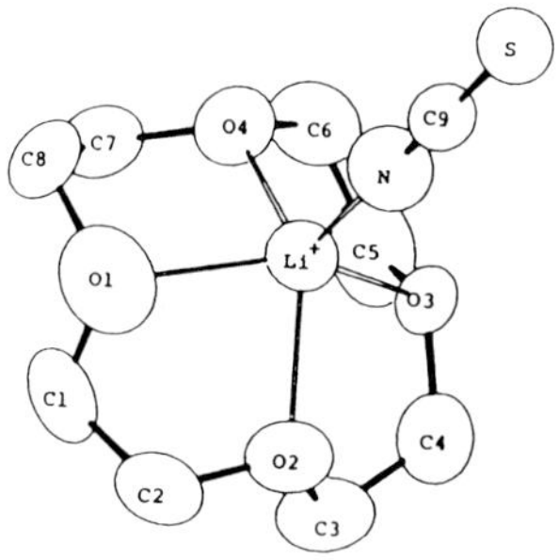

Fig. 3-109. Crystal structure of the LiSCN-[12-O-4] crown ether complex (taken from P. Groth, *Acta Chim. Scand.,* **A35,** 463, 1981).

This ligand is not specific for the Li^+ cation. The stability constants, determined by an electrochemical method and collected in Table 3-15, show that the macrocycle strongly complexes the sodium cation [258 b].

Two slightly larger macrocycles, 1,5,9,13-tetraoxacyclohexadecane, [16-O-4], and its 3,3,7,7,11,11,15,15-octamethyl derivative, have been synthesized (Fig. 3-110a). In both these componds, two successive oxygens are separated by three carbon atoms instead of the two found in almost all the macrocyclic polyethers we have met with up until now.

Table 3-15. Stability constants (log K_s) of alkali metal complexes of [12-O-4] crown ether in propylene carbonate.

$L:M^+$	Li^+	Na^+	K^+	Rb^+	Cs^+
1 : 1	2.93	3.60	2.15	1.69	1.43
2 : 1		6.31			

(Taken from J. F. Desreux et al., *J. Chem. Soc., Dalton,* 865, 1980).

(a)

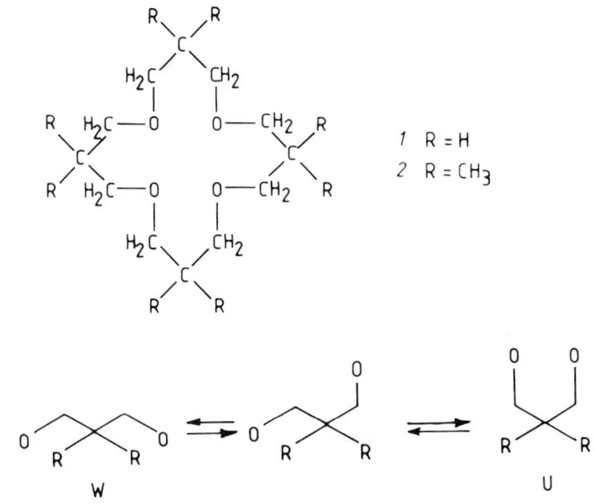

1 R = H
2 R = CH_3

(b)

Fig. 3-110. (a) [16-O-4] Tetraoxacyclohexadecane macrocycles, (b) conformational equilibria of the OCCCO subunit.

The OCCCO subunit in ligands *1* and *2* has specific conformational properties and can exist in three forms in equilibrium with each other. In the U form, the distance between the two oxygens is close to that in normal crown ethers having the chain sequence $-O-(CH_2)_2-O-$. Consequently, if the four subunits in crowns *1* or *2* are in the U form, the cavity size can be close to that in [12-O-4] (Fig. 3-110b).

The X-ray crystal structures of ligand *1* and of its complex with Li^+ cation have been determined [258 c, d]. The large difference in conformation between the free and complexed forms can be noted (Fig. 3-111 a and b).

The lithium cation is once again pentacoordinated, and it is displaced from the mean plane by ca. 0.55 Å (Fig. 3-111 c).

NMR studies have shown that in solution the same conformational changes occur on complexation of Li^+ by *1*, and similar behavior is observed with ligand *2* [258 e].

These two macrocycles have been incorporated into membranes in order to determine their ability to act as neutral carriers in membrane electrodes [258 f]. For both compounds, the response of the electrode is linear for Li^+, NH_4^+, Na^+ and K^+ chloride solutions over the concentration range 1 to 10^{-4} mol/l. The selectivity coefficients (Table 3-16) indicate that in the case of the alkali metal cations, the electrode response is at a maximum for Li^+ cation.

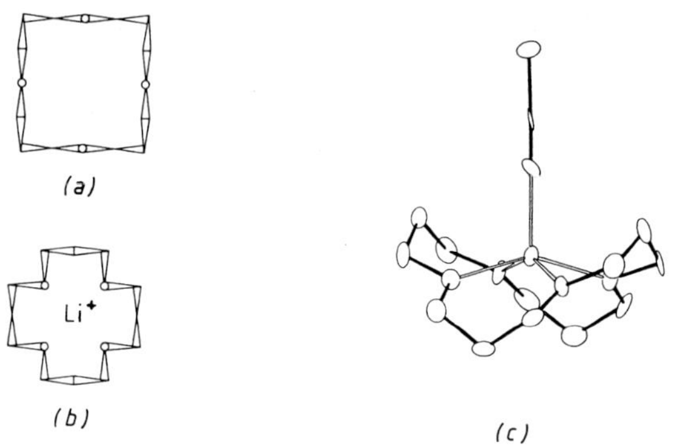

Fig. 3-111. Conformation of [16-O-4] crown ether in (a) the free form and (b) complexed; (c) structure of the complex with LiSCN (taken from P. Groth, *Acta Chim. Scand.,* **A 35**, 460, 1981).

Table 3-16. Electrode selectivity coefficients.

	Carrier *(1)* K_{LiM}	$1/K_{LiM}$	Carrier *(2)* K_{LiM}	$1/K_{LiM}$
Li^+	1.0	1.0	1.0	1.0
H^+	4.7	0.2	3.5	0.3
NH_4^+	3.4×10^{-1}	3	3.2×10^{-1}	3
Na^+	1.0×10^{-1}	10	9.9×10^{-1}	10
K^+	2.0×10^{-2}	50	2.9×10^{-2}	35
Rb^+	1.4×10^{-2}	72	1.5×10^{-2}	67
Cs^+	1.3×10^{-2}	77	1.4×10^{-2}	71
Mg^{++}	1.7×10^{-3}	590	3.5×10^{-3}	286
Ca^{++}	1.4×10^{-3}	715	9.5×10^{-3}	105
Sr^{++}	1.2×10^{-3}	830	1.4×10^{-3}	715
Ba^{++}	1.2×10^{-3}	830	1.6×10^{-3}	625

(Taken from J. Krane et al., *Acta Chem. Scand.,* **A 36**, 227, 1982).

The selectivities towards NH_4^+ and Na^+ are low, but increase for the larger alkali metal cations, and are high for divalent cations. H^+ interference only becomes a problem under acid conditions. Macrocycles *1* and *2* are thus efficient carriers of Li^+.

These compounds fulfil some of the criteria suggested as being necessary for an efficient carrier of Li^+. (a) Substitution of the water molecules in $Li(H_2O)_n$ should take place stepwise, the ligand interaction sites gradually replacing the water molecules (page 139 in [259 a]). This process assumes quite high conformational flexibility, which is the case for compounds *1* and *2* due to the presence of three carbon atoms between the heteroatoms. (b) Complexation of Li^+ should induce the macrocycle to contract in order to favor efficient interactions between Li^+ and the ether groups.

Because of the importance of Li^+ in the treatment of certain mental disorders [259 a, b], numerous ligands and potential carriers of this cation have been synthesized (Fig. 3-112). Thus, dibenzo-[14-O-4] (*1*) [259c], benzo-[14-O-4] (*2*) [259d], didecalino-[14-O-4] (*3*) [259e], and benzo-[13-O-4] (*4*) [259f] are all efficient and selective complexing agents for lithium cation. The continuing interest in complexing agents for lithium cation is demonstrated by the large number of studies covering the various aspects of its complexation [259g].

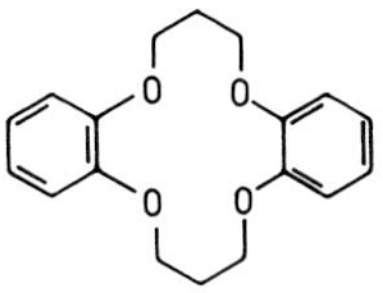

1

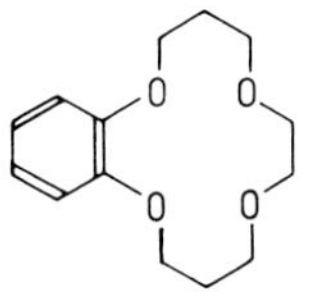

2

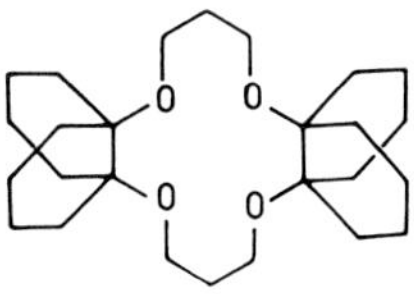

3

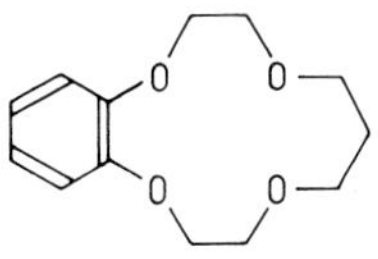

4

Fig. 3-112. Macrocycles which complex Li^+ cation.

f) Medium effects

The solvent has a very great effect on stabililty, selectivity, rate of exchange and also, in certain cases, stoichiometry.

The stability constant of a complex increases as the solvating power of the solvent towards cations decreases. Thus, complex formation competes with cation solvation [253, 260].

Examples of the effect of change in solvent MeOH:H_2O composition on stability constant are to be found in Table 3-17, and are illustrated in Fig. 3-113. For both crown ethers, replacement of water by methanol gives rise to a ca. 300-fold increase in stability constant [261, 262].

An NMR study of the complexes of Na^+ with [15-O-5], benzo-[15-O-5], and [18-O-6] crown ethers in water and in a number of nonaqueous solvents has shown that in complexation reactions the donor ability of the solvent is much more important than is its dielectric constant [263].

The selectivity generally increases with the stability of the complex, but cation size also plays a part [255, 264, 265]. Small cations are more sensitive than are large ones to change in solvating power of the solvent. Thus, M_i^+/M_j^+ selectivity increases with decreasing solvating power if the M_i^+ cation is smaller than the M_j^+ cation, and the selectivity decreases if the M_i^+ cation is larger.

Table 3-17. Sodium cation binding. Variation in stability constants as a function of MeOH : H_2O ratio.

		15-O-5		18-O-6	
wt. % of McOH	Mol fraction of MeOH	K_s	log K_s	K_s	log K_s
0	0	6.2	0.79	63	1.80
20	0.123	30.9	1.49	151	2.18
40	0.273	51.0	1.71	293	2.47
60	0.458	164	2.21	644	2.81
80	0.692	448	2.65	1759	3.25
90	0.835	926	2.97	5378	3.73
100	1.000	1780	3.25	22580	4.35

(Taken from G. W. Gokel et al., *J. Org. Chem.*, **47**, 147, 1982).

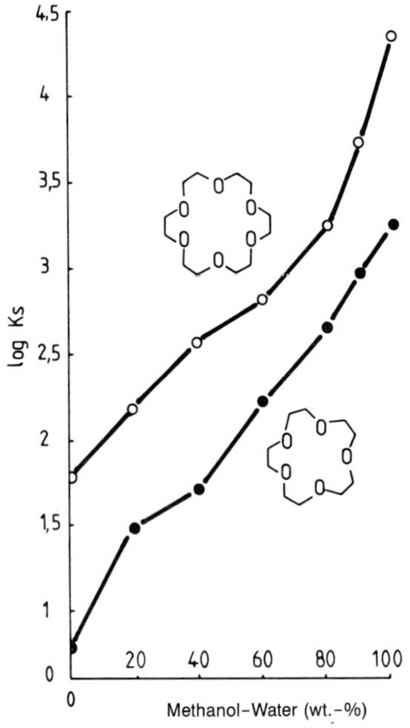

Fig. 3-113. Variation of the stability constants of complexes of Na^+ ion with 15-crown-5 and 18-crown-6 as a function of the proportion of methanol in water.

With 3,3-dimethyldibenzo[18-crown-6] and the two cations K^+ and Na^+, the selectivity favors K^+ in methanol and in water, whereas it favors Na^+ in tetrahydrofuran [264a] and in acetonitrile [264b]. These results show the importance of solvent-cation interactions, and it is thus always necessary to compare stability constants and selectivities of complexes in the same solvent.

g) Stereochemical control of complex stability. Effect of macrocycle conformation

A remarkable effect of the ligand conformation and stereochemistry on its complexing ability is shown by comparing the stability constants of the potassium complexes of macrocycles *1* and *2* (Fig. 3-114). The values are $K_s \sim 30000$ for *1* and less than 10 for *2* in 95:5 methanol: water [266].

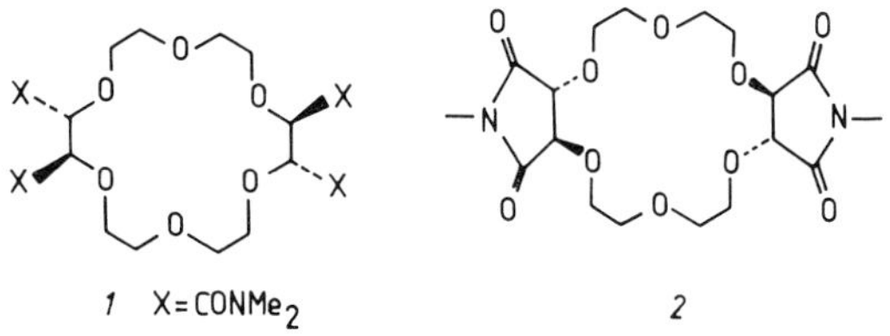

Fig. 3-114. Tetracarboxamide *(1)* and bis-imide *(2)* derivatives of [18-O-6] crown.

The distortion of the macrocycle conformation resulting from the presence of two imide rings prevents simultaneous interactions of all the oxygens with the cation. The distortion of compound 2 is clearly visible in Fig. 3-115b.

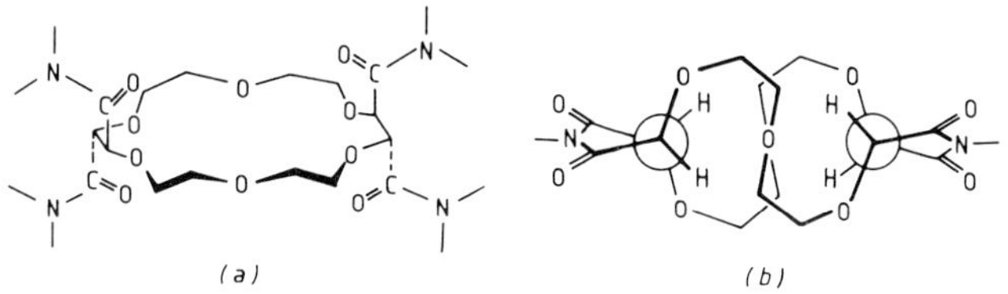

Fig. 3-115. Idealized conformations for macrocycles *1* and *2* in Fig. 3-114.

The conformation of the macrocycle determines not only the arrangement of the interaction sites but also the size and shape of the cavity, and thus plays a very important role in the stability and selectivity of complexation.

h) Effect of lateral appendages

In substituted macrocycles of type *1* (Fig. 3-114), the lateral groups are in a diaxial position on each side of the mean plane of the ring [37, 267, 268]. Under these conditions, the shape of the central cavity is conserved even when the lateral groups are changed or when they bear a charge. Some examples of charged R substituents, together with the arrangement of these substituents with respect to the macrocycle, are shown in Fig. 3-116.

The lateral groups greatly influence the stability of the complexes as demonstrated in Fig. 3-117 showing the change in K_s for the complex with potassium ion in water as a function of the nature of R [269, 270].

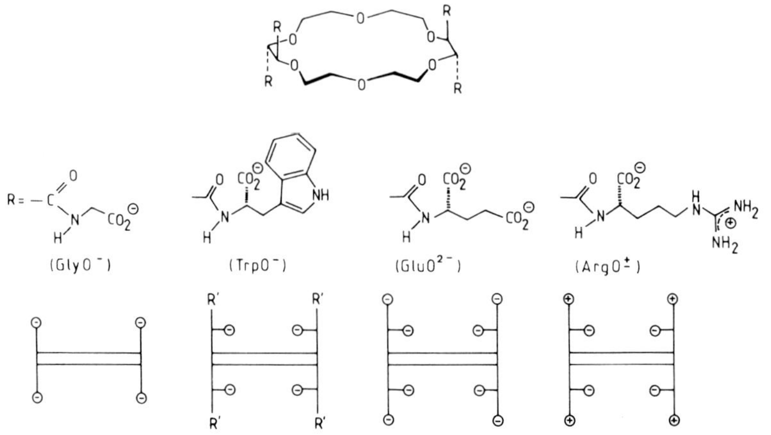

Fig. 3-116. Structures of macrocycles containing charged lateral groups (R' = (3-indolyl)CH_2).

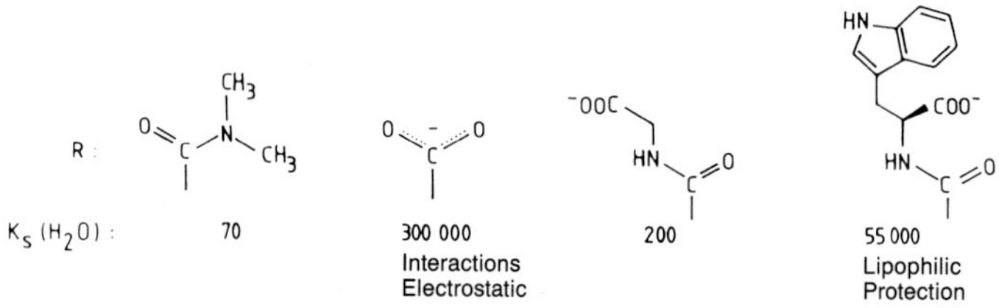

Fig. 3-117. Effect of lateral groups in the tetrasubstituted macrocycle (see Fig. 3-116 for general formula) on the stability of complexes formed with K^+ ion.

These effects are due to various factors. *Electrostatic interactions*, when present, dominate the binding strength which leads to the following sequence with respect to the nature of the R side chain: cationic < neutral ≪ anionic. The observed stability constants and their dependence on the nature of R are shown in Fig. 3-118.

The most pronounced effect is observed when the anionic groups are directly bound to the ring (R = CO_2^-). These complexes are the most stable among all known polyether macrocyclic ligands. The stability of the potassium complex is ca. 4000-fold greater than that of the complex of potassium with [18-O-6] (compound *2* in Fig. 3-118), and is comparable to that of the most stable alkali metal cryptates. This tetracarboxylate macrocycle binds the NH_4^+ cation just as strongly.

The involvement of *lipophilic shielding* groups which prevent solvation of the polar groups explains the remarkable increase in complex stability when the glycinate groups (GlyO⁻) are replaced by amino acid residues with side chains of increasing lipophilicity: prolinate (PrO⁻), phenylalaninate (PheO⁻), tryptophanate (TrpO⁻).

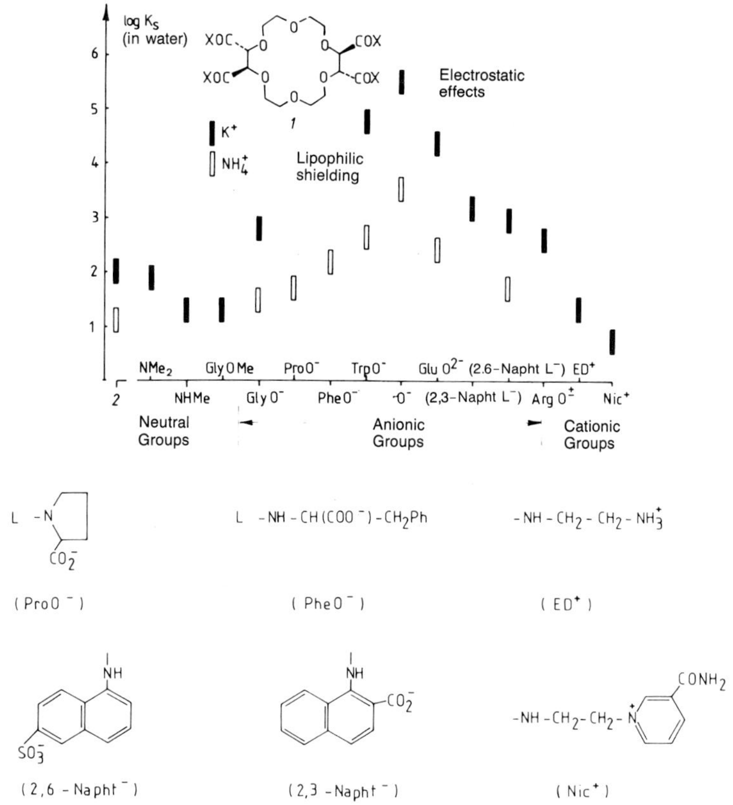

Fig. 3-118. Variation in stability of the K^+ and NH_4^+ complexes of macrocycle *1* in aqueous solution as a function of the nature of the side chain, X. Compound *2* is the macrocycle [18-O-6]. The detailed formulae are given here for those substituents not shown in Fig. 3-116.

The introduction of lipophilic groups into crowns of type *1* thus reinforces the electrostatic interactions giving rise to more efficient ligands. The effects observed are similar to those which appear to be involved in substrate binding to biological receptors, and also in active sites containing hydrophobic residues in proteins.

A series of macrocyclic polyethers with a carboxylic acid group and a monoamide group bearing a long chain (Fig. 3-119) have been synthesized. These compounds possess several of the properties necessary for efficient carriers of cations across membranes:

- they contain complexation sites (the ether groups in the crown);
- they, as well as their complexes, are lipophilic, and are thus soluble in chloroform;

- the carboxylic acid side group plays a double role: firstly, under basic conditions, it provides an additional binding site, and secondly it allows transport of cations against the concentration gradient of the transported ions as a result of a coupled counterflow of protons (see below) [271].

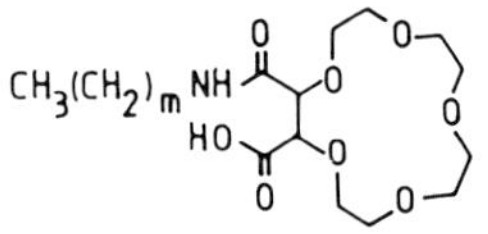

1 a m = 7
1 b m = 13
1 c m = 17

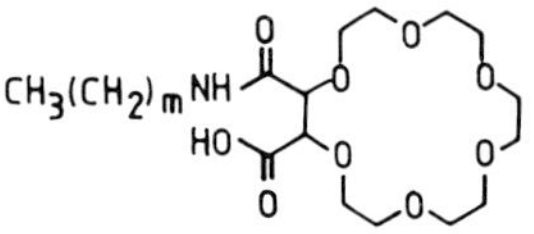

2 a m = 7
2 b m = 13
2 c m = 17
2 d m = 3

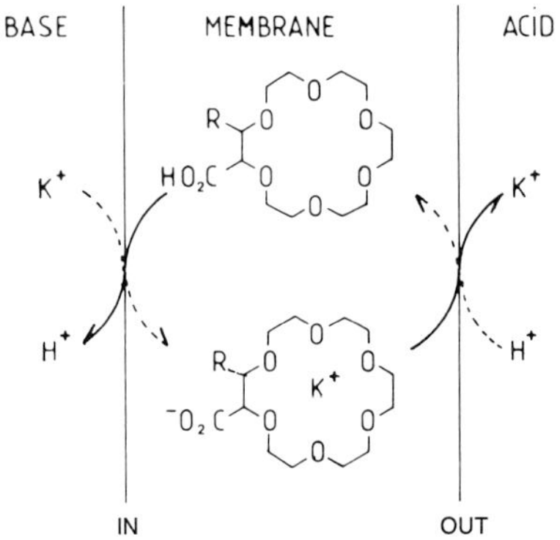

Fig. 3-119. [15-O-5] and [18-O-6] macrocycles with a carboxylic acid group and a monoamide group bearing a long-chain alkyl group; mechanism of K^+ transport and H^+ countertransport. The following reactions occur at the in and out interfaces, respectively:
$K^+(aq) + HCrn(org) \rightleftharpoons KCrn(org) + H^+(aq)$
$KCrn(org) + H^+(aq) \rightleftharpoons HCrn(org) + K^+(aq)$

The stability constants of alkali metal complexes formed with the above ligands in both the protonated, LH (carboxylic acid), and deprotonated, L (carboxylate anion), forms have been determined in methanol/water (9 : 1 v/v). The two constants are defined by the following equilibria:

$$K_1 = c(LHM)/c(LH) \cdot c(M), \quad K_2 = c(LM)/c(L) \cdot c(M)$$

The stability constants were determined in the case of ligands *1a* and *2d* using ion selective electrode (ISE) or pH-metric techniques (Table 3-18). Both methods gave very similar results.

The introduction of a charge on the ligand increases the stabilities of the sodium and potassium complexes about fivefold for [15-O-5] crown ($\Delta \log K = 0.7$ with Na^+ and 0.8 with K^+), and about 20-fold with [18-O-6] crown ($\Delta \log K = 1.1$ with Na^+ and 1.3 with K^+). Nevertheless, the selectivities are not greatly affected by the introduction of a charge.

[15-O-5]: $\Delta \log K_{(Na^+/K^+)} = 0.5$ with LH and 0.4 with L.
[18-O-6]: $\Delta \log K_{(K^+/Na^+)} = 1.2$ with LH and 1.4 with L.

Table 3-18. Stability constants, in methanol/water (9 : 1 v/v), for lipophilic ligands bearing a carboxylic acid group.

	Stability Constant	Na^+		K^+		Rb^+	
		ISE	pH	ISE	pH	ISE	pH
1a	log K_1	2.7	2.7	2.2	2.1	–	1.8
	log K_2	3.4	3.4	3.0	2.9	2.6	2.7
2d	log K_1	3.3	3.1	4.5	4.2	4.0	3.8
	log K_2	4.4	4.2	5.8	5.5	5.0	4.8

(Taken from T. M. Fyles et al., *Canad. J. Chem.*, **59**, 1724, 1981).

All these lipophilic ligands are efficient carriers of alkali metal cations across a chloroform membrane [272]. The schematic representation in Fig. 3-119 shows clearly that the transport of the cation in one direction is coupled with countertransport of H^+ in the other direction. The processes occurring at the two interfaces are given in the caption to the figure.

Other lipophilic ligands with two acid functional groups on an [18-O-6] crown have been synthesized. They were designed for the transport of divalent cations [273, 274]. Depending on the pH of the starting aqueous phase, one of these ligands transports either a univalent cation (acid pH) or a divalent cation (basic pH) [273].

i) Case of onium cations

Macrocyclic polyethers of the [18-O-6] type form complexes with $NH_4{}^+$ and RNH_3^+ primary ammonium salts [275–278]. In the case of RNH_3^+, the charged end is attached to the ligand, the R group being located above the macrocycle. Binding occurs via N-H$^+$---O electrostatic interactions (Fig. 3-120).

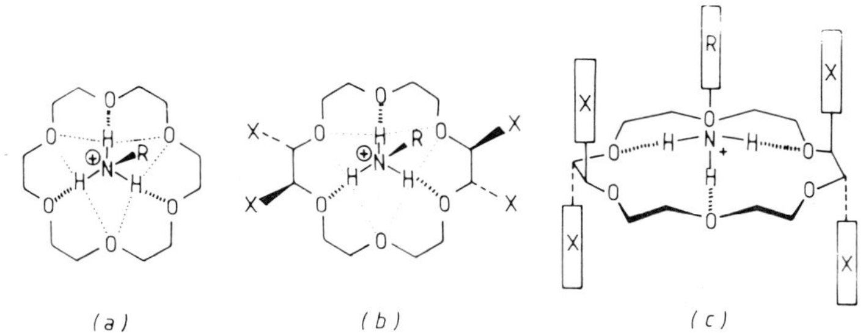

Fig. 3-120. Electrostatic interactions in complexes formed between [18-O-6] type macrocycles and RNH_3^+ primary ammonium salts.

In the complexes formed by RNH_3^+ salts with [18-O-6] macrocycles bearing lateral substituents, the R residues are probably located between the X groups (Fig. 3-120 c) with which they are able to interact [270].

The X-ray crystal structure shows that the ammonium ion is in the perching position, i.e., about 1 Å out of the median plane of the macrocycle oxygens (Fig. 3-121). The displacement from the median plane is 1 Å in the case of NH_4^+ [278 a] and 0.84 Å with $CH_3NH_3^+$ [278 b]. The perching position is quite general for ammonium cations [278 b].

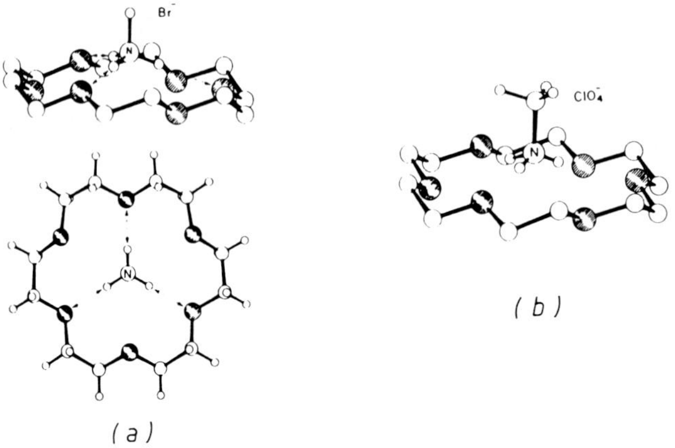

Fig. 3-121. Crystal structures of the complexes formed between [18-O-6] crown and (a) ammonium bromide and (b) methylammonium perchlorate (taken from O. Nagano et al., *Bull. Chem. Soc. Japan,* **51,** 790, 1978 and K. N. Trueblood et al., *J. Am. Chem. Soc.,* **104,** 1355, 1982).

Macrocyclic polyethers of the [27-crown-9] type (Fig. 3-122) complex guanidinium salts [279, 280]. The presence of carboxylate groups in the chiral compound *2* greatly enhances the complexing properties. The stability constant of the complex in aqueous solution is ca. 9000 $l\,mol^{-1}$ whereas that of *3* is too low to be measured [269, 280 a].

In the case of substituted guanidinium salts, the stability constants decrease very rapidly with increasing degree of substitution. For methyl-, ethyl- and tetramethylguanidinium salts they are 450, 150 and $<10\ l\ mol^{-1}$, respectively [280 a].

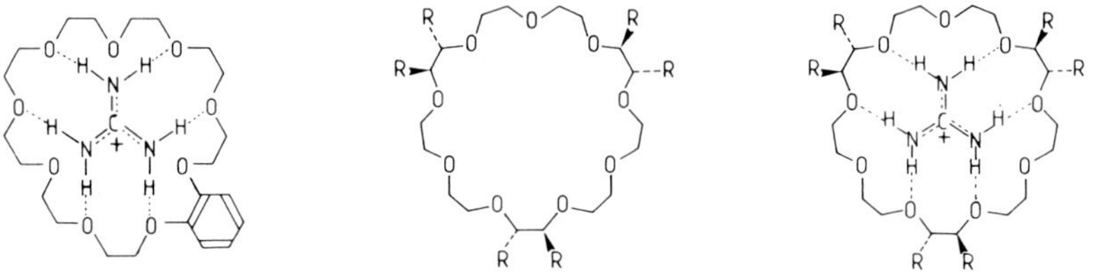

1

2 R = $CO_2^-\ {}^+HN(CH_2CH_2OH)_3$
3 R = $CONMe_2$

4

Fig. 3-122. [27-O-9] type macrocyclic polyethers and examples of complexes formed with a guanidinium salt.

The X-ray crystal structure of complex *1* has confirmed both the good receptor-substrate complementarity, and also the network of hydrogen bonds depicted in Fig. 3-122 [280b].

j) Introduction of binding sites other than oxygen

Replacement of oxygen by nitrogen. The stability of potassium complexes decreases greatly as a result of replacement of oxygen by nitrogen [253]. For each NH nitrogen site introduced, the stability constant decreases by a factor close to 100 (Table 3-19) [253].

However, when the nitrogen bears an alkyl group, the decrease in stability is much lower being only by a factor of six when the two [18-crown-6] oxygens are replaced by two NMe groups, and by a factor of ca. eight when the dibenzo[18-crown-6] oxygen is replaced by an $nC_8H_{17}N$ group [265, 253].

The presence of nitrogen atoms gives rise to stronger solvation of the macrocycle because of hydrogen bonding to the protic solvent, and hence relative destabilization of the potassium complex. In addition, the NH groups form, with the ring oxygens, internal hydrogen bonds which must be broken in order to form the complex. The difference in stability between the ligands $O_4(NMe)_2$ and $O_4(NH)_2$, $\log K_s = 5.3 - 2.0 = 3.3$, represents the energy necessary to break four NH---O bonds, that is 4 kcal/mol, or 1 kcal/mol for each hydrogen bond.

Table 3-19. Effect of replacement of oxygen by nitrogen and sulfur on stability constants of 18-crown-6 derivatives.

	Polyether[a)]		Log K_1'	
Type	A	B	K^+ in methanol	Ag^+ in water
1	O	O	6.10	1.60
1	NMe	NMe	5.30	
2	O	O	5.00	
2	NR[b)]	O	4.10	
1	NH	O	3.90	3.3
2	NH	O	3.20	
1	NH	NH	2.04	7.8
2	NH	NH	1.63	
1	S	S	1.15	4.34

a) Type 1 = $A(CH_2CH_2OCH_2CH_2OCH_2CH_2)_2B$;
Type 2 = $A[CH_2CH_2O(o\text{-}C_6H_4)OCH_2CH_2]_2B$;
b) R = $n\text{-}C_8H_{17}$
(Taken from H. K. Frensdorf, *J. Am. Chem. Soc.*, **93**, 600, 1971).

Replacement of oxygen by sulfur. The replacement of two oxygen atoms by two sulfur atoms in [18-crown-6] decreases the stability constant of the potassium complex by a factor of 10^5 (Table 3-19) [253, 254].

This large decrease is explained chiefly by the low affinity of S (soft donor) for K^+ (hard cation). In addition, the size of the cavity is modified, and there is a substantial change in the conformation of the macrocycle, the sulfur atoms having a marked tendency to point outside the ring [281] (e. g., see Fig. 1-49 giving the structure of the [18-crown-6] monothia dcrivativc).

Conversely, in the case of silver cation, a different order of stability of the complexes as a function of the nature of the sites is observed: $O_6 < O_4S_2 < O_4(NH)_2$ (log K_s is equal to 1.6, 4.3 and 7.8, respectively). In this case, because of the appreciable covalent character of the Ag^+---N and Ag^+---S bonds, electrostatic interactions do not predominate [253].

This difference in behavior between the cations can be utilized in their selective extraction.

k) Photocontrol of ion extraction and ion transport via combined complexation and irradiation

Shinkai et al. [282] set out control ion transport by means of light, by using a ligand possessing complexing properties capable of being modified by irradiation (Fig. 3-123). These workers synthesized a bis(crown ether) which, on photoirradiation, isomerized from the *trans* to the *cis* form.

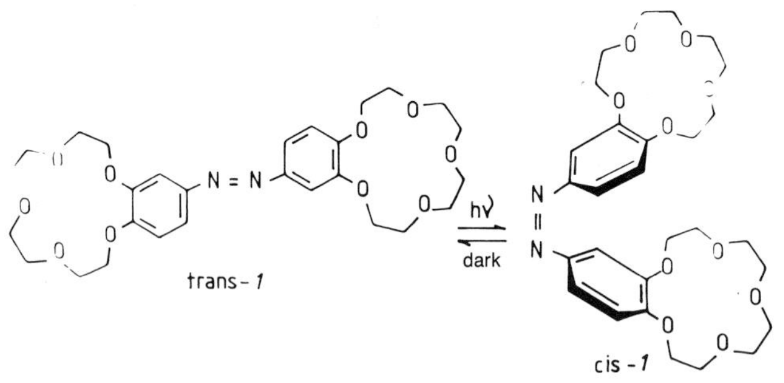

Fig. 3-123. Photoisomerization of an azo-bis(crown ether).

It was hoped that the *cis* form would complex large cations more efficiently than would the *trans* form, which indeed turned out to be so. In the presence of this ligand, extraction of alkali metal salts of methyl orange from water to *o*-dichlorobenzene showed that the *trans-1* isomer extracts Na^+ 5.6 times more efficiently than does *cis-1*, whereas *cis-1* extracts K^+ 42.5 times more efficiently than does *trans-1* (Table 3-20).

The selectivity, which is expressed by the ratio *trans/cis* of extractability (Ex) for Na^+ against that for K^+, is 238-fold. Rb^+ and Cs^+ cations are also extracted by *cis-1* more efficiently than by *trans-1*, but with lower selectivity than for K^+.

Table 3-20. Extraction of alkali metal cations (% extracted).

1	Na^+	K^+	Rb^+	Cs^+
Trans-1	29.6	1.3	29.3	24.1
photoirradiated *1*	17.1	29.0	49.6	29.9
cis-1 [a]	5.3	55.2	68.8	36.2

a) Values calculated from the composition of the photostationary state.
cis 1/trans 1 = 51.4/48.6: $Ex_{cis\text{-}1} = (Ex_{photo\text{-}1} - 0.486\ Ex_{trans\text{-}1})/0.514$
(Taken from S. Shinkai et al., *J. Am. Chem. Soc.*, **103**, 111, 1981).

Na^+ cation is extracted with a cation: crown stoichiometry of 1:1, that is 2 Na^+ per bis(crown ether) ligand. The larger cations are extracted in the form cation: crown = 1:2, corresponding to one cation per bis(crown ether) ligand (sandwich-type complex). These cation: crown = 1:1 and 1:2 associations are in complete agreement with the observations made for simple crown ethers (see Table 3-11).

The rate of cation transport across a membrane is also affected by irradiation of ligand *1*. However, the rate is strongly dependent on the anion present [282].

l) Complexing properties of spherands

We have mentioned in Part I the synthesis and some properties of these compounds (cf. Section 3.4.4f, Part I).

Spherands are macrocyclic ligands with the distinctive feature of possessing the same conformational arrangement as that in their complexes [283a–c].

Compound 1 (Fig. 3-124a) is the archetype for the family of spherands. Complexation of Li^+ occurs without substantial change in the macrocycle conformation (Fig. 3-124b).

Since the first spherand, *1*, was reported in 1979 by Cram et al., dozens of such compounds have been prepared. We shall comment briefly on the complexing properties of four spherands, and compare them to cryptands *5* and *6* (Fig. 3-125).

For the following reasons, the determination of stability constants for spherands is not straightforward: (a) spherands are soluble in only a limited number of solvents, (b) the very high stability constants prevent the use of ion selective electrodes, and (c) the absence of acid-base sites precludes the use of potentiometric methods.

Various methods for determining stability constants have been used (extraction, measurement of k_1 and k_{-1}). The extraction method is based on the study of the distribution between D_2O and $CDCl_3$ of the metal picrate M^+Pic^- ($M^+ = Li^+$, Na^+, etc.; Pic^- = picrate) both in the absence and in the presence of a ligand [283d], giving rise to the following equilibrium:

$$L + MPic \underset{k_{-1}}{\overset{k_1}{\rightleftharpoons}} LMPic$$

and hence $K = c(LMPic)/c(L) \cdot c(MPic) = k_1/k_{-1}$.

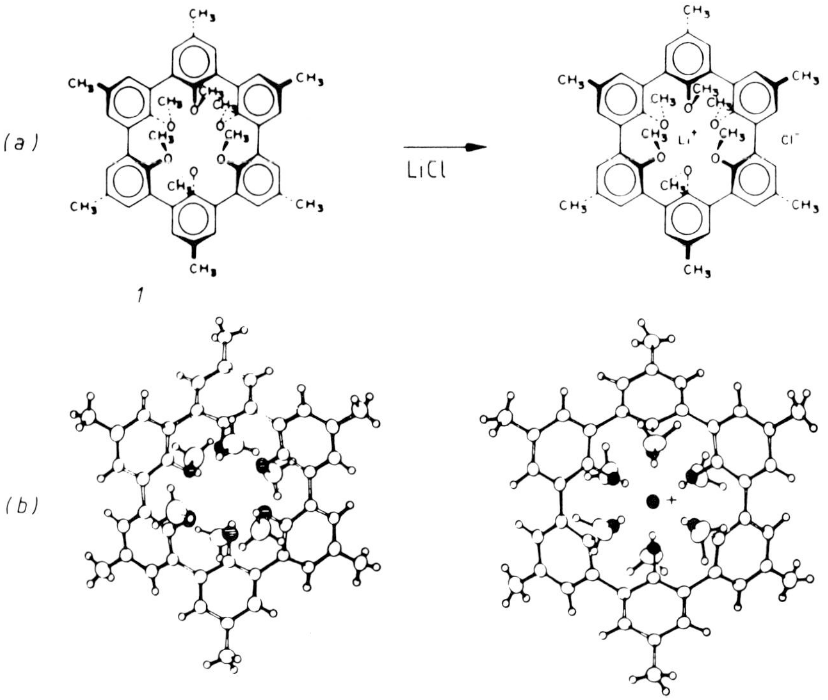

Fig. 3-124. Structures of spherand *1* and of its complex with Li^+ cation (taken from D. J. Cram, *Angew. Chem., Int. Ed. Engl.,* **25**, 1039, 1986).

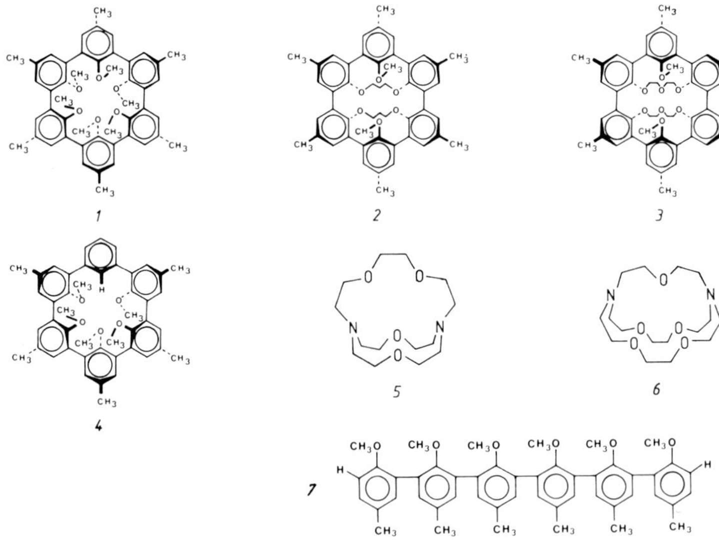

Fig. 3-125. Structures of some representative spherands *(1–4).*

Using UV and NMR spectroscopy, the concentrations of the species can be determined and the equilibrium constant K calculated. In certain cases, this equilibrium constant can also be obtained by measurement of k_1 and k_{-1} (in $CDCl_3$ saturated with D_2O in order that the results can be compared to those obtained using the extraction method). The equilibrium constants, K, and the rate constants, k_1 and k_{-1}, for the complexes of the compounds shown in Fig. 3-125 are collected in Table 3-21.

Table 3-21. Spherand stability constants. Complexation and decomplexation rate constants at 25 °C in $CDCl_3$ saturated with D_2O.

		1	*2*	*3*	*4*	*5*	*6*	*7*
Li$^+$	(a) K	$>7 \times 10^{16}$	2×10^{12}	4.4×10^{11}	4.1×10^{7}	1.7×10^{12}	–	$<2.5 \times 10^{4}$
	(b) k_1	7.5×10^{4}	3.8×10^{5}	3×10^{5}	–	4.8×10^{5} (d)	–	–
	(c) k_{-1}	$<10^{-12}$	1.9×10^{-7}	6.7×10^{-7}	–	4.4×10^{-3} (d)	–	–
Na$^+$	K	1.2×10^{14}	5.5×10^{9}	5.4×10^{13}	6.8×10^{4}	–	9.0×10^{11}	$<2.5 \times 10^{4}$
	k_1	4.1×10^{5}	1.2×10^{6}	8.6×10^{4}	–	–	1.7×10^{8} (d)	–
	k_{-1}	3.4×10^{-9}	2.2×10^{-4}	1.6×10^{-9}	–	–	2.35×10^{-2} (d)	–

a) K in l mol^{-1}; b) k_1 in l mol^{-1} sec^{-1}; c) k_{-1} in sec^{-1}; d) measurements carried out in methanol [283e].
(Taken from D. J. Cram et al., *J. Am. Chem. Soc.* **107**, 3657, 1985).

Compound *1* forms extremely stable complexes with Li^+ and Na^+. The complex (*1*, Li^+) is ca. 5×10^4-fold more stable than the complex of the cryptand selective for Li^+ (*5*), and (*1*, Na^+) is ca. 10^2-fold more stable than the complex of the cryptand selective for Na^+ (*6*).

Spherand *2* is slightly smaller than *1*. The decrease in K for (*2*, Li^+) is explained by a poorer distribution of interaction sites. The decrease in the size of the cavity greatly affects the stability constant in the case of Na^+.

Spherand *3* possesses a larger cavity giving rise to better complexation of Na^+ than of Li^+.

The subtle differences between spherands *1*, *2* and *3* results in substantial changes in Li^+/Na^+ selectivity (600 for *1*, ~360 for *2*, 1/120 for *3*). These ligands are very specific for Li^+ and Na^+. Other cations such as K^+, Rb^+, Cs^+, Ca^{2+} and Mg^{2+} do not form any complexes detectable by the method used. It follows that in the case of *1*, for example, Na^+/K^+ selectivity is estimated to be greater than 10^{10}.

Finally, the absence of one OCH_3 group in spherand *1* (compound *4*) very greatly affects the complexing power of the corresponding ligand. A decrease by a factor of ca. 10^{10} is observed both for Li^+ and for Na^+.

Comparison of spherand *1* with its open analogue *7* reveals in a dramatic way how worthwhile, from the point of view of complexation, were the synthetic efforts made to obtain *1*. The stabilities of the complexes of *1* with Li^+ and Na^+ are, respectively, ca. 10^{12}- and 10^{10}-fold greater than those observed for *7*. These differences illustrate well the influence of preorganization of the ligand, a concept which Cram has particularly emphasized [283 c]. The same principles have of course been applied to the elaboration of cryptands [162]. Because of facile rotation about the Ar – Ar and CH_3O-Ar bonds, the open compound *7* possesses numerous possible conformations (more than 1000 according to Cram). The conformation giving rise to octahedral coordination of the cation undergoing complexation is thus highly improbable and difficult to attain energetically (bond angle deformation, intraligand repulsions). Conformations affording a favorable arrangement of two or three complexation sites are more readily attained, but the resulting interaction types are, of course, of low efficiency giving rise to poor stability of the complexes ($< 2.5 \times 10^4$). It should also be noted that the open conformation of *7* permits extensive solvation of the OCH_3 groups thus making an overall negative contribution to the process of complexation.

Unlike compound *7*, spherand *1* is structured with very limited conformational mobility. In the free state, the cavity is already present and the dimensions well adapted to the Li^+ and Na^+ cations being complexed. The OCH_3 sites are not solvated. Intraligand repulsions already exist and so do not further develop on complexation. All these features explain the exceptional complexing properties of this compound and its analogues.

The rate constants k_1 and k_{-1} have been elegantly commented by Cram. For the majority of the spherands studied (to be found in Table 3-21 and also in [283 d]), the k_1 complexation rate constants fall into a narrow range (from 7.5×10^4 to 1.2×10^6 l mol^{-1} sec^{-1}). The same is not so for k_{-1} decomplexation rate constants which range from 10^{-4} to 10^{-12} sec^{-1}. It emerges from these observations that the equilibrium constants, K, are almost entirely governed by the rate constants, k_{-1}, for dissociation of the complexes. Consequently, the plot of log k_{-1} vs. log K obtained from the constants measured for various spherands is close to being linear [283 d].

Analysis of the processes of complexation and decomplexation provides an explanation for the values of k_1, and k_{-1}. On complexation, the solvated cation approaches the ligand which presents an almost entirely lipophilic region (the oxygen atoms pointing towards the cavity center). The cation therefore has to overcome this lipophilic barrier and become desolvated before it can enter the cavity. During the stages of approach and crossing of the lipophilic barrier, the nature and arrangement of the binding sites inside the cavity do not play a large part, thus explaining the fact that k_1 varies relatively little with the spherand under consideration. The opposite is so for decomplexation. If the cavity is well adapted to the cation and if the binding sites are favorably arranged (and so strongly binding), it is evident that the cation will have great difficulty in overcoming the lipophilic barrier leading to the exterior, the more so since this barrier greatly reduces participation by the solvent in the process of decomplexation. The large variations in k_{-1} and the reason why K is controlled by dissociation of the complex, can thus be readily understood. Theoretical studies have also been devoted to this interesting class of complexing agents [283 f].

m) Stability of the complexes formed by natural macrocycles

The stability constants of some natural antibiotics are collected in Table 3-22. The first three compounds (nonactin, monactin and valinomycin) exhibit high K^+/Na^+ selectivity, particularly valinomycin ($\sim 10^4$).

Table 3-22. Stability constants of complexes of natural macrocycles. Comparison with [18-O-6] crowns (MeOH solvent).

Ligands	Log K_s						
	Na^+	K^+	Rb^+	Cs^+	Ca^{2+}	Sr^{2+}	Ba^{2+}
Nonactin[a)]	2.43	3.70	3.62	2.97	–	–	1.72
Monactin[a)]	2.62	4.15	3.62	3.15	–	–	2.15
Valinomycin[b)]	0.67	4.90	5.26	4.41	2.70	2.23	3.34
Enniatin B[b)]	2.41	2.92	2.74	2.34	2.95	2.65	2.93
[18-O-6][b)]	4.32	6.10	5.35	4.79	3.90	>5.5	7.04
B [18-O-6][b)]	4.03	5.27	4.62	3.66	3.50	4.92	5.35
DB [18-O-6][b)]	4.36	5.00	4.23	3.55	–	3.55	4.28

– Temperature of measurement a) 30 °C, b) 25 °C.
– The stability constants in this Table originate from both reviews in [284].
– The formulas of the four antibiotics are given in Figs. 2-5 and 2-32.

With the larger cations, Rb^+ and Cs^+, these ligands form complexes possessing stabilities close to those observed with K^+ (the K^+/Rb^+ and K^+/Cs^+ selectivities are thus low).

These three antibiotics are characterized by the large size of the macrocycle (32-membered ring for nactins, and 36-membered ring for valinomycin). The X-ray crystal structures of the complexes (Figs. 2-11, 2-12, 2-34, 2-35) show that the ligand wraps itself round the cation. This arrangement provides an explanation for the very high K^+/Na^+ selectivity of these antibiotics as well as their low K^+/Rb^+ and K^+/Cs^+ selectivities.

For efficient complexation of the Na^+ cation (in other words in order to attain optimum Na^+---O distances for efficient interactions with all the donor sites) the ligand has to contract very greatly. This process is limited by steric factors, thus giving rise to weak complexation of Na^+. For complexation of larger cations (Rb^+, Cs^+) the ligand must expand slightly, a process which does not result in any appreciable steric hindrance thus explaining the fact that the stability constants remain large with these cations (see the discussion in Section 2.5.1 concerning the structures of nactin complexes). As regards selectivities, it should be noted that in biological systems, only the K^+/Na^+ selectivity is sizeable since Rb^+ and Cs^+ cations are only present in extremely low concentrations in nature, and so do not interfere with K^+. The low K^+/Rb^+ and K^+/Cs^+ selectivities are therefore without practical consequences.

Enniatin B (18-membered ring) forms relatively unstable complexes, and no noteworthy selectivity is observed. The properties of this antibiotic approach those of [18-O-6] crowns (with, however, lower stabilities).

Another distinctive feature for the first three antibiotics is to be noted. They all show high K^+/Ba^{2+} selectivity. This is not the case for either enniatin B or [18-O-6] crowns, which form complexes of comparable stability with K^+ and Ba^{2+}. The factors controlling M^+/M^{2+} selectivity are discussed below.

3.3.5 Stability and Selectivity of Cryptates (Macrobicyclic Complexes)

a) Diazapolyoxa macrobicyclic ligands: cryptates of alkali and alkaline earth metal cations

The stability constants of cryptates have been determined in water and in the mixture $MeOH:H_2O$ (95 : 5) using a potentiometric method, and in methanol using an ion selective electrode method (Table 3-23).

Table 3-23. Stability constants (log K_s) of cryptates in H_2O.

		Ionic radius (Å)							
	Radius of the	Li^+	Na^+	K^+	Rb^+	Cs^+	Ca^{2+}	Sr^{2+}	Ba^{2+}
Ligand	Cavity	0.60	0.95	1.33	1.48	1.69	0.99	1.13	1.35
[1.1.1]	0.5	2.2	–	–	–	–	–	–	–
[2.1.1]	0.8	5.5	3.2	<2	<2	<2	2.5	<2	<2
[2.2.1]	1.1	2.5	5.40	3.95	2.55	<2	6.95	7.35	6.30
[2.2.2]	1.4	<2	3.9	5.4	4.35	<2	4.4	8.0	9.5
[3.2.2]	1.8	<2	1.65	2.2	2.05	2.0	2.0	3.4	6.0
[3.3.2]	2.1	<2	<2	<2	<0.7	<2	2.0	2.0	3.65
[3.3.3]	2.4	<2	<2	<2	<0.5	<2	<2	<2	–

(Taken from J. M. Lehn, *Structure and Bonding,* **16**, 1, 1973).

The cryptates of alkali metal cations are amongst the most stable complexes known [162, 163, 265, 269].

These high stabilities can be attributed to the following factors:

- the preservation of ligand structure,
- the difficulty in cation solvation,
- the partial preorganization of the binding sites.

The highest stabilities are observed for the [$Li^+ \subset$ 2.1.1], [$Na^+ \subset$ 2.2.1], and [$K^+ \subset$ 2.2.2] cryptates, in each case corresponding to complementarity between the cavity and cation sizes.

Rigid-type ligands such as [2.1.1], [2.2.1] and [2.2.2] display *peak selectivity* for the optimal cation, whereas the more flexible ligands such as [3.2.2] and the larger cryptands display *plateau selectivity*. In this latter case, the discrimination between Na^+ and K^+ is high, but the stability is approximately the same with K^+, Rb^+ and Cs^+. The two types of selectivity are clearly visible in Fig. 3-126.

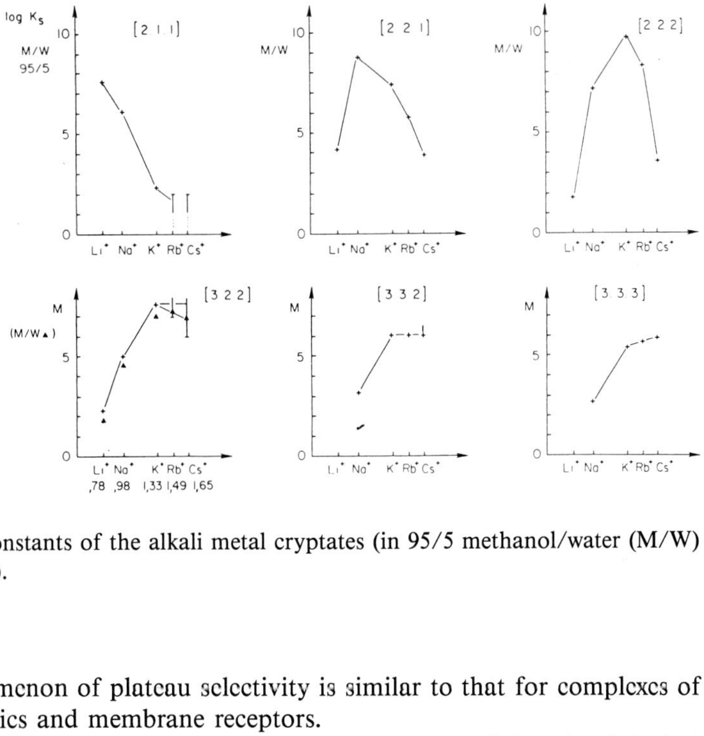

Fig. 3-126. Stability constants of the alkali metal cryptates (in 95/5 methanol/water (M/W) or neat methanol (M)).

The phenomenon of plateau selectivity is similar to that for complexes of macrocyclic antibiotics and membrane receptors.

This type of selectivity is, as we have already seen, sufficient in biological systems since the only discrimination necessary is Na^+/K^+, these two alkali metal cations being the only ones occurring to any great extent in nature. This selectivity is readily obtainable for various reasons:

- The difference in ionic radius for the above two cations is great (~0.35 Å).
- Adequate complexation of K^+ can be obtained with an uncharged polar cavity, e.g., one having a peptide structure, of low rigidity and sufficiently large for K^+, which can resist contraction for accepting Na^+ better than it can resist expansion for complexing Rb^+ and Cs^+, the latter two ions being of low abundance.

- Solvation plays a determining role; the sodium cation having a free enthalpy of solvation much greater than that of K^+ (by about 18 kcal/mol) is desolvated by an uncharged ligand with much greater difficulty than is K^+.

The high selectivity of rigid cryptands towards bulky cations arises from their denser structural frameworks and greater number of rings than for biological receptors.

The organization of the ligand is the keystone of the chemistry of molecular recognition. We have already seen the spectacular results in the case of spherands (see compounds *1* and *7* in Fig. 3-125). We have also mentioned the *macrocyclic effect* which is recalled in Fig. 3-127. A similar effect is observed for cryptands. Compounds *3* and *4* (Fig. 3-127) possess an identical number of interaction sites, but in *3* one of the strands is not connected, resulting in a decrease in stability for this compound by a factor of the order of 10^5 with respect to that for cryptand *4*.

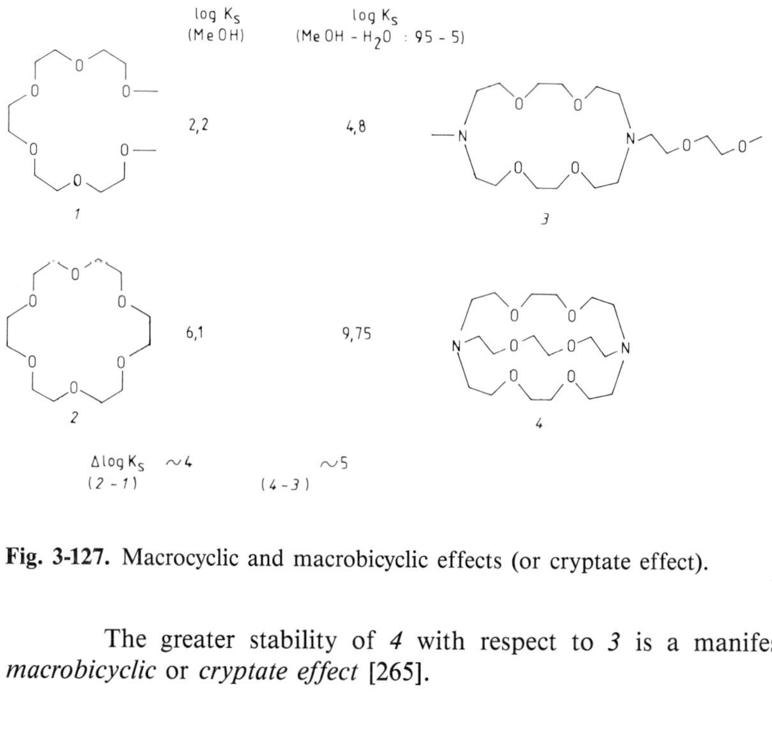

Fig. 3-127. Macrocyclic and macrobicyclic effects (or cryptate effect).

The greater stability of *4* with respect to *3* is a manifestation of the *macrobicyclic* or *cryptate effect* [265].

b) Polyaza-polyoxa macrobicyclic ligands: cryptates of transition metal and toxic heavy metal cations

Stability of the complexes. The complexes of [2.2.2], [2.2.1] and [2.1.1] cryptands with transition metals are of relatively low stability, but those with heavy metal cations such as Pb^{2+} and Cd^{2+} are very stable [163, 265, 285].

Whereas the "hard" alkali and alkaline earth metal cations are strongly bound to electron-rich polar sites, the binding of transition metal and group B metal cations is favored by "softer" nitrogen and sulfur sites. A series of cryptands has been synthesized in which some of the oxygens in [2.2.2] and [2.1.1] cryptands are replaced by nitrogen atoms (Fig. 3-128, see also figure 3.10 and ref. 174), and their complexing properties have been studied [286].

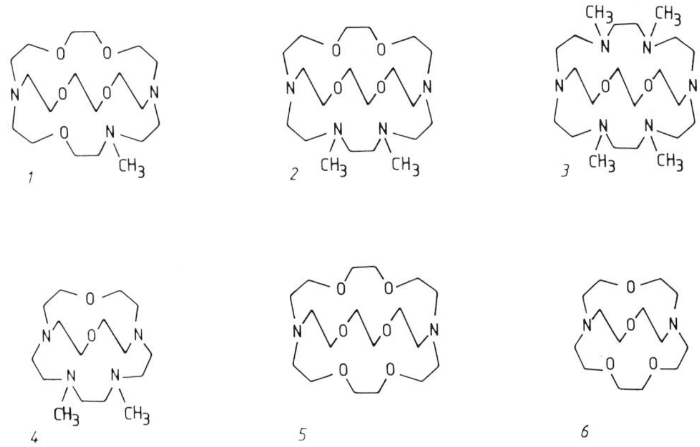

Fig. 3-128. Macrobicycles analogous to [2.2.2] and [2.2.1] cryptands, in which various oxygen atoms have been replaced by nitrogens.

The stability constants of the alkali and alkaline earth metal complexes of ligands *1* and *4* found to be lower than those for the corresponding complexes of cryptands *5* and *6*. In the series of cryptand derivatives of [2.2.2] the stabilities of the complexes fall off very rapidly (Fig. 3-129) especially in the case of compound *3*.

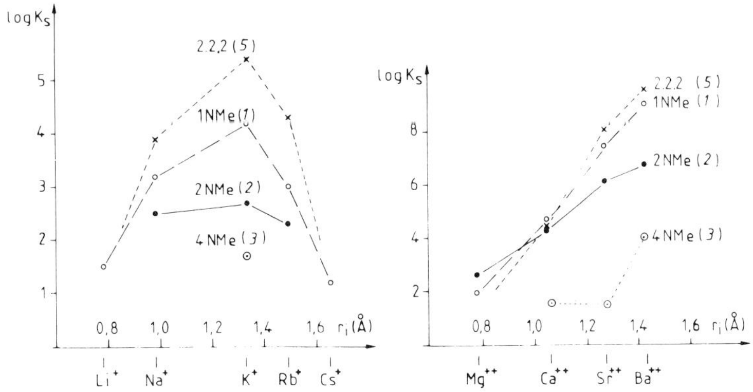

Fig. 3-129. Stability constants of the cryptates formed with alkali and alkaline earth metal cations as a function of the number of nitrogen atoms in the chains. In the case of the 4NMe macrobicycle *(3)*, the stability constants with Na^+ and Rb^+ are too small to be measurable.

Transition metal cations form the most stable complexes with the [2.1.1] analogue, tetraaza-cryptand *4*. The polyaza derivatives *1-3* of [2.2.2] in which the cavities are too large give complexes of lower stability with these cations.

In the case of transition and heavy metals, the stability depends on the number of nitrogens and the size of the cavity. It increases with the number of nitrogen sites, the effects of the first, or first two, sites introduced being the most important. The small cations Co^{2+}, Ni^{2+}, Cu^{2+} and Zn^{2+} form complexes which are appreciably more stable with ligand *4* having a small cavity than with ligands *1–3* possessing cavities which are too large. Ligands *1–4* all form very stable complexes with the heavy metal cations Cd^{2+}, Pb^{2+} and Hg^{2+} (Fig. 3-130).

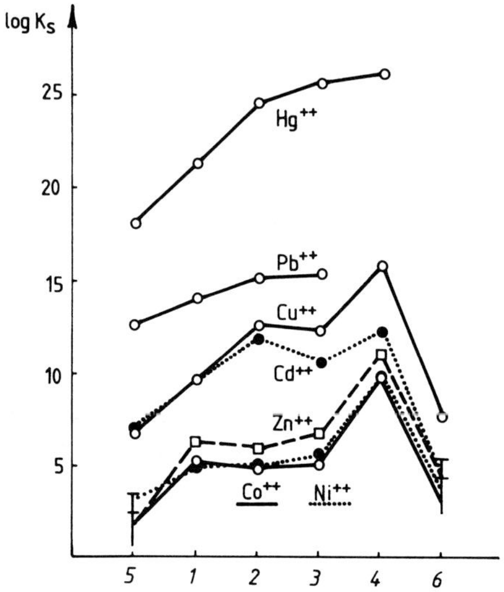

Fig. 3-130. Stability constants of the complexes formed by cryptands *1–6* with the cations Co^{2+}, Ni^{2+}, Zn^{2+}, Cd^{2+}, Cu^{2+}, Pb^{2+} and Hg^{2+}.

Selectivity of complexation. The selectivities discussed up until now principally concern cryptates of the same family of cations (alkali, alkaline earth or transition metal). With a view to various applications, it is interesting to compare the selectivities between cations of different families, and to assess the influence of ligand structure on this type of discrimination.

The "soft" cation/"hard" cation complexation selectivity increases with the number of nitrogen sites in the ligands. Some examples have led to quantification of this trend [286].

Thallium/potassium selectivity. In the study of various biological systems, K^+ has been replaced by Tl^+ which is usually bound more strongly and which provides a very useful UV and NMR spectroscopic probe. Furthermore, Tl^+ is very toxic and its specific complexation is of great interest. The Tl^+/K^+ complexation selectivities are 130, 630, 250 and 8 for the ligands *1, 2, 3* and *5*, respectively (see Fig. 3-128). The conversion of *5* into *2* via introduction of two NCH_3 sites thus gives rise to a ca. 80-fold increase in selectivity.

Cadmium/zinc and cadmium/calcium selectivity. Industrial development leads to very intensive use and wide dissemination of extremely toxic heavy metals (lead, mercury, cadmium, etc.) resulting in both acute and chronic poisoning, and hence the need for the availability of a large selection of suitable therapeutic agents. The principal drawback to the chelating agents currently used in *chelatotherapy* for cadmium. lead and mercury poisoning (ethylenediaminetetraacetic acid, D-penicillamine and 2,3-dimercaptopropan-1-ol) is their low selectivity. It should be possible to greatly improve selectivity by using more specific ligands such as cryptands, for example. Indeed, for effective treatment of heavy metal poisoning it is essential that the compound being used should complex very strongly Cd^{2+}, Hg^{2+} and Pb^{2+}, whereas the biologically important cations, Na^+, K^+, Mg^{2+}, Ca^{2+} and Zn^{2+}, should be complexed much less strongly in order to avoid the latter valuable cations being "washed out" and eliminated from the organism.

We shall illustrate this process for cadmium. The Cd^{2+} cation is much larger than Zn^{2+} but the same size as Ca^{2+}. Also, Cd^{2+} and Zn^{2+} both prefer nitrogen rather than oxygen sites, which is not the case for Ca^{2+}. Thus, the most efficient cryptand should have a cavity which is too large for Zn^{2+}, and sufficient nitrogen sites to strongly favor Cd^{2+} over Ca^{2+}.

Ligands *2* and *3* (Figs. 3-128 to 3-130) have the required properties and indeed show extremely high selectivities, these being, respectively, 10^6 and 10^4 for Cd^{2+}/Zn^{2+}, and 10^8 and 10^9 for Cd^{2+}/Ca^{2+} [286].

c) Lipophilic cryptands. Control of M^{2+}/M^+ selectivity

As we have already pointed out (see Table 3-22), the vast majority of natural ionophoric macrocycles complex more strongly K^+ than Ba^{2+} (it is legitimate to compare the stabilities of these two cations since their ionic radii are very similar, 1.33 and 1.35 Å). A very detailed study by Morf and Simon has shown the role played by the lipophilicity of these antibiotics in the control of M^{2+}/M^+ selectivity [287a].

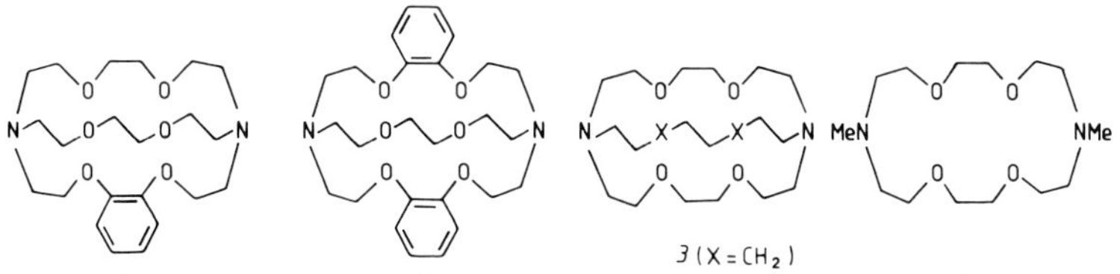

1 *2* *3* (X = CH_2) *4* (X = O) *5*

Fig. 3-131. Lipophilic cryptands *(1–3)*.

The synthetic lipophilic cryptands *1–3* (Fig. 3-131) provide support for these conclusions (with certain limits as will be seen below).

It should be noted that the lipophilicity of the ligand and of its complexes plays a very important part especially in the transfer of cations across lipid membranes. The lipophilicity of a ligand can be estimated using the lipophilicity increments determined by Hansch [287 b]. The hydrophobic constant π_x for a substituent X is given by the relationship:

$$\pi_X = \log P_X - \log P_H,$$

where P_X and P_H are, respectively, the partition coefficients for the compound RX and a reference compound RH in the solvent system 1-octanol/water.

The lipophilicity and thickness of the compounds in Fig. 3-131 are given in Table 3-24.

Table 3-24. Thickness and lipophilicity of various ligands.

	Ligand	Thickness (Å)[a)]	Lipophilicity log P
5	[2.2 NCH_3]	3.7	+0.5
4	[2.2.2]	4.1	+0.5
1	[2B.2.2]	4.25	+1.7
2	[2B.2B.2]	4.4	+2.8
3	[2.2.C8]	4.2	+3.5
	Valinomycin	6.7	+2.0
	Enniatin B	5.5	+3.7

a) Calculated from the van der Waals volumes for the subunits from which the ligand is composed. (Same source as Table 3.23).

It can be noted that: (a) compounds *5* and *4* are of low lipophilicity, (b) log *P* increases steadily in the sequence *4*<*1*<*2*<*3*, and (c) compounds *2* and *3* display lipophilicities which are comparable to those of natural antibiotics.

Influence of ligand thickness and number of interaction sites on divalent cation/univalent cation selectivity

Solvation sphere effect: The determination of thermodynamic functions ($\Delta G, \Delta H$ and ΔS) associated with transfer of an ion from the gaseous state into solution is complex [287a]. The establishment of the first coordination sphere is accompanied by multiple interactions (ion-dipole, ion-induced dipole, repulsive forces, etc.) about which we shall not enter into detail. Instead, we shall look at, in particular, the step involving transfer of the solvated ion from the gaseous state into solution:

$$(M^+, nH_2O) \text{ (gaseous)} \rightarrow (M^+, nH_2O) \text{ (solution)}.$$

Interactions between the cation and the *second* solvation sphere arise during this transfer, and the energy involved is given by the Born equation:

$$\Delta G_{\mathrm{B}} = \frac{-Z^2 e^2}{2\,a}\left(1 - \frac{1}{\varepsilon}\right)$$

where Z is the charge on the ion, ε is the dielectric constant of the solvent (78.3 in the case of water), a is the sum of the cation radius (r_c) and the thickness of the first solvation sphere (s), that is $a = r_c + s$. In the case of a complex, the s term can also be the thickness of the ligand since, as we have seen, on complexation of a cation, the first solvation sphere is replaced by the ligand. The process of cation complexation is portrayed in Fig. 3-132.

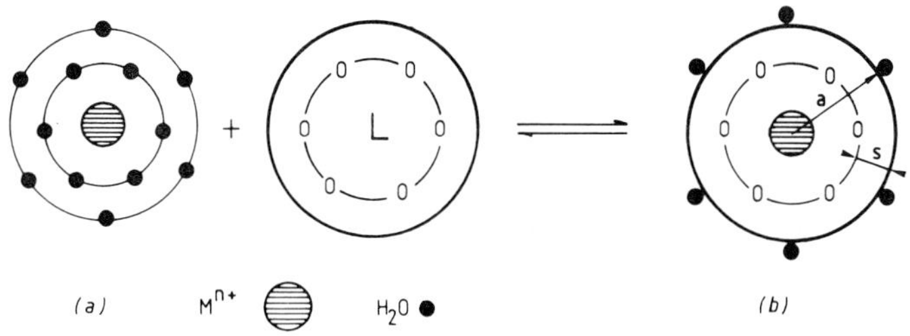

Fig. 3-132. Schematic representation of cation complexation by a ligand; the distance between the cation and the solvation sphere in the complex (b) depends on the thickness of the ligand.

It can be clearly seen that in the case of a cation in solution (Fig. 3-132 a), the distance between the cation and the second solvation sphere depends on the nature of the cation and on that of the solvent. In complexes (Fig. 3-132 b), this distance is dependent on the thickness, s, of the ligand. Depending on the value of s, the second solvation sphere will be closer to or further from the cation. Increase in the distance of the second solvation sphere from the cation (increase in a) gives rise to a decrease in the contribution of ΔG_{D} to the overall energy of the reaction, and hence a decrease in the stability of the complex. However, because of the Z^2 term, this decrease will be more marked for divalent than for univalent cations. The thickness of the ligand therefore enables M^{2+}/M^{+} selectivity to be controlled.

Cryptands *4*, *1* and *2* in Fig. 3-131 possess the same number of atoms in their chains and are of increasing lipophilicity: *4* < *1* < *2* (Table 3-24) [287c]. The trend in Ba^{2+}/K^{+} selectivity is given in Table 3-25.

The incorporation of a first benzene ring into [2.2.2] cryptand (compound *1*) does not greatly affect Ba^{2+}/K^{+} selectivity, probably because approach of the solvent at one side of the bicyclic system is not hampered. However, the presence of two benzene rings greatly decreases the selectivity ratio. The stabilities of the cryptates of Ba^{2+} and K^{+} become equal in the case of *2*, thus demonstrating the effect of ligand thickness.

Table 3-25. Stability constants (log K_s) of lipophilic cryptates in 95:5 methanol:water.

Ligand	*(4)*	*(1)*	*(2)*	*(3)*	*(5)*	Nonactin
K^+	9.45	9.05	8.6	4.35	4.38	3.6
Ba^{2+}	11.5	11.05	8.5	<2.0	6.67	1.7
Ba^{2+}/K^+	110	100	ca. 1	1/>220	200	1/80

(Taken from J. M. Lehn et al., *J. Chem. Soc., Chem. Commun.,* 15, 1973).

Several comments are, however, pertinent. The introduction of aromatic rings into the chains of [2.2.2] gives rise to various other changes apart from the increase in lipophilicity: (a) the oxygens bonded to the aromatic rings are of lower basicity and therefore of lower donor power, (b) rotation about the carbon-carbon bond in O–C–C–O is prevented when this unit is part of the aromatic system. The two oxygen atoms are therefore in the *cis* configuration. This gives rise to a decrease in the size of the macrobicycle and limited conformational flexibility, that is to say difficulty in modifying the size of the cavity to fit the cation undergoing complexation [287d].

A recent and more complete study by Buschmann illustrates well the effect of various modifications resulting from the introduction of aromatic rings [287e]. In the case of Ba^{2+}/K^+ selectivity, the same trends as those described above are observed (Table 3-26). In this study, just as in that performed in methanol/water solvent system, complexes of cryptand *2* with Ba^{2+} and K^+ are found to have similar stabilities.

Table 3-26. Stability constants (log K_s) of lipophilic cryptates in methanol.

		Na^+	K^+	Rb^+	Ca^{2+}	Sr^{2+}	Ba^{2+}	Ba^{2+}/K^+
4	[2.2.2]	7.97	10.49	9.10	8.16	11.75	12.9	260
1	[2_B.2.2]	7.50	9.21	7.19	7.04	10.32	10.99	60
2	[2_B.2_B.2]	7.60	8.74	5.91	5.96	8.83	8.85	~1
	$\frac{K_s\ [2.2.2]}{K_s\ [2_{BB}]}$	~2	~55	~1550	~160	~830	~11200	

(Taken from H. J. Buschmann, *Inorg. Chim. Acta,* **134,** 225, 1987).

The decrease in cavity size in the dibenzo compound *2*, makes this ligand more capable of complexing small cations. As a result, the decrease in stability with Na^+, and also to a certain extent with Ca^{2+}, is lower than that observed with larger ions. In the [$Na^+ \subset$ *2*] and [$Ca^{2+} \subset$ *2*] complexes, the destabilizing effects (lipophilicity, lower donor power) are compensated by the good fit of the size of the cation to that of the cavity.

The importance of the second solvation sphere has been shown by calorimetric measurements allowing calculation of $\Delta G, \Delta H$ and ΔS for the complexes of cryptands *4*, *1* and *2* with K^+ and Ba^{2+} [287e].

For the complexes of these ligands with K^+, the enthalpies and entropies of complexation vary only slightly (Table 3-27). Both ΔH and ΔS contribute to the decrease in stability, which is relatively small, accompanying the increase in lipophilicity.

Table 3-27. Ba^{2+}/K^+ selectivity[a)].

		$-\Delta G_{K^+}$	$-\Delta G_{Ba^{2+}}$	$-\Delta H_{K^+}$	$-\Delta H_{Ba^{2+}}$	$T\Delta S_{K^+}$	$T\Delta S_{Ba^{2+}}$	$T\Delta S_{Ba^{2+}} - T\Delta S_{K^+}$
[2.2.2]								
	K^+	14.27		17.94		−3.67		
4								4.73
	Ba^{2+}		17.54		16.48		1.06	
[2_B.2.2]								
	K^+	12.52		15.62		−3.10		
1								5.15
	Ba^{2+}		14.94		12.89		2.05	
[2_B.2_B.2]								
	K^+	11.89		15.82		−3.93		
2								7.95
	Ba^{2+}		12.04		8.02		4.02	

a) Thermodynamic parameters (ΔG, ΔH, $T\Delta S$ in kcal mol^{-1}) for [2.2.2], [2_B.2.2] and [2_B.2_B.2] cryptands at 25 °C in methanol.
(Same source as Table 3.26).

On the other hand, for the complexes with Ba^{2+}, the deciding factor is the enthalpy, whereas the entropy varies only slightly. The large decrease in ΔH in the case of [$Ba^{2+} \subset$ *2*] is explained by the disruption of the second solvation sphere. This can be understood by calculating, for each ligand, the difference $T\Delta S_{Ba^{2+}} - T\Delta S_{K^+}$ (Table 3-27). In the case of ligand [2.2.2], this difference is calculated to be 4.73 kcal/mol indicating that complexation of Ba^{2+} liberates four more solvent molecules than in the case of complexation of K^+ by this ligand, in agreement with the fact that Ba^{2+} is solvated to a greater extent than is K^+. With compound [2_B.2.2], the difference (5.15 kcal/mol) is more or less the same, hence incorporation of an aromatic ring only slightly disrupts the solvation. In the case of the dibenzo compound, the $T\Delta S$ difference is substantially greater (7.95 kcal/mol) corresponding to the liberation of two additional solvent molecules with respect to the previous cases. These solvent molecules therefore no longer interact with Ba^{2+}, and hence there is a large decrease in ΔH [287e].

The number of complexation sites. The replacement of two oxygen atoms in [2.2.2] cryptand by two methylene groups gives macrobicycle *3* (see Fig. 3-131). The thickness of these two ligands is approximately the same, but the lipophilicity of *3* is appreciably larger than that of *4* (see Table 3-24). Replacement of two oxygens in a bridge by two carbons decreases enormously the stability of the cryptates (Table 3-25). Because of the double charge, the complexes of alkaline earth metal cations are much more destabilized than are those of alkali metal cations. As a result, the Ba^{2+}/K^{+} selectivity is reversed by a factor of more than 10^4. Comparison of the stability constants of *3* and *5* shows that the large destabilization of the barium complex of *3* is due to its *cryptate* structure. The three bridges in the bicyclic ligand isolate the cation from the solvent much more efficiently than in the case of macrocycle *5* for which the approach of solvent to the cation can take place from either above or below the crown. Consequently, the cations in the complexes of ligand *5* can complete their solvation sphere with solvent molecules thus giving a Ba^{2+}/K^{+} ratio close to that for cryptand *4* [287c].

d) Effect of ligand size on the redox potential

As we have already pointed out for the sepulchrates (see discussion of Fig. 3-61), the ligand size has an effect on the redox potential of an $M^{n+}/M^{(n-1)+}$ couple.

On complexation of the metal by [2.2.1] and [2.2.2] cryptands, the redox potential of the Eu^{3+}/Eu^{2+} couple increases by 190 mV and 400 mV, respectively [288a]. These results indicate that the +2 oxidation state is strongly stabilized (the values of log Eu^{2+}/Eu^{3+} are equal to 3.4 with [2.2.1], and 7.1 with [2.2.2]). This stabilization can be ascribed both to a significant increase in the ionic radius of europium through reduction to the Eu^{2+} state, resulting in a better fit of the cation into the cryptate cavity, and also to the fact that greater desolvation is needed for complexation of Eu^{3+}, the latter being very strongly hydrated. The application of the complexing properties of crown ethers and cryptands to the rare earths has been very extensively studied [288b].

e) Thermodynamics of complexation by cryptands

The enthalpies and entropies of complexation of various cations by macrocyclic and macrobicyclic ligands in water have been determined [285a, 289].

Some examples are given in Table 3-28. For complexes with K^{+}, it can be noted that in all cases the stability is enthalpic in origin, the entropy term being always unfavorable.

A negative change in entropy on complexation can be the result of rearrangement in the structure of the water molecules surrounding the complexed cation. The formation of a cryptate converts a small inorganic cation into a large hydrophobic organic cation in which the water of solvation is more highly structured [290]. The NBu_4^+ cation can be taken as a model for hydrophobic cations. The change in entropy for the process $M^+ \rightarrow NBu_4^+$ may be estimated from the difference in the entropy of transfer to water of M^+ and NBu_4^+. Thus, the difference in the entropy of transfer is −24 e.u. for $K^+ \rightarrow NBu_4^+$, K^+ being structure-breaking and NBu_4^+

structure-making [291]. Another factor which may also contribute to the decrease in ΔS, is the loss in degrees of freedom of the ligand on complexation.

Table 3-28. Free energies, enthalpies and entropies of complexation[a)], at 25 °C in water.

Ligand		Cation						
		Li^+	Na^+	K^+	Rb^+	Ca^{2+}	Sr^{2+}	Ba^{2+}
[2.2.1][b)]	$-\Delta G_c$	3.4	7.2	5.4	3.45	9.5	10.0	8.6
	$-\Delta H_c$	0.0	5.35	6.8	5.4	2.9	6.1	6.3
	$T\Delta S_c$	3.4	1.85	− 1.4	− 1.95	6.6	3.9	2.3
	$+\Delta S_c$	11.4	6.2	− 4.7	− 6.5	22	13.1	7.7
[2.2.2][b)]	$-\Delta G_c$		5.3	7.2	5.9	6	10.9	12.9
	$-\Delta H_c$		7.4	11.4	11.8	0.2	10.3	14.1
	$T\Delta S_c$		−2.1	− 4.2	− 5.9	5.8	0.6	− 1.2
	$+\Delta S_c$		−7	−14.1	−19.8	19.5	2	− 4.0
[18-C-6][c)]	$-\Delta G_c$			2.8			3.7	5.3
	$-\Delta H_c$			6.2			3.6	7.6
	$T\Delta S_c$			− 3.4			0.1	− 2.3
	$+\Delta S_c$			−11.4			0.3	− 7.7

a) ΔG_c, ΔH_c, $T\Delta S_c$ in kcal mol^{-1}; ΔS in entropy units
b) (Taken from J. M. Lehn et al., *Helv. Chim. Acta,* **59,** 1099, 1976).
c) (Taken from R. M. Izatt et al., *J. Am. Chem. Soc.* **98,** 7620, 1976).

Selectivity of cryptands. Plots of the enthalpy of complexation of cryptands as a function of the cation show selectivity peaks, and the curves are approximately the same as those for the free energies (Fig. 3-133). Conversely, the entropies of complexation decrease steadily in always the same sequence becoming less positive or more negative as the cation becomes larger or more weakly charged. This is in agreement with the greater gain in entropy expected for small cations of high charge density for which the entropies of hydration are large [289].

The selectivity peaks observed for the stability constants of cryptates are therefore almost entirely enthalpic in origin. However, with small cations (Li^+, Ca^{2+}) the entropy contribution is large ([$Ca^{2+} \subset$ 2.2.1]) or even determining ([$Li^+ \subset$ 2.2.1], [$Ca^{2+} \subset$ 2.2.2]), the enthalpy term being close to zero for these latter two complexes.

The M^{2+}/M^+ selectivity is completely different in terms of free energy as compared to enthalpy alone. Whereas cryptates of divalent cations are much more stable than those of univalent cations with the same radii, their enthalpies of complexation are similar or smaller so that the enthalpic M^{2+}/M^+ selectivities may be reversed.

Effect of enthalpy-entropy compensation. Several different studies have shown that for a given ligand the stability ($-\Delta G$) of the complexes formed with various cations

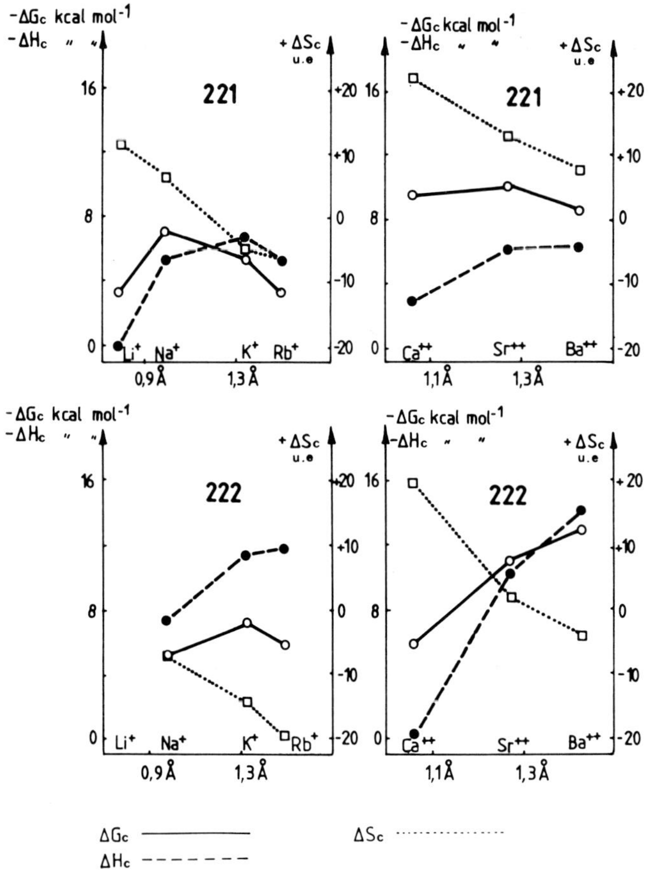

Fig. 3-133. Free energies $-\Delta G_c$ (solid line), enthalpies $-\Delta H_c$ (dashed line) and entropies $+\Delta S_c$ (dotted line) of cryptate formation by alkali and alkaline earth metal cations with [2.2.1] and [2.2.2] ligands in water at 25 °C.

arises from enthalpy-entropy compensation [292 a, b]. This phenomenon is also observed for cryptates. Thus, in the case of [2.2.1] cryptand, the complexes with Li^+ and Rb^+ have similar stabilities, but for different reasons (Table 3-28). With Li^+, $-\Delta H = O$, and $T\Delta S$ (favorable) is equal to $+3.4$ kcal/mol. With Rb^+, the high enthalpy, $-\Delta H = +5.4$ kcal/mol, is counterbalanced by an unfavorable entropy term, $T\Delta S = -1.95$ kcal/mol. The same is so with the alkaline earth cations, the increase in $-\Delta$H proceeding from Ca^{2+} to Ba^{2+} being compensated by a decrease in $T\Delta S$. Similar trends are observed with [2.2.2] complexes.

By analyzing the literature data, Inoue et al. have established that there is a linear relationship between $T\Delta S$ and ΔH [292 c]. For each type of ligand (glyme, podand, crown ether, cryptand, macrocyclic antibiotic) the straight line, $T\Delta S = f(\Delta H)$, is of characteristic slope, α (Fig. 3-134).

The three straight lines shown are those for podands (P), macrocyclic systems (CE), and cryptands (C). It can be seen that the unfavorable contribution

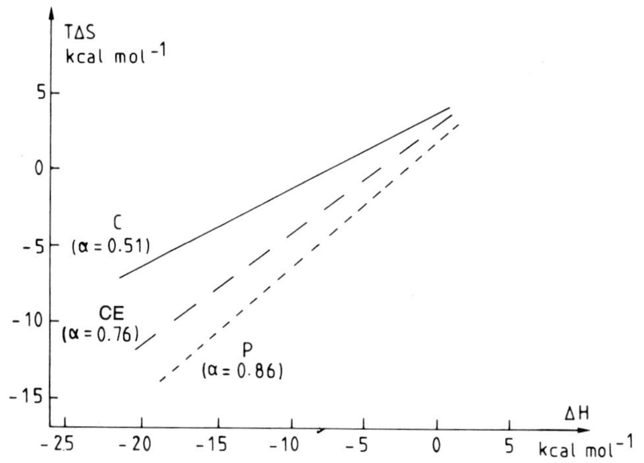

Fig. 3-134. Plots of $T\Delta S$ vs. ΔH for cryptands (C), crown ethers (CE) and podands (P). The three straight lines were obtained by linear regression analysis of experimental results for the three types of compound (39, 207 and 23 points for C, CE and P, respectively). Complexes formed with a wide variety of cations were considered: alkali metals, Ca^{2+}, Sr^{2+}, Ba^{2+}, Ag^{+}, Pb^{2+}, Tl^{+}, NH_4^+, Hg^{2+}, Co^{2+}, Eu^{3+}, Cd^{2+}, etc. NB.: Fewer than 5% of the complexes show severe deviations from the regression line.

($T\Delta S < 0$) of the entropy term decreases in the order P > CE > C. This sequence is attributed to preorganization of the ligand; the degrees of freedom in an acyclic system must be considerably reduced when it forms the cyclic structure required for complexation. This decrease in degrees of freedom is evidently less important for macrocyclic compounds, and even more so for bicyclic ligands, since in these systems the conformation and arrangement of interaction sites in the free state are more similar to those in the complexed state. In all three cases (P, CE, C), it can also be noted that, for $\Delta H = 0$, the $T\Delta S$ term is positive, the formation of a complex thus being due to a $T\Delta S > 0$ entropy contribution (see examples in Table 3-28).

f) Applications of selectivity of cation complexation by macrobicyclic cryptands

The recovery and purification of metals from waste materials is becoming of increasing importance because of pollution problems and increases in the costs of raw materials. In certain specific cases, cryptands can be of use. Thus, [2.2.2] cryptand provides an extraction method for the purification of cesium, in particular ^{137}Cs [163]. It has also been used for the decorporation of radioactive strontium [293].

Isotopic separation can be envisaged, and remarkable results have been obtained for $^{22}Na/^{24}Na$ with [2.2.1] cryptand in methanol [294a]. Other separations such as $^{6}Li/^{7}Li$ and $^{44}Ca/^{40}Ca$ have also been carried out [294b].

Further information on various aspects of macrocyclic and macrobicyclic chemistry is to be found in a variety of reviews, namely those concerning kinetics [68, 295], thermodynamics [295], and complexation of ammonium cations [296]. Several reviews and works on complexation by macrocycles have appeared recently [297].

There are numerous applications of macrocyclic and macrobicyclic complexing agents in the following fields: ion transport [298], analytical chemistry [299], chemical reactivity [300], and stabilization of anionic species (electrides, alkalides and polymetallic anions) [301].

As a general rule, addition of an efficient complexing agent to a system containing a metal cation gives rise to substantial changes in the system, thus explaining the wide extent of applications of macrocyclic complexing agents.

References for Part II

1. G. R. Painter, B. C. Pressman, *"Host Guest Complex Chemistry II"*, Ed. F. Vögtle Springer Verlag, 1982, p. 83.
2. C. J. Pedersen, *J. Am. Chem. Soc.,* **89,** 2495 and 7017, 1967; **92,** 391, 1970; *J. Org. Chem.,* **36,** 1690, 1971. F. Vögtle, E. Weber, *Kontakte (Merck),* **1,** 11, 1977; C. J. Pedersen. *"Synthetic Multidentate Macrocyclic Componds"* Eds. R. M. Izatt, J. J. Christensen, Academic Press New-York, 1978, p. 1.
3. D. Bright, M. R. Truter, *J. Chem. Soc.,* B, 1544, 1970; Nature, **225,** 176, 1970.
4. R. Hilgenfeld, W. Saenger, *Angew. Chem. Int. Ed. Engl.,* **20,** 1045, 1981.
5. M. A. Bush, M. R. Truter, *J. Chem. Soc. Perkin II,* 345, 1972.
6. J. Hašek, K. Huml, D. Hlavatá, *Acta Cryst,* **B35,** 330, 1979.
7. J. D. Owen, M. R. Truter, J. N. Wingfield, *Acta Crst.,* **C40,** 1515, 1984.
8. a) W. Klyne, V. Prelog, *Experientia,* **16,** 521, 1960; b) J. D. Dunitz, P. Seiler, *Acta Cryst,* **B30,** 2739, 1974.
9. J. D. Dunitz, M. Dobler, P. Seiler, R. P. Phizackerley, *Acta Cryst.,* **B30,** 2733, 1974.
10. A. Knöchel, J. Kopf, J. Oehler, G. Rudolph, *J. Chem. Soc., Chem. Comm.,* 595, 1978.
11. P. Groth, *Acta Chem. Scand.,* **A36,** 109, 1982.
12. M. A. Bush, M. R. Truter, *J. Chem. Soc., Perkin II,* 341, 1972.
13. P. R. Mallinson, M. R. Truter, *J. Chem. Soc., Perkin II,* 1818, 1972.
14. a) F. P. van Remoortere, F. P. Boer, *Inorg. Chem.,* **13,** 2071, 1974; b) F. P. Boer, M. A. Neuman, F. P. van Remoortere, E. C. Steiner, *Inorg, Chem.,* **13,** 2826, 1974; c) E. Mason, H. A Eick, *Acta Cryst.,* **B38,** 1821, 1982.
15. M. R. Truter, *Structure and Bonding,* **16,** 71, 1973.
16. a) M. Bourgoin, K. H. Wong, J. Y. Hui, J. Smid, *J. Am. Chem. Soc.,* **97,** 3462, 1975; b) K. H Wong, H. L. Ng, *Tetrahedron Letters,* 4295, 1979.
17. a) M. J. Calverley, J. Dale, *J. Chem. Soc., Chem. Comm.,* 684, 1981; b) P. Groth, *Acta Chem. Scand.,* **A38,** 337 and 342, 1984.
18. M. Dobler, R. P. Phizackerley, *Acta Cryst.,* **B30,** 2748 and 2746, 1974.
19. M. Dobler, J. D. Dunitz, P. Seiler, *Acta Cryst.,* **B30,** 2741, 1974.
20. W. S. Sheldrick, J. Kroner, F. Zwaschka, A. Schmidpeter, *Angew. Chem. Int. Ed. Engl.,* **18,** 934, 1979.
21. J. Hašek, D. Hlavatá, R. Huml, *Acta Cryst.,* **B36,** 1782, 1980.
22. D. L. Hughes, *J. Chem. Soc., Dalton,* 2374, 1975,
23. D. E. Fenton, M. Mercer, N. S. Poonia, M. R. Truter, *J. Chem. Soc., Chem. Comm.,* 66, 1972.
24. J. D. Owen, *Acta Cryst.,* **C40,** 951, 1984.
25. N. S. Poonia, *J. Am. Chem. Soc.,* **96,** 1012, 1974; N. S. Poonia, A. V. Bajaj, *Chem. Rev.,* **79,** 389, 1979.
26. C. Riche, C. Pascard-Billy, C. Cambillau, G. Bram, *J. Chem. Soc., Chem. Comm.,* 183, 1977.
27. D. L. Ward, A. I. Popov, N. S. Poonia, *Acta Cryst.,* **C40,** 238, 1984.
28. J. D. Owen, *J. Chem. Soc., Dalton,* 1418, 1978.
29. P. D. Cradwick, N. S. Poonia, *Acta Cryst.,* **B33,** 197, 1977.
30. D. L. Ward, A. I. Popov, N. S. Poonia, *Acta Cryst.,* **C40,** 1183, 1984.
31. C. M. Means, N. C. Means, S. G. Bott, J. L. Atwood, *J. Am. Chem. Soc.,* **106,** 7627, 1984.

32. V. W. Bhagwat, H. Manohar, N. S. Poonia, *Inorg. Nucl. Chem. Letters,* **16,** 373, 1980 and references therein.
33. M. Czugler, E. Weber, *J. Chem. Soc., Chem. Comm.,* 472, 1981.
34. a) D. L. Hughes, J. N. Wingfield, *J. Chem. Soc., Chem. Comm.,* 804, 1977; b) D. L. Hughes, C. L. Mortimer, M. R. Truter, *Acta Cryst.,* **B34,** 800, 1978.
35. M. A. Neuman, E. C. Steiner, F. P van Remoortere, F. P. Boer, *Inorg. Chem.,* **14,** 734, 1975.
36. J. P. Behr, J. M. Lehn, D. Moras, J. C. Thierry, *J. Am. Chem. Soc.,* **103,** 701, 1981.
37. a) J. P. Behr, J. M. Lehn, A. C. Dock, D. Moras, *Nature,* **295,** 526, 1982; b) A. C. Dock, D. Moras, J. P. Behr, J. M. Lehn, *Acta Cryst.,* **C39,** 1001, 1983.
38. a) R. A. Schultz, D. M. Dishong, G. W. Gokel, *Tetrahedron Letters,* 2623, 1981; F. R. Fronczek, V. J. Gatto, R. A. Schultz, S. J. Jungk, W. J. Colucci, R. D. Gandour, G. W. Gokel, *J. Am. Chem. Soc.,* **105,** 6717, 1983; b) R. A. Schultz, B. D. White, D. M. Dishong, K. A. Arnold, G. W. Gokel, *J. Am. Chem. Soc.,* **107,** 6659, 1985; R. D. Gandour, F. R. Fronczek, V. J. Gatto, C. Minganti, R. A. Schultz, B. D. White, K. A. Arnold, D. Mazzocchi, S. R. Miller, G. W. Gokel, *J. Am. Chem. Soc.,* **108,** 4078, 1986.
39. S. Buoen, J. Dale, P. Groth, J. Krane, *J. Chem. Soc., Chem. Comm.,* 1172, 1982.
40. G. Shoham, D. W. Christianson, R. A. Bartsch, G. S. Heo, V. Olsher, W. N. Lipscomb, *J. Am. Chem. Soc.,* **106,** 1280, 1984.
41. M. Dobler *"Ionophores and their Structures",* John Wiley and Sons, New York 1981; R. Hilgenfeld, W. Saenger, *Topics Current Chem.,* **101,** 1, 1982; *"Host Guest Complex Chemistry. Macrocycles"*, Eds. F. Vögtle, E. Weber, Springer-Verlag Berlin 1985, p. 43.
42. H. M. Colquhoun, J. F. Stoddart, D. J. Williams, *J. Chem. Soc., Chem. Comm.,* 847, 1981.
43. H. M. Colquhoun, J. F. Stoddart, D. J. Williams, *J. Chem. Soc., Chem. Comm.,* 849, 1981.
44. H. M. Colquhoun, J. F. Stoddart, D. J. Williams, J. B Wolstenholme, R. Zarzycki, *Angew. Chem. Int. Ed. Engl.* **20,** 1051, 1981; H. M. Colquhoun, S. M. Doughty, J. M. Maud, J. F. Stoddart, D. J. Williams, J. B. Wolstenholme, *Israel J. of Chem.,* **25,** 15, 1985.
45. H. M. Colquhoun, J. F. Stoddart, D. J. Williams, *Angew. Chem. Int. Ed. Engl.,* **25,** 487, 1986; J. F. Stoddart, *Pure and Appl. Chem.,* **60,** 467, 1988.
46. J. P. Behr, P. Dumas, D. Moras, *J. Am. Chem. Soc.,* **104,** 4540, 1982.
47. D. Britton, M. K. Chantooni, W. J. Wang, I. M. Kolthoff, *Acta Cryst.,* **C40,** 1584, 1984.
48. D. L. Hughes, M. R. Truter, *J. Chem. Soc., Chem. Comm.,* 727, 1982.
49. H. Stetter, J. Marx, *Ann. Chem.,* **607,** 59, 1957.
50. B. Dietrich, J. M. Lehn, J. P. Sauvage, *Tetrahedron Letters,* 2885, 1969.
51. J. E. Richman, T. J. Atkins, *J. Am. Chem. Soc.,* **96,** 2268, 1974.
52. A. E. Martin, J. E. Bulkowski, *J. Org. Chem.,* **47,** 415, 1982; B. Dietrich, M. W. Hosseini, J. M. Lehn, R. B. Sessions, *Helv. Chim. Acta,* **66,** 1262, 1983; M. W. Hosseini, J. M. Lehn, *J. Am. Chem. Soc.,* **104,** 3525, 1982; *Helv. Chim. Acta,* **69,** 587, 1986.
53. B. L. Shaw, *J. Am. Chem. Soc.,* **97,** 3856, 1975.
54. J. Comarmond, B. Dietrich, J. M. Lehn, R. Louis, *J. Chem. Soc., Chem. Comm.,* 74, 1985; A. Bencini, A. Bianchi, E. Garcia-Espana, M. Giusti, M. Micheloni, P. Paoletti, *Inorg. Chem.,* **26,** 681, 1987; J. E. Bulkowski, W. E. Summers III *"Copper Coordination Chemistry: Biochemical and Inorganic Perspectives"*. Eds K. D. Karlin, J. Zubieta, Adenine Press New-York, 1983, p. 445.
55. a) B. Dietrich, M. W. Hosseini, J. M. Lehn, R. B. Sessions, *J. Am. Chem. Soc.,* **103,** 1282, 1981, see also [141], Part I; B. Dietrich, *"Inclusion Compounds",* Eds. J. L. Atwood, J. E. D. Davies, D. D. MacNicol, Academic Press, vol. 2, 1984, p. 337; b) K. E. Krakowiak, J. S. Bradshaw, D.J. Zamecka-Krakowiak, *Chem. Rev.,* **89,** 929, 1989.

56. M. Herceg, R. Weiss, *Bull. Soc. Chim. Fr.,* 549, 1972.
57. D. Moras, B. Metz, M. Herceg, R. Weiss, *Bull. Soc. Chim. Fr.,* 551, 1972.
58. M. Herceg, R. Weiss, *Acta Cryst.,* **B29,** 542, 1973.
59. J. S. Bradshaw, J. Y. Hui, Y. Chan, B. L. Haymore, R. M. Izatt, J. J. Christensen, *J. Heterocycl. Chem.,* **11,** 45, 1974.
60. J. R. Dann, P. P. Chiesa, J. W. Gates Jr., *J. Org. Chem.,* **26,** 1991, 1961.
61. M. L. Campbell, N. K. Dalley, R. M. Izatt, J. D. Lamb, *Acta Cryst.,* **B37,** 1664, 1981.
62. J. C. Huffman, M. L. Campbell, N. K. Dalley, S. B. Larson, *Acta Cryst.,* **B37,** 1739, 1981.
63. M. L. Campbell, S. B. Larson, N. K. Dalley, *Acta Cryst.,* **B37,** 1741, 1744, 1750, 1981; M. L. Campbell, N. K. Dalley, S. H. Simonsen, *ibid,* 1747, 1981.
64. N. K. Dalley, R. M. Izatt, J. J. Christensen, *Fourth Int. Congress of Heterocyclic Chemistry Proceedings,* 211 Salt Lake City, Utah, 1973.
65. L. Rossa, F. Vögtle, *Topics in Current Chem.,* **113,** 1, 1983.
66. A. J. Blake, R. O. Gould, A. J. Lavery, M. Schröder, *Angew. Chem. Int. Ed. Engl.,* **25,** 274, 1986; J. A. R. Hartman, E. J. Hintsa, S. R. Cooper, *J. Am. Chem. Soc.,* **108,** 1208, 1986; J. A. R. Hartman, S. R. Cooper, *J. Am. Chem. Soc.,* **108,** 1202, 1986; M. N. Bell, A. J. Blake, M. Schröder, T. A. Stephenson, *J. Chem. Soc., Chem. Comm.,* 471, 1986; G. Reid, M. Schröder, *Chem. Soc. Rev.,* **19,** 239, 1990.
67. a) A. E. Martin, S. J. Lippard, *"Copper Coordination Chemistry: Biochemical and Inorganic Perspectives"* Eds. K. D. Karlin, J. Zubieta, Adenine Press, New York, 1983, p. 395.; Y. L. Agnus, *ibid,* p. 371; S. R. Cooper, *Acc. Chem. Res.,* **21,** 141, 1988; D. B. Rorabacher, M. M. Bernardo, A. M. Q. van de Linde, G. H. Leggett, B. C. Westerby, M. J. Martin, L. A. Ochrymowycz, *Pure and Appl. Chem.,* **60,** 501, 1988; T. Yoshida, T. Adachi, T. Ueda, *ibid,* **62,** 1127, 1990; M. Schröder, *Pure and Appl. Chem.,* **60,** 517, 1988; b) K. Wieghardt, *Pure and Appl. Chem.,* **60,** 509, 1988; R. Bhula, P. Osvath, D. C. Weatherburn, *Coord. Chem. Rev.,* **91,** 89, 1988.
68. W. Burgermeister, R. Winkler-Oswatitsch, *Topics in Current Chem.,* **69,** 91, 1977.
69. R. Hilgenfeld, W. Saenger, *Topics in Current Chem.,* **101,** 1, 1982.
70. C. Ghélis, J. Yon *"Protein Folding"* Eds. B. Horecker, N. O. Kaplan, J. Marmur, H. A. Scheraga, Academic Press, 1982.
71. G. D. Rose, L. M. Gierasch, J. A. Smith, *Advances in Protein Chemistry,* **37,** 1, 1985.
72. H. Brockmann, G. Schmidt-Kastner, *Chem. Ber.,* **88,** 57, 1955.
73. M. M. Shemyakin, N. A. Aldanova, E. I. Vinogradova, M. Y. Feigina, *Tetrahedron Letters,* 1921, 1963.
74. W. McMurray, R. W. Begg, *Arch. Biochem. Biophys.,* **84,** 546, 1959.
75. C. Moore, B. C. Pressman, *Biochem. Biophys. Res. Comm.,* **15,** 562, 1964.
76. B. C. Pressman, *Proc. Natl. Acad. Sci. U.S.A.,* **53,** 1076, 1965.
77. P. Mueller, D. O. Rudin, *Biochem. Biophys. Res. Comm.,* **26,** 398, 1967.
78. B. C. Pressman, E. J. Harris, W. S. Jagger, J. H. Johnson, *Proc. Natl. Acad. Sci. U.S.A.,* **58,** 1949, 1967.
79. R. L. Harned, P. H. Hidy, C. J. Corum, K. L. Jones, *Antibiot. Chemotherap.,* 594, 1951.
80. a) L. A. R. Pioda, H. A. Wachter, R. E. Dohner, W. Simon, *Helv. Chim. Acta,* **50,** 1373, 1967; b) B. T. Kilbourn, J. D. Dunitz, L. A. R. Pioda, W. Simon, *J. Mol. Biol.,* **30,** 559, 1967.
81. D. B. Davies, M. Abu Khaled, *J. Chem. Soc., Perkin II,* 1327, 1976.
82. Y. A. Ovchinnikov, V. T. Ivanov, *Tetrahedron,* **30,** 1871, 1974.
83. D. J. Patel, A. E. Tonelli, *Biochem.,* **12,** 486, 1973.
84. D. J. Patel, *Biochem.,* **12,** 496, 1973.

85. J. D. Glickson, S. L. Gordon, T. P. Pitner, D. G. Agresti, R. Walter, *Biochem.,* **15,** 5721, 1976.
86. K. L. Servis, D. J. Patel, *Tetrahedron,* **31,** 1359, 1975.
87. K. J. Rothschild, I. M. Asher, H. E. Stanley, E. Anastassakis, *J. Am. Chem. Soc.,* **99,** 2032, 1977.
88. G. W. Feigenson, P. R. Meers, *Nature,* **283,** 313, 1980.
89. E. Grell, T. Funck, *J. Supramol. Struct.,* **1,** 307, 1973.
90. V. T. Ivanov, T. A. Laine, N. O. Abdulaev, L. B. Senyavina, E. M. Popov, Y. A. Ovchinnikov, M. M. Shemyakin, *Biochem. Biophys. Res. Comm.,* **34,** 803, 1969.
91. B. Maigret, B. Pullman, *Theoret. Chem. Acta,* **37,** 17, 1975.
92. W. L. Duax, H. Hauptman, C. M. Weeks, D. A. Norton, *Science,* **176,** 911, 1972.
93. W. L. Duax, H. Hauptman, *Acta Cryst.,* **B28,** 2912, 1972.
94. I. L. Karle, *J. Am. Chem. Soc.,* **97,** 4379, 1975.
95. G. D. Smith, W. L. Duax, D. A. Langs, G. T. DeTitta, J. W. Edmonds, D. C. Rohrer, C. M. Weeks, *J. Am. Chem. Soc.,* **97,** 7242, 1975.
96. M. Pinkerton, L. K. Steinrauf, P. Dawkins, *Biochim. Biophys. Res. Comm.,* **35,** 512, 1969.
97. K. Neupert-Laves, M. Dobler, *Helv. Chim. Acta,* **58,** 432, 1975.
98. Y. A. Ovchinnikov, V. T. Ivanov, A. M. Shkrob, *"Membrane-active complexones",* BBA Library, vol. XII, Elsevier, Amsterdam 1974.
99. J. A. Hamilton, M. N. Sabesan, L. K. Steinrauf, *J. Am. Chem. Soc.,* **103,** 5880, 1981.
100. L. K. Steinrauf, J. A. Hamilton, M. N. Sabesan, *J. Am. Chem. Soc.,* **104,** 4085, 1982.
101. O. S. Estrada, C. Gomez-Louero, M. Montal, *Bioenergetics,* **3,** 417, 1972.
102. R. W. Roeske, S. Isaac, T. E. King, L. K. Steinrauf, *Biochem. Biophys. Res. Comm.,* **57,** 554, 1974.
103. R. C. Prince, A. R. Crofts, L. K. Steinrauf, *Biochem. Biophys. Res. Comm.,* **59,** 697, 1974.
104. B. Braden, J. A. Hamilton, M. N. Sabesan, L. K. Steinrauf, *J. Am. Chem. Soc.,* **102,** 2704, 1980.
105. J. A. Hamilton, L. K. Steinrauf, B. Braden, *Biochim. Biophys. Res. Comm.,* **64,** 151, 1975.
106. Y. A. Ovchinnikov, V. T. Ivanov, A. V. Evstratov, V. F. Bystrov, N. D. Abdullaev, E. M. Popov, G. M. Lipkind, S. F. Arkhipova, E. S. Efremov, M. M. Shemyakin, *Biochem. Biophys. Res. Comm.,* **37,** 668, 1969.
107. Y. A. Ovchinnikov, *Febs Letters,* **44,** 1, 1974.
108. E. Grell, F. Eggers, Th. Funck, *Chimia,* **26,** 632, 637, 1972.
109. M. Dobler, J. D. Dunitz, J. Krajewski, *J. Mol. Biol.,* **42,** 603, 1969.
110. V. T. Ivanov, A. V. Evstratov, L. V. Sumskaya; E. I. Melnik, T. S. Chumburidze, S. L. Portnova, T. A. Balashova, Y. A. Ovchinnikov, *Febs Letters,* **36,** 65, 1973; V. T. Ivanov, *Ann. N. Y. Acad. Sci.,* **264,** 221, 1975.
111. T. Wieland, G. Lüben, H. Ottenheym, J. Faesel, J. X. de Vries, W. Konz, A. Prox, J. Schmid, *Angew. Chem. Int. Ed. Engl.,* **7,** 204, 1968.
112. a) I. L. Karle, T. Wieland, D. Schermer, H. C. J. Ottenheym, *Proc. Natl. Acad. Sci. USA,* **76,** 1532, 1979; b) I. L. Karle, J. Karle, T. Wieland, W. Burgermeister, B. Witkop, *Proc. Natl. Acad. Sci. USA,* **73,** 1782, 1976; I. L. Karle, *J. Am. Chem. Soc.,* **99,** 5152, 1977; *Int. J. Peptide Protein Res.* **28,** 6, 1986; I. L. Karle, T. Wieland, *Int. J. Peptide Protein Res.,* **29,** 596, 1987; H.L. Lotter, *Z. Naturforsch.,* **39C,** 535, 1984.
113. I. L. Karle, J. Karle, T. Wieland, W. Burgermeister, H. Faulstich, B. Witkop, *Proc. Natl. Acad. Sci. USA,* **70,** 1836, 1973.

114. I. L. Karle, *J. Am. Chem. Soc.,* **96,** 4000, 1974; *Biochem.,* **13,** 2155, 1974.
115. a) Y. A. Ovchinnikov, V. T. Ivanov, *Tetrahedron,* **31,** 2177, 1975; *"The Proteins"* Vol. V, Eds. H. Neurath, R. L. Hill, Academic Press 1982, p. 307; b) I. L. Karle, *Proc. Natl. Acad. Sci. USA,* **82,** 7155, 1985.
116. a) M. B. Sankaram, K. R. K. Easwaran, *Biopolymers,* **21,** 1557, 1982; b) D. C. Hodgkin, B. M. Oughton, *Biochem. J.,* **65,** 752, 1957; S. E. Hull, R. Karlsson, P. Main, M. M. Woolfson, E. J. Dodson, *Nature,* **275,** 206, 1978; b) M. R. Pincus, H. A. Scheraga, *Acc. Chem. Res.,* **18,** 372, 1985.
117. a) E. R. Blout, C. M. Deber, L. G. Pease, *"Peptides, polypeptides and proteins",* John Wiley and Sons, 1974, p. 266; C. M. Deber, V. Madison, E. R. Blout, *Acc. Chem. Res.,* **9,** 106, 1976; b) V. Madison, M. Atreyi, C. M. Deber, E. R. Blout, *J. Am. Chem. Soc.,* **96,** 6725, 1974; c) G. Kartha, K. I. Varughese, S. Aimoto, *Proc. Natl. Acad. Sci. USA,* **79,** 4519, 1982; d) *Int. J. Peptide Protein Res.* **27,** 118, 1986; e) B. Bartman, C. M. Deber, E. R. Blout, *J. Am. Chem. Soc.,* **99,** 1028, 1977.
118. B. F. Gisin, R. B. Merrifield, *J. Am. Chem. Soc.,* **94,** 6165, 1972.
119. B. F. Gisin, D. G. Davis, *Biophys. J.,* **13,** 288 a, 1973.
120. H. P. Ting-Beall, M. T. Tosteson, B. F. Gisin, D. C. Tosteson, *J. Gen. Physiol.,* **63,** 492, 1974.
121. D. G. Davis, B. F. Gisin, D. C. Tosteson, *Biochem.,* **15,** 768, 1976.
122. R. Benz, B. F. Gisin, H. P. Ting-Beall, D. C. Tosteson, P. Laüger, *Biochem. Biophys. Acta,* **455,** 665, 1976.
123. J. A. Hamilton, M. N. Sabesan, L. K. Steinrauf, *Acta Cryst.,* **B36,** 1052, 1980.
124. D. Baron, L. G. Pease, E. R. Blout, *J. Am. Chem. Soc.,* **99,** 8299, 1977.
125. J. P. Degelaen, P. Pham, E. R. Blout, *J. Am. Chem. Soc.,* **106,** 4882, 1984.
126. C. M. Deber, P. D. Adawadkar, J. Tom-Kun, *Biochem. Biophys. Res Comm.,* **81,** 1357, 1978.
127. K. Titlestad, *Acta Chem. Scand.,* **B29,** 153, 1975.
128. P. Groth, *Acta Chem. Scand.,* **A30,** 840, 1976; **A31,** 232, 1977.
129. J. Dale, P. Groth, K. Titlestad, *Acta Chem. Scand.,* **B31,** 523, 1977.
130. M. Dobler, *Helv. Chim. Acta,* **55,** 1371, 1972; Y. Nawata, T. Sakamaki, Y. Iitaka, *Acta Cryst.,* **B30,** 1047, 1974.
131. Y. Nawata, T. Hayashi, Y. Iitaka, *Chem. Letters,* 315, 1980.
132. Y. Kyogoku, M. Ueno, H. Akutsu, Y. Nawata, *Biopolymers,* **14,** 1049, 1975; M. J. O. Anteunis, A. De Bruyn, *Bull. Soc. Chim. Belg.,* **86,** 445, 1977.
133. W. Simon, W. E. Morf, P. Ch. Meier, *Structure and Bonding,* **16,** 113, 1973.
134. T. Sakamaki, Y. Iitaka, Y. Nawata, *Acta Cryst.,* **B32,** 768, 1976.
135. M. Dobler, J. D. Dunitz, B. T. Kilbourn, *Helv. Chim. Acta,* **52,** 2573, 1969.
136. M. Dobler, R. P. Phizackerley, *Helv. Chim Acta,* **57,** 664, 1974.
137. T. Sakamaki, Y. Iitaka, Y. Nawata, *Acta Cryst.,* **B33,** 52, 1977.
138. K. Neupert-Laves, M. Dobler, *Helv. Chim. Acta,* **59,** 614, 1976.
139. Y. Nawata, T. Sakamaki, Y. Iitaka, *Chem. Letters,* 151, 1975.
140. Y. Nawata, T. Sakamaki, Y. Iitaka, *Acta Cryst.* **B33,** 1201, 1977.
141. A. Samat, M. El Malouli Bibout, M. Chanon, J. Elguero, *Nouv. J. Chimie,* **6,** 483, 1982.
142. A. Samat, M. El Malouli Bibout, J. Elguero, J. *Chem. Soc., Perkin I,* 1717, 1985.
143. K. Cooper, E. Jakobsson, P. Wolynes, *Prog. Biophys. Molec. Biol.,* **46,** 51, 1985; P. Laüger, *Angew. Chem. Int. Ed. Engl.,* **24,** 905, 1985.
144. R. Sarges, B. Witkop,, *Biochem.,* **4,** 2491, 1965.
145. D. W. Urry, *Proc. Natl. Sci. USA,* **68,** 672, 1971; *ibid,* **69,** 1610, 1972.
146. E. Bamberg, P. Laüger, J. *Membr. Biol.,* **11,** 177, 1973.

147. a) J. F. Hinton, W. L. Whaley, D. Shungu, R. E. Koeppe II, F. S. Millett, *Biophys. J.,* **50,** 539, 1986; b) D. W. Urry, T. L. Trapane, J. T. Walker, K. U. Prasad, *J. Biol. Chem.,* **257,** 6659, 1982.
148. W. R. Veatch, E. T. Fossel, E. R. Blout, *Biochem.,* **13,** 5249, 1974; W. R. Veatch, E. R. Blout, *ibid,* 5257.
149. W. R. Veatch, R. Mathies, M. Eisenberg, L. Stryer, J. *Mol. Biol.,* **99,** 75, 1975.
150. D. W. Urry, M. C. Goodall, J. D. Glickson, D. F. Mayers, *Proc. Natl. Acad. Sci. USA,* **68,** 1907, 1971.
151. S. Weinstein, B. A. Wallace, E. R. Blout, J. S. Morrow, W. Veatch, *Proc. Natl. Acad. Sci. USA,* **76,** 4230, 1979; D. W. Urry, T. L. Trapane, K. U. Prasad, *Science,* **221,** 1064, 1983; D. W. Urry, *Topics in Current Chemistry,* **128,** 175, 1985.
152. R. E. Koeppe II, K. O. Hodgson, L. Stryer, *J. Mol. Biol.,* **121,** 41, 1978; R. E. Koeppe II, J. M. Berg, K. O. Hodgson, L. Stryer, *Nature,* **279,** 723, 1979; R. E. Koeppe II, B. P. Schoenborn, *Biophys. J.,* **45,** 503, 1984; R. E. Koeppe II, M. Kimura, *Biopolymers,* **23,** 23, 1984.
153. J. Kamegai, S. Kimura, U. Imanishi, *Biophys. J.,* **49,** 1101, 1986.
154. D. R. Martin, R. J. P. Williams, *Biochem. J.,* **153,** 181, 1976.
155. G. Jung, W. A. König, D. Leibfritz, T. Ooka, K. Janko, G. Boheim, *Biochim. Biophys. Acta,* **433,** 164, 1976; G. Boheim, K. Janko, D. Leibfritz, T. Ooka, W. A. König, G. Jung, *Biochim. Biophys. Acta,* **433,** 182, 1976.
156. S. B. Hladky, L. G. M. Gordon, D. A. Haydon, *Ann. Rev. Phys. Chem.,* **25,** 11, 1974.
157. G. Baumann, P. Mueller, *J. Supramol. Struct.,* **2,** 538, 1974.
158. R. Nagaraj, P. Balaram, *Acc. Chem. Res.,* **14,** 356, 1981.
159. R. O. Fox Jr., F. M. Richards, *Nature,* **300,** 325, 1982.
160. R. Bosch, G. Jung, H. Schmitt, W. Winter, *Biopolymers,* **24,** 979, 1985.
161. A. J. M. van Beijnen, R. J. M. Nolte, J. W. Zwikker, W. Drenth, *Rec. Trav. Chim. Pays-Bas,* **101,** 409, 1982; J. H. van der Linden, J. Schoonman, R. J. M. Nolte, W. Drenth, *Rec. Trav. Chim. Pays-Bas,* **103,** 260, 1984.
162. J. M. Lehn, *Structure and bonding,* **16,** 1, 1973.
163. J. M. Lehn, *Acc. Chem. Res.,* **11,** 49, 1978.
164. H. E. Simmons, C. H. Park, *J. Am. Chem. Soc.,* **90,** 2428, 1968.
165. H. Stetter, J. Marx, *Ann.,* **607,** 59, 1957.
166. B. Dietrich, J. M. Lehn, J. P. Sauvage, *Tetrahedron Letters,* 2885, 1969.
167. B. Dietrich, J. M. Lehn, J. P. Sauvage, J. Blanzat, *Tetrahedron,* **29,** 1629, 1973.
168. J. L. Dye, M. T. Lok, F. J. Tehan, J. M. Ceraso, K. J. Voorhees, *J. Org. Chem.,* **38,** 1773, 1973.
169. C. H. Park, H. E. Simmons, *J. Am. Chem. Soc.,* **90,** 2429, 2431, 1968.
170. J. Cheney, J. M. Lehn, *J. Chem. Soc., Chem. Comm.,* 487, 1972.
171. H. E. Simmons, C. H. Park, R. T. Uyeda, M. F. Habibi, *Trans. N.Y. Acad. Sci., Ser. II,* **32,** 521, 1970.
172. B. Metz, D. Moras, R. Weiss, *J. Chem. Soc., Perkin II,* 423, 1976; B. Metz, R. Weiss, *Nouv. J. Chimie,* **2,** 615, 1978.
173. B. Dietrich, J. M. Lehn, J. P. Sauvage, *J. Chem. Soc., Chem. Comm.,* 1055, 1970.
174. a) J. M. Lehn, F. Montavon, *Tetrahedron Letters,* 4557, 1972; *Helv. Chim. Acta,* **59,** 1556, 1976; b) B. Dietrich, J.-M. Lehn, J. Guilhem, C. Pascard, *Tetrahedron Letters,* **30,** 4125, 1989.
175. D. Landini, F. Montanari, F. Rolla, *Synthesis,* 223, 1978; F. Montanari, D. Landini, F. Rolla, *Topics in Current Chem.,* **101,** 147, 1982.

176 a) A. P. Bell, C. D. Hall, *J. Chem. Soc., Chem. Comm.,* 163, 1980; P. J. Hammond, P. D. Beer, C. D. Hall, *ibid,* 1161, 1983; b) C. D. Hall, N. W. Sharpe, I. P. Danks, Y. P. Sang, *J. Chem. Soc., Chem. Comm.,* 419, 1989; c) P.D. Beer, *Chem. Soc. Rev.,* **18,** 409, 1989.
177. T. Asano, T. Okada, S. Shinkai, K. Shigematsu, Y. Kusano, O. Manabe, *J. Am. Chem. Soc.,* **103,** 5161, 1981.
178. a) G. R. Newkome, V. K. Majestic, F. R. Fronczek, *Tetrahedron Letters,* **22,** 3039, 1981; b) K. N. Raymond, T. J. McMurry, T. M. Garrett, *Pure and Appl. Chem.,* **60,** 545, 1988; K. N. Raymond and T. M. Garrett, *ibid.,* **60,** 1807, 1988.
179. J. P. Konopelski, F. Kotzyba-Hibert, J. M. Lehn, J. P. Desvergne, F. Fagès, A. Castellan, H. Bouas-Laurent, *J. Chem. Soc., Chem. Commun.,* 433, 1985.
180. G. Guinand, P. Marsau, J. M. Lehn, F. Kotzyba-Hibert, J. P. Konopelski, J. P. Desvergne, F. Fagès, A. Castellan, H. Bouas-Laurent, *Acta Cryst.,* **C42,** 715, 1986.
181. G. Jenner, *Nouv. J. Chimie,* **3,** 329, 1979; K. Matsumoto, A. Sera, T. Uchida, *Synthesis,* 1, 1985; S. Suvachittanont, *J. Chem. Ed.,* **60,** 150, 1983.
182. J. Jurczak, M. Pietraszkiewicz, *Topics in Current Chem.,* **130,** 183, 1985.
183. M. Pietraszkiewicz, P. Salanski, J. Jurczak, *J. Chem. Soc., Chem. Comm.,* 1184, 1983; *Tetrahedron,* **40,** 2971, 1984.
184. A.H. Haines, P. Karntiang, *J. Chem. Soc., Perkin I,* 2577, 1979.
185. C. H. Park, H. E. Simmons, *J. Am. Chem. Soc.,* **94,** 7184, 1972.
186. D. G. Parsons, *J. Chem. Soc., Perkin I,* 451, 1978.
187. J. D. Owen, *J. Chem. Soc., Perkin II,* 12, 1981.
188. A. C. Coxon, J. F. Stoddart, *J. Chem. Soc., Chem. Comm.,* 537, 1974; *J. Chem. Soc., Perkin I,* 767, 1977.
189. B. L. Allwood, S. E. Fuller, P. C. Y. K. Ning, A. M. Z. Slawin, J. F. Stoddart, D. J. Williams, *J. Chem. Soc., Chem. Commun.,* 1356, 1984.
190. Y. Nakatsuji, T. Mori, M. Okahara, *J. Chem. Soc., Chem. Commun.,* 1045, 1984.
191. B. Son, B. P. Czech, R. A. Bartsch, *Tetrahedron Letters,* **26,** 1787, 1985.
192. a) C. G. Krespan, *J. Org. Chem.,* **45,** 1177, 1980; b) *ibid.,* **39,** 2351, 1974.
193. G. R. Newkome, A. Nayak, *J. Org. Chem.,* **43,** 409, 1978.
194. F. Vögtle, G. Hohner, E. Weber, *J. Chem. Soc., Chem. Comm.,* 366, 1973.
195. F. Vögtle, G. Hohner, *Angew. Chem, Int. Ed. Engl.,* **14,** 497, 1975.
196. A. Ricci, R. Danieli, S. Rossini, *J. Chem. Soc., Perkin I,* 1691, 1976.
197. a) M. Kanishi, J. I. Kunizaki, J. Inanaga, M. Yamaguchi, *Bull. Chem. Soc. Japan,* **54,** 3828, 1981; b) R. A. Pascal Jr., J. Spergel, D. Van Engen, *Tetrahedron Letters,* **27,** 4099, 1986.
198. B. Dietrich, M. W. Hosseini, J. M. Lehn, R. B. Sessions, *Helv. Chim. Acta,* **68,** 289, 1985.
199. M. W. Hosseini, Thèse de Doctorat d'Etat, Université Louis Pasteur, Strasbourg 1983; B. Dietrich, J. Guilhem, J. M.Lehn, C. Pascard, E. Sonveaux, *Helv. Chim. Acta,* **67,** 91, 1984.
200. D. Heyer, J. M. Lehn, *Tetrahedron Letters,* **27,** 5869, 1986.
201. a) J. Franke, F. Vögtle, *Angew. Chem. Int. Ed. Engl.,* **24,** 219, 1985; b) F. Vögtle, W. M. Müller, W. H. Watson, *Topics in Current Chemistry,* **125,** 131, 1984; c) Y. Murakami, J.-I. Kikuchi, *Pure and Appl. Chem.,* **60,** 549, 1988; d) F. Diederich, *Angew. Chem. Int. Ed. Engl.,* **27,** 362, 1988; F. Diederich, *Cyclophanes,* Ed. J. F. Stoddart, *Royal Society of Chemistry,* 1991.
202. D. O'Krongly, S. R. Denmeade, M. Y. Chiang, R. Breslow, *J. Am. Chem. Soc.,* **107,** 5544, 1985.
203. G. R. Newkome, V. K. Majestic, F. R. Fronczek, *Tetrahedron Letters,* **22,** 3035, 1981.

204. G. R. Newkome, V. Majestic, F. Fronczek, J. L. Atwood, *J. Am. Chem. Soc.,* **101,** 1047, 1979.
205. J. Jazwinski, J. M. Lehn, D. Lilienbaum, R. Ziessel, J. Guilhem, C. Pascard, *J. Chem. Soc., Chem. Comm.,* 1691, 1987.
206. a) R. W. Alder, R. B. Sessions, J. M. Mellor, M. F. Rawlins, *J. Chem. Soc., Chem. Comm.,* 747, 1977; b) R. W. Alder, R. B. Sessions, *J. Am. Chem. Soc.,* **101,** 3651, 1979; c) R. W. Alder, R. B. Sessions, A. J. Bennet, R. E. Moss, *J. Chem. Soc., Perkin I,* 603, 1982.
207. a) R. W. Alder, R. J. Arrowsmith, A. Casson, R. B. Sessions, E. Heilbronner, R. Kovač, H. Huber, M. Taagepera, *J. Am. Chem. Soc.,* **103,** 6137, 1981; b) S.F. Nelsen, P.M. Gannett, *J. Am. Chem. Soc.* **104,** 4698, 1982; c) R. W. Alder, *Acc. Chem. Res.,* **16,** 321, 1983; d) S. F. Nelsen, R. W. Alder, R. B. Sessions, K. D. Asmus, K. O. Hiller, M. Göbl, *J. Am. Chem. Soc.,* **102,** 1429, 1980.
208. R. Hoffmann, *Acc. Chem. Res.,* **4,** 1, 1971; R. Hoffmann, A. Imamura, W. J. Hehre, *J. Am. Chem. Soc.,* **90,** 1499, 1968; E. Heilbronner, K. A. Muszkat, *J. Am. Chem. Soc.,* **92,** 3818, 1970; E. C. Baird, *J. Chem. Ed.,* **54,** 291, 1977; K. D. Asmus, *Acc. Chem. Res.,* **12,** 436, 1979; T. Clark, *J. Am. Chem. Soc.,* **110,** 1672, 1988.
209. a) R. W. Alder, A. G. Orpen, R. B. Sessions, *J. Chem. Soc., Chem. Comm.* 999, 1983; b) R. W. Alder, A. G. Orpen, J. M. White, *J. Chem. Soc., Chem. Comm.* 949, 1985; c) O. Ermer, R. Gerdil, J. D. Dunitz, *Helv. Chim. Acta,* **54,** 2476, 1971; d) B. Kirste, R. W. Alder, R. B. Sessions, M. Bock, H. Kurreck, S. F. Nelsen, *J. Am. Chem. Soc.,* **107,** 2635, 1985.
210. R. Annunziata, F. Montanari, S. Quici, M. T. Vitali, *J. Chem. Soc., Chem. Comm.,* 777, 1981; P. L. Anelli, F. Montanari, S. Quici, *J. Org. Chem.,* **50,** 3453, 1985.
211. a) I. I. Creaser, J. MacB. Harrowfield, A. J. Herlt, A. M. Sargeson, J. Springborg, R. J. Geue, M. R. Snow, *J. Am. Chem. Soc.,* **99,** 3181, 1977; b) A. M. Sargeson, *Pure Appl. Chem.,* **50,** 905, 1978; c) J. MacB. Harrowfield, A. J. Herlt, A. M. Sargeson, *Inorg. Syntheses,* **20,** 85, 1980; d) I. I. Creaser, R. J. Geue, J. MacB. Harrowfield, A. J. Herlt, A. M. Sargeson, M. R. Snow, J. Springborg, *J. Am. Chem. Soc.,* **104,** 6016, 1982.
212. A. M. Sargeson, *Chem. in Britain,* **15,** 23, 1979.
213. a) A. M. Bond, G. A. Lawrance, P. A. Lay, A. M. Sargeson, *Inorg. Chem.,* **22,** 2010, 1983; b) T. W. Hambley, *Inorg. Chem.,* **27,** 2496, 1988; c) R. Nagao, F. Marumo, Y. Saito, *Acta Cryst.,* **B29,** 2438, 1973; d) K. Wieghardt, W. Schmidt, W. Herrmann, H. J. Küppers, *Inorg. Chem.,* **22,** 2953, 1983; e) N. F. Curtis, G. F. Gainsford, T. W. Hambley, G. A. Lawrance, K. R. Morgan, A. Siriwardena, *J. Chem. Soc., Chem. Comm.,* 295, 1987; f) Y. Hung, L. Y. Martin, S. C. Jackels, A. M. Tait, D. H. Busch, *J. Am. Chem. Soc.,* **99,** 4029, 1977.
214. a) A. M. Sargeson, *Pure Appl. Chem.,* **56,** 1603, 1984; b) A. M. Bond, T. W. Hambley, M. R. Snow, *Inorg. Chem.,* **24,** 1920, 1985; c) R. J. Geue, T. W. Hambley, J. M. Harrofield, A. M. Sargeson, M. R. Snow, *J. Am. Chem. Soc.,* **106,** 5478, 1984.
215. a) L. R. Gahan, T. W. Hambley, A. M. Sargeson, M. R. Snow, *Inorg. Chem.,* **21,** 2699, 1982; b) L. R. Gahan, G. A. Lawrance, A. M. Sargeson, *Inorg. Chem.,* **23,** 4369, 1984; c) R. V. Dubs, L. R. Gahan, A. M. Sargeson, *Inorg. Chem.,* **22,** 2523, 1983.
216. a) G. J. Gainsford, R. J. Geue, A. M. Sargeson, *J. Chem. Soc., Chem. Comm.,* 233, 1982; b) H. A. Boucher, G. A. Lawrance, P. A. Lay, A. M. Sargeson, A. M. Bond, D. F. Sangster, J.C. Sullivan, *J. Am. Chem. Soc.,* **105,** 4652, 1983; c) M. P. Suh, W. Shin, D. Kim, S. Kim., *Inorg. Chem.,* **23,** 618, 1984; d) P. Comba, L. M. Engelhardt, J. MacB. Harrowfield, G. A. Lawrance, L. L. Martin, A. M. Sargeson, A. H. White, *J. Chem.*

Soc., Chem. Comm., 174, 1985; e) P. Comba, I. I. Creaser, L. R. Gahan, J. M. Harrowfield, G. A. Lawrance, L. L. Martin, A. W. H. Mau, A. M. Sargeson, W. H. F. Sasse, M. R. Snow., *Inorg. Chem.,* **25,** 384, 1986.

217. P. A. Lay, A. W. H. Mau, W. H. F. Sasse, I. I. Creaser, L. R. Gahan, A. M. Sargeson, *Inorg. Chem.,* **22,** 2347, 1983; I. I. Creaser, L. R. Gahan, R. J. Geue, A. Launikonis, P. A. Lay, J. D. Lydon, M. G. McCarthy, A. W. H. Mau, A. M. Sargeson, W. H. F. Sasse, *Inorg. Chem.,* **24,** 2671, 1985; V. Balzani, N. Sabbatini, F. Scandola, *Chem. Rev.,* **86,** 319, 1986.
218. J. C. Rodriguez-Ubis, B. Alpha, D. Plancherel, J. M. Lehn, *Helv. Chim. Acta,* **67,** 2264, 1984.
219. A. Caron, J. Guilhem, C. Riche, C. Pascard, B. Alpha, J. M. Lehn, J. C. Rodriguez-Ubis, *Helv. Chim. Acta,* **68,** 1577, 1985.
220. B. Alpha, J. M. Lehn, G. Mathis, *Angew. Chem. Int. Ed. Engl.* **26,** 266, 1987.
221. a) D. H. Busch, *Pure Appl. Chem.,* **52,** 2477, 1980; D. H. Busch, C. Cairns, *Progress in Macrocyclic Chemistry,* vol. 3, Eds. R. M. Izatt, J. J. Christensen, John Wiley and Sons, New York, 1981; b) J. C. Stevens, P. J. Jackson, W. P. Schammel, G. G. Christoph, D. H. Busch, *J. Am. Chem. Soc.,* **102,** 3283, 1980; c) J. C. Stevens, D. H. Busch, *J. Am. Chem. Soc.,* **102,** 3285, 1980; d) N. Herron, D. H. Busch, *J. Am. Chem. Soc.,* **103,** 1236, 1981; e) D. H. Busch, S. C. Jackels, R. C. Callahan, J. J. Grzybowski, L. L. Zimmer, M. Kojima, D. J. Olszanski, W. P. Schammel, J. C. Stevens, K. A. Holter, J. Mocak, *Inorg. Chem.,* **20,** 2834, 1981; f) D. H. Busch, G. C. Christoph, L. L. Zimmer, S. C. Jackels, J. J. Grzybowski, R. C. Callahan, M. Kojima, K. A. Holter, J. Mocak, N. Herron, M. Chavan, W. P. Schammel, *J. Am. Chem. Soc.,* **103,** 5107, 1981.
222. J. H. Cameron, M. Kojima, B. Korybut-Daszkiewicz, B. K. Coltrain, T. J. Meade, N. W. Alcock, D. H. Busch, *Inorg. Chem.,* **26,** 427, 1987.
223. a) J. S. Lindsey, D. C. Mauzerall, *J. Am. Chem. Soc.,* **104,** 4498, 1982. b) J. S. Lindsey, D. C. Mauzerall, H. Linschitz, *J. Am. Chem. Soc.,* **105,** 6528, 1983.
224. a) J. Weiser, H. A. Staab, *Angew. Chem. Int. Ed. Engl.,* **23,** 623, 1984; B. Morgan, D. Dolphin, *Angew. Chem. Int. Ed. Engl.,* **24,** 1003, 1985; b) J. E. Baldwin, P. Perlmutter, *Topics in Current Chem.,* **121,** 181, 1984; E. Tsuchida, H. Nishide, *Topics in Current Chem.,* **132,** 63, 1986.
225. J. C. Tolle, M. A. Staples, E. R. Blout, *J. Am. Chem. Soc.,* **104,** 6883, 1982; *"Conformation in Biology"* Eds. R. Srinivasan, R. H. Sarma, Adenine Press N. Y. 1983, p. 147.
226. a) M. Eigen, *Angew. Chem. Int. Ed. Engl.,* **3,** 1, 1964; b) S. Scheiner, *Acc. Chem. Res.,* **18,** 174, 1985.
227. a) J. J. Delpuech, G. Serratrice, A. Strich, A. Veillard, *J. Chem. Soc., Chem. Comm.,* 817, 1972, *Molecular Physics,* **29,** 849, 1975; b) S. Scheiner, L. B. Harding, *J. Am. Chem. Soc.,* **103,** 2169, 1981; S. Scheiner, *J. Phys. Chem.,* **86,** 376, 1982.
228. a) J. Rozière, C. Belin, M. S. Lehman, *J. Chem. Soc., Chem. Comm.,* 388, 1982; b) D. J. Jones, I. Brach, J. Rozière, *J. Chem. Soc., Dalton,* 1795, 1984,
229. A. J. Kresge, *Acc. Chem. Res.,* **8,** 354, 1975; *Pure Appl. Chem.,* **53,** 189, 1981.
230. J. R. de la Vega, *Acc. Chem. Res.,* **15,** 185, 1982 and references therein; J.H. Busch, J. R. de la Vega, *J. Am. Chem. Soc.,* **99,** 2397, 1977.
231. J. M. Lehn, *Topics Current Chem.,* **15,** 311, 1970 and references therein.
232. a) R. W. Alder, P. S. Bowman, W. R. S. Steele, D. R. Wintermann, *J. Chem. Soc., Chem. Comm.* 723, 1968; b) H. Einspahr, J. B. Robert, R. E. Marsh, J. D. Roberts, *Acta Cryst.,* **B29,** 1611, 1973; c) D. E. Fenton, M. R. Truter, B. L. Vickery, *J. Chem. Soc., Chem. Comm.,* 93, 1971; M. R. Truter, B. L. Vickery, *J. Chem. Soc., Dalton,* 395, 1972.
233. R. W. Alder, M. R. Bryce, N. C. Goode, N. Miller, J. Owen, *J. Chem. Soc., Perkin I,* 2840, 1981; H. A. Staab, T. Saupe, *Angew. Chem. Int. Ed. Engl.,* **27,** 865, 1988.

234. F. Hibbert, *J. Chem. Soc., Perkin II,* 1862, 1974; A. Awwal, F. Hibbert, *J. Chem. Soc., Perkin II,* 1589, 1977; R.W. Alder, N. C. Goode, N. Miller, F. Hibbert, K. P. P. Hunte, H. J. Robbins, *J. Chem. Soc., Chem. Comm.* 89, 1978; F. Hibbert, K. P. P. Hunte, *J. Chem. Soc., Perkin II,* 1895, 1983.
235. F. Hibbert, *Acc. Chem. Res.,* **17,** 115, 1984; *Adv. Phys. Org. Chem.,* **22,** 113, 1986.
236. E. Haselbach, A. Henriksson, F. Jachimowicz, J. Wirz, *Helv. Chim. Acta,* **55,** 1757, 1972.
237. a) J. Clark, D. D. Perrin, *Quart. Rev.,* **18,** 295, 1964; b) E. Q. Adams, *J. Am. Chem. Soc.,* **38,** 1503, 1916; c) N. Bjerrum, *Zeit. Phys. Chem.,* **106,** 219, 1923; d) L. Meites, *J. Chem. Ed.,* **49,** 682, 1972; e) H. Ogino, *Inorg. Chem.,* **19,** 1619, 1980; f) G. Soula, *J. Org. Chem.,* **50,** 3717, 1985; g) G. Wipff, J. M. Wurtz, *New. J. Chem.,* **13,** 807, 1989; h) G. Anderegg, *Helv. Chim. Acta,* **64,** 1790, 1981; J. H. Burns, C. F. Baes Jr., *Inorg. Chem.,* **20,** 616, 1981.
238. a) R. C. Lord, M. N. Siamwiza, *Spectrochim. Acta,* **31A,** 1381, 1975; b) G. Wipff, P. A. Kollman, J. M. Lehn, *J. Mol. Struct.,* **93,** 153, 1983; c) G. Wipff, P. Kollman, *Nouv. J. Chimie,* **9,** 457, 1985.
239. a) B. G. Cox, D. Knop, H. Schneider, *J. Am. Chem. Soc.,* **100,** 6002, 1978; b) R. Pizer, *J. Am. Chem. Soc.,* **100,** 4239, 1978.
240. a) B. G. Cox, J. Murray-Rust, P. Murray-Rust, N. van Truong, H. Schneider, *J. Chem. Soc., Chem. Comm.,* 377, 1982; b) B. G. Cox, N. van Truong, H. Schneider, *J. Chem. Soc., Perkin II,* 515, 1983; c) B. Metz, D. Moras, R. Weiss, *J. Chem. Soc., Perkin II,* 423, 1976.
241. J. Cheney, J. P. Kintzinger, J. M. Lehn, *Nouv. J. Chimie,* **2,** 411, 1978.
242. P. B. Smith, J. L. Dye, J. Cheney, J. M. Lehn, *J. Am. Chem. Soc.,* **103,** 6044, 1981.
243. H. J. Brügge, D. Carboo, K. von Deuten, A. Knöchel, J. Kopf, W. Dreissig, *J. Am. Chem. Soc.,* **108,** 107, 1986.
244. R. Geue, S. H. Jacobson, R. Pizer, *J. Am. Chem. Soc.,* **108,** 1150, 1986.
245. R. W. Alder, A. G. Orpen, R. B. Sessions, *J. Chem. Soc., Chem. Comm.,* 999, 1983.
246. a) R. W. Alder, R. E. Moss, R. B. Sessions, *J. Chem. Soc., Chem. Comm.,* 1000, 1983; b) R. W. Alder, *Chem. Rev.,* **89,** 1215, 1989.
247. a) B. Dietrich, J. M. Lehn, J. P. Sauvage, *Tetrahedron Letters,* **34,** 2885, 2889, 1969; b) *Tetrahedron,* **29,** 1647, 1973.
248. B. Metz, D. Moras, R. Weiss, *J. Chem. Soc., Chem. Comm.,* 444, 1971.
249. a) B. Metz, D. Moras, R. Weiss, *J. Am. Chem. Soc.,* **93,** 1806, 1971; b) L. Pauling *"The nature of the Chemical Bond",* Cornell University Press, 1960.
250. L. G. Sillen, A. E. Martell, *Chem. Soc. Spec. Publ. n°* 17, 1964, n° 25, 1971.
251. N. K. Dalley, "Structural studies of synthetic macrocyclic molecules and their cation complexes", In: *"Synthetic multidentate macrocyclic compounds",* R. M. Izatt, J. J. Christensen eds, N.Y. Acad. Press, 1978.
252. a) R. M. Izatt, D. J. Eatough, J. J. Christensen, *Structure and bonding,* **16,** 161, 1973; R. M. Izatt, R. E. Terry, B. L. Haymore, L. D. Hansen, N. K. Dalley, A. G. Avondet, J. J. Christensen, *J. Am. Chem. Soc.,* **98,** 7620, 1976; R. M. Izatt, L. D. Hansen, D. J. Eatough, J. S. Bradshaw, J. J. Christensen, *"Metal.Ligand Interactions in Organic Chemistry and Biochemistry".* Eds. B. Pullman, N. Goldblum, D. Reidel, Publishing Company 1976, p. 337; b) N. K. Dalley, D. E. Smith, R. M. Izatt, J. J. Christensen, *J. Chem. Soc., Chem. Comm.,* 90, 1972.
253. H. K. Frensdorf, *J. Am. Chem. Soc.,* **93,** 600, 1971.
254. J. D. Lamb, R. M. Izatt, C. S. Swain, J. J. Christensen, *J. Am. Chem. Soc.,* **102,** 475, 1980.
255. R. Ungaro, B. El Haj, J. Smid, *J. Am. Chem. Soc.,* **98,** 5198, 1976.

256. F. De Jong, D. N. Reinhoudt, *Advances in Physical Organic Chemistry,* vol. 17, 1980, p. 279. Academic Press; R. M. Izatt, J. S. Bradshaw, S. A. Nielsen, J. D. Lamb, J. J. Christensen, *Chem. Rev.,* **85,** 271, 1985; R. M. Izatt, K. Pawlak, J. S. Bradshaw, R. L. Bruening, *Chem. Rev.,* **91,** 1721, 1991.
257. G. Wipff, P. A. Kollman, J. M. Lehn, *J. Mol. Struct.,* **93,** 153, 1983.
258. a) P. Groth, *Acta Chem. Scand.* **A35,** 463, 1981; b) J. Massaux, J. F. Desreux, G. Duyckaerts, *J. Chem. Soc., Dalton,* 865, 1980; c) P. Groth, *Acta Chem. Scand.,* **25,** 725, 1971; d) P. Groth, *Acta Chem. Scand.,* **A35,** 460, 1981; e) K. M. Aalmo, J. Krane, *Acta Chem. Scand.,* **A36,** 219, 1982; f) *ibid,* 227.
259. a) W. E. Bunney, Jr., D. L. Murphy, "The Neurobiology of Lithium", *Neurosciences Research Program Bull.,* **14,** 1976; b) M. T. Doig III, M. G. Heyl, D. F. Martin, *J. Chem. Ed.,* **50,** 343, 1973; c) G. Shoham, W. N. Lipscomb, U. Olsher, *J. Chem. Soc., Chem. Comm.,* 208, 1983; d) E. M. Holt, G. D. Malpass Jr., R. G. Ghirardelli, R. A. Palmer, B. Rubin, *Acta Cryst.,* **C40,** 394, 1984; *ibid,* **C40,** 396; e) K.Kobiro, T. Matsuoka, S. Takada, K. Kakiuchi, Y. Tobe, Y. Odaira, *Chem. Letters,* 713, 1986; f) U. Olsher, J. Jagur-Grodzinski, *J. Chem. Soc., Dalton,* 501, 1981; g) J. Dale, S. B. Fredriksen, *Pure and Appl. Chem.,* **61,** 1587, 1989; L. J. Rodriguez, M.Xu. E. M. Eyring, S. Petrucci, *ibid,* **61,** 1593, 1989; P. A. Kollman, P. D. J. Grootenhuis, M. A. Lopez, *ibid,* **61,** 593, 1989; E. M. Eyring, S. Petrucci, M.Xu,L.J. Rodriguez, D. P. Cobranchi, M. Masiker, P. Firman, *ibid,* **62,** 2237, 1990; U. Olsher, R. M. Izatt, J. S. Bradshaw, N. K. Dalley, *Chem. Rev.,* **91,** 137, 1991; P. P Power, *Acc. Chem. Res.,* **21,** 147, 1988.
260. C. J. Pedersen, *J. Am. Chem. Soc.,* **89,** 7017, 1967; C. J. Pedersen, H. K. Frensdorff, *Angew. Chem. Int. Ed. Engl.,* **11,** 16, 1972.
261. R. M. Izatt, R. E. Terry, D. P. Nelson, Y. Chan, D. J. Eatough, J. S. Bradshaw, L. D. Hansen, J. J. Christensen, *J. Am. Chem. Soc.,* **98,** 7626, 1976.
262. D. M. Dishong, G. W. Gokel, *J. Org. Chem.,* **47,** 147, 1982.
263. J. D. Lin, A. I. Popov, *J. Am. Chem. Soc.,* **103,** 3773, 1981.
264. a) K. H. Wong, G. Konizer, J. Smid, *J. Am. Chem. Soc.,* **92,** 666, 1970; b) A. Hofmanova, J. Koryta, M. Brezina, M. L. Mittal, *Inorg. Chim. Acta* **28,** 73, 1978; I.M. Kolthoff, M.K. Chantooni, *Anal. Chem.* **52,** 1039, 1980.
265. J. M. Lehn, J. P. Sauvage, *J. Am. Chem. Soc.,* **97,** 6700, 1975.
266. J. M. Lehn, *Pure Appl. Chem.,* **51,** 979, 1979.
267. J. J. Daly, P. Schönholzer, J. P. Behr, J. M. Lehn, *Helv. Chim. Acta,* **64,** 1444, 1981.
268. J. P. Behr, J. M. Lehn, D. Moras, J.C. Thierry, *J. Am. Chem. Soc.,* **103,** 701, 1981.
269. J. M. Lehn, *Pure Appl. Chem.,* **50,** 871, 1978.
270. J. P. Behr, J. M. Lehn, P. Vierling, *Helv. Chim. Acta,* **65,** 1853, 1982.
271. L. A. Frederick, T. M. Fyles, N. P. Gurprasad, D. M. Whitfield, *Canad. J. Chem.,* **59,** 1724, 1981.
272. T. M. Fyles, V. A. Malik-Diemer, D. M. Whitfield, *Canad. J. Chem.,* **59,** 1734, 1981; T. M. Fyles, V. A. Malik-Diemer, C. A. McGavin, D. M. Whitfield, *Canad. J. Chem.,* **60,** 2259, 1982.
273. A. Hriciga, J. M. Lehn, *Proc. Natl. Acad. Sci. U.S.A.,* **80,** 6426, 1983.
274. T. M. Fyles, C. A. McGavin, D. M. Whitfield, *J. Org. Chem.,* **49,** 753, 1984.
275. D. J. Cram, J. M. Cram, *Acc. Chem. Res.,* **11,** 8, 1978 and references therein.
276. D. A. Laidler, J. F. Stoddart, *J. Chem. Soc., Chem. Comm.,* 481, 1977.
277. J. P. Behr, J. M. Lehn, P. Vierling, *J. Chem. Soc., Chem. Comm.,* 621, 1976.
278. a) O. Nagano, A. Kobayashi, Y. Sasaki, *Bull. Chem. Soc. Japan,* **51,** 790, 1978; b) K. N. Trueblood, C. B. Knobler, D. S. Lawrence, R. V. Stevens, *J. Am. Chem. Soc.,* **104,** 1355,

1982; c) I. Goldberg, *"Inclusion Compounds"*, Eds. J. L. Atwood, J. E. D. Davies, D. D. MacNicol, Academic Press, vol. 2, 1984, p. 261.

279. K. Madan, D. J. Cram, *J. Chem. Soc., Chem. Comm.*, 427, 1975; E. P. Kyba, R. C. Hegelson, K. Madan, G. W. Gokel, T. L. Tarnowski, S. S. More, D. J. Cram, *J. Am. Chem. Soc.*, **99.** 2564, 1977.
280. a) J. M. Lehn, P. Vierling, R. C. Hayward, *J. Chem. Soc., Chem. Comm.*, 296, 1979; b) J. W. H. M. Uiterwijk, S. Harkema, J. Geevers, D. N. Reinhoudt, *J. Chem. Soc., Chem. Comm.* 200, 1982.
281. S. R. Cooper, *Acc. Chem. Res.*, **21**, 141, 1988.
282. S. Shinkai, T. Nagaji, T. Ogawa, K. Shigematsu, O. Manabe, *J. Am. Chem. Soc.*, **103**, 111, 1981; S. Shinkai, O. Manabe, *Topics in Current Chem.*, **121**, 67, 1984.
283. a) D. J. Cram, T. Kaneda, R. C. Helgeson, G. M. Lein, *J. Am. Chem. Soc.*, **101**, 6752, 1979; b) K. N. Trueblood, C. B. Knobler, E. Maverick, R. C. Helgeson, S. B. Brown, D. J. Cram, *J. Am. Chem. Soc.*, **103**, 5594, 1981; c) D. J. Cram, K. N. Trueblood, *Topics in Current Chem.*, **98**, 43, 1981; D. J. Cram, *Angew. Chem. Int. Ed. Engl.*, **25**, 1039, 1986; d) D. J. Cram, G. M. Lein, *J. Am. Chem. Soc.*, **107**, 3657, 1985; e) B.G. Cox, J. Garcia-Rosas, H. Schneider, *J. Am. Chem. Soc.*, **103**, 1054, 1981; f) P. Kollman, *Acc. Chem. Res.*, **18**, 105, 1985.
284. E. Grell, T. Funck, F. Eggers, *"Membranes"* vol. 3, Ed. G. Eisenman, Marcel Dekker Inc. New York 1975, p. 1; R. M. Izatt, J. S. Bradshaw, S. A. Nielsen, J. D. Lamb, J. J. Christensen, D. Sen, *Chem. Rev.*, **85**, 271, 1985.
285. a) G. Anderegg, *Helv. Chim. Acta,* **58**, 1218, 1975; b) F. Arnaud-Neu, B. Spiess, M. J. Schwing-Weill, *Helv. Chim. Acta,* **60**, 2633, 1977.
286. J. M. Lehn, F. Montavon, *Helv. Chim. Acta,* **59**, 1566, 1976; **61**, 67, 1978.
287. a) W. E. Morf, W. Simon, *Helv. Chim. Acta,* **54**, 794 et 2683, 1971; S. Goldman, R. G. Bates, *J. Am. Chem. Soc.*, **94**, 1476, 1972; b) C. Hansch, *Acc. Chem. Res.*, **2**, 232, 1969; c) B. Dietrich, J. M. Lehn, J. P. Sauvage, *J. Chem. Soc., Chem. Comm.*, 15, 1973; d) N. L. Ott, C. L. Barnes, R. W. Taylor, D. van der Helm, *Acta Cryst.*, **B38**, 2277, 1982; e) H. J. Buschmann, *Inorg. Chim Acta,* **134**, 225, 1987.
288. a) O. A. Gansow, A. R. Kausar, K. M. Triplett, M. J. Weaver, E. L. Yee, *J. Am. Chem. Soc.*, **99**, 7087, 1977, E. L. Yee, O. A. Gansow, M. J. Weaver, *J. Am. Chem. Soc.*, **102**, 2278, 1980; b) J. C. Bünzli, D. Wesser, *Coord. Chem. Rev.*, **60**, 191, 1984; D. E. Fenton, P. A. Vigato, *Chem. Soc. Rev.*, **17**, 69, 1988.
289. E. Kauffmann, J. M. Lehn, J. P. Sauvage, *Helv. Chim. Acta,* **59**, 1099, 1976.
290. H. G. Hertz, *Angew. Chem. Int. Ed. Engl.*, **9**, 124, 1970.
291. B. G. Cox, A. J. Parker, *J. Am. Chem. Soc.*, **95**, 6879, 1973.
292. a) R. M. Izatt, R. E. Terry, B. L. Haymore, L. D. Hansen, N. K. Dalley, A. G. Avondet, J. J. Christensen, *J. Am. Chem. Soc.*, **98**, 7620, 1976; b) G. Michaux, J. Reisse, *J. Am. Chem. Soc.*, **104**, 6895, 1982; c) Y. Inoue, T. Hakushi, *J. Chem. Soc., Perkin II,* 935, 1985.
293. W. H. Müller, *Naturwiss.*, **57**, 248, 1970.
294. a) A. Knöchel, R. D. Wilken, *J. Radioanal. Chem.*, **32**, 345, 1976; b) K. G. Heumann, *Topics Current Chem.* **127**, 77, 1985.
295. B. G. Cox, *Annual Reports on the Progress of Chemistry,* Section C, **81**, 43, 1984.
296. I. O. Sutherland, *Chem. Soc. Rev.*, **15**, 63, 1986; *J. Chem. Soc., Faraday I,* **82**, 1145, 1986.
297. J. M. Lehn, *Science,* **227**, 849, 1985; B. Dietrich, *J. Chem. Ed.*, **62**, 954, 1985; P. G. Potvin, J.-M. Lehn, *"Synthesis of Macrocycles: The Design of Selective Complexing Agents"*, Eds. R M. Izatt, J. J. Christensen, John Wiley and Sons Inc. 1987, p. 167; J. M. Lehn, *Angew. Chem. Int. Ed. Engl.*, **27**, 89, 1988; D. J. Cram, *ibid,* 1009; C. J. Pedersen,

ibid 1021. G. W. Gokel, S. H. Korseniowski, *"Macrocyclic Polyether Syntheses"*, Springer-Verlag 1982; L. F. Lindoy, *"The chemistry of macrocyclic ligand complexes"*, Cambridge University Press 1989; R. D. Hancock, A. E. Martell, *Chem. Rev.,* **89,** 1875, 1989; A. V. Bajaj, N. S. Poonia, *Coord. Chem. Rev.,* **87,** 55, 1988; D. H. Busch, N. A. Stephenson, *ibid,* **100,** 119, 1990.

298. M. Okahara, Y. Nakatsuji, *Topics Current Chem.* **128,** 37, 1985; T. M. Fyles, *"Bioorganic Chemistry Frontiers",* vol. 1, Springer-Verlag, 1990, p. 71; E. Pretsch, M. Badertscher, M. Welti, T. Maruizumi, W. E. Morf, W. Simon, *Pure and Appl. Chem.* **60,** 567, 1988; L. F. Lindoy, D. S. Baldwin, *ibid,* **61,** 909, 1989; A. Pullman, *Chem. Rev.,* **91,** 793, 1991.
299. H. G. Löhr, F. Vögtle, *Acc. Chem. Res.* **18,** 65, 1985; D. B. Gomis, E. F. Alonso, E. A. Garcia, P. A. Abrodo, *Talanta,* **36,** 1237, 1989; M. Takagi, K. Ueno, *Topics in Current Chemistry,* **121,** 39, 1984; S. Misumi, *Pure and Appl. Chem.* **62,** 493, 1990.
300. W. P. Weber, G. W. Gokel, *"Phase Transfer Catalysis in Organic Synthesis",* Springer-Verlag 1977; R. M. Kellogg, *Angew. Chem. Int. Ed. Engl.* **23,** 782, 1984.
301. J. L. Dye, M. G. Debacker, *Ann. Rev. Phys. Chem.* **38,** 271, 1987; J. L. Dye, *Pure and Appl. Chem.,* **61,** 1555, 1989; J. D. Corbett, *Chem. Rev.,* **85,** 383, 1985.

Illustration Acknowledgements

The following figures have been reproduced or redrawn:

I, *1,* 2	(*J. Chem. Soc. B.,* 1544 © 1970)
I, *1,* 5	(Houben Weyl, Georg Thieme Verlag, Stuttgart (1955), p. 729)
I, *1,* 9	(*Acc. Chem. Res.,* **14,** 95 © 1981)
I, *2,* 4, 5, 6	(*Tetrahedron,* **27,** 2561 © 1971)
I, *2,* 8	(*Z. Phys. Chem.,* **101,** 267 © 1976)
I, *2,* 12	(*J. Am. Chem. Soc.,* **101,** 5736 © 1979)
I, *2,* 18, 19	(*J. Am. Chem. Soc.,* **97,** 3313, 3321, 5831, 5834 © 1975)
I, *3,* 61	(*J. Am. Chem. Soc.,* **102,** 3283 © 1980)
I, *3,* 64	(*J. Chem. Soc., Chem. Comm.,* 244 © 1985)
I, *3,* 88	(*J. Chem. Soc., Chem. Comm.,* 533 © 1981)
I, *3,* 89	(*J. Chem. Soc., Chem. Comm.,* 1005 © 1979)
I, *3,* 97	(*J. Am.. Chem. Soc.,* **103,** 5594 © 1981)
I, *3,* 121	(*J. Am. Chem. Soc.,* **103,** 2780 © 1981)
II, *1,* 3	(*Angew. Chem., Int. Ed. Engl.,* **20,** 1041 © 1981)
II, *1,* 4	(*J. Chem. Soc., Perkin II,* 345 © 1975)
II, *1,* 5	(*Acta Cryst.,* **B30,** 2739 © 1974)
II, *1,* 6	(*Acta Chem. Scand.,* **A36,** 109 © 1982)
II, *1,* 7	(*J. Chem. Soc., Perkin II,* 341 © 1972)
II, *1,* 8	(*J. Chem. Soc., Perkin II,* 1818 © 1972)
II, *1,* 9	(*Inorg. Chem.,* **13,** 2826 © 1974)
II, *1,* 10	(*Structure and Bonding,* **16,** 71 © 1973)
II, *1,* 12	(*Acta Chem. Scand.,* **A38,** 337 © 1984)
II, *1,* 13	(*Acta Cryst.,* **B30,** 2748 © 1974)
	(*Acta Cryst.,* **B30,** 2741 © 1974)
II, *1,* 14	(*Angew. Chem., Int. Ed. Engl.,* **18,** 934 © 1979)
II, *1,* 15	(*Topics in Current Chem.,* **101,** 1 © 1982)
II, *1,* 16	(*J. Chem. Soc., Dalton,* 2374 © 1975)
II, *1,* 17	(*Acta Cryst.,* **C40,** 951 © 1984)

II, *1,* 17 (*Acta Cryst.,* **C40,** 951 © 1984)

II, *1,* 18	(*J. Chem. Soc., Chem. Comm.,* 183 © 1977)
II, *1,* 19	(*J. Chem. Soc., Dalton,* 1418 © 1978)
	(*Acta Cryst.,* **C40,** 1183 © 1984)
II, *1,* 20	(*J. Am. Chem. Soc.,* **106,** 7627 © 1984)
II, *1,* 21, 22	(*Topics in Current Chem.,* **101,** 1 © 1982)
II, *1,* 22	(*J. Chem. Soc., Chem. Comm.,* 472 © 1981)
II, *1,* 23	(*J. Chem. Soc., Chem. Comm.,* 804 © 1977)
	(*Acta Cryst.,* **B34,** 800 © 1978)
II, *1,* 24	(*Inorg. Chem.,* **14,** 734 © 1975)
II, *1,* 26	(*J. Am. Chem. Soc.,* **103,** 701 © 1981)
II, *1,* 27	(*Nature,* **295,** 526 © 1982)
II, *1,* 28	(*J. Am. Chem. Soc.,* **105,** 6717 © 1983)
II, *1,* 29	(*J. Chem. Soc., Chem. Comm.,* 1172 © 1982)
II, *1,* 30	(*J. Am. Chem. Soc.,* **106,** 1280 © 1984)
II, *1,* 31, 32	(*J. Chem. Soc., Chem. Comm.,* 847 © 1981)
II, *1,* 33	(*J. Chem. Soc., Chem. Comm.,* 849 © 1981)
II, *1,* 35	(*Angew. Chem., Int. Ed. Engl.,* **20,** 1051 © 1981)
II, *1,* 36	(*J. Am. Chem. Soc.,* **104,** 4540 © 1982)
II, *1,* 37	(*Acta Cryst.,* **C40,** 1584 © 1984)
	(*Acta Cryst.,* **44,** 303 © 1988)
II, *1,* 38	(*J. Chem. Soc., Chem. Comm.,* 727 © 1982)
II, *1,* 43	(*Bull. Soc. Chim. Fr.,* 549 © 1972)
II, *1,* 44	(*Bull. Soc. Chim. Fr.,* 551 © 1972)
II, *1,* 45	(*Acta Cryst.,* **B29,** 542 © 1973)
II, *1,* 49	(*Acta Cryst.,* **B37,** 1739 © 1981)
II, *1,* 50	(*Acta Cryst.,* **B37,** 1741, 1744, 1747 © 1981)
II, *1,* 51	(*Acta Cryst.,* **B37,** 1750 © 1981)
II, *1,* 52	(*J. Heterocycl. Chem.,* **11,** 45 © 1974)
II, *2,* 6, 7	(*Tetrahedron,* **30,** 1871 © 1974)
II, *2,* 8	(*J. Am. Chem. Soc.,* **97,** 4379 © 1975)
II, *2,* 9, 10	(*Science,* **176,** 911 © 1972)
II, *2,* 11	(*Helv. Chim. Acta,* **58,** 432 © 1975)
II, *2,* 12	(*J. Am. Chem. Soc.,* **103,** 5880 © 1981)
II, *2,* 13	(*J. Am. Chem. Soc.,* **104,** 4085 © 1982)
II, *2,* 15	(*J. Am. Chem. Soc.,* **102,** 2704 © 1980)
II, *2,* 17, 18	(*FEBS Letters,* **44,** 1 © 1974)
II, *2,* 20	(*Z. Naturforsch.,* **39C,** 535 © 1984)
II, *2,* 21	(*Proc. Natl. Acad. Sci.,* **70,** 1836 © 1973)
II, *2,* 22	(*Proc. Natl. Acad. Sci.,* **82,** 7155 © 1985)
II, *2,* 24, 25	(*Acc. Chem. Res.,* **9,** 106 © 1976)
	(*Proc. Natl. Acad. Sci.,* **79,** 4519 © 1982)
II, *2,* 26	(*Acta Cryst.,* **B36,** 1052 © 1980)
II, *2,* 29	(*Acta Chem. Scand.,* **B29,** 153 © 1975)
	(*Acta Chem. Scand.,* **A30,** 840 © 1976)

	(*Acta Chem. Scand.,* **A31,** 232 © 1977)
II, *2,* 31	(*Acta Chem. Scand.,* **B31,** 523 © 1977)
II, *2,* 33	(*Helv. Chim. Acta,* **55,** 1371 © 1972)
	(*Acta Cryst.,* **B30,** 1047 © 1974)
II, *2,* 34	(*J. Mol. Biol.,* **30,** 559 © 1967)
II, *2,* 36	(*Helv. Chim. Acta,* **59,** 614 © 1976)
II, *2,* 41	(*Proc. Natl. Acad. Sci.,* **68,** 1907 © 1971)
II, *2,* 42	(*Biochem.,* **13,** 5249 © 1974)
II, *2,* 44	(*J. Supramol. Struct.,* **2,** 538 © 1974)
II, *2,* 45	(*Acc. Chem. Res.,* **14,** 356 © 1981)
II, *2,* 46, 47, 48	(*Nature,* **300,** 325 © 1982)
II, *2,* 49	(*Rec. Trav. Chim. Pays-Bas,* **103,** 260 © 1984)
II, *3,* 16	(*Tetrahedron Letters,* **22,** 3039 © 1981)
II, *3,* 27	(*J. Chem. Soc., Perkin II,* 12 © 1981)
II, *3,* 49	(*Tetrahedron Letters,* **22,** 3035 © 1981)
II, *3,* 51	(*J. Am. Chem. Soc.,* **101,** 1047 © 1979)
II, *3,* 56	(*J. Chem. Soc., Chem. Comm.,* 949 © 1985)
II, *3,* 60	(*Chem. in Britain,* **15,** 23 © 1979)
II, *3,* 67	(*Helv. Chim. Acta,* **68,** 1577 © 1985)
II, *3,* 79	(*J. Chem. Soc., Chem. Comm.,* 388 © 1982)
II, *3,* 80-84	(*Acc. Chem. Res.,* **15,** 185 © 1982)
II, *3,* 88	(*Helv. Chim. Acta,* **55,** 1757 © 1972)
II, *3,* 91	(*Spectrochim. Acta,* **31A,** 1381 © 1975)
II, *3,* 93	(*J. Chem. Soc., Chem. Comm.,* 377 © 1982)
II, *3,* 98	(*J. Am. Chem. Soc.,* **108,** 107 © 1986)
II, *3,* 99	(*J. Chem. Soc., Chem. Comm.,* 999 © 1983)
II, *3,* 103	(*J. Am. Chem. Soc.,* **93,** 1806 © 1971)
II, *3,* 107	(*J. Am. Chem. Soc.,* **102,** 475 © 1980)
II, *3,* 109	(*Acta Chem. Scand.,* **A35,** 463 © 1981)
II, *3,* 111	(*Acta Chem. Scand.,* **A35,** 460 © 1981)
II, *3,* 121	(*Bull. Chem. Soc. Japan,* **51,** 790 © 1978)
	(*J. Am. Chem. Soc.,* **104,** 1355 © 1982)
II, *3,* 124, 125	(*J. Am. Chem. Soc.,* **101,** 6752 © 1979)
	(*J. Am. Chem. Soc.,* **103,** 5594 © 1981)

By kind permission of the American Chemical Society, the American Association for the Advancement of Science (AAAS), MacMillan Ltd., Pergamon Press.

We have made every attempt to obtain permission to reproduce the figures which appear in this work. Any inadvertent omission which is pointed out to us, will be rectified in the next edition.

Index